A Specialist Periodical Report

Spectroscopic Properties of Inorganic and Organometallic Compounds

Volume 1

A Review of the Literature Published during 1967

Senior Reporter

N. N. Greenwood, *Department of Inorganic Chemistry, University of Newcastle upon Tyne*

Reporters

J. W. Akitt, W. Errington, T. C. Gibb, and B. P. Straughan

The Chemical Society
Burlington House, London, W1V 0BN

Organic formulae composed by Wright's Symbolset method

PRINTED IN GREAT BRITAIN BY JOHN WRIGHT AND SONS LTD., AT THE STONEBRIDGE PRESS, BRISTOL

Foreword

This is the first volume of a series which will appear annually on the spectroscopic properties of inorganic and organometallic compounds. It is being published by the Chemical Society as part of their policy to provide specialist periodical reports in several areas of chemistry to supplement and extend *Annual Reports on the Progress of Chemistry*. Since 1904, when the first volume of *Annual Reports* appeared, chemical publications have expanded enormously both in number and in size and the treatment of the subject in *Annual Reports* has necessarily become increasingly compressed and selective. There is still a real need for such a publication which defines and reports 'the principal definite steps in advance which have been accomplished in the preceding year for the benefit of all workers, students, or teachers of chemistry, or those chemists who are engaged in technical or manufacturing applications of chemistry, in order that specialists in any one department of the science may obtain without difficulty information as to the nature and extent of progress in other branches of the subject to which they have not paid special attention'.* For this reason the Chemical Society intends to continue publishing *Annual Reports* though from Volume 64 (1967) it will be published in two separate parts.

Equally, there is a place for specialist reports on various aspects of chemistry which aim at a more comprehensive coverage. The present volume is one of first of these new *Specialist Periodical Reports* and covers the year 1967. Spectroscopic techniques are central to the study of chemical compounds and we have endeavoured to give a balanced account of the results published during the year. The book is divided into 7 chapters each dealing with a particular technique and arranged in order of increasing energy of the transition involved. It has been our aim not only to record results but to indicate the wide variety of ways in which spectroscopic information is being used. Each chapter is self-contained and its length and the number of references it contains indicate the extent to which each technique is being applied to the study of inorganic and organometallic compounds. There are, of course, many other applications of these techniques but they fall outside the scope of the Report. For example, e.s.r. finds extensive use in the study of free radicals and of doped crystalline solids but inclusion of such topics which are peripheral to the main theme would have resulted in a book of unmanageable proportions. Similarly the borderline between organic and organometallic chemistry is somewhat arbitrary but we have tried to include all spectroscopic information which might be of interest to people working primarily with inorganic or organometallic compounds.

* Quotation from the Introduction to Vol. 1 (1904) of *Annual Reports*.

Each chapter states the range of work covered and the areas which have been excluded. It has been the intention to include all significant advances published during 1967. This has involved an extensive literature survey since spectroscopic techniques are so all-pervasive that it is insufficient to scan *Current Chemical Papers* or even the pages of *Chemical Abstracts*. These sources have, of course, been used but the primary source of data has been the page-by-page scanning of a very large number of scientific periodicals which span the full range from nuclear physics to preparative chemistry. This has resulted in the ordered presentation and discussion of over 2800 references, a figure which can be compared with some 2250 references appearing in the Sections covering the whole of General, Physical, and Inorganic Chemistry in *Annual Reports* for 1966. It would be idle to suppose that no paper published during 1967 has passed unnoticed but we believe that the vast majority of papers containing information on the spectra of inorganic and organometallic compounds have been covered. We have also endeavoured to reduce the time for preparing and publishing the manuscript to a minimum.

The uses of such a compilation are manifold. We hope that it will serve both as a source of ideas and information and also as an indication of the pattern of development of the subject. For this reason we have tried to keep a balance between numerical data on the one hand and scientific results and conclusions on the other. For the non-specialist reader there are indications of the scope of each technique and its possible applications to his own particular problems. For the specialist reader the review should serve as a ready source of references to specific compounds.

The authors would welcome comments on the design of this new type of annual review and would appreciate suggestions which might make the publication more useful in the future.

Note. In Chapter 1 the following reference numbers, 215–217, 220, 269 and 416 give the periodical as *Chem. Ber.* They should read *Ber. Bunsengesellschaft Phys. Chem.*

Contents

Ligand abbreviations

The following abbreviations have been used

acac	acetylacetone
bipy	2,2′-bipyridyl
cod	cyclo-octadiene
cot	cyclo-octatetraene
diars	*o*-phenylenebisdimethylarsine
dien	diethylenetriamine
diphos (or dpe)	1,2-bisdiphenylphosphinoethane
dipyam	di-2-pyridylamine
DMF	dimethylformamide
dmg	dimethyl glyoxime
DMSO	dimethyl sulphoxide
dpa	di-2-pyridylamine
dpte	1,2-di(phenylthio)ethane
en	ethylenediamine
ffars	1,2-bis(dimethylarsino)tetrafluorocyclobutene
ffos	1,2-bis(diphenylphosphino)tetrafluorocyclobutene
H_4edta	ethylenediaminetetracetic acid
hfac	hexafluoroacetylacetone
ibn	isobutylenediamine
lut	3,4-lutidine
Me_4daeo	bis-(2-dimethylaminoethyl)oxide
Me_4daes	bis-(2-dimethylaminoethyl)sulphide
ox	oxalato
paphy	pyridine-2-aldehyde-2′-pyridylhydrazone
pc	phthalocyanine
phen (*o*-phen)	1,10-phenanthroline
pic	β- (or γ-)picoline
pmt	pentamethylenetetrazole
pn	propylenediamine
py	pyridine
tdt	toluene-3,4-dithiolate
terpy	2,2′,2″-terpyridine
tetren	tetraethylenepentamine
tpa	tri-2-pyridylamine
tpt	2,4,6-tri(2-pyridyl)-1,3,5-triazine
tren	2,2′,2″-triaminotriethylamine
trien	triethylenetetramine
tripyam	tri-2-pyridylamine
tu	thiourea

Conversion factors

	cm.$^{-1}$	joule mole^{-1}	ev	kcal. mole^{-1}	Mc. sec.$^{-1}$
cm.$^{-1}$	1	11·957	$1{\cdot}2394 \times 10^{-4}$	$2{\cdot}8584 \times 10^{-3}$	$2{\cdot}9979 \times 10^{4}$
joule mole^{-1}	$8{\cdot}3626 \times 10^{-2}$	1	$1{\cdot}0364 \times 10^{-5}$	$2{\cdot}3904 \times 10^{-4}$	2506·2
ev	8068·3	$9{\cdot}6484 \times 10^{4}$	1	23·063	$2{\cdot}4188 \times 10^{8}$
kcal. mole^{-1}	349·83	4183·3	$4{\cdot}3359 \times 10^{-2}$	1	$1{\cdot}0487 \times 10^{7}$
Mc. sec.$^{-1}$	$3{\cdot}3356 \times 10^{-5}$	$3{\cdot}9903 \times 10^{-4}$	$4{\cdot}1344 \times 10^{-9}$	$9{\cdot}5345 \times 10^{-8}$	1

Mössbauer spectra

For ^{57}Fe ($E\gamma$ = 14·413 kev): 1 mm. sec.$^{-1}$ = $3{\cdot}879 \times 10^{-4}$ cm.$^{-1}$ = $4{\cdot}638 \times 10^{-3}$ joule mole^{-1}
= $4{\cdot}809 \times 10^{-8}$ ev = $1{\cdot}109 \times 10^{-6}$ kcal. mole^{-1}
= 11·63 Me. sec.$^{-1}$

For other nuclides multiply the above conversion factors by $E\gamma$ (kev)/14·413.

1
Nuclear Magnetic Resonance Spectroscopy

1 Introduction

Nuclear magnetic resonance spectroscopy, n.m.r., is playing an increasingly important part in the study of inorganic and organometallic chemistry. Some idea of its importance relative to the other spectroscopic techniques can be obtained from the fact that this chapter contains some 610 references which places it next in size after the chapters on vibrational and electronic spectroscopy. As would be expected, proton n.m.r. provides much of the experimental information but, in contrast to the organic field, many significant contributions have come from studies involving directly the nuclei ^{2}D, ^{7}Li, ^{9}Be, ^{11}B, ^{13}C, ^{14}N, ^{17}O, ^{19}F, ^{23}Na, ^{27}Al, ^{31}P, ^{35}Cl, ^{55}Mn, and ^{59}Co. Indirect double-resonance techniques have, in addition, embraced ^{29}Si, ^{195}Pt, and ^{199}Hg.

The subject breaks down naturally into a series of topics which are dealt with in the order:

Stereochemistry of complexes; dynamic systems; adducts and solvent effects; bonding problems and contact shifts; solid-state n.m.r.; boron and Group III elements; fluorine and phosphorus; less-common resonances. There is also an Appendix which lists compounds which have been characterised by their n.m.r. spectra but which are not discussed elsewhere in the chapter. As the list indicates, technique-oriented topics are discussed first and these are followed by those in which the chemistry of the element forms the main cohesive theme. On this basis there is inevitably some overlap between sections and some arbitrary decisions have been made about where a particular paper should go. In general, papers were placed where the most coherent n.m.r story would result. The order within sections follows the order of Groups in the periodic table.

Although this section has necessarily been written so as to emphasise the applications of n.m.r. to each particular problem, it should be borne in mind that other techniques frequently play an important part and all of these should be considered in a balanced appraisal of the work.

Only a few books and reviews have been published during the year on the application of n.m.r. to inorganic and organometallic chemistry. Volume 2 of 'Progress in Nuclear Magnetic Resonance Spectroscopy' has appeared [1]

[1] *Prog. N.M.R. Spectroscopy*, 1967, **2**.

and this contains articles on 'Chemical Shift Calculations,' by D. E. O'Reilly; 'High Resolution Nuclear Magnetic Resonance in Partially Oriented Molecules,' by A. D. Buckingham and K. A. McLauchlan; 'Nuclear Magnetic Resonance of Paramagnetic Systems,' by E. de Boer and H. von Willigen; and 'The Cause and Calculation of Proton Chemical Shifts in Non-conjugated Organic Compounds,' by A. F. Zürcher. The relevance of the first three of these articles to this review will be obvious; the principles outlined in the last one, where it is shown that shifts are due mainly to neighbour-anisotropy effects and not to inductive effects are also of general importance. Recent work on the anisotropies of the C=C and C=O bonds are relevant to this.[2, 3a] Volume 3 of the same series also appeared at the end of the year;[3b] it contains articles on 'Sub-spectral Analysis,' by P. Diehl, R. K. Harris, and R. G. Jones, 'The Isotope Shift,' by H. Batiz Hernandez and R. A. Bernheim, 'Nuclear Spin Relaxation Studies of Molecules absorbed on Surfaces,' by K. J. Packer, 'Relaxation Processes in Systems of Two Non-identical Spins,' by E. L. Mackor and C. MacLean, 'Microdynamic Behaviour of Liquids as studied by N.m.r. Relaxation Times,' by H. G. Hertz and 'Solvent Effects and N.m.r.,' by P. Laszlo. The last two articles are of particular relevance to work covered in this chapter. Chemical applications of ^{17}O nuclear and electron spin resonance have been reviewed,[4] an extensive paper has appeared dealing with 'Nuclear Magnetic Resonance Studies of Ions in Pure and Mixed Solvents,'[5a] and the study of ligand-substitution reactions of paramagnetic species using n.m.r. has been reviewed.[5b]

It is becoming increasingly clear that double-resonance experiments can provide full assignments in quite complex inorganic molecules though the technique is still used infrequently (see refs. 39, 61, 141, 151, and 203, later in the text). More attention could usefully be devoted to such experiments which reduce considerably the ambiguity of the n.m.r. experiment. Deuteron magnetic resonance spectra are now being produced with good resolution despite the quadrupole moment of the deuteron and it seems that this may also become an important technique in future years.[6]

It is also appropriate to review here papers which have dealt with the more theoretical or general aspects of n.m.r.; these tend to be distinct from the main bulk of papers dealing with applications to inorganic and organometallic compounds though they are relevant to these applications and often involve typically inorganic compounds.

[2] J. W. Apsimon, W. S. Craig, P. V. Demarco, D. W. Mathieson, L. Saunders, and W. B. Whalley, *Tetrahedron*, 1967, **23**, 2357.
[3a] G. J. Karabatsos, G. C. Sonnichsen, N. Hsi, and D. J. Fenoglio, *J. Amer. Chem. Soc.*, 1967, **89**, 5067.
[3b] *Progr. N.M.R. Spectroscopy*, 1967, **3**.
[4] B. L. Silver and Z. Luz, *Quart. Rev.*, 1967, **21**, 458.
[5a] J. F. Hinton and E. S. Amis, *Chem. Rev.*, 1967, **67**, 367.
[5b] T. R. Stengle and C. H. Langford, *Co-ord. Chem. Rev.*, 1967, **2**, 349.
[6] L. K. Montgomery, A. O. Clouse, A. M. Crellier, and L. E. Applegate, *J. Amer. Chem. Soc.*, 1967, **89**, 3453.

Coupling Constants and Chemical Shifts.—By looking at the proton spectra of the compounds Me_nSnX_{4-n} using double-resonance techniques it was shown that the reduced coupling constant $K(^{119}Sn-^{13}C)$ is positive whereas $K(^{119}Sn-^{1}H)$ is negative; there is a linear relation between $J(Sn-C)$ and $J(Sn-H)$, but the line does not pass through the origin and it is suggested that terms other than the Fermi contact term are important in determining the magnitude of K.[7] A similar conclusion was reached for compounds of type Me_4M (M = C, Si, Sn, and Pb).[8] The sign of $J(^{195}Pt-^{1}H)$ in $(Et_3P)_2PtHCl$ has been found to be positive as predicted.[9] In an experiment which measured $J(^{13}C-H)$, $J(^{13}C-Se)$, and $J(^{13}C-^{125}Te)$ by double resonance in the compounds Me_2S, Me_2Te, Me_3S^+, Me_3Te^+, and Me_2Se_2 it was concluded that the hybridisation of Se or Te was an important factor in determining the magnitude of the coupling constants.[10] Similarly the coupling constants $J(C-H)$ and $J(Si-F)$ in the compounds MeSiFXY depend on the electronegativities of X and Y.[11] The signs of the couplings in PH_3, PH_4^+, PF_3, P_2, and SiH_4 have been calculated successfully and the correct order of magnitude obtained though the method did not succeed with SiF_4.[12] The effects of the electronegativity of substituents on the coupling constants $J(^{13}C-H)$ and $J(^{29}Si-H)$ in a series of substituted methanes and silanes has also been calculated.[13] The coupling constants in $(Me_3SiNMe)_nSnMe_{4-n}$ all increase with n.[14]

It has been suggested that the sign of $J(^{11}B-^{1}H)$ might be determined from the differences in width of the four components of the proton spectrum of $(CD_3O)_2BH$ though such differences could not be detected.[15]

The relationship between $J(^{119}Sn-^{1}H)$ and Mössbauer chemical shifts in the series Me_nSnH_{4-n} ($n = 1-4$) has been investigated but it seems that the Mössbauer data are not yet sufficiently accurate to make a valid comparison.[16]

Full analyses have been given of the 'deceptively simple' proton and ^{31}P spectra of a number of phosphorus compounds of general types $R_2P(CH_2)PR_2$ and $R_2P(X)(CH_2)_nP(X)R_2$ (n = 0, 1, or 2; R = Ph or Me; X = O, S, or Se).[17, 18] The proton spectrum of pyridine has been analysed fully using double irradiation of the ^{14}N.[19]

[7] W. McFarlane, *J. Chem. Soc.* (*A*), 1967, 528.
[8] W. McFarlane, *Mol. Phys.*, 1967, **13**, 587.
[9] W. McFarlane, *Chem. Comm.*, 1967, 772.
[10] W. McFarlane, *Mol. Phys.*, 1967, **12**, 243.
[11] S. G. Frankiss, *J. Phys. Chem.*, 1967, **71**, 3418.
[12] A. H. Cowley, W. D. White, and S. L. Manatt, *J. Amer. Chem. Soc.*, 1967, **89**, 6433.
[13] R. Ditchfield, M. A. Jensen, and J. W. Murrell, *J. Chem. Soc.* (*A*), 1967, 1674.
[14] O. J. Scherer and P. Hornig, *J. Organometallic Chem.*, 1967, **8**, 465.
[15] N. Boden, H. S. Gutowsky, J. R. Hansen, and T. C. Farrar, *J. Chem. Phys.*, 1967, **46**, 2849 (L).
[16] C. May and J. J. Spijkerman, *J. Chem. Phys.*, 1967, **46**, 3272 (L).
[17] A. J. Carty and R. K. Harris, *Chem. Comm.*, 1967, 234.
[18] E. G. Finer and R. K. Harris, *Mol. Phys.*, 1967, **12**, 457.
[19] S. Castellano, C. Sun, and R. Kostelnik, *J. Chem. Phys.*,1967, **46**, 327.

The Table gives references to other compounds for which coupling data have also been obtained.

Compound	*Ref.*	*Compound*	*Ref.*
$MeSiHCl_2$	20	$(p\text{-}MePh)_4Pb$, $(p\text{-}MePh)_6Pb_2$	25
$(CF_3)_nPX_{3-n}$ (X = H, F, Cl)	21	$Me_3SnCF_2\cdot CFH\cdot CF_3$ and related compounds	26
$PhCH_2SeH$	22	GeH_3SMe,SiH_3SMe	27
Si_3F_8	23	$(H_2SiCCl_2)_3SiH_3CCl_2$	28
$MeSiF_3$,$HSiF_3$, HPF_2	24		

If, in a double-resonance experiment, the irradiated line is broad and the observed line normally narrow, then some broadening is carried over into the narrow line by the double irradiation. Thus by observing the ^{13}C spin satellites in the proton spectrum of chloroform while irradiating ^{13}C it was found that the ^{13}C line is 1·9 c./sec. wide owing to chlorine relaxation. This has allowed $J(^{13}C—^{35}Cl)$ to be estimated as about 49 c./sec.[29]

A nuclear–nuclear Overhauser experiment has been carried out using weak irradiation which has allowed simultaneous observation of both ^1H and ^{31}P nuclei. Intensity enhancements of twenty times have been observed on spin-coupled phosphorus while irradiating ^1H. If a highly degenerate line is irradiated, such as in $(MeO)_3P$, then both positive and negative enhancements can be seen indicating that transverse relaxation times are different for different symmetry classes.[30]

Several papers have dealt with factors affecting chemical shifts in specific compounds. The chemical shifts of ^{199}Hg and ^{13}C in Me_2Hg have been determined by a double-resonance method relative to tetramethylsilane (TMS).[31] The SH proton in $(\pi\text{-}C_5H_5)_2M(SH)_2$ (M = Mo or W) is at a very high field (τ 12·36 or 11·46, respectively); it is suggested that this is due either to shielding by the metal *d* orbitals or to a mixing with the π orbitals on sulphur which leads to magnetic anisotropy.[32]

In a study of ^{13}C, ^{14}N, and ^{17}O chemical shifts of ligands in metal complexes it was found that two effects appear to operate: (*a*) constraints

20 W. McFarlane, *J. Chem. Soc.* (*A*), 1967, 1275.
21 S. L. Manatt, D. D. Ellemann, A. H. Cowley, and A. B. Burg, *J. Amer. Chem. Soc.*, 1967, **89**, 4544.
22 W. McFarlane, *Chem. Comm.*, 1967, 963.
23 R. B. Johannenesen, *J. Chem. Phys.*, 1967, **47**, 955.
24 R. B. Johannenesen, *J. Chem. Phys.*, 1967, **47**, 3088.
25 W. Kitching, V. G. Kumar Das, and P. R. Wells, *Chem. Comm.*, 1967, 356.
26 H. C. Clark, N. Cyr, and J. M. Tsai, *Canad. J. Chem.*, 1967, **45**, 1073.
27 J. T. Wang and C. H. Van Dyke, *Chem. Comm.*, 1967, 612.
28 G. Fritz, H. Frohlich, and D. Kummer, *Z. anorg. Chem.*, 1967, **353**, 34.
29 R. Freeman, N. N. Ernst, and W. A. Anderson, *J. Chem. Phys.*, 1967, **46**, 1125.
30 D. D. Elleman, S. L. Manatt, A. J. R. Bourn, and A. H. Cowley, *J. Amer. Chem. Soc.*, 1967, **89**, 4542.
31 R. R. Dean and W. McFarlane, *Mol. Phys.*, 1967, **13**, 343.
32 M. L. H. Green and W. F. Lindsell, *J. Chem. Soc.* (*A*), 1967, 1455.

forced on loosely bound ligand-electrons by the bonding to the metal, and (*b*) delocalisation of metal *d*-electrons leading to a low-field shift.[33] In the complex $[(\pi\text{-}C_5H_5)_2Ti\ acac]^+$ (acac = acetylacetone), it is found that the CH resonance of L is at the lowest recorded value for any acac complex ($\delta = -6{\cdot}33$ p.p.m. from TMS) due to the net positive charge on the complex.[34]

The hydride proton in the series of stannanes R_nSnH_{4-n} (R = alkyl or aryl) is always to low field of SnH_4 and the magnitude of the shift increases with *n*. It is proposed that this is due to the magnetic anisotropy of the Sn—C bond. $J(^{119}Sn—^1H)$ varies smoothly with *n* and its affected by the electron-donor properties of R.[35]

Proton shifts in the compounds (1) and (2) indicate that the phthalocyanine ring in (1) is aromatic whereas the hemiporphyrazine ring in (2)

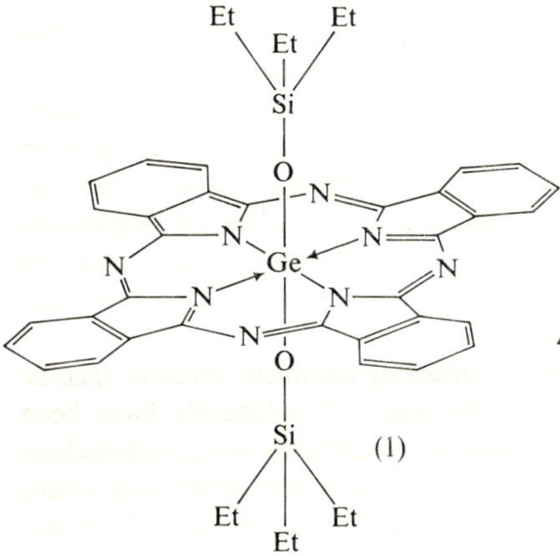

(1)

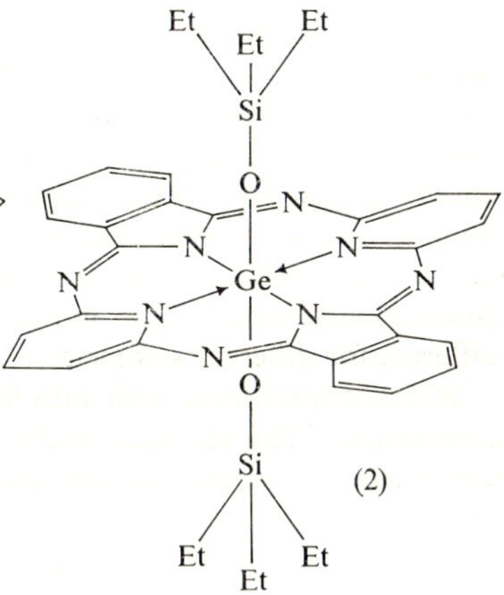

(2)

is not. When M = Ge the shift of the CH_2 groups of the $SiEt_3$ groups in (1) is τ 12·42 and in (2) is τ 9·5.[36] Similar conclusions were drawn for the compounds when M = Sn.[37]

A theoretical interpretation has been given of the well known n.m.r. shift additivity rules for substituents using McWeeney group functions for compounds of the type CHXYZ, BXYZ, $CF_nH_mX_j$, and SiH_xX_y. The proton shifts are pairwise additive in CHXYZ, *i.e.* $\delta = \eta_{xy}+\eta_{yz}+\eta_{xz}$ and in BXYZ whereas the CH_3 shifts in CH_3CXYZ are directly additive; $\delta = \eta_x+\eta_y+\eta_z$. Values are given for the shift coefficients η.[38]

[33] R. Bramley, B. F. Figgis, and R. S. Nyholm, *J. Chem. Soc.* (*A*), 1967, 861.
[34] G. Doyle and R. S. Tobias, *Inorg. Chem.*, 1967, **6**, 1111.
[35] J. Dufermont and J. C. Maire, *J. Organometallic Chem.*, 1967, **7**, 415.
[36] J. N. Esposito, L. E. Sutton, and M. E. Kenny, *Inorg. Chem.*, 1967, **6**, 1116.
[37] L. E. Sutton and M. E. Kenney, *Inorg. Chem.*, 1967, **6**, 1869.
[38] T. Vladimiroff and E. R. Malinowski, *J. Chem. Phys.*, 1967, **46**, 1830.

2 Stereochemistry of Complexes

It is often possible to obtain detailed knowledge of the arrangement of ligands in a complex using quite simple n.m.r. data. For this reason stereochemical studies (often linked with other techniques) form an important section of the n.m.r. literature of inorganic and organometallic compounds and comprise nearly 15% of the references abstracted. The section falls naturally into a series of sub-sections:

π Complexes, treated here in order of decreasing ring size; complexes with phosphine ligands; other transition-metal complexes; and compounds containing tin.

π Complexes.—It is a general feature of π complexes that the olefinic proton signal of the complexed ligand occurs to high field of the uncomplexed ligand, since the complexing removes some of the anisotropy of the double bond.

Three tricarbonyl-iron or -ruthenium complexes of 1,2-substituted cyclooctatetraene have been prepared. The maintenance of a cyclic triene arrangement after substitution was proved by a series of double-resonance experiments.[39] The bisnaphthalene π complex of ruthenium $[Ru(C_{10}H_8)_2]^{2+}(PF_6^-)_2$ has been prepared; its proton n.m.r. shows two sets of A_2B_2 spectra, indicating that only one ring of each naphthalene nucleus is involved in bonding to the metal.[40] In the case of unsymmetrically substituted naphthalene complexes of tricarbonyl chromium it was possible to distinguish two isomers (3) and (4), since the protons in the ring which is bonded to Cr are shifted upfield and give different patterns for each complex.[41] Acetylpyrrolyl and indolyl π complexes have also been prepared and their proton spectra recorded.[42] Phenyl π complexes of type (5)

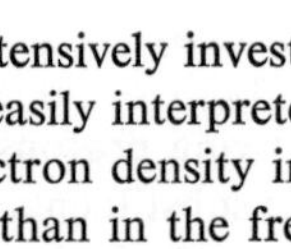

(3)

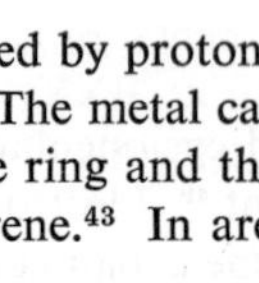

(4)

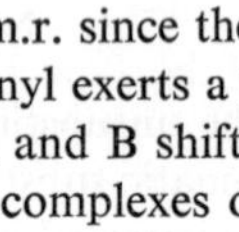

(5)

have been extensively investigated by proton n.m.r. since the A_2B_2 pattern produced is easily interpreted. The metal carbonyl exerts a levelling effect upon the electron density in the ring and the A and B shifts are closer in the complex than in the free arene.[43] In arene complexes of $Ru_6C(CO)_{14}$

[39] M. Green and D. C. Wood, *Chem. Comm.*, 1967, 1062.
[40] E. O. Fischer, C. Elschenbroich, and C. G. Kreiter, *J. Organometallic Chem.*, 1967, 7, 481.
[41] B. Denbzer, H. P. Fritz, C. G. Kreiter, and K. Ofele, *J. Organometallic Chem.*, 1967, 7, 289.
[42] P. L. Pauson, A. R. Quazi, and B. W. Rockett, *J. Organometallic Chem.*, 1967, 7, 325.
[43] H. P. Fritz and C. G. Kreiter, *J. Organometallic Chem.*, 1967, 7, 427.

the arene ring appears to act as a six-electron donor. The individual carbon atom is also of interest in these compounds.[44]

Phosphorus–proton coupling has been observed in a series of complexes (π-C_5H_5)Mo(CO)$_2$L$_n$I$_{2-n}$ (n = 1 or 2; L a phosphine ligand).[45] It was found by correlating n.m.r. and i.r. data that the π-C_5H_5 resonance was a singlet if L was *trans* to (CO), a doublet if L was *trans* to I, and a triplet if L was *trans* to another L (n = 2). It was possible to follow *cis–trans* isomerisation by n.m.r. in these systems. Ferrocenes have also received attention. The methylene groups in (6) give rise to a singlet though it is not

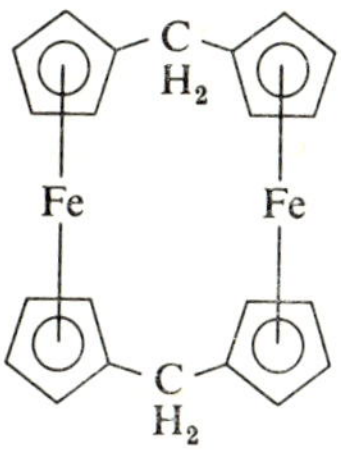

(6)

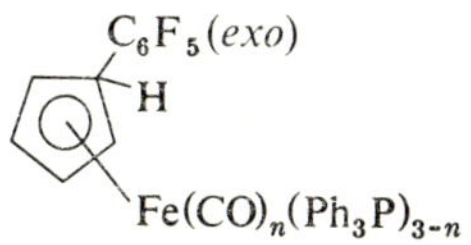

(7)

certain whether the rotation necessary to give this equivalence is possible. The ring protons give the expected pair of triplets.[46] Protonation of ethyl-substituted ferrocenes has been followed by n.m.r. and several carbonium ions characterised in strong acid solutions. The protonation reduces the shielding difference between the ring protons α and β to the substituent.[47] Nine complexes of type (7) are reported. The proton n.m.r. shows that only one type of complex is present and evidence is given that the C_6F_5 substituent is directed away from the iron.[48] Quinone substituted ferrocenes (8) have also been made. The CH_2 of the CH_2NMe_2 group is an AB

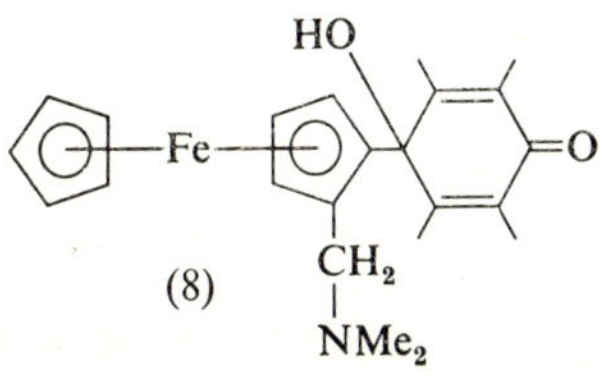

(8)

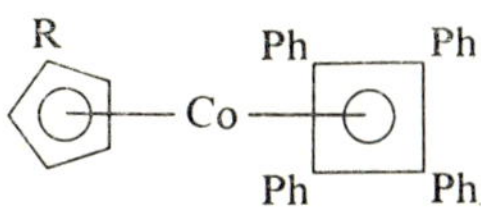

(9)

quartet and the unsubstituted cyclopentadienyl ring protons suffer a shift due to the aromatic substituent.[49] Electrophylic substituents of the π-C_5H_5 ring have been reported in some cobalt complexes (9); [50] the protons of the

[44] B. F. G. Johnson, R. D. Johnston, and J. Levis, *Chem. Comm.*, 1967, 1057.
[45] A. K. Manning, *J. Chem. Soc.* (*A*), 1967, 1984.
[46] W. E. Watts, *J. Organometallic Chem.*, 1967, **10**, 191.
[47] W. L. Horspool and R. G. Sutherland, *Chem. Comm.*, 1967, 786.
[48] P. M. Treichel and R. L. Shubkin, *Inorg. Chem.*, 1967, **6**, 1328.
[49] R. E. Moore, B. W. Rockett, and D. G. Brown, *J. Organometallic Chem.*, 1967, **9**, 141.
[50] M. D. Rausch and R. A. Genetti, *J. Amer. Chem. Soc.*, 1967, **89**, 5502.

substituent R are also shifted high field by the phenyl groups (R = HgCl, I^+, CN^+, $SiMe_3$, COMe, CHO, $CH_2 \cdot OH$, $CH_2 \cdot NMe_2$, and CH_2NMe_3I). Me_3Si substituents have been added to the rings of $(\pi\text{-}C_5H_4) \cdot CH_2 \cdot NMe_2$ and substituent cyclopentadienyl rings have been linked by Me_2Si groups.[51] The complex (10) is σ bonded to the nitrogen of the pyrrolyl ring. The shift between the α, β protons (0·15 p.p.m.) is smaller than in the parent pyrrole (0·5 p.p.m.) whereas in a π-complexed pyrrole the shift should be increased.[52]

A series of transition-metal complexes containing π-cyclopentadienyl and bis-(trifluoromethyl)ethylenedithiolate ligands are also reported (M = Mo, W, Co, Rh, or Ir) and a variety of structures given on the basis of proton and fluorine n.m.r. and other spectroscopic techniques.[53]

The bonding in bis-(6,6′-diphenylfulvene) π complexes of Co and Rh (11) has been shown to be asymmetric to the two rings since H_3 and H_4 have different chemical shifts.

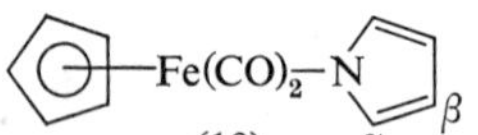

(10)

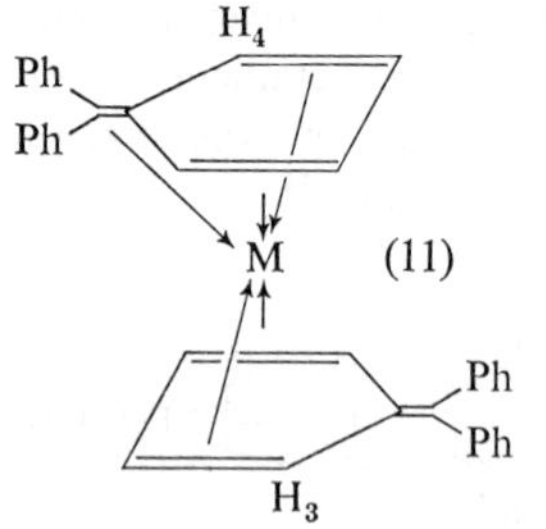

(11)

If dimethyl fulvene is used as starting material this re-arranges to give the bisisopropyl π-cyclopentadienyl complex.[54] Tricarbonyldiphenylfulvenechromium is π bonded to the fulvene ring and can be reduced to π-cyclopentadienyl complexes.[55] Data have been given which suggest that it is not possible to differentiate between possible structures of certain metallocenes by using proton chemical shift data for the substituents alone.[56]

Exchange mechanisms involving π-ring complexes (see refs. 146–154) are dealt with in section 2.

The spectrum of tricarbonylcyclobutadieneiron oriented in a liquid-crystal solution indicates that the protons are in a square configuration. The square is slightly oblong, to the extent that the ratio of the lengths of the sides is 0·9977, suggesting some interaction of the molecule with the nematic solvent.[57]

Spectra are recorded for several π-olefinic complexes. The reactions between 3,3,3-trifluoropropyne and metal carbonyls yield a series of

[51] G. Marr, *J. Organometallic Chem.*, 1967, **9**, 147.
[52] P. L. Pauson and A. R. Quazi, *J. Organometallic Chem.*, 1967, **7**, 321.
[53] R. B. King and M. B. Bisnette, *Inorg. Chem.*, 1967, **6**, 469.
[54] E. O. Fischer and U. J. Weimann, *J. Organometallic Chem.*, 1967, **8**, 535.
[55] R. L. Cooper, E. O. Fischer, and W. Semmlinger, *J. Organometallic Chem.*, 1967, **9**, 333.
[56] T. G. Traylor and J. C. Ware, *J. Amer. Chem. Soc.*, 1967, **89**, 2304.
[57] C. S. Yannoni, G. P. Ceasar, and B. P. Dailey, *J. Amer. Chem. Soc.*, 1967, **89**, 2833.

complexes which have had structures assigned on the basis of proton and fluorine resonances; with $Fe(CO)_5$ a substituted cyclopentane-2,4-dien-1-one is formed while with $Co_2(CO)_8$ both a σ and a π complex are formed.[58] The diolefin complexes (12), (13), (14), and (15) have been prepared[59-62] and characterised by proton and fluorine n.m.r. The olefinic protons of (12) are equivalent and indicate symmetrical bonding of the iron. For (13) the methyl peaks are shifted downfield from uncomplexed 'Dewar' benzene. Excess of $PdCl_2$ causes isomerisation to normal hexamethylbenzene. In the case of (14) the magnitudes of the proton–proton coupling constants were investigated by double resonance and were found to be consistent with the structure given. In (15) all the olefinic protons are equivalent. The CH_2 protons of the uncomplexed ligand are broad due to ring flexing but the ring is constrained in the complex and the lines are sharper. There is evidence for $Rh—CH_2$ interaction at the asterisked positions. Proton n.m.r. spectra have given evidence for the *trans*-acetoxy addition of palladium to cyclo-octadiene.[63] Silver perchlorate forms complexes with $\alpha\omega$-diolefins ($\alpha\omega$D) of type $Ag_2\cdot(\alpha\omega D)_3$ for short-chain and $Ag\cdot(\alpha\omega D)$ for long-chain olefins.[64]

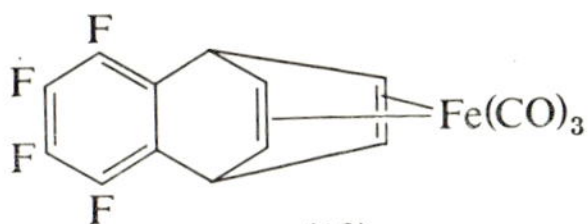

(12)

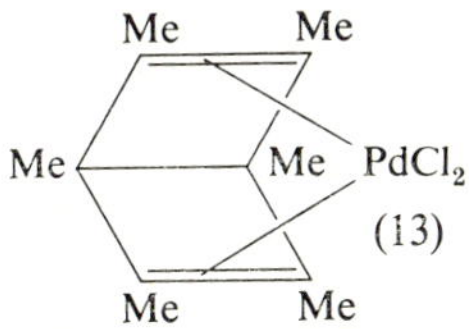

(13)

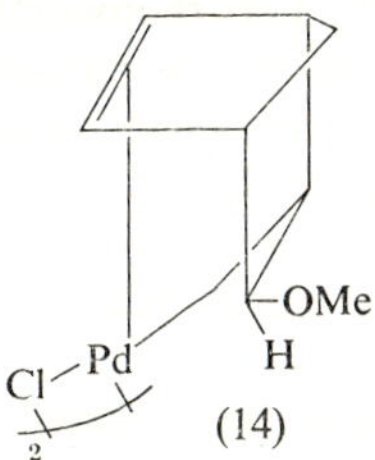

(14)

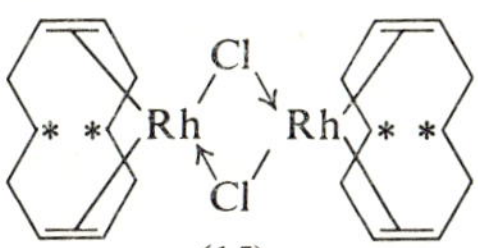

(15)

Extensive proton n.m.r. data are given for a series of π-allylphosphine complexes of rhodium and platinum,[65] and these were useful in confirming that the complexes were indeed π-allylic with the phosphine *trans* to the

[58] R. S. Dickson and D. B. W. Yawney, *Austral. J. Chem.*, 1967, **20**, 77.
[59] A. J. Tomlinson and A. G. Massey, *J. Organometallic Chem.*, 1967, **8**, 321.
[60] H. Dietl and P. M. Maitlis, *Chem. Comm.*, 1967, 759.
[61] M. Green and R. I. Hancock, *J. Chem. Soc.* (*A*), 1967, 2054.
[62] J. C. Trebellas, J. R. Olechowski, H. B. Jonassen, and D. W. Moore, *J. Organometallic Chem.*, 1967, **9**, 153.
[63] C. B. Anderson and B. J. Burreson, *J. Organometallic Chem.*, 1967, **7**, 181.
[64] G. Bresson, R. Broggi, M. P. Lachi, and A. C. Segre, *J. Organometallic Chem.*, 1967, **9**, 355.
[65] M. C. Volger and K. Vrieze, *J. Organometallic Chem.*, 1967, **9**, 527.

π-allyl group in the six-co-ordinated rhodium complexes. The π-allyl, π-cyclohexyl, and π-cyclo-octenyl complexes of tetrakis(triphenylphosphine)rhodium show Rh—H coupling only to the central proton of the π-system whereas P—H coupling occurs only to the terminal protons. A simple MO theory is given.[66] A variable temperature study of bisallylrhodium chloride favours the structure in which both the carbon–carbon bonds have partial double-bond character.[67] π-Allyl complexes of Zr, Ni, and Pd (AM_2X_2 spectra), Hg (AX_4), and Pt ($ABCM_2$ and $ABCM_2X$, showing unsymmetrical bonding) are also reported.[68]

Fluoro(mono-olefin)triethylphosphine complexes are reported with nickel, platinum, and palladium. Spin coupling is observed between Pt and F and between P and F. The platinum–fluorine coupling constants are smaller in these complexes than in π-perfluorobenzene complexes and it is suggested that this indicates weaker Pt—C π bonding.[69] In some platinum complexes the terminal methyl groups of the triethylphosphine are quintets, indicating that the phosphines are *trans*. The olefinic fluorine spectra are complex and are interpreted.[70] The proton spectra of some hydrogenic olefin and alkyne complexes of platinum are also reported.[71] It is noted that in the case of monosubstituted-olefin platinum complexes with pyridine-*N*-oxide the coupling constants between each geminal proton and the platinum are not the same, being about 77 and 67·5 c./sec. This suggests that the olefin is twisted.[72] A similar situation exists for the complex (16) where H_2 and H_3 are shifted upfield by different amounts on

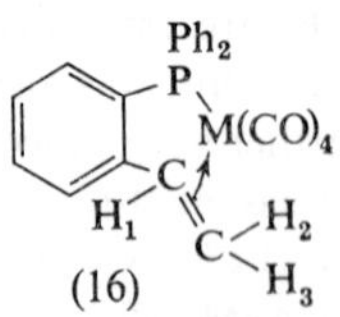

(16)

complexing (1·68 and 1·63 p.p.m. respectively) though the changes in coupling constants indicate that the olefin remains planar. This is interpreted as showing that the olefin–metal bond is not perpendicular to the plane of the olefin and so affects H_2 and H_3 differently.[73] The reaction of ethylene and mercuric ion leads to a substance with an A_2B_2 spectrum showing $J(^{199}Hg—^{1}H)$ of 247 c./sec. which is possibly a mercurinium

66 C. A. Reilly and H. Thyret, *J. Amer. Chem. Soc.*, 1967, **89**, 5144.
67 W. B. Wise, D. C. Lini, and K. C. Ramey, *Chem. Comm.*, 1967, 463.
68 J. K. Becconsall, B. E. Job, and S. O'Brien, *J. Chem. Soc.* (*A*), 1967, 423.
69 A. J. Rest, D. T. Rosevear, and F. G. A. Stone, *J. Chem. Soc.* (*A*), 1967, 66.
70 H. C. Clark and W. S. Tsang, *J. Amer. Chem. Soc.*, 1967, **89**, 533.
71 S. Cenini, R. Ugo, F. Bonati, and G. La Monica, *Inorg. Nuclear Chem. Letters*, 1967, **3**, 191.
72 P. D. Kaplan and M. Orchin, *Inorg. Chem.*, 1967, **6**, 1096.
73 M. A. Bennett, R. S. Wyholm, and J. D. Saxby, *J. Organometallic Chem.*, 1967, **10**, 301.

π complex.[74] The spectra of

$$RCO\cdot CH_A{=}CH_B\cdot Cl \quad \text{and} \quad RCO\cdot CH_A{=}CH_BCl \text{ (with } Fe(CO)_4 \text{ bonded to the C=C)}$$

have been compared. The iron is bonded to the π system and the olefinic protons shift to high field while J(AB) increases from about 8·5 to 13·5 c./sec. on complexing, indicating a reduced bond-order.[75] The complexes formed by rhodium and palladium chlorides with olefins, in solvents which cause dimerisation of the olefins, have been identified by proton n.m.r. as π complexes.[76] The proton n.m.r. of $[MoCl(CO)_2(ol)_3]^-$ (ol = maleimide or maleic anhydride) gives an AB pattern in the olefinic region. The olefinic protons are non-equivalent and several isomeric structures are given, though proton n.m.r. alone cannot differentiate between them.[77]

Complexes with Phosphine (and Arsine) Ligands.—A number of such complexes have already been dealt with under 'π complexes' (see refs. 45, 48, 66, and 69) and some are also dealt with in dynamic systems (see refs. 143, 144, and 154 later in the text). It is well known that the n.m.r. spectra of the methyl groups of methylphosphine ligands (L) which are *trans* to each other show triplets due to strong P—P spin–spin coupling whereas, if they are *cis*, doublets are obtained. Several examples of the use of such data have appeared during the year and an interesting double-resonance experiment has been performed on *cis* and *trans* $Cl_2Pt(Et_3P)_2$. It is found that the signs of the couplings are the same in the two forms and only their magnitudes change. The values of J(P—P) are estimated to be: *cis* <5 c./sec.; *trans* 90 c./sec. The platinum chemical shift was also measured and is 552 p.p.m. between the two isomers, a further very powerful diagnostic tool.[78]

Stereochemistries have been assigned to complexes of the type $RhCl_3L_3$ and $RhCl_2L_3X$ [79] and $IrY_nMe_{3-n}L_3$ [80] [X = Br, I, NCO, NCS, NO_2, or N_3; Y = halogen; L = PMe_2Ph or $AsMe_2Ph$]. In tetramethylplatinum derivatives, which were recently prepared for the first time, two phosphine or arsine ligands are *cis* to one another (17). This is demonstrated by the

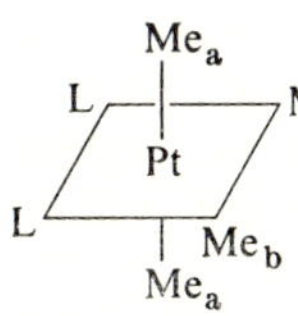

(17)

[74] Yasukazu Saito and Masashi Matsuo, *Chem. Comm.*, 1967, 961.
[75] A. N. Nesmeyanov, Khalil Ahmed, L. V. Rybin, M. J. Rybinskaya, and Yu. A. Ustynyuk, *J. Organometallic Chem.*, 1967, **10**, 121.
[76] A. D. Ketley, L. P. Fisher, A. J. Berlin, C. R. Morgan, E. H. Gorman, and T. R. Steadman, *Inorg. Chem.*, 1967, **6**, 657.
[77] H. D. Murdoch, K. Henzi, and F. Calderazzo, *J. Organometallic Chem.*, 1967, **7**, 441.
[78] W. McFarlane, *J. Chem. Soc.* (*A*), 1967, 1922.
[79] P. R. Brookes and B. L. Shaw, *J. Chem. Soc.* (*A*), 1967, 1079.
[80] B. L. Shaw and A. C. Smithies, *J. Chem. Soc.* (*A*), 1967, 1047.

coupling patterns of the phosphines and by the presence of two sorts of platinum methyl groups for the arsines.[81] The data for the methyl groups are τ_a 9·66, τ_b 8·33; $J(Pt-H_a)$ 44, and $J(Pt-H_b)$ 66 c./sec.

For triethylphosphine ligands, *trans* stereochemistry leads to a methyl quintet. This is seen in the complex (18) which also has a 1 : 4 : 1 triplet for the Me_3Ge group due to Pt—H coupling. By contrast the compound (19) gives a very complex ethyl spectrum which has not been interpreted, but the Me_3Ge triplet is further split into doublets with $J(P-H)$ 1·4 c./sec. Data are also given for several complexes of type $(Et_3P)_2Pt(Me_3M)X$ (M = Ge or Si; X = Cl, Br, I, CN, NCS, or Ph).[82]

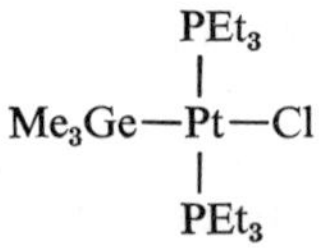

(18)

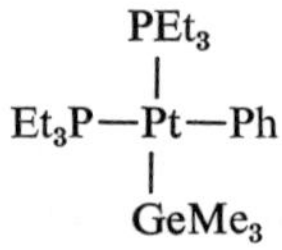

(19)

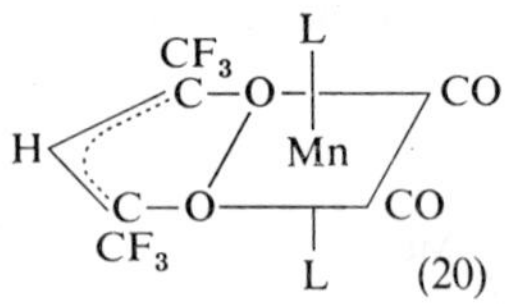

(20)

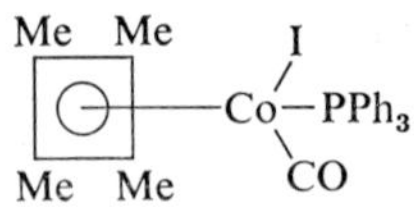

(21)

The phosphine ligands are *trans* in $(Et_3P)_2PdHCl$[83] and in (20),[84] the information being obtained in both cases using proton spectra. There is some suggestion, in the case of the complex $Mo(CO)_4(Ph_2PH)_2$, that the *trans* effect may not be diagnostic in some cases. Here the dipole moment suggests that the phosphines are *cis* and an A_2X_2 proton half-spectrum is taken as confirmatory evidence. However, the spectrum (which, in fact, is presumably AA′XX′) indicates a large phosphorus–phosphorus coupling constant such as might be associated with a *trans* structure and it seems that this complex merits further investigation.[85] For the case where no group is attached to phosphorus, which is capable of giving a simply interpreted proton spectrum, the ^{31}P resonance can be utilised. It has been found for the complexes $(R_nPh_{3-n}P)_2PtCl_2$ (R = alkyl) that $J[Pt-P(cis)] >$ $J[Pt-P(trans)]$ and $\delta[P(cis)]$ is upfield from $\delta[P(trans)]$. The formation of the ionic compound RPh_2PCl^+ with an excess of phosphine is also demonstrated.[86]

The complexes $H_2MPt(Ph_3P)_2$ (M = S or Se) which are formed from the reaction of H_2M and $Pt(Ph_3P)_2$ have been shown by proton n.m.r. to be

81 J. D. Ruddick and B. L. Shaw, *Chem. Comm.*, 1967, 1135.
82 F. Glockling and K. A. Hooton, *J. Chem. Soc.* (*A*), 1967, 1066.
83 E. H. Brooks and F. Glockling, *J. Chem. Soc.* (*A*), 1967, 1030.
84 F. A. Hartman, M. Kilner, and A. Wojcicki, *Inorg. Chem.*, 1967, **6**, 34.
85 J. G. Smith and D. T. Thompson, *J. Chem. Soc.* (*A*), 1967, 1694.
86 S. O. Grim, R. C. Keiter, and W. McFarlane, *Inorg. Chem.*, 1967, **6**, 1133.

of two types in each case, namely the simple adduct and a complex in which one M-hydrogen has transferred to the platinum. The two sorts of hydrogen in the complex are differentiated by their chemical shifts and coupling constants; thus for the H_2S complex J(Pt—H) is 932 and J(Pt—S—H) 43·8 c./sec.[87] and for the H_2Se complex J(Pt—H) is 993 and J(PtSe—H) 44·6 c./sec. Two isomers of the complex $(Ph_3P)_2Rh(Br)H\cdot Si(OEt)_3$ have been detected from the n.m.r. spectrum of the hydride proton.[88] Phosphorus couples to the methyl protons in the compound (21) with J(P—H) 3·8 c./sec.[89] The ^{31}P shifts are given for twentyfive compounds of type $(R_nPh_{3-n}P)M(CO)_5$ (M = Cr, Mo, or W) and it is noted that $J(^{183}W—^{31}P)$ correlates linearly with the C=O stretching frequencies.[90] In the compound *trans*-$(Bu_3P)Mo(CO)_4(PPh_3)$ J(P—P) is measured as 50 c./sec. which is in good agreement with values calculated from the proton spectrum of $(R_3P)_2Mo(CO)_4$ (60 c./sec.) and is of the same order of magnitude as found for platinum complexes (90 c./sec.).[78]

Other Transition-metal Complexes. *First-row Transition-metals.*—In this section, the stereochemistry of complexes of the elements Ti, V, Co, and Ni is considered.

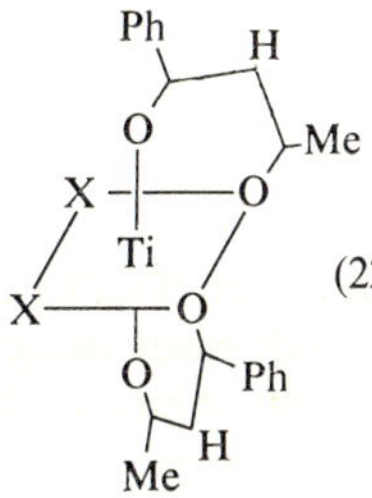

(22)

The stereochemistry and lability of dihalogenobis-(β-diketonato)-titanium(IV) complexes (22) have been confirmed by low-temperature proton n.m.r. spectra.[91] The halogen atoms were shown to be *cis* and the chemical shifts of the β-diketonato-groups are accounted for both in terms of the magnetic anisotropy of phenyl substituents and of the electric-field effects generated by the large dipole moment of the complex. Thus, the CH protons are at low field. The complex can twist about its C_3 axis and all three isomers made possible by re-arranging the unsymmetrical β-diketonato-units are present in statistical amounts. Low-temperature ^{19}F studies of TiF_4 bis-adducts have also shown the *cis* arrangement of the two ligands; an A_2X_2 spectrum is obtained if the ligands are identical and a more complex ABX_2 pattern if the ligands are different.[92]

[87] D. Morelli, A. Segre, R. Ugo, G. La Monica, S. Genini, F. Conti, and F. Bonati, *Chem. Comm.*, 1967, 524.
[88] R. N. Haszeldine, R. V. Parish, and D. J. Parry, *J. Organometallic Chem.*, 1967, **9**, P 13.
[89] E. O. Greaves, R. Bruce, and P. M. Maitlis, *Chem. Comm.*, 1967, 860.
[90] S. O. Grim, D. A. Wheatland, and W. McFarlane, *J. Amer. Chem. Soc.*, 1967, **89**, 5573.
[91] N. Serpone and R. C. Fay, *Inorg. Chem.*, 1967, **6**, 1835.
[92] D. S. Dyer and R. O. Ragsdale, *Inorg. Chem.*, 1967, **6**, 8.

The stereochemistry of six-co-ordinate complexes $TiX_2\,acac_2$ as deduced from n.m.r. data is compared with that of the corresponding tin compounds (see page 18).

The paramagnetic vanadium complexes VL_3, [L is (23), (24), (25), or (26)] show contact shifts of the ligands. Complexes with (23), (24), or (25) are all *trans*, each ring substituent giving three signals [93, 94] whereas for (26) both *cis*- (one signal) and *trans*-forms are found.[95]

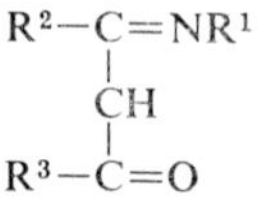

(23)

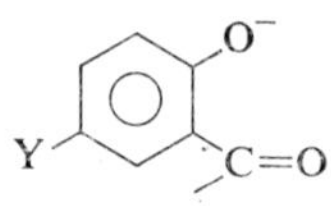

(24)

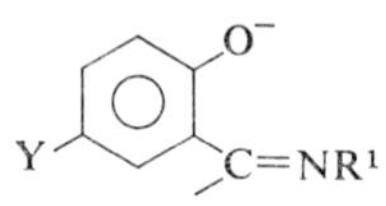

(25)

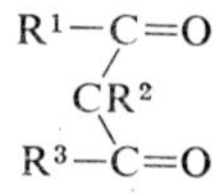

(26)

Triethylenetetramine complexes of cobalt have been investigated using a variety of techniques. The α- and β-forms of the complex (27) can be distinguished by their NH patterns, the resonance of the α-form having three and the β-form five lines [96] (NH proton exchange in similar complexes has also been studied [192]). In the case of the 2,9-dimethyltriethylenetetramine complex with an amino-acid adduct there is a slight difference in the methyl group patterns of the α- and β-complexes and this can be correlated with a positive and negative Cotton effect respectively.[97] The N—H protons in *cis*- and *trans*-octahedral bisethylenediamine complexes have also been examined in detail and it is shown that protons on nitrogen *trans* to a chloride or an acidic ligand resonate to high field.[98]

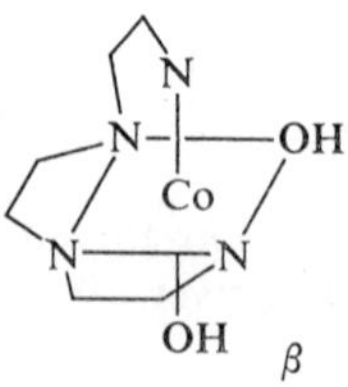

(27)

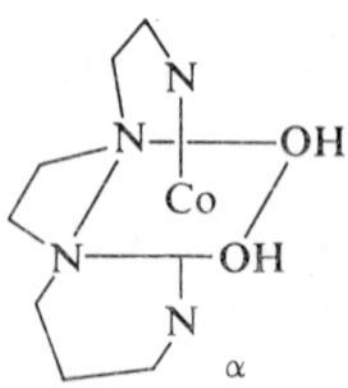

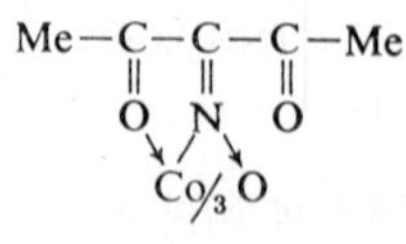

(28)

Asymmetric bonding in the isonitrosoacetylacetonate complex (28) through N and O is indicated by a methyl doublet with a shift between the two peaks of 0·5 p.p.m. The ^{59}Co shift is similar to that of the corresponding acac complex.[99] The structure of a series of related cobalt complexes with unsymmetrical bidentate ligands has been demonstrated by proton n.m.r.

[93] F. Rohrschied, R. E. Ernst, and R. H. Holm, *Inorg. Chem.*, 1967, **6**, 1607.
[94] F. Rohrscheid, R. E. Ernst, and R. H. Holm, *J. Amer. Chem. Soc.*, 1967, **89**, 6472.
[95] F. Rohrscheid, R. E. Ernst, and R. H. Holm, *Inorg. Chem.*, 1967, **6**, 1315.
[96] A. M. Sargeson and G. M. Searle, *Inorg. Chem.*, 1967, **6**, 787.
[97] R. G. Asperger and C. F. Liu, *J. Amer. Chem. Soc.*, 1967, **89**, 708.
[98] I. R. Lantzke and D. W. Watts, *Austral. J. Chem.*, 1967, **20**, 35.
[99] N. J. Patel and B. C. Haldar, *Inorg. Nuclear Chem.*, 1967, **29**, 1037.

studies, to be *trans* and to have no symmetry elements,[100-102] *e.g.* the methyl protons in compound (29) give rise to three resonances. In the case of the ligand (30) the Bu^t and Me groups give singlets, indicating the symmetrical *cis* structure (31); the benzylic protons, however, give an AB quartet.[103]

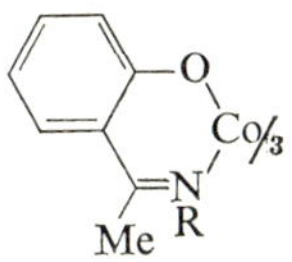

(29)

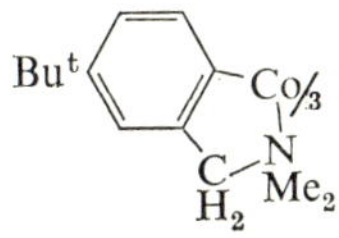

(30)

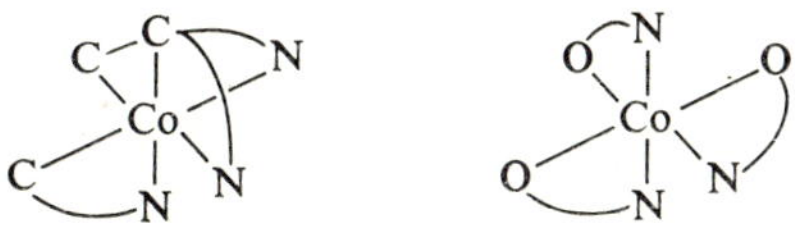

cis (31) *trans*

The cobalt is σ bonded to the aryl-ring carbon. It has been suggested that optically active complexes of 1,2-diaminopropane can be differentiated by proton n.m.r. since the shift of the methyl group depends on the orientation of the NC—CN bond axis in the complex. Shifts of 1·5 c./sec. are reported between (+) and (−)-forms.[104] Ammine complexes with a 5-nitrosalcylato-ligand can be made by the reaction:

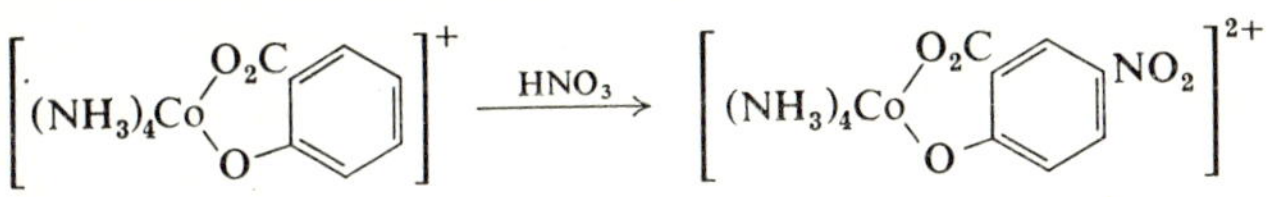

The ring is nitrated and Co^{III} oxidised to Co^{IV}. The *cis*- and *trans*-ammine groups are well separated and resonate at τ 5·9 and 6·5 respectively.[105]

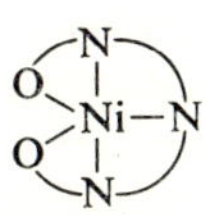

(32)

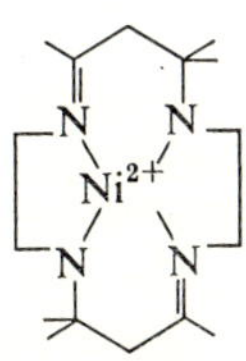

(33)

[100] A. Chakravorty and K. C. Kalia, *Inorg. Chem.*, 1967, **6**, 690.
[101] A. Chakravorty and B. Behera, *Inorg. Chem.*, 1967, **6**, 1812.
[102] A. Chakravorty and K. C. Kalia, *Inorg. Nuclear Chem. Letters*, 1967, **3**, 319.
[103] A. C. Cope and R. N. Gourley, *J. Organometallic Chem.*, 1967, **8**, 527.
[104] J. G. Brushmiller and L. G. Stadtherr, *Inorg. and Nuclear Chem. Letters*, 1967, **3**, 525.
[105] Y. Yamamoto and K. Ito, *Bull. Chem. Soc. Japan*, 1967, **40**, 2580.

The contact shifts present in pentadentate Schiff's base complexes of nickel give double lines for many protons and show non-equivalence of ring protons, because the complex (32) is neither exactly a square pyramid nor a trigonal bipyramid. Not all lines are doubled and this eliminates the possibility that two species are present. On addition of pyridine an adduct is formed and almost all the splitting disappears.[106] Paramagnetic nickel complexes, NiL_2, of the ligands 5-methyl salicylate, MeC(NR)CH·COPh and CH(NR)CH·COPh have three asymmetric centres, two on the ligand and one on the metal. The contact shifts of the ligands distinguish the active and meso forms of the complexes. The energies of configurational change were measured and it was found that tetrahedral nickel is slightly favoured in these particular complexes and that ligand exchange occurs with no overall stereoselectivity.[107] Changes in the stereochemistry of the complex cation (33) in water and acetone have been followed by n.m.r. spectroscopy.[108] The structure of some nickel–sulphur chelates has been clarified with the aid of proton n.m.r.[109]

Second- and Third-row Transition-metals (*and Uranium*).—The sequence of metals followed in this section is Mo, (U); Rh; Os, Ir; Pd, Pt.

The diamagnetism of molybdenum chelates indicates that these contain an Mo_2O_4 unit with the chelating agents wrapped around the two Mo atoms giving an AB and singlet spectrum. The chelates $(EDTA)Mo_2O_4$ and $Mo(NTA)_2$ are described (EDTA = ethylenediaminetetra-acetic acid; NTA = nitrilotriacetic acid).[110]

The extraction of uranyl ion by >N·CO·CO·N< and >N·CO·$[CH_2]_n$·CO·N< ($n = 1$ or 2) is described and proton n.m.r. data compared with the extraction data. The former are poor extractors though they should be good chelating agents and it is shown that the C=O groups are skew.[111]

The cationic rhodium(III) complex $[RhCl_2L_2]^+$ where L is $MeSCH_2{\cdot}CH_2{\cdot}SMe$ shows five closely spaced Me doublets with J(Rh—H) *ca.* 3 c./sec. This indicates that the five possible stereo-isomers of the complex had been detected in the mixture;[112] one arrangement of the methyl groups is shown in structure (34).

A rhodium chelate of hexafluoroacetylacetone (HFA), $RhCl_2(HFA)_2$, exhibits only one proton and two fluorine resonances. This suggests that structure (35) is one of the two possible structures.[113]

The two hydrogens in dihydrotetracarbonylosmium appear to be *cis* from i.r. and proton n.m.r. evidence. Substitution of one carbonyl by triphenylphosphine gives a compound in which the two protons and the phosphorus

[106] G. N. La Mar and L. Sacconi, *J. Amer. Chem. Soc.*, 1967, **89**, 2282.
[107] R. E. Ernst, M. J. O'Connor, and R. H. Holm, *J. Amer. Chem. Soc.*, 1967, **89**, 6104.
[108] L. G. Warner, N. J. Rose, and D. H. Busch, *J. Amer. Chem. Soc.*, 1967, **89**, 703.
[109] J. P. Fackler and D. Coucouvanis, *J. Amer. Chem. Soc.*, 1967, **89**, 1747.
[110] L. V. Haynes and D. T. Sawyer, *Inorg. Chem.*, 1967, **6**, 2146.
[111] T. H. Siddall, jun., and M. L. Good, *Inorg. Nuclear Chem.*, 1967, **29**, 149.
[112] R. A. Walton, *J. Chem. Soc.* (*A*), 1967, 1852.
[113] S. C. Chattoraj and R. E. Sievers, *Inorg. Chem.*, 1967, **6**, 408.

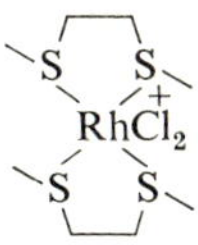

(34)

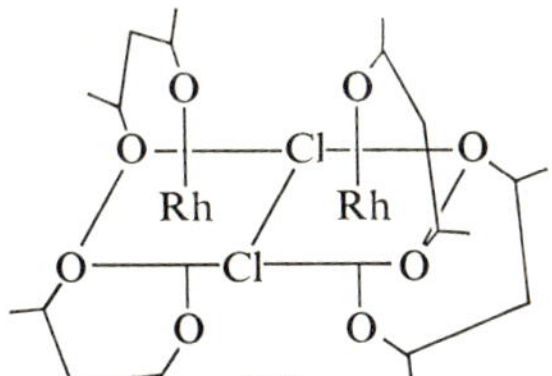

(35)

are all on one triangular face of the octahedron. The protons are at τ 18·0 with J(P—Os—H) 24 c./sec.[114]

The ester groups of the acetylene complex of iridium (36) have been shown to be chemically different.[115]

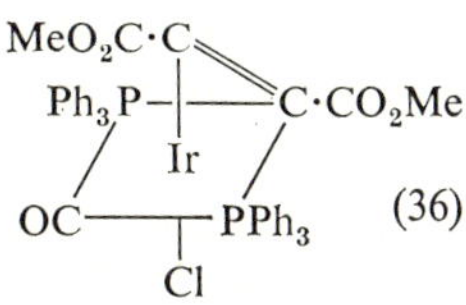

(36)

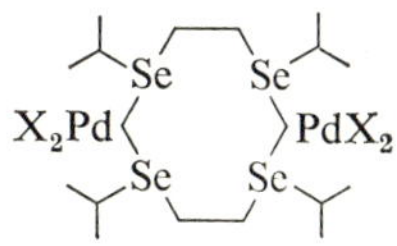

(37)

The bis-(2-pyrrolealdimine)chelates of palladium are probably mostly *trans* square planar.[116] Similar complexes of mercury are also described.

The ligand 1,2-bis-(isopropylseleno)ethane forms 1 : 1 complexes with PdX_2 and PtX_2 which can be obtained as monomers or dimers. The $—SeCH_2 \cdot CH_2Se—$ group gives a singlet proton resonance in the free ligand and an A_2B_2 pattern in the complex showing that it has become part of a ring structure. The $SeCH_2$ shifts are different in the monomer and dimer (τ monomer 9·05; τ dimer 8·5) and this is taken as additional evidence for a bridged structure (37) for the dimers.[117] Proton n.m.r. and i.r. spectroscopy have been used to determine the structure of HPt $acac_2X$. One acac group is bonded normally while the other is in the enol form and is bonded as a π complex.[118] In the octahedral trimethylplatinum complex Me_3PtX_3 it has been found that J(Pt—C—H) depends on the nature of X [X = H_2O, NH_3, $MeNH_2$, py (= pyridine), SCN^-, NO_2^-, CN^-]. By making $Me_3Pt\,py_n(H_2O)_{3-n}$ (n = 0–3) it was shown that J(Pt—C—H) depends upon the nature of the group *trans* to the methyl and is either 80 (X_{trans} = H_2O) or 67 c./sec. (X_{trans} = py). No exchange takes place.[119] The structure of the trimethyl platinum complex with X = NH_3 has been shown to be as in (38) from both i.r. and proton n.m.r. spectra;[120] J is 71 c./sec. and only one methyl triplet is seen.

[114] F. L'Eplattenier and F. Calderazzo, *Inorg. Chem.*, 1967, **6**, 2092.
[115] J. P. Collman and Jung W. Kang, *J. Amer. Chem. Soc.*, 1967, **89**, 844.
[116] Kwan-Nan Yeh and R. H. Barker, *Inorg. Chem.*, 1967, **6**, 830.
[117] N. N. Greenwood and G. Hunter, *J. Chem. Soc.* (*A*), 1967, 1520.
[118] D. Gibson, J. Lewis, and C. Oldham, *J. Chem. Soc.* (*A*), 1967, 72.
[119] D. E. Clegg and J. R. Hall, *Austral. J. Chem.*, 1967, **20**, 2025.
[120] D. E. Clegg and J. R. Hall, *Spectrochim. Acta*, 1967, **23A**, 263.

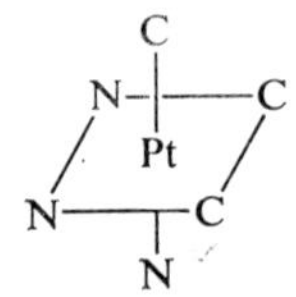

(38)

Compounds containing Tin.—The proton n.m.r. spectra of a number of heterocyclic tin compounds have been reported. For the compound (39) two methyl signals are seen and it is suggested that this occurs because of intermolecular Sn → O co-ordination.[121] The 1:3-dithio-2-stannacyclopentenes (40) have been shown to have planar rings, since if R = Me only one signal is seen.[122]

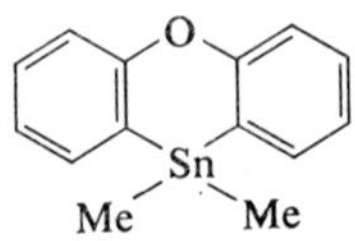

(39)

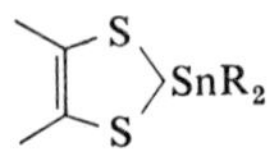

(40)

It has been suggested that the tin acetylacetonato-complex SnX_2 $acac_2$ is *trans* and exists in two forms which are differentiated by bond tautomerism or by a distorted octahedral form. The proton resonance, however, is best interpreted by a *cis* configuration in which the ligand methyl-groups are non-equivalent but can exchange. Several possible mechanisms are suggested which could cause coalescence of the lines at high temperatures.[123] More extensive results including i.r. and dipole moment data confirm that such complexes are *cis*.[124] Confirmatory evidence is also available from the similar titanium compounds TiX_2 $acac_2$ which are also shown to be *cis*.[125] The activation energies for exchange, determined by variable temperature proton n.m.r., are Ti 11·6[125] and Sn 5·4 kcal. mole^{-1}.[126] The value for the tin complex is lower than for the titanium complex but is still fairly high, certainly much higher than one would expect for simple bond tautomerism or positional vibration of the X atoms.

3 Dynamic Systems

The study of dynamic chemical processes is one of the major areas in which n.m.r. has contributed to chemistry; certainly this section contains the largest single group of related references in this review, covering 20% of the total. The general topic is considered under the headings:

[121] E. J. Kupchik, J. A. Ursino, and P. R. Bondjok, *J. Organometallic Chem.*, 1967, **10**, 269.
[122] E. W. Abel and C. R. Jenkins, *J. Chem. Soc.* (*A*), 1967, 1344.
[123] J. W. Faller and A. Davidson, *Inorg. Chem.*, 1967, **6**, 182.
[124] W. H. Nelson, *Inorg. Chem.*, 1967, **6**, 1509.
[125] R. C. Fay and R. N. Lowry, *Inorg. Chem.*, 1967, **6**, 1512.
[126] Y. Kawasaki and T. Tanaka, *Inorg. Nuclear Chem. Letters*, 1967, **3**, 13.

Rotational and conformational exchange; the dynamics of π complexed systems and of some σ complexed rings; ligand exchange; equilibria in multi-component systems; monitoring the course of reactions (including slow ligand-exchanges); and ionic solutions in both aqueous and non-aqueous media.

Rotational and Conformational Exchange.—The mechanism of fluorine positional interchange in SF_4 has been discussed.[127, 128] It is not yet possible to decide whether S—F bonds are broken or not during the interchange and experiments to decide this point by searching for ^{33}S satellites using ^{33}S-enriched SF_4 are suggested. In the case of PF_5 the fluorine resonance is a doublet so that no P—F bond breaking can occur. Linewidth studies show that this exchange rate increases with increasing PF_5 concentration and decreasing temperature and it is suggested that the intramolecular exchange is caused by intermolecular association.[129] This work suggests a further way to tackle the SF_4 problem.

The titanium tetrafluoride complex (41), where L is dimethylformamide, gives an A_2X_2 spectrum at low temperatures. The coalescence temperature is affected by an excess of L, which slows the rate of exchange. Fluorine positional exchange occurs via loss of one L, fluorine scrambling, and reassociation.[130]

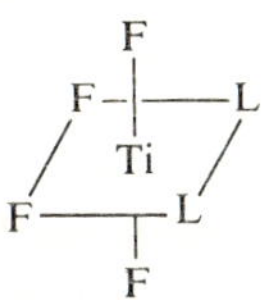

(41)

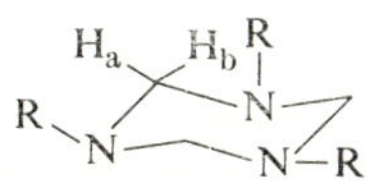

(42)

The geometry about the sulphur in the square planar sulphur–platinum complexes $(R^1R^2S)_2PtCl_2$ is shown to be pyramidal.[131] When $R^1 = R^2 =$ $PhCH_2$ the CH_2 protons are non-equivalent and give an ABX spectrum with ^{195}Pt as X, J_{AX} 29·3, J_{BX} = 55·7 c./sec. Pyramidal inversion occurs at S which hops between its lone pairs. The rate is greater in *trans*- than in *cis*-complexes.[132]

The pentacyclic ring compound $(P{\cdot}CF_3)_5$ exhibits temperature-dependent ^{19}F and ^{31}P spectra. The ^{19}F spectrum consists of two bands which come near to coalescence only at 202°. The motion is ascribed to torsional vibration of the ring giving the impression of pseudo-rotation.[133]

Ring inversion in hexahydro-1,3,5-triazines (42) has been studied by following changes in the AB proton pattern for the CH_2 groups with

[127] E. C. Muetterties and W. D. Phillips, *J. Chem. Phys.*, 1967, **46**, 2861.
[128] R. L. Redington and C. V. Berney, *J. Chem. Phys.*, 1967, **46**, 2862.
[129] S. Brownstein, *Canad. J. Chem.*, 1967, **45**, 1711.
[130] D. S. Dyer and R. O. Ragsdale, *J. Amer. Chem. Soc.*, 1967, **89**, 1528.
[131] P. Haake and P. C. Turley, *J. Amer. Chem. Soc.*, 1967, **89**, 4611.
[132] P. Haake and P. C. Turley, *J. Amer. Chem. Soc.*, 1967, **89**, 4617.
[133] E. J. Wells, H. P. K. Lee, and L. K. Peterson, *Chem. Comm.*, 1967, 894.

temperature. The *N*-alkyl groups give only one signal so that inversion at N must be rapid.[134] Hindered rotation of perfluorophenyl substituents has been observed in complexes containing sterically bulky ligands. At low temperatures the *ortho*-fluorines in $[(C_6F_5)_3P]_2MX_2$ (MX_2 = $PdBr_2$, $PtBr_2$, or PtI_2) give a 1 : 1 : 1 triplet or, when MX_2 = $PtCl_2$, a 1 : 2 doublet. In the latter case the rings are rotating freely about the P—C bond but no rotation occurs around the P—Pt bond axis.[135] A similar situation occurs for $Ph_nM(C_6F_5)_{4-n}$ (M = Ge or Sn) where if $n < 3$ the ^{19}F lines are broadened.[136] Restricted rotation has also been demonstrated about the >N—C bond in (43)[137] and about the N—N bond in (44).[138] The latter compound has been investigated in both the liquid and vapour states and it is found that in the liquid state about 2 kcal. mole^{-1} is added to the energy barrier to rotation.

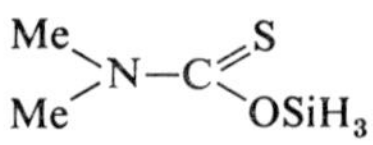

(43)

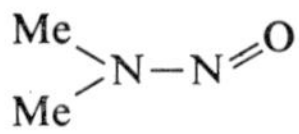

(44)

π Complexes.—The complex $(\pi\text{-}C_5H_5)Rh(C_2H_4)SO_2$ has been made from the bisethylene complex by electrophilic attack of SO_2. Restricted rotation of the ethylene is observed below −10° leading to an A_2B_2 spectrum at −50°.[139] Several papers have dealt with the problem of the exchange of π-allyl ligands and the mechanism which leads to temperature-dependent spectra. In the complex (45) two isomers are present in unequal amounts at −10° but these rapidly interconvert at 130°. This may be due to rotation about an axis through the plane of the π-allyl ligand.[140] The bis-(π-allyl) complex (46) has two forms below 5° [(46a) and (46b)] which give rise to two A_2M_2X spectra. As the temperature is raised the CH_2 peaks broaden due to both the a ⇌ b transformation and the H_2–H_3 interchange. The existence of this transformation is neatly shown by a double-resonance experiment in which irradiation of the H_2 peak of either (46a) or (46b) perturbs equally the H_3 peaks of both (46a) and (46b).[141] A more detailed analysis of the mechanisms which lead to equivalence of H_2 and H_3 in π-allyl complexes has led to the conclusion that exchange occurs via σ-allylic complexes; it is shown that a basic ligand is needed in the molecule for exchange to occur [142] and a very detailed discussion is given of several

[134] J. M. Lehn, F. G. Riddell, B. J. Price, and I. O. Sutherland, *J. Chem. Soc.* (*B*), 1967, 387.
[135] R. D. W. Kemmitt, D. I. Nichols, and R. D. Peacock, *Chem. Comm.*, 1967, 599.
[136] D. E. Fenton, A. G. Massey, K. W. Jolley, and L. H. Sutcliffe, *Chem. Comm.*, 1967, 1097.
[137] E. A. V. Ebsworth, G. Rocktaschel, and J. C. Thompson, *J. Chem. Soc.* (*A*), 1967, 362.
[138] R. K. Harris and R. A. Spragg, *Chem. Comm.*, 1967, 362.
[139] R. Cramer, *J. Amer. Chem. Soc.*, 1967, **89**, 5377.
[140] A. Davison and W. C. Rode, *Inorg. Chem.*, 1967, **6**, 2124.
[141] J. K. Becconsall and S. O'Brien, *J. Organometallic Chem.*, 1967, **9**, P 27.
[142] F. A. Cotton, J. W. Faller, and A. Musco, *Inorg. Chem.*, 1967, **6**, 179.

fast rate processes occurring in solutions containing $PdX_2(\pi\text{-allyl})_2$ and phosphine, arsine, or stibine ligands.[143] The effect on the proton n.m.r. spectra of the ligand : complex ratio is discussed. The effect of increased electron-donor capacity of the ligand in π-allylic complexes of Rh^{III} and Pt^{II} is to decrease the activation energy of exchange, some E_A values being $9{\cdot}2 \pm 1{\cdot}0$ (Ph_3P), $18{\cdot}0 \pm 1{\cdot}0$ (Ph_3As), and 20 kcal. mole^{-1} (Ph_3Sb).[144]

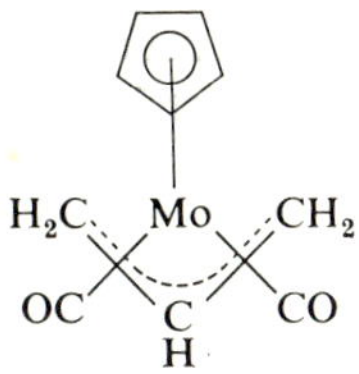

(45)

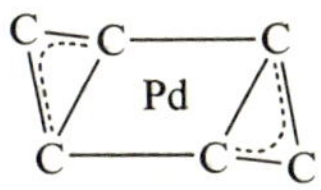

(46a)

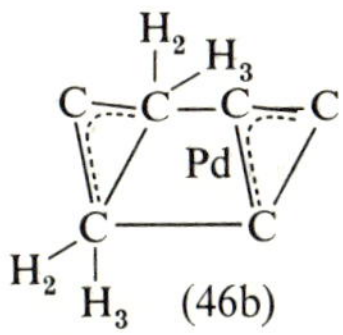

(46b)

It has been found that the *trans* and *cis* geometries of the two starting materials (47a) and (47b) used to prepare 1,3-diphenylallyl-lithium are not maintained despite the π bonding and the same product is obtained from both.[145]

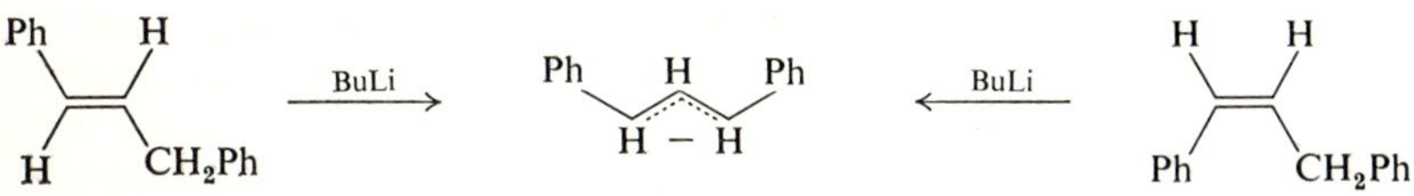

(47a) (47b)

Ring-whizzing is also a popular subject. In many cyclic polyolefin complexes a single sharp proton resonance is obtained even though the ring protons ought not to be all equivalent and some sort of ligand positional interchange is proposed. Low-temperature studies have enabled the non-equivalence to be observed in a number of cases. The C_8H_8 rings in $(C_8H_8)M(CO)_3$ (M = Cr, Mo, W) and the C_7H_7 rings in $(C_7H_7)M(CO)_2(C_5H_5)$ (M = Mo or W) show such behaviour with activation energies for interchange of 6–7 kcal. mole^{-1}; these activation energies increase in the order Mo < Cr ≪ W.[146] The structure in solution of cyclooctatetraene complexes of tricarbonyl-iron and -ruthenium has previously

[143] J. Powell and B. L. Shaw, *J. Chem. Soc.* (*A*), 1967, 1839.
[144] K. Vrieze and H. C. Volger, *J. Organometallic Chem.*, 1967, **9**, 537.
[145] H. H. Freedman, V. R. Sandel, and B. P. Thill, *J. Amer. Chem. Soc.*, 1967, **89**, 1762.
[146] R. B. King, *J. Organometallic Chem.*, 1967, **8**, 129.

been the subject of a controversy which has been clarified during the year.[147-151] Both sets of compounds are instantaneous 1 : 3 dienes and valence tautomerism occurs via 1 : 2 hops. The tub structure undergoing 1 : 5 hops is ruled out.[149] Supporting evidence is obtained from better resolved low-temperature spectra, which enable all four types of proton to be distinguished below −130°, and from the behaviour of substituted ring complexes. The disubstituted compounds (48) and (49) show no dynamic behaviour and have been shown by double-resonance experiments, by u.v., and by Mössbauer spectroscopy to be 1 : 3 diene complexes.[151]

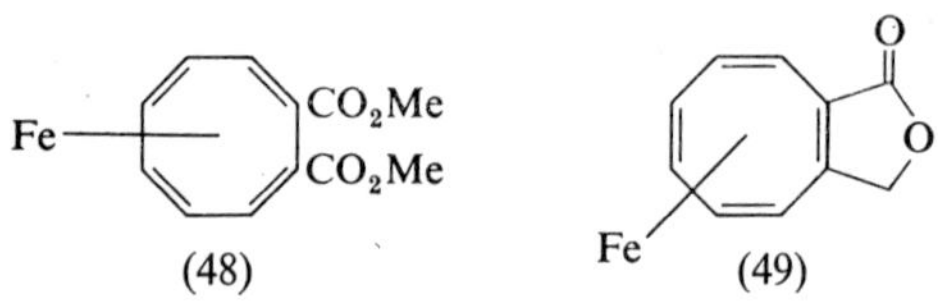

(48) (49)

That exchange is blocked is evidence in favour of 1 : 2 hops round a puckered ring with bond shifting rather than 1 : 5 jumps with the bonds static. The n.m.r. variable-temperature behaviour of the monomethyl-substituted ring compound has also been explained on the basis of a 1 : 2 hop mechanism, the iron preferring a position adjacent to the methyl group. Proof has been provided using the deuteriated compound (50)

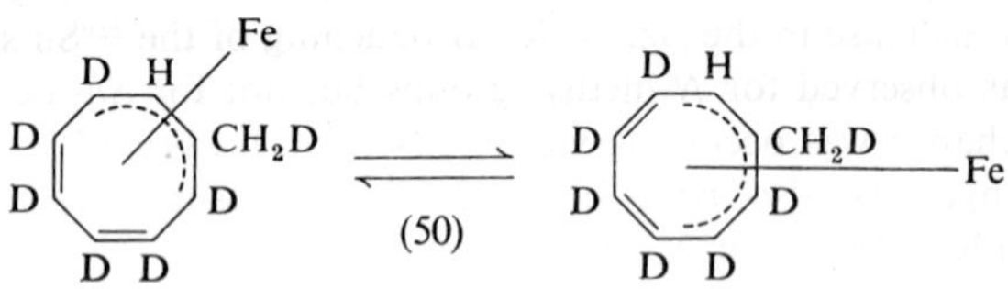

(50)

which gives two ring proton signals below −125°.[150] The compound $(C_8H_8)Ru_2(CO)_6$ gives an $A_2B_2C_2X_2$ spectrum which indicates that one olefinic bond is free and the others are statically bonded to the two ruthenium atoms.[148] The compounds $(C_8H_8)Ru_2(CO)_5$ and $(C_8H_8)_2Ru_3(CO)_4$, however, show rapid valence tautomerism. Judging from the crystal structure of the latter [152] 1 : 5 hops would be difficult in this case. A 1 : 2 shift mechanism is also proposed for the σ-cyclopentadienyl ring of $(\pi\text{-}C_5H_5)Cr(NO)_2(\sigma\text{-}C_5H_5)$ which gives an A_2B_2X spectrum only below −88·5°. In the case of the related complex (51) further evidence in support of a 1 : 2 shift is obtained since no exchange occurs up to +70°.

147 M. J. Bruce, M. Cooke, M. Green, and F. G. A. Stone, *Chem. Comm.*, 1967, 523.
148 F. A. Cotton, A. Davison, and A. Musco, *J. Amer. Chem. Soc.*, 1967, **89**, 6796.
149 F. A. L. Anet, M. D. Kaesz, A. Maasbol, and S. Winstein, *J. Amer. Chem. Soc.*, 1967, 89, 2489.
150 F. A. L. Anet, *J. Amer. Chem. Soc.*, 1967, **89**, 2492.
151 R. Grubbs, R. Breslow, R. Herber, and S. J. Lippard, *J. Amer. Chem. Soc.*, 1967, **89**, 6864.
152 M. J. Bennett, F. A. Cotton, and P. Legzdins, *J. Amer. Chem. Soc.*, 1967, **89**, 6797.

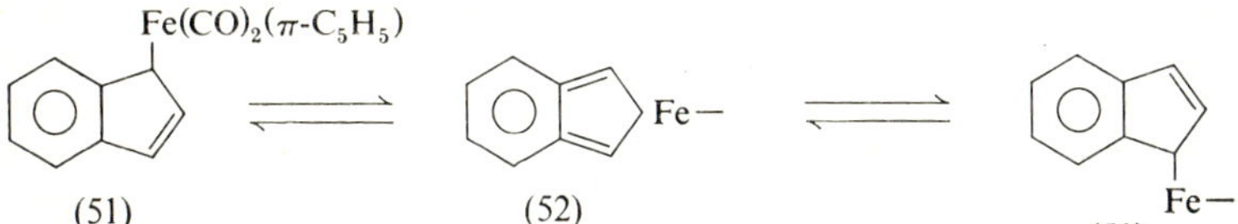

(51) (52) (51)

In this case a 1 : 2 shift is inhibited by the necessity to pass through the high-energy state (52) whereas a 1 : 3 hop mechanism would not be inhibited.[153] The σ-cyclopentadienyl complex (53) gives a well-resolved A_2B_2X spectrum at $-70°$. As the temperature is increased the A and B proton peaks fuse asymmetrically and this can be used to differentiate between a 1 : 2 or a 1 : 3 hop mechanism provided the A and B chemical shifts can be assigned correctly. In this case a 1 : 3 mechanism is tentatively postulated.[154]

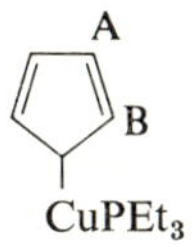

(53)

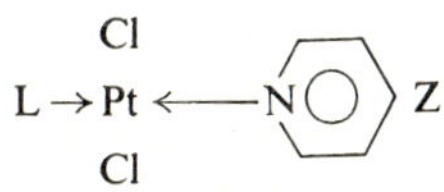

(54)

Ligand Exchange.—Studies of exchange in stannylamines $R_nSn[NRR]_{4-n}$ (R = alkyl or phenyl) have shown that exchange is favoured by a decrease in *n* and an increase in the size of R. Broadening of the ^{119}Sn satellites on heating was observed for *N*-methyl groups but not for *Sn*-methyl groups so that exchange was occurring via Sn—N bond breaking.[155] Association and exchange has also been noted in dithiophosphinates of the type $R_{3-n}M(SSPR_2)_n$ (n = 1 or 2; M = Sn or Tl).[156]

The exchange of 2-picoline (2-pic) with the paramagnetic complex $Co(2\text{-pic})_2Cl_2$ has been studied by analysing line shapes. The signal due to the excess of 2-picoline is broadened due to exchange with the paramagnetic complex. The reaction is speeded by an increase in free 2-picoline, and an activation energy of $5{\cdot}34 \pm 0{\cdot}3$ kcal mole^{-1} was obtained.[157] Ligand lability of the pyridine and olefin in (54), where L = ethylene, or *cis*- or *trans*-but-2-ene, has been studied by finding the temperatures at which the platinum–proton spin coupling can be observed in the olefin and pyridine. It was found that both were labile.[158]

The cyclo-octene–irridium complex $IrCl(CO)(C_8H_{14})_3$ will take up reversibly two molecules of ethylene, but in the presence of an excess of ethylene, rapid exchange of ethylene and cyclo-octene occurs.[159]

[153] F. A. Cotton, A. Musco, and G. Yagnpsky, *J. Amer. Chem. Soc.*, 1967, **89**, 6136.
[154] G. M. Whitesides and J. S. Fleming, *J. Amer. Chem. Soc.*, 1967, **89**, 2855.
[155] E. W. Randall, C. H. Yoder, and J. J. Zuckerman, *J. Amer. Chem. Soc.*, 1967, **89**, 3438.
[156] F. Bonati, S. Cenini, and R. Ugo, *J. Organometallic Chem.*, 1967, **9**, 395.
[157] S. S. Zumdahl and R. S. Drago, *J. Amer. Chem. Soc.*, 1967, **89**, 4319.
[158] P. D. Kaplan, P. Schmidt, and M. Orchin, *J. Amer. Chem. Soc.*, 1967, **89**, 4537.
[159] B. L. Shaw and E. Singleton, *J. Chem. Soc.* (*A*), 1967, 1683.

Many examples of ligand exchange occur in ionic solutions and these are discussed under that heading.

Equilibria in Multicomponent Systems.—Exchange between methyl-lithium and lithium bromide in ether has been demonstrated [160] and occurs via a mixed aggregate. Exchange reactions of organic groups in mixtures of organo-lithium and organo-metal compounds have been investigated by ^1H and ^7Li n.m.r. In the systems $PhLi–Ph_2M$ (M = Mg or Zn), the aggregates Li_2MPh_4 are formed when $[Li]/[M] \geqslant 2$ and $LiMPh_3$ at lower stoicheiometric ratios of lithium. Exchange of both lithium and phenyl groups occurs; the rate of the former determined by solvent-separated ion-pair formation whereas the faster phenyl exchange rate is determined by dissociation of the complex.[161] In the case of magnesium only, and with both methyl and phenyl groups present, the complexes $Li_2MgMe_{4-n}Ph_n$ are produced which at low temperatures give five ^7Li signals corresponding to $n = 0, 1, 2, 3$, and 4. In mixtures of MeLi and PhLi the complexes $[Li_2MePh]_n$ and Li_4Me_3Ph are formed.[162] The systems $EtLi+Et_2M$ in Et_2O (M = Cd, Hg, or Zn), which exhibit fast exchange, have been investigated by following the ethyl-group shift as a function of composition of the mixtures. Breaks in the curves thus obtained indicate the existence of 1 : 1 complexes (M = Cd or Zn) and also demonstrate the formation of the etherates $(EtLi)_2,Et_2O$, $LiZnEt_3,Et_2O$, $LiZnEt_3,2THF$, and $LiCdEt_3,2THF$ (THF = tetrahydrofuran).[163]

The reaction:

$$Me_2Be+BeBr_2 \underset{}{\overset{Et_2O}{\rightleftharpoons}} 2MeBrBe$$

has been shown to occur;[164] the Me_2Be and MeBrBe proton signals are separated by 0·05 p.p.m. Fluorine magnetic resonance of solutions of perfluoroaryl Grignard reagents at low temperatures has identified both RMgX and R_2Mg, and equilibrium constants have been estimated. Mixing R_2Mg and MgI rapidly gives the conventional Grignard reagent. The shifts of the R_2Mg peaks are solvent- and concentration-dependent in these mixtures and there are probably rapid equilibria involving various solvent and halide adducts.[165] $Et_2N·CH_2·CH_2·NEt_2$ forms a stable adduct with the Grignard reagent p-FC_6H_4MgBr. Exchange is slow and free and complexed amine can be distinguished; the amine slows the exchange of the aryl groups.[166] Exchange has also been demonstrated in the system $EtZnX–Et_2Zn$ by following changes in proton chemical shifts with composition. With an excess of Et_2Zn and X = I, a complex $EtZnI(Et_2Zn)_2$ is formed.[167]

160 R. Waack, M. A. Doran, and E. B. Baker, *Chem. Comm.*, 1967, 1291.
161 L. M. Seitz, and T. L. Brown, *J. Amer. Chem. Soc.*, 1967, **89**, 1602.
162 L. M. Seitz and T. L. Brown, *J. Amer. Chem. Soc.*, 1967, **89**, 1607.
163 S. Toppet, G. Slinckx, and G. Smets, *J. Organometallic Chem.*, 1967, **9**, 205.
164 E. C. Ashby, R. Sanders, and J. Carter, *Chem. Comm.*, 1967, 997.
165 D. F. Evans and M. S. Khan, *J. Chem. Soc.* (*A*), 1967, 1643.
166 D. F. Evans and M. S. Khan, *J. Chem. Soc.* (*A*), 1967, 1648.
167 J. Boersma and J. G. Noltes, *J. Organometallic Chem.*, 1967, **8**, 551.

The exchange and relaxation processes in methylmercuric iodide have received considerable attention.[168–170] It is shown that the broadening of the ^{199}Hg satellites of the methyl group in the presence of Me_nAlCl_{3-n} is not due to methyl exchange but to ^{199}Hg relaxation induced by the rapid quadrupole relaxation of the iodide, presumably via spin coupling; fast anion exchange occurs.[168, 169] The system $MeHgX + CN^- \rightleftharpoons MeHgCN + X^-$ exhibited relatively slow exchange and rate constants were calculated from changes in line shape with temperature.[170] The previously reported broadening of the ^{199}Hg satellites in MeHgSCN were found not to occur in the purified compound.[169]

Slow exchange takes place in organo-mercury compounds and equilibria were established over several days for the systems

$$R^1{}_2Hg + R^2{}_2Hg \rightleftharpoons 2R^1R^2Hg.$$

Proton n.m.r. spectroscopy was used to analyse the systems where diphenylmercury was a component and, in general, non-random distribution of products was found.[171]

Rates of optical inversion have been measured for aluminium diketonato-complexes of type $Al(AA)_2(BB)$ and $Al(AA)(BB)_2$ where (AA) and (BB) are symmetrical bidentate ligands. Racemic mixtures are obtained, but the study can be carried out without resolving the mixtures by following the collapse of the $B{-}CH_3$ and $B^*{-}CH_3$ resonances as the temperature is increased (55). Rates of ligand exchange were also studied.[172]

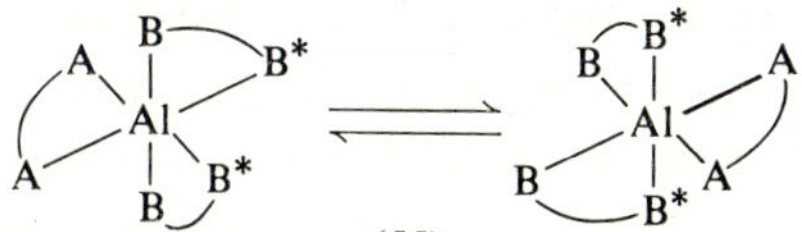

(55)

Studies of the methyl proton resonances in a solution of mixtures of Me_3Al,OEt_2 and Me_2EtAl,OEt_2 have shown that exchange takes place between the etherates as well as between the free trialkyls though at a slower rate. The rate is decreased by decreasing the aluminium concentration or by adding an excess of ether.[173] Donor exchange on the adducts of Me_3Ga with Me_2NH (i), $MeNH_2$ (ii), NH_3 (iii), and Me_2O (iv) has also been studied. With (i) exchange occurs via a dissociation step while for (ii) and (iii) electrophilic displacement is required; exchange of (iv) was too fast to study.[174] In $(PhAlMe_2)_2$ the phenyl groups act as the bridging

[168] N. S. Ham, E. A. Jeffery, T. Mole, and S. N. Stuart, *Chem. Comm.*, 1967, 254.
[169] D. N. Ford, P. R. Wells, and P. C. Lauterbur, *Chem. Comm.*, 1967, 616.
[170] R. B. Simpson, *J. Chem. Phys.*, 1967, **46**, 4775.
[171] G. F. Reynolds and S. R. Daniel, *Inorg. Chem.*, 1967, **6**, 480.
[172] J. J. Fortman and R. E. Sievers, *Inorg. Chem.*, 1967, **6**, 2022.
[173] N. S. Ham, E. A. Jeffery, T. Mole, and J. K. Saunders, *Austral. J. Chem.*, 1967, **18**, 2641.
[174] J. B. De Roos and J. P. Oliver, *J. Amer. Chem. Soc.*, 1967, **89**, 3970.

groups, since only one Me signal is seen, even at low temperatures; phenyl groups also act as bridging groups in Me_5Al_2Ph.[175]

Redistribution reactions in monomethyl-silicon and -germanium compounds, in silanes and germanes, and in mixtures of these two, have been studied and shifts assigned to the Me groups of $MeMZ_{3-n}T_n$ (M = Ge or Si; Z, T = halogen or donor ligands) for n = 0–3. If M = Si the ligands Z and T are distributed at random whereas if M = Ge the mixed products are preferred.[176-178]

Exchange has been demonstrated in chloroform and methylene chloride solutions of dimethyltin 8-hydroxyquinolatetropolonate [(ox)(T)], three methyl peaks being seen below 87°:[179]

$$2Me_2Sn(ox)(T) \rightleftharpoons Me_2Sn(ox)_2 + Me_2Sn(T)_2.$$

Similarly the complex $(Me_3SbS)_2SnMe_2Cl_2$ in chloroform solution gives $Me_3SbS + Me_3SbCl_2 + (Me_2SnS—)_2$ and can be prepared from a solution of these three components.[180] Trimethyltin formate and acetate polymers have been obtained in soluble forms which show concentration-dependent tin–methyl proton coupling constants whose magnitude indicates that in the formate, four-co-ordinate tin monomers are in equilibrium with small five-co-ordinate polymer units, whereas in the acetate the four-co-ordinate form predominates.[181] The chloroacetates were also studied and were found to be associated in CCl_4 but not in $CHCl_3$, which may hydrogen-bond to the -CO_2^- groups.[182]

Information has been obtained about the conformational interchange of diamine ligands in $Co^{III}X_4$ en complexes (en = ethylenediamine).[183] The hydrolysis equilibria of *trans*-dinitrobis(acetylacetonato)cobalt(III) have also been worked out.[184] The interaction between the Co^{II} *meso*-porphyrin complex and nitrobenzene has been studied by observing the changes in pseudo-contact-shift of the nitrobenzene protons with concentration.[185]

The extent of the formation of derivatives during the competitive Friedel–Crafts acetylation of π-cyclopentadienyl–metal complexes has been determined by proton n.m.r. spectroscopy and has enabled a series of complexes to be placed in an order of decreasing reactivity.[186]

Monitoring the Course of Reactions.—Reactions of amine complexes of cobalt have been extensively investigated by n.m.r. techniques. The sub-

[175] E. A. Jeffery, T. Mole, and J. K. Saunders, *Chem. Comm.*, 1967, 696.
[176] K. Meodritzer and J. R. Van Wazer, *J. Inorg. Nuclear Chem.*, 1967, **29**, 1571.
[177] K. Meodritzer and J. R. Van Wazer, *J. Inorg. Nuclear Chem.*, 1967, **29**, 1851.
[178] S. Cradock and E. A. V. Ebsworth, *J. Chem. Soc.* (*A*), 1967, 1226.
[179] M. Komura, T. Tanaka, and T. Mukai, *Inorg. Nuclear Chem. Letters*, 1967, **3**, 17.
[180] Mizuo Shindo and Rokuro Okawara, *Inorg. Nuclear Chem. Letters*, 1967, **3**, 75.
[181] P. B. Simons and W. A. G. Graham, *J. Organometallic Chem.*, 1967, **8**, 479.
[182] P. B. Simons and W. A. G. Graham, *J. Organometallic Chem.*, 1967, **10**, 457.
[183] D. A. Buckingham, L. Durham, and A. M. Sargeson, *Austral. J. Chem.*, 1967, **20**, 257.
[184] B. P. Cotsoradis and R. D. Archer, *Inorg. Chem.*, 1967, **6**, 800.
[185] H. A. O. Hill, B. E. Mann, and R. J. P. Williams, *Chem. Comm.*, 1967, 906.
[186] E. O. Fischer, M. van Foerster, C. G. Kreiter, and K. E. Schwarzhans, *J. Organometallic Chem.*, 1967, **7**, 113.

stitution reactions of the ligand Y in $(RNH_2)_5CoY$ (R = H or Me) have been followed and the effect of the Y upon the lability of the amine protons found. Chloride ligand is replaced by water in acid solution (R = Me). The methyl proton spectrum of $[(MeNH_2)_5CoCl]^{2+}$ consists of a triplet and a singlet, which indicates that the protons of the $MeNH_2$ *trans* to the chloride are labilised. Once the chloride is replaced by water these protons are no longer labile and can no longer be deutierated.[187] The replacement of NO_2 by water in $[(NH_3)_5CoNO_2]^{2+}$ has also been followed by proton resonance and additional tracer studies using ^{18}O indicate that the water oxygen is derived principally from the NO_2^- ligand. The suggested mechanism is

$$[(NH_3)_5Co(NO_2)]^{2+} + H^+ \longrightarrow$$

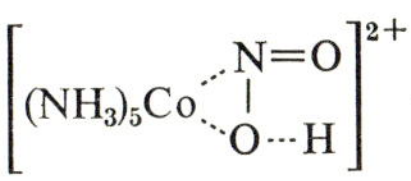

$$\longrightarrow [(NH_3)_5Co(OH)]^{2+} + NO^+$$

$$[(NH_3)_5Co(OH)]^{2+} + H^+ \longrightarrow [(NH)_3Co(H_2O)]^{3+}$$

The water can be replaced by HSO_4^- in concentrated sulphuric acid.[188]

The stereochemistry of base hydrolysis of the ^{15}N-labelled cation $[Co(^{14}NH_3)_4(^{15}NH_3)X]^{2+}$ with the $^{15}NH_3$ group *trans* to X has been studied; X^- is replaced by OH^- to give a product in which 50% of the OH^- is *trans* and 50% *cis* to the $^{15}NH_3$ group, implying a common intermediate during the hydrolysis. The $^{15}NH_3$ group gives a doublet; one doublet is obtained before hydrolysis and two equal doublets after hydrolysis.[189] Racemisation and proton exchange have been followed in the complex ions (56) and (57)

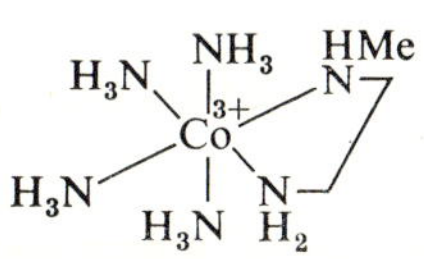

(56)

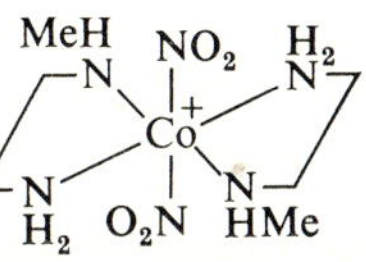

(57)

by following changes in the *N*-methyl spectrum upon deuteriation. In the case of (56) slow deuterium exchange occurs (detected after time of the order of 300 min.) and racemisation is slow[190] whereas for (57) it was

[187] M. Parris, *J. Chem. Soc.* (*A*), 1967, 583.
[188] A. D. Harris, R. Stewart, D. Hendrickson, and W. L. Jolly, *Inorg. Chem.*, 1967, **6**, 1052.
[189] D. A. Buckingham, I. I. Olsen, and A. M. Sargeson, *J. Amer. Chem. Soc.*, 1967, **89**, 5129.
[190] D. A. Buckingham, L. G. Marzilli, and A. M. Sargeson, *J. Amer. Chem. Soc.*, 1967, **89**, 825.

found to be much faster than the rate of racemisation.[191] For the triethylenetetramine complexes (58) it is found that when X = Cl^-, no NH proton exchange with D_2O solvent occurs if there is no change in stereochemistry when Cl^- is replaced by water, but if the complex changes to the β-form [see (27)] then NH proton exchange does occur. NH Proton exchange also occurs when X = H_2O and the stereochemistry changes to the β-form.[192] With the bidentate ligands glycine or sarcosine, in place of the two monodentate ligands X in (58), the NH protons are stable in acid solution but are lost to solvent in neutral or alkaline solution with sharpening of the methyl resonances.[193] The rate of α-proton exchange in the L-(+)-valine and L-(+)-alanine complexes D- and L-[Co en_2(AA)]$^{2+}$ have been studied by proton n.m.r.[194]

The hexa-amminenickel(II) cation reacts with acetone to give the complexes $[NiL_3]^{2+}$ and $[NiL_2]^{2+}$, in which L was shown to be $NH_2{\cdot}CMe_2{\cdot}CH_2{\cdot}CMe{:}NH$ on the basis of proton n.m.r. and other techniques; the structure of $[NiL_2]^{2+}$ is shown in (59).[195]

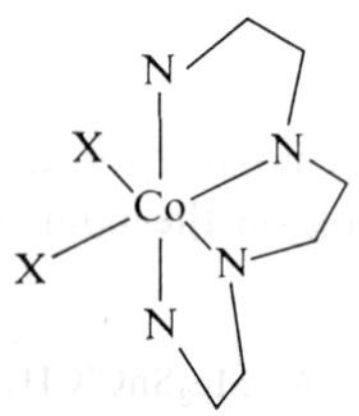

(58)

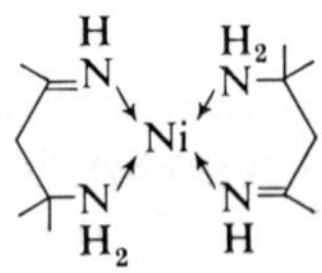

(59)

The reactions of chlorotris(triphenylphosphine)rhodium(I):

$$(s)RhCl(PPh_3)_3 \rightleftharpoons (s)RhCl_2H(PPh_3)_2 \longrightarrow RhCl_2R(PPh_3)_2$$

where s is a solvent molecule, have been followed by proton n.m.r. The PPh_3 groups are *cis* and equivalent. Carbon monoxide can be inserted to give, *e.g.* $RhCl_2(CO{\cdot}CH{:}CH_2)(PPh_3)_2$.[196]

A miscellaneous series of reactions followed by n.m.r. spectroscopy includes the insertion of sulphur dioxide into the π-allyl complex $(CO)_5Mn(\pi\text{-}C_3H_5)$ which leads to re-arrangement of the π-allyl group to $(CO)_5MnSO_2(CH_2{\cdot}CH{:}CH_2)$.[197] The olefinic complex (60) re-arranges to a

[191] D. A. Buckingham, L. G. Marzilli, and A. M. Sargeson, *J. Amer. Chem. Soc.*, 1967, **89**, 3428.
[192] D. A. Buckingham, L. G. Marzilli, and A. M. Sargeson, *Inorg. Chem.*, 1967, **6**, 1032.
[193] L. G. Marzilli and D. A. Buckingham, *Inorg. Chem.*, 1967, **6**, 1042.
[194] D. A. Buckingham, L. G. Marzilli, and A. M. Sargeson, *J. Amer. Chem. Soc.*, 1967, **89**, 5133.
[195] N. J. Rose, M. S. Elder, and D. M. Busch, *Inorg. Chem.*, 1967, **6**, 1924.
[196] M. C. Baird, J. T. Magne, J. A. Osborn, and G. Wilkinson, *J. Chem. Soc.* (*A*), 1967, 1347.
[197] F. A. Hartman, P. J. Pollick, R. L. Downs, and A. Wojcicki, *J. Amer. Chem. Soc.*, 1967, **89**, 2493.

mixture of two π-allylic complexes as shown.[198] Methyl cyanide is lost when $W_3Cl_6\,py_4$,MeCN is heated and W_2Cl_6(4-isopropylpyridine)$_4$ exists in two forms.[199] Hydroxymercuriated propene $^+HgCH_2\cdot CH(OH)Me$ decomposes to acetone.[200] A series of monosodium acetonitriles have been made and it is shown that $CH_2(Na)CN$ is a mixture of tautomers, containing 75% of the product together with olefinic, acetylenic- and amino-protons.[201]

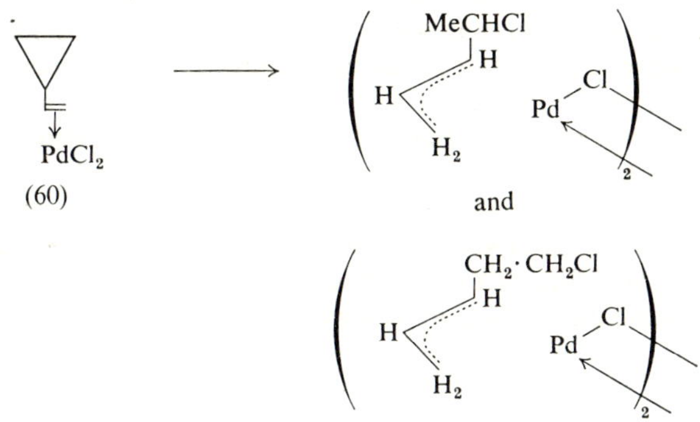

(60)

The reaction between trimethyltin methoxide and β-propiolactone can proceed in either of two ways depending on the point of cleavage of the lactone:

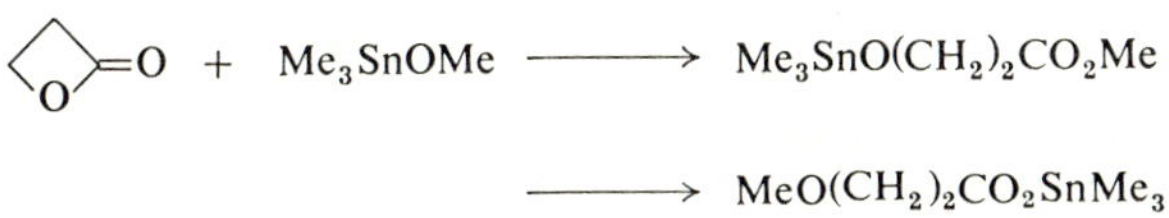

The dependence of the ratio of the two products on solvent has been determined using proton n.m.r. and it was shown that the transition state for *O*-alkyl bond cleavage (to give the second product) has the greater dipole character.[202] The proton n.m.r. spectrum of bis(t-butyl)stannane shows that it decomposes either by disproportionation or coupling to give a new species:

$$2Bu^t_2SnH_2 \rightarrow Bu^t_3Sn^+ + Bu^tSnH_3$$

$$2Bu^t_2SnH_2 \rightarrow Bu^t_2Sn(H){-}Sn(H)Bu^t_2 + H_2$$

Double irradiation of the Bu^t groups allows observation of the tin-hydrogen atoms and indicates the presence of a long-distance H—C—C—Sn—H coupling.[203]

[198] A. D. Ketley and J. A. Braatz, *J. Organometallic Chem.*, 1967, **9**, P 5.
[199] R. Saillant, J. L. Hayder, and R. A. D. Wentworth, *Inorg. Chem.*, 1967, **6**, 1497.
[200] Y. Saito and M. Matsuo, *J. Organometallic Chem.*, 1967, **10**, 524.
[201] C. Kruger, *J. Organometallic Chem.*, 1967, **9**, 125.
[202] K. Itoh, S. Kobayashi, S. Sakai, and Y. Ishii, *J. Organometallic Chem.*, 1967, **10**, 451.
[203] J. C. Maire and J. Dufermont, *J. Organometallic Chem.*, 1967, **10**, 373.

In aqueous solutions the ^{35}Cl resonance of the chloride ion broadened in the presence of mercury(II) chloride because of chlorine exchange with the $[HgCl_4]^{2-}$ ion. Poly-L-glutamate will complex with the mercury(II) chloride and its addition causes changes in the ^{35}Cl linewidth; this has enabled the pH-dependent helix–random-coil transition of the poly-L-glutamate chain to be followed.[204] The hydrolytic polymerisation of iron(III) citrate has been followed in a similar way. As excess of citrate (cit) is added at high pH the bulk-solvent proton relaxation times decrease because an anionic chelate $[Fe\,cit_2]^-$ is formed which is more effective in inducing relaxation than is the iron bound in the polymer.[205]

The mechanism of the formation of P_2I_4 from its elements in carbon disulphide has been elucidated by following the reaction using ^{31}P n.m.r.[206] PI_3 is formed first (all P—P bonds broken) and this undergoes slow conversion to P_2I_4 with excess of phosphorus. Growth and decay curves are given for PI_3 and P_2I_4. The ^{31}P shifts of PI_3 and P_2I_4 are reassigned as -178 and -108 p.p.m. respectively from H_3PO_4.

The extent of the reaction:

$$R^1R^2PMe + Bu^tLi \longrightarrow R^1R^2PCH_2Li + Bu^tH$$

was followed by ^{31}P n.m.r. and the ^{31}P spectra are given for many derivatives of the product, *e.g.* $Ph_2P(S)CH_2C(OH)Ph_2$, etc.[287]

Ionic Solutions.—This section deals with weak solvation complexes of ions, which invariably show exchange at ordinary temperatures. The situation is dominated by the ionic nature of the constituents rather than their ability to form stable complexes.

(*a*) *Aqueous Solutions.*—By studying a sufficiently concentrated solution of aluminium trichloride at $-47°$ it has been possible to obtain separate proton resonances for the bulk- and solvation-water and to show that the aquated aluminium ion is $[Al(H_2O)_6]^{3+}$; ΔH, for proton exchange, is calculated to be 24 ± 3 kcal. mole^{-1}, though this is felt to be too large.[208] This may not be so since it has been demonstrated that the Al—O—H bonding in the hydrated aluminium cation has sufficient covalent character to transmit ^{27}Al–^{1}H spin coupling[209] and must be quite strong. The low lability of the Al—O bond of the complex has also been demonstrated using ^{27}Al n.m.r. though it is found that rapid proton exchange does take place.[210] The hydration number of Ga^{3+} has been estimated by two methods to be $5{\cdot}9 \pm 0{\cdot}1$; one method used an empirical proton line-broadening effect due to added manganese(II) ions, and the other involved comparing the ^{17}O signals from complexed- and bulk-water, the latter being shifted by

[204] R. G. Bryant, *J. Amer. Chem. Soc.*, 1967, **89**, 2497.
[205] J. G. Spiro, L. Pape, and P. Saltman, *J. Amer. Chem. Soc.*, 1967, **89**, 5555.
[206] R. L. Carroll and R. P. Carter, *Inorg. Chem.*, 1967, **6**, 401.
[207] D. J. Peterson, *J. Organometallic Chem.*, 1967, **8**, 199.
[208] R. E. Schuster and A. Fratiello, *J. Chem. Phys.*, 1967, **47**, 1554.
[209] L. D. Supran and N. Sheppard, *Chem. Comm.*, 1967, 832.
[210] J. W. Akitt, N. N. Greenwood, and D. Lester, Chem. Soc. Autumn Meeting, Durham, Sept. 1967.

adding a paramagnetic ion (Dy^{3+}).[211] This paper led to some controversy.[212, 213] It was also concluded that thorium forms the complex $[Th(H_2O)_{10}]^{4+}$ and that H^+ and NH_4^+ do not form long-lived hydration spheres.[211] A similar conclusion was reached for Li^+ from proton relaxation times of ^7LiCl and ^6LiCl solutions.[214] A theoretical treatment of proton relaxation times in aqueous solutions also suggests that dynamically well defined hydration spheres are absent from alkali-halide solutions.[215–217] Anion–cation–solvent interactions in sodium salt solutions have been inferred from ^{23}Na relaxation measurements.[218] Proton chemical shifts in melts of $Ca(NO_3)_2(H_2O)_4$ with added salts indicate that the Ca^{2+} ion is selectively hydrated in the presence of K^+, NMe_4^+, and NO_3^-; Mg^{2+}, however, competes successfully with Ca^{2+} for water and alters the position of the water proton resonance when its salts are added.[219]

The structure of aqueous solutions has been investigated by ^{17}O n.m.r. spectroscopy [220] and the ^{17}O shifts in water at various temperatures and with respect to steam have been used to investigate hydrogen-bonding interactions in water.[221] The ^{17}O shifts behave similarly to the proton shifts. Proton exchange in water has also been studied by ^{17}O n.m.r. spectroscopy.[222] The hydration of the complex $[PtMe_3(H_2O)_n]^+$ has been investigated using ^{17}O n.m.r. in the presence of Dy^{3+} whence $n = 3{\cdot}0 \pm 0{\cdot}1$; the water exchanges rapidly.[223]

Three papers have dealt with the hydration of the vanadyl ion (61). The equatorial water molecules exchange slowly (τ 1·35 × 10^{-3} sec.) whereas the water *trans* to oxygen exchanges rapidly (τ *Ca.* 10^{-11} sec.). The number of equatorial water molecules involved were estimated by adding a paramagnetic ion (Dy^{3+}) to pure water and to a VO^{2+} solution and comparing the solvent shifts.[224, 225]

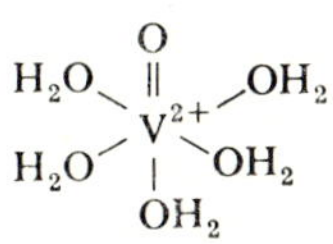

(61)

[211] T. J. Swift, O. G. Fritz, jun., and T. A. Stephenson, *J. Chem. Phys.*, 1967, **46**, 406.
[212] S. Meiboom, *J. Chem. Phys.*, 1967, **46**, 410.
[213] T. J. Swift and W. G. Sayre, *J. Chem. Phys.*, 1967, **46**, 410.
[214] B. F. Fabricand and S. S. Goldberg, *Mol. Phys.*, 1967, **13**, 323.
[215] H. G. Hertz, *Chem. Ber.*, 1967, **71**, 979.
[216] H. G. Hertz, *Chem. Ber.*, 1967, **71**, 999.
[217] L. Endom, H. G. Hertz, B. Thül, and M. D. Zeidler, *Chem. Ber.*, 1967, **71**, 1008.
[218] M. Eisenstadt and H. L. Friedman, *J. Chem. Phys.*, 1967, **46**, 2182.
[219] C. T. Moynihan and A. Fratiello, *J. Amer. Chem. Soc.*, 1967, **89**, 5546.
[220] F. Fister and H. G. Hertz, *Chem. Ber.*, 1967, **71**, 1032.
[221] A. E. Florin and Mohammed Alei, *J. Phys. Chem.*, 1967, **47**, 4268.
[222] S. W. Rabideau and H. C. Hecht, *J. Chem. Phys.*, 1967, **47**, 544.
[223] G. E. Glass and R. S. Tobias, *J. Amer. Chem. Soc.*, 1967, **89**, 6371.
[224] J. Reuben and D. Fiat, *Inorg. Chem.*, 1967, **6**, 579.
[225] K. Wuthrich and R. E. Connick, *Inorg. Chem.*, 1967, **6**, 583.

The transfer of protons from (61) and from the aquated chromic ion to bulk-water have been measured by proton spin relaxation studies.[226, 227] The transfer of complete water molecules from the uranyl ion to bulk water has been followed by monitoring the change in ^{17}O signal intensity of a ^{17}O-enriched UO_2^{2+} sample. This ion, which has the extremely large chemical shift of -1115 p.p.m. from water, has two ^{17}O environments.[228] The pseudo-contact ^{17}O shifts of water in the presence of lanthanide ions are altered in magnitude by the presence of nitrate and acetate anions, which indicates that they displace water from the inner co-ordination sphere of the lanthanides.[229–231]

The hydration number of Co^{2+} is maintained at six over the range $-10°$ to $+183°$. The rate of exchange at 27° is given by τ $4{\cdot}2 \times 10^{-7}$ sec. and ΔH 10·4 kcal. mole^{-1}.[232] Proton n.m.r. investigations of Co^{2+} solutions containing monodentate ligands indicate that the bulk-water shift does not necessarily decrease as water is displaced from the co-ordination sphere. Magnetically anisotropic complexes can provide a pseudo-contact shift which differs markedly from the isotropic shift of the hexa-aquo ion. There is, therefore, danger in interpreting water chemical shifts in the presence of ligands in terms of hydration numbers of paramagnetic ions.[233] This is illustrated in the data shown:

Ion	$[Co(H_2O)_6]^{2+}$	$[Co(CNS)(H_2O)_5]^+$
Shift of bound water (c./sec., H_2O std.)	4470	6400

Ion	$[Co(CNS)_2(H_2O)_4]$	$[Co(CNS)_3(H_2O)_3]^-$
Shift of bound water (c./sec., H_2O std.)	6800	14400

Relaxation times for water protons in 0·05M-Cu^{2+} solutions lengthen as ethylenediamine is added and indicate that a bis-chelate is formed. If an excess of ethylenediamine is added the relaxation times shorten due to N—H proton exchange and this effect can be suppressed if the ligand is *N*-alkylated.[234] The chelating abilities of various derivatives of nitrilotriacetic acid and its *N*-oxide have been determined using ^1H and ^{31}P n.m.r. at different pH values.[235] The proton data offer strong evidence for nitrogen involvement in the final deprotonation step of nitrilotriacetic acid.[236] The addition of ions to $MeN(CH_2CO_2^-)_2$ or $HO{\cdot}CH_2{\cdot}CH_2{\cdot}N(CH_2CO_2^-)_2$ leads to a high-field proton shift rather than the low-field shift expected on

226 T. J. Swift, T. A. Stephenson, and G. R. Stein, *J. Amer. Chem. Soc.*, 1967, **89**, 1611.
227 C. E. Manley and V. L. Pollak, *J. Chem. Phys.*, 1967, **46**, 2106.
228 S. W. Rabideau, *J. Phys. Chem.*, 1967, **71**, 2747.
229 I. Abrahamer and Y. Marcus, *Inorg. Chem.*, 1967, **6**, 2103.
230 J. Reuben and D. Fiat, *Israel. J. Chem.*, 1967, **5**, 33P.
231 J. Reuben and D. Fiat, *Chem. Comm.*, 1967, 729.
232 A. M. Chmelnick and D. Fiat, *J. Chem. Phys.*, 1967, **47**, 3986.
233 W. de W. Horrocks and J. R. Hutchinson, *J. Chem. Phys.*, 1967, **46**, 1703.
234 M. Griffel, *J. Phys. Chem.*, 1967, **71**, 3284.
235 R. P. Carter, M. M. Crutchfield, and R. R. Irani, *Inorg. Chem.*, 1967, **6**, 943.
236 R. P. Carter, R. C. Carroll, and R. R. Irani, *Inorg. Chem.*, 1967, **6**, 939.

electron-donor shift or electrostatic grounds and it is concluded that complex formation profoundly affects electron densities in the molecule.[237]

(*b*) *Non-aqueous Solutions.*—References in this section are ordered according to the solvent used, *i.e.* NH_3, dimethyl formamide (DMF), dimethyl acetamide (DMA), dimethyl sulphoxide (DMSO), MeCN, $CDCl_3$, MeOH.

Liquid ammonia solvates many ions. Magnesium gives the ion $[Mg(NH_3)_5]^{2+}$ with rapid intramolecular exchange of ammonia molecules [238] whereas aluminium forms $[Al(NH_3)_6]^{3+}$, this latter figure being obtained from ^{14}N measurements in the presence of Cu^{2+} which broadened the bulk-solvent line to zero intensity.[239] A series of hydride anions X^- ($X^- = OH^-$, SH^-, SeH^-, PH_2^-, AsH_2^-, and SbH^-) show slow proton exchange with NH_3 but fast exchange with X^-, MX, and NH_4^+.[240]

Below 0° bulk and solvating DMF can be distinguished in solutions of Be^{II} in DMF. The solvate $[Be(DMF)_4]^{2+}$ is present and reacts with $[Be\ acac_2]^{2+}$ to give the mixed complex $[Be\ acac(DMF)_2]^{2+}$ and lines for all three components can be distinguished below +5°.[241] Solutions of $AlCl_3$ in dimethylformamide show that the ion $[Al(DMF)_6]^{3+}$ is formed and that exchange with bulk solvent is slow. The ^{27}Al line is fairly narrow indicating a symmetrical species. The *N*-methyl lines remain nonequivalent in the ion and the HCO proton is shifted most on complexing so that co-ordination probably occurs through the oxygen.[242] ΔH for DMF exchange is less for that of water being about 15–18 kcal. mole^{-1}.[242, 243] Co-ordination numbers (in parentheses) and ΔH values are also reported for Ga^{3+} (6), $\Delta H = 21$; $SbCl_5$ (1), $\Delta H = 10$; and $TiCl_4$ (2), $\Delta H = 28$ kcal. mole^{-1}.[242–244] It has also been shown that DMF will complex with Ln(2,2,6,6-tetramethyl-3,5-heptanedione)$_3$. Complexed and bulk DMF signals were seen and the methylene protons of the ketone are shifted upfield.[245] *N*-Methylacetamide–water mixtures also show promising results.[246] In this case, however, it is the N—H protons which suffer the largest shifts. The tin complexes $[Me_2SnL_4]^{2+}$ (L = DMF or DMSO) give two SnMe and ligand peaks possibly due to different ion-pairs. The signals from $[Me_3SnL_2]^+$ change with time and this may also be an ion-pair effect.[247]

In DMF, DMA, or DMSO *cis*- and *trans*-$[CoXY\ en_2]^+$ form ion-pairs with anions, though pairing is stronger in the *cis*-complex because of its

[237] G. H. Noncollas and A. C. Park, *J. Phys. Chem.*, 1967, **71**, 3678.
[238] T. J. Swift and H. H. Lo, *J. Amer. Chem. Soc.*, 1967, **89**, 3988.
[239] H. H. Glaeser, H. W. Dodgen, and J. P. Hunt, *J. Amer. Chem. Soc.*, 1967, **89**, 3065.
[240] G. H. Sheldrick, *Trans. Faraday Soc.*, 1967, **63**, 1065.
[241] N. A. Matwiyoff and W. G. Movius, *J. Amer. Chem. Soc.*, 1967, **89**, 6077.
[242] A. Fratiello, D. P. Miller, and R. Schuster, *Mol. Phys.*, 1967, **12**, 111.
[243] A. Fratiello and R. Schuster, *J. Phys. Chem.*, 1967, **71**, 1948.
[244] W. G. Movius and N. A. Matwiyoff, *Inorg. Chem.*, 1967, **6**, 847.
[245] J. E. Schwarberg, D. R. Gere, R. E. Sievers, and K. J. Eisentrant, *Inorg. Chem.*, 1967, **6**, 1933.
[246] J. F. Hinton and E. S. Amis, *Chem. Comm.*, 1967, 100.
[247] V. G. Kumar Das and W. Kitching, *J. Organometallic Chem.*, 1967, **10**, 59.

larger dipole moment. The proton n.m.r. spectra of the *cis*-complex indicates that the NH protons play a part in the association.[248] Ni^{II} and Co^{II} dissolved in DMSO give a single peak at room temperature, the position of which depends upon metal concentration due to exchange between bulk solvent and complexed DMSO, $[M(DMSO)_6]^{2+}$; kinetic parameters are derived for the nickel complex. The shifts are pure contact shifts.[249] The exchange rate of DMSO in a series of complexes *cis*-$[CoX(DMSO)\,en_2]^{2+}$ (X = Cl, Br, NO_2, or DMSO), when dissolved in fully deuteriated DMSO has been measured by comparing areas of solvent and solvate DMSO peaks as exchange proceeded with the deuteriated solvent.[250]

Different N—H proton resonances have been seen in $[Co\,en_3]^{3+}$ which is thought to be due to the 'freezing' of the ethylenediamine ring by hydrogen bonding to the solvent. The activation energy for ring flexing is $10{\cdot}5 \pm 0{\cdot}5$ kcal. $mole^{-1}$.[251]

In concentrated aqueous ethylenediamine solutions of Cu^{II} the ^{14}N linewidths are affected both by relaxation times and residence times in the first solvation sphere. The diamine exchanges as a whole between bulk solution and the co-ordination sphere.[252] The kinetic parameters for solvent exchange around Ni^{II} and Co^{II} in methyl cyanide have also been measured. At low temperatures solvent bound to cobalt could be distinguished and it was found that $n = 5{\cdot}3 \pm 0{\cdot}3$ in $[Co(MeCN)_n]^{2+}$. Probably only one complex is formed for each metal and these do not interact with either BF_4^- or ClO_4^-.[253–255]

Ion-pairing has been detected in $CDCl_3$ solution for $[Ph_4As]^+[Ph_3PCoI_3]^-$ which shows large cationic phenyl proton chemical shifts due to pseudo-contact interaction with the anion. Calculations suggest that the cation lies along a C_3 axis of the anion with a C_2 axis concurrent with the anion C_3 axis and at a distance of $9{\cdot}0 \pm 0{\cdot}8$ Å.[256]

The exchange rate of methanol between bulk solvent and $[Mg(MeOH)_n]^{2+}$ has been followed by measuring OH line broadening.[257] It is shown that bulk-ligand exchange and not proton exchange occurs and the value $n = 6$ was measured. Addition of water produces mixed solvates of form $[Mg(H_2O)_n(MeOH)_{6-n}]^{2+}$. No spin coupling was observed between ^{25}Mg and MeOH and this, together with the temperature dependence of the complexed-methanol shift, suggests that this methanol may be hydrogen bonded into a cage-like structure. No separated solvent–ligand signals

[248] W. A. Millen and D. W. Watts, *J. Amer. Chem. Soc.*, 1967, **89**, 6858.
[249] S. Thomas and W. L. Reynolds, *J. Chem. Phys.*, 1967, **46**, 4164.
[250] I. R. Lantzke and D. W. Watts, *Austral. J. Chem.*, 1967, **20**, 173.
[251] B. M. Fung, *J. Amer. Chem. Soc.*, 1967, **89**, 5788.
[252] M. Alei, W. B. Lewis, A. B. Denison, and L. O. Morgan, *J. Chem. Phys.*, 1967, **47**, 1062.
[253] J. F. O'Brien and W. L. Reynolds, *Inorg. Chem.*, 1967, **6**, 2110.
[254] N. A. Matwiyoff and S. V. Hooker, *Inorg. Chem.*, 1967, **6**, 1127.
[255] K. D. Ravage, T. R. Stengle, and C. H. Langford, *Inorg. Chem.*, 1967, **6**, 1252.
[256] G. N. Lamar, R. H. Fischer, and W. De W. Horrocks, *Inorg. Chem.*, 1967, **6**, 1798.
[257] S. Nakamura and S. Meiboom, *J. Amer. Chem. Soc.*, 1967, **89**, 1765.

were seen for Li^+, Ca^{2+}, Sr^{2+}, or Ba^{2+}.[257] The extent to which a series of solvents compete with water to solvate aluminium, titanium, and cobalt cations have been determined in binary solvent–water mixtures.[258] The solvation of ion pairs has been studied for Li^+, Na^+, and Ag^+ in various solvents,[259] and for tetra-alkylammonium ions in water and chloroform.[260]

The interaction between Me_3SnBr and I^- or Br^- in several non-aqueous solvents has also been studied by observing the effect of halide ions on *J*(Sn—C—H).[261]

4 Adducts and Solvent Effects

This section inevitably overlaps with the preceding section on dynamic processes and with the section on boron and other Group III metals. It was felt, however, that it should be kept separate from these two since it reflects a distinct section of the literature.

In past years considerable use has been made of n.m.r. data to measure donor base strengths. Recent work, however, has suggested that such correlations may be fortuitous.[262] The Et_2O adducts of BF_3, BCl_3, and BBr_3 in ether, at $-45°$ give separate complexed and uncomplexed ether signals, which shows that a 1 : 1 complex is formed and enables the shift of the adducts to be obtained exactly; the CH_2 downfield shifts, with respect to bulk ether, are 54 (BF_3), 82 (BCl_3), and 89 c./sec. (BBr_3). This trend is opposite to that expected from the halogen electronegativities.[263] It does, however, roughly parallel the known increase in acceptor strengths from BF_3 to BBr_3. By contrast, the thermochemical data for the M—S bond in sulphide adducts of AlX_3 and GaX_3 do not correlate well with n.m.r. shifts and it is suggested that the former data are more reliable.[264] Similarly, ethers give inconsistent n.m.r. results.[265] When diphenylketimine is used as donor to triaryl- or trialkyl-aluminium or -gallium it is found that the donor protons all shift to high field on complexing, which is in the opposite direction to that expected from the inductive effect.[266, 267] The CH_2 proton shifts in Et_2AlX complexed with amine ligands are shifted downfield in the order $I < Br < Cl < F$; I is the furthest downfield and this could be explained if the electronegativity of the aluminium increases from X = F through to X = I. The acceptor properties of Et_2AlX increase in the order $F < Cl < Br < I$.[268]

[258] A. Fratiello, R. E. Lee, D. P. Miller, and Von Mori Nishida, *Mol. Phys.*, 1967, **13**, 349.
[259] D. Nicholls and M. Szwarc, *J. Phys. Chem.*, 1967, **71**, 2727.
[260] E. W. Randall and D. Shaw, *Spectrochim. Acta*, 1967, **23**, *A*, 1235.
[261] M. Gieler and J. Nasielski, *J. Organometallic Chem.*, 1967, **7**, 273.
[262] N. N. Greenwood and S. Srivastava, *J. Chem. Soc.*, 1966, 703.
[263] R. E. Schuster, A. Fratiello, and T. P. Onak, *Chem. Comm.*, 1967, 1038.
[264] R. L. Richards and A. Thompson, *J. Chem. Soc.* (*A*), 1967, 1248.
[265] R. L. Richards and A. Thompson, *J. Chem. Soc.* (*A*), 1967, 1244.
[266] K. Wade and B. K. Wyatt, *J. Chem. Soc.* (*A*), 1967, 1339.
[267] J. R. Jennings, I. Pattinson, K. Wade, and B. K. Wyatt, *J. Chem. Soc.* (*A*), 1967, 1608.
[268] C. A. Smith and M. G. H. Wallbridge, *J. Chem. Soc.* (*A*), 1967, 7.

Measurements of basicities of donors and acidities of acceptors have been made by comparing shifts in an inert and in an active solvent. The chemical-shift difference for the NH protons of amines at infinite dilution in CCl_4 and DMSO gives a measure of the acidity [269] and the same principle has been used for silylcarbinols where the OH proton shift is measured in DMSO and is found to correlate with the Hammett σ-constants of substituent Y for several different R groups in $Ph_3SiC(OH)(R)(C_6H_4Y)$.[270] Basicities can be found by measuring the $CHCl_3$ shift at infinite dilution in the base. In the case of silicon and germanium *gem*-diamine derivatives it is found that the silicon derivative is always the least basic member but that the n.m.r. results and i.r. measurements do not compare well in detail.[271] It has been found, however, that it is important to compare only isostructural series of bases, *e.g.* the series Me_3SiNHR or the series Me_3MNMe_2 (M = Si, Ge, or Sn). When this is done the n.m.r. and i.r. results correlate well.[272]

In liquid HF, HCN has been shown to act as a ligand towards Ag^+. Proton n.m.r. spectroscopy has shown that a 2 : 1 complex is formed in the reaction: [273]

$$2AgCN + 2HF \rightleftharpoons 2[HCN,Ag^+] + 2F^-$$

$$2[HCN,Ag^+] + 2F^- \rightleftharpoons [(HCN)_2Ag]^+ + Ag^+ + 2F^-$$

Tetracarbonyl nickel will co-ordinate to $P_4(NMe)_6$ or $As_4(NMe)_6$ at phosphorus or arsenic. The multiplicities of the methyl resonances indicate the formation of $As_4(NMe)_6,Ni(CO)_3$ (1 : 1 doublet) and $As_4(NMe)_6,2Ni(CO)_3$ (1 : 4 : 1 triplet). Addition of more nickel gives a crystalline product $As_4(NMe)_6,4Ni(CO)_3$. In the phosphorus system 1 : 1, 1 : 2, and 1 : 3 complexes are formed.[274]

Two tertiary butyl phosphate (TBP) resonances are obtained when uranyl nitrate is dissolved in TBP and peak-area measurements show that the complex $UO_2(NO_3)_2,2TBP$ is formed. It was possible to measure relative adduct strengths in mixed phosphate solutions.[275] It has been shown that the donor 5,6-benzoquinoline must migrate between the two aluminium atoms in $Et_2AlOAlEt_2$ [276] since only a single ethyl-group resonance is seen.

On a negative note it has been demonstrated that the tin, in compounds such as $Me_nSn[Re(CO)_5]_{4-n}$, does not co-ordinate with pyridine and that the $Co(CO)_4$ group is more electronegative than is $Re(CO)_5$.[277]

[269] H. Suhr, *Chem. Ber.*, 1967, **71**, 1104.
[270] A. G. Brook and K. H. Pannell, *J. Organometallic Chem.*, 1967, **8**, 179.
[271] E. W. Randall, C. H. Yoder, and J. J. Zuckerman, *Inorg. Chem.*, 1967, **6**, 744.
[272] E. W. Abel, D. A. Armitage, and S. P. Tyfield, *J. Chem. Soc.* (*A*), 1967, 554.
[273] M. F. A. Dove and J. G. Hallett, *Chem. Comm.*, 1967, 571.
[274] J. G. Riess and J. R. van Wazer, *J. Organometallic Chem.*, 1967, **8**, 347.
[275] T. H. Siddall and W. E. Stewart, *Inorg. Nuclear Chem. Letters*, 1967, **3**, 279.
[276] H. Tani, T. Araki, N. Oguni, and N. Ugyama, *J. Amer. Chem. Soc.*, 1967, **89**, 173.
[277] J. A J. Thompson and W. A. G. Graham, *Inorg. Chem.*, 1967, **6**, 1365.

Perfluoroacetone does not complex with Group IV elements. No changes were seen in the proton n.m.r. spectra when $Me_3SiOCH(CF_3)_2$, $Me_3SiOC(CF_3)_2OCH(CF_3)_2$, and $(CF_3)_2CO$ were mixed.[278]

Solvent effects have been accounted for in terms of collision complexes (which provide an anisotropy shift) rather than in terms of a reaction-field effect, and it is claimed that the new model gives better results.[279] Solvent effects for Me_2SnCl_2, Me_3SbCl, and Me_3PbCl have been measured for thirty-two organic solvents which have been classified on the basis of the results into four groups consisting of (*a*) isotropic inert solvents, (*b*) isotropic polar solvents which can co-ordinate to the metal, (*c*) anisotropic polar donating solvents such as pyridine, and (*d*) anisotropic non-donating solvents such as alkyl- or halogen-substituted aromatics. The solvent interaction can affect τ (Me) or J(Sn(Pb)—H) and it was found that J–τ plots follow different lines for each group, depending on the relative importance of collision- and donor-complexing, the latter affecting J more than the former.[280] The proton-tin coupling constants of the dithiolene complexes (62) are also solvent dependent, indicating co-ordination at Sn.

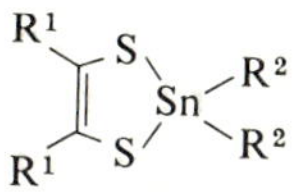

(62)

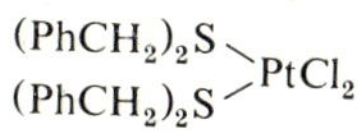

(63)

Several workers have contrasted benzene with other more isotropic solvents. The GeH_3 and CH_2 protons of $GeH_3CH_2OMe_3$ are coincident in cyclohexane, but are separated in benzene so that a complex A_3B_2 pattern results.[281] The same two solvents affect the methyl protons of $(MeBeNEt_2)_3$ differently, giving a doublet in benzene but not in cyclohexane.[282] Benzene is thought to form a weak complex with metal clusters in $[M(CO)_4SeMe]_n$, where M = Mn, or Re, and $n \geqslant 2$, a large solvent effect (0·4–0·8 p.p.m.) being observed between this solvent and CCl_4.[283] The β-protons of tetrahydrofuran in $AlH_nX_{3-n},2THF$ suffer a large benzene solvent-effect.[284] The proton spectrum of PH_3 shifts to high field in benzene and a complex is formed. In a series of solvents J(P—H) increases, with increase in dielectric constant, due to changes in the angle H—P—H; AsH_3 and SbH_3 behave similarly though couplings cannot be observed.[285] A similar mechanism may cause the chemical shift in the AB spectrum of the CH_2 protons of (63) to be solvent and temperature dependent.[131]

[278] F. Janzer and J. Willis, *Inorg. Chem.*, 1967, **6**, 1900.
[279] I. D. Quntz, jun., and M. D. Johnston, *J. Amer. Chem. Soc.*, 1967, **89**, 6008.
[280] G. Matsubayashi, Y. Kawasaki, T. Tanaka, and R. Okawara, *Bull. Chem. Soc. Japan*, 1967, **40**, 1566.
[281] G. A. Gibbon, J. T. Wang, and C. H. van Dyke, *Inorg. Chem.*, 1967, **6**, 1989.
[282] G. E. Coates and A. H. Fishwick, *J. Chem. Soc.* (*A*), 1967, 1199.
[283] E. W. Abel, B. C. Crosse, and G. V. Hutson, *J. Chem. Soc.* (*A*), 1967, 2014.
[284] D. C. Schmidt and E. E. Flagg, *Inorg. Chem.*, 1967, **6**, 1262.
[285] E. A. V. Ebsworth and G. M. Sheldrick, *Trans. Faraday Soc.*, 1967, **63**, 1071.

5 Bonding Problems and Contact Shifts

π Interactions.—There is considerable interest in the problem of whether p_π–d_π bonding is present in complexes or compounds, in certain situations, and several different approaches have been undertaken, silicon compounds receiving the most attention. π-Bonding effects are suggested as the reason for non-random equilibration [286] in the reaction:

$$Me_2SiX_2 + Me_3SiY \rightleftharpoons Me_2SiXY + Me_3SiX$$

Oxygen–silicon p_π–d_π bonding is also invoked to account for the variation of chemical shifts with substituents in $R_{n-4}Si(OR)_n$ as n is varied.[287] In a series of compounds of type $R_3MC{:}CH$ and $R_3MCH{:}CHMR_3$ (M = Si or Sn) it is found that the acetylenic and olefinic protons are considerably de-shielded. This is ascribed to increased acidity resulting from electron withdrawal from the double bond by p_π–d_π back-bonding to Sn and Si.[288] Most papers suggest either that π bonding does not occur in these compounds to any great extent or that n.m.r. may not be a good tool for its detection. Thus the imidazolidine ring-protons of (64; M = C or Si) suffer only a very small chemical shift if the nature of M is changed. A single-line spectrum is seen so that the ring must be inverting rapidly and the nitrogen must be pyramidal much of the time with no p_π–d_π bonding.[289]

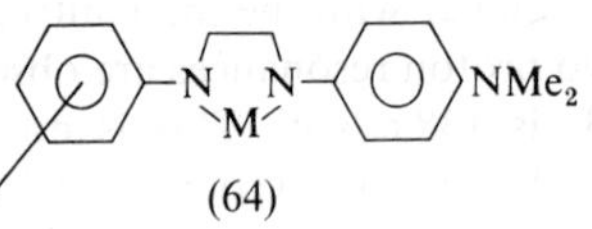

(64)

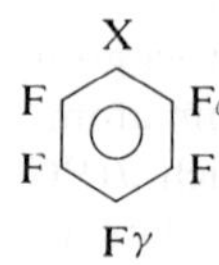

(65)

The problem has also been tackled by utilising the properties of substituted pentafluorobenzenes (65) for which it has been shown that $J_{\alpha-\gamma}$ and ϕ_γ depend on the π-donating properties of X. Thus π donation is seen when X = NH_2 or NHR. If the group R π-bonds to N then the effect is observed in the fluorine spectra. The results indicate that π bonding occurs for R = BPh_2 but not for R = $SiMe_3$.[290] An attempt has been made to detect conjugation via Si—C p_π–d_π bonding using contact shifts in the complexes (66). It is known that aromatic substituents on the cyclo-heptatriene ring can delocalise electron spin over a considerable distance. It was found that though the β-proton is shifted some 3000 c./sec. in the complex, the phenyl protons on M were shifted by only 40 c./sec., indicating only slight σ-delocalisation and no conjugation.[291] A search for restricted

[286] K. Moedritzer and J. R. van Wazer, *Inorg. Chem.*, 1967, **6**, 93.
[287] P. T. Ostdick and P. A. McCusker, *Inorg. Chem.*, 1967, **6**, 98.
[288] C. S. Kraihanzel and M. L. Losee, *J. Organometallic Chem.*, 1967, **10**, 427.
[289] C. H. Yoder and J. J. Zuckerman, *Inorg. Chem.*, 1967, **6**, 103.
[290] M. G. Hogben, A. J. Oliver, and W. A. G. Graham, *Chem. Comm.*, 1967, 1183.
[291] D. R. Eaton and W. R. McClellan, *Inorg. Chem.*, 1967, **6**, 2134.

rotation about the Si—N bonds in solid $(Me_3Si)_2NSiMe_nCl_{3-n}$, $(Me_3Si)_2NH$, $(Me_2ClSi)_2NH$, and $[Me_2ClSiNSiMe_2]_2$ was not successful and the results could be interpreted in terms of intermolecular barriers to rotation.[292] The ^{15}N—H coupling constant in aniline is unaffected if a hydrogen is replaced by -$SiMe_3$ and this, it is suggested, indicates that the nitrogen remains pyramidal. A theoretical treatment indicates that p_π–d_π bonding can occur between silicon and the phenyl ring which flattens the nitrogen pyramid without affecting the electronic structure around N.[293]

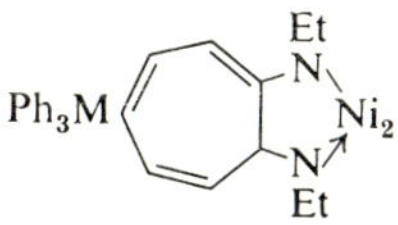

(M = Si, Ge, Sn) (66)

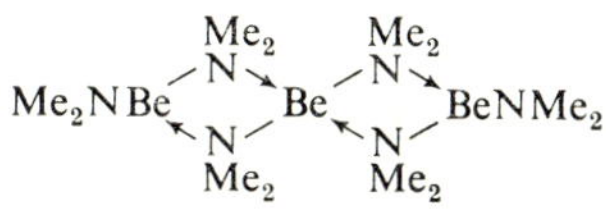

(67)

The chemical bonding in $(SiH_3)_3P$ and $(GeH_3)_3P$ on the basis of ^{31}P n.m.r. shifts are not very different, but the Raman spectra show the latter to be pyramidal in contrast with the planar silicon analogue.[294] No evidence of unusual electronic states has been found in the compounds SiH_3OR (R = SiH_3, Me, CHO, COMe, and $COCF_3$).[295]

Attention has been paid to p_π–d_π bonding in a miscellaneous range of compounds. The compound (67) is shown by its crystal structure to have the terminal methyl groups co-planar with the three nitrogens around each terminal beryllium atom. Two proton resonances are observed in solution, for each of which $J(^{13}C$—H) is 138 c./sec. This is normal for bridging *N*-methyl groups but not for the terminal groups and is taken as further evidence of Be—N π bonding.[296]

In a paper discussing mainly i.r. results it is noted that the ^{11}B shifts of BF_3, $B(OMe)_3$, and $B(NMe_2)_3$ suggest a bond order of 1·33, whereas Me_3B seems to have very little double-bond character.[297]

In the aminoboranes $R^1R^2NBR^3R^4$ proton resonance shows the presence of structural isomers and Hückel calculations of relative π-bond orders have been carried out and found to correlate with the B—N stretching frequency. The ^{11}B shift is related to the π density on B.[298] In the three-co-ordinate boron compounds $(XBNEt_3)_3$ and $PhB(X)NMe_2$ B—X π-bonding occurs and diminishes in the order

$$X = F > NCO > NCS > Cl > Br.^{299}$$

[292] H. Levey, *J. Inorg. Nuclear Chem.*, 1967, **29**, 1859.
[293] P. G. Perkins, *Chem. Comm.*, 1967, 268.
[294] S. Cradock, E. A. V. Ebsworth, G. Davidson, and L. A. Woodward, *J. Chem. Soc.* (*A*), 1967, 1229.
[295] E. A. V. Ebsworth and J. C. Thompson, *J. Chem. Soc.* (*A*), 1967, 69.
[296] J. L. Atwood and G. D. Stucky, *Chem. Comm.*, 1967, 1169.
[297] J. E. de Moor, G. P. van der Kelen, and Z. Eeckhaut, *J. Organometallic Chem.*, 1967, **9**, 31.
[298] M. R. Chakrabarty, C. C. Thompson, and W. S. Brey, *Inorg. Chem.*, 1967, **6**, 518.
[299] M. F. Lappert and H. P. Pyszora, *J. Chem. Soc.* (*A*), 1967, 854.

The ^{19}F and ^{1}H n.m.r. of the compounds $(p\text{-}FC_6H_4)_nSnCl_{4-n}$ suggest that, when $n \geqslant 2$, p_π–d_π interaction occurs between the π orbitals on the phenyl ring and the empty 4*d* tin orbitals.[300] In the compounds R_3SnH (R = alkyl or H), *J*(Sn—H) correlates with the sum of the Taft constants for the substituents, but it is suggested that this is due not to C—Sn p_π–d_π interaction but to hyperconjugation.[301]

The CF_3 spectra of $Me_2NNHAs(CF_3)_2$ is a sharp singlet so that the CF_3 groups must be rotating rapidly around the N—As bond which thus can have no p_π–d_π bond character.[302] On the other hand, ^{31}P shifts in aminophosphonium compounds $[(R^1R^2)_nP(NR_2{}^3)_{4-n}]^+$ are interpreted in terms of both electronegativity and π-bonding effects.[303]

The ^{19}F shifts in the complex (68) are very sensitive to the nature of X. It is suggested that this is due to F—Ti p_π–d_π interactions modified by the predominating resonance structure of the pyridine *N*-oxide, the nature of which is determined by the electron withdrawing properties of X. On the other hand, the fluorine coupling constants are very insensitive to the nature of the ligands or to X and this is taken as evidence for direct through-space interaction of the fluorines.[304] In the reaction:

$$(\pi\text{-}C_5H_5)Mn(CO)_3 \xrightarrow[L]{h\nu} (\pi\text{-}C_5H_5)Mn(CO)_2L \text{ or } (\pi\text{-}C_5H_5)Mn(CO)L_2$$

charge redistribution does not affect the cyclopentadienyl ring, only the carbonyl groups and the ligands L (Ph_3P, Ph_3As, Ph_3Sb and Ph_3Bi).[305]

A series of nickel(0) complexes, NiL_4, in which L is a substituted fluorophosphine show low-field ligand–fluorine shifts on complexing. The ligands are also stabilised in the complex which suggests Ni—P d_π–d_π back-bonding.[306] This is also indicated in the related complexes $Ni(CO)_nL_{4-n}$ where n = 0, 1, or 2; the ^{31}P shifts depend upon a paramagnetic term introduced by π bonding and upon unknown bond-angle changes. The increase in the co-ordination number of phosphorus reduces *J*(P—F).[307]

The contact shifts of the phenyl groups of aniline ligands co-ordinated to Ni^{II} acac$_2$ complexes indicate that π delocalisation takes place from nickel to the ring. It is suggested that the σ bond carries the electron spin to the sp^3 lobe of the nitrogen which can there mix with the ring π-orbitals.[308]

Little Pt—CO p_π–d_π bonding is detected in the complexes shown in (69).[309] The proportions of σ and π bonding from the hydrocarbon ligand to platinum in the complexes C_8H_8PtX and $C_8H_{12}PtX_2$ (X = Cl, Br, I or Me)

[300] J. C. Maire, *J. Organometallic Chem.*, 1967, **9**, 271.
[301] K. Kawakami, T. Saito, and R. Okawara, *J. Organometallic Chem.*, 1967, **8**, 377.
[302] C. K. Peterson and K. I. Thé, *Chem. Comm.*, 1967, 1056.
[303] S. R. Jain, W. S. Brey, and H. H. Sisler, *Inorg. Chem.*, 1967, **6**, 515.
[304] D. S. Dyer and R. O. Ragsdale, *J. Phys. Chem.*, 1967, **71**, 2309.
[305] C. Barbeau, *Canad. J. Chem.*, 1967, **45**, 161.
[306] J. F. Nixon, *J. Chem. Soc.* (*A*), 1967, 1136.
[307] G. S. Reddy and R. Schmutzler, *Inorg. Chem.*, 1967, **6**, 823.
[308] R. W. Kluiber and W. De W. Horrocks, jun., *Inorg. Chem.*, 1967, **6**, 431.
[309] A. R. Brause, M. Rycheck, and M. Orchin, *J. Amer. Chem. Soc.*, 1967, **89**, 6500.

have been estimated; σ-bonding is the most important when X = halogen but π bonding is of more importance when X = Me.[310]

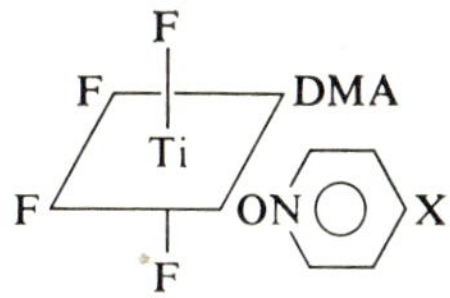

(68)

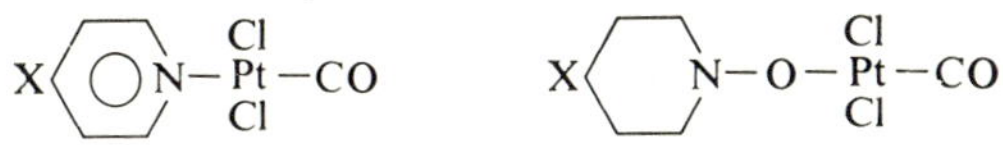

(69)

The bond orders in the dichromate ion have been determined by ^{17}O n.m.r. spectroscopy; the ^{17}O chemical shifts of (70) is -1460 from H_2O (both oxygens are π bonded) and of (71) is -835 p.p.m. (average Cr—O bond-order 1·5). The dichromate ion, on the other hand, shows two resonances (at -1127 and -345 p.p.m.) and these can be assigned to Cr—O bond-orders of 1·67 and 1 respectively; these are in accordance with the formulation (72) with a bent Cr—O—Cr bond.[311]

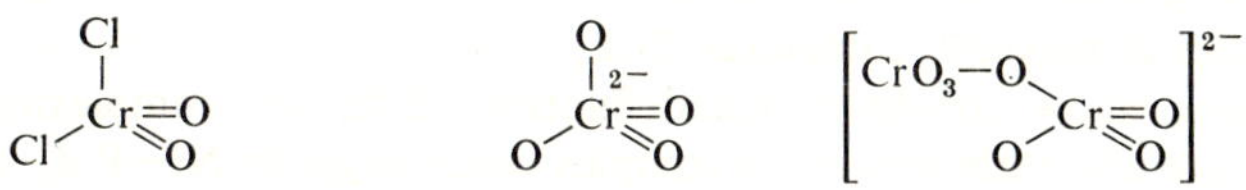

(70) (71) (72)

Solvent effects on the proton n.m.r. spectra of $(EtO)_3AsO$ show this to be a polar molecule and it is concluded that the As—O bond is $\sigma+\pi$.[312]

Other Bonding Problems.—The values of *J*(C—H) in (73) have been used to check the hybridisation of the ring carbon-atoms.[313] The chemical shifts, coupling constants $J(Sn-CH_3)$, and Sn—O and C=O stretching frequencies (ν) have been compared for a series of acetylacetonates XYM $acac_2$ (X, Y = alkyl, aryl, or halogen; M = Sn, Pb, Ti, Ga, or Sb). The n.m.r. and i.r. data may be correlated; increasing ν and *J* are consistent with increasing covalency of the Sn—O bond.[314] As the electron attracting power of X and Y increases the τ values of the acac protons decrease and this is explained in terms of an inductive effect.[315] In the complexes

[310] H. P. Fritz and D. Sellmann, *Spectrochim. Acta*, 1967, **23**, *A*, 1991.
[311] R. G. Kidd, *Canad. J. Chem.*, 1967, **45**, 605.
[312] J. P. Laurent, M. Durand, and F. Gallais, *Compt. rend.*, 1967, **264**, *C*, 1005.
[313] M. E. Vol'pin, V. G. Dulova, Yu. T. Struchkov, N. K. Bokiy, and D. N. Kursanov, *J. Organometallic Chem.*, 1967, **8**, 87.
[314] Y. Kawasaki, *Inorg. Nuclear Chem.*, 1967, **29**, 840.
[315] Y. Kawasaki, T. Tanaka, and R. Okawara, *Bull. Chem. Soc. Japan*, 1967, **40**, 1562.

Me_2PbL_2 (L = Cl, ClO_4, OH, $\frac{1}{2}$ acac, $\frac{1}{2}$ oxalate) the coupling constant J(Pb—CH) is more than twice that in Me_4Pb. It is suggested from this that L—Pb bonds are highly ionic.[316] It has been shown that selenium has the same electronegativity in Et_2Se and Et_2SH.[317]

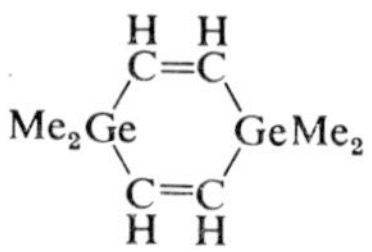

(73)

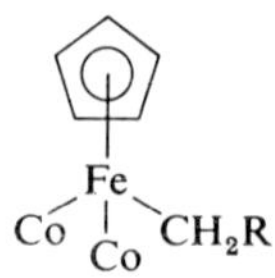

(74)

Comparison of the proton shifts in $MeMn(CO)_5$, $CH_2F{\cdot}Mn(CO)_5$, $MeCO{\cdot}Mn(CO)_5$, and $CH_2F{\cdot}CO{\cdot}Mn(CO)_5$ suggests that the C—Mn bond is stronger in the fluorinated compound.[318] The CH_2R proton shifts in the complex (74) have been compared with those of the compounds CH_3R. There is a larger range of shifts in (74) and it is suggested that this is because R effects the electronegativity of the iron via the iron *d*-orbitals.[319] The equilibrium:

$$[Co^{III}(corrin)X] + H_2O \rightleftharpoons [Co^{III}(corrin)X(H_2O)]$$

has been studied by proton n.m.r. and the results suggest that there is no rigid distinction between formally five- or six-co-ordinate cobalt in the ground state.[320] The methyl groups of dimethylglyoxime (DG) in the complexes $Co(DG)_2L_2$, where L is a phosphine ligand, are split due to coupling with the phosphorus (*J ca.* 3 c./sec.) so that the bonds to cobalt in these compounds must be covalent.[321] In $RCo(DG)_2PPh_3$, τ of the dimethylglyoxime methyl groups is correlated with Hammett's σ constants for R. It is suggested that this may be due to electronic transmission.[322] The reaction of complex (75a) with molecular oxygen gives a product which may be aromatic. The methyl groups c and d are nonequivalent in (75a) but become equivalent in (75b) owing to the molecule becoming planar. Protons a and b shift to the aromatic region.[323]

Paramagnetic Species.—The n.m.r. of paramagnetic species has proved useful in investigations of many exchanging systems, particularly those containing solvated ions because of the large chemical shifts caused by the paramagnetic centre, either by the contact or pseudo-contact mechanisms.[93–95, 106, 107, 151, 185, 224–227, 229–234, 248, 249, 256] The ways in which these

[316] Y. Kawasaki, *J. Organometallic Chem.*, 1967, **9**, 549.
[317] J. P. Mila and J. P. Laurent, *Bull. Soc. chim. France*, 1967, 2735.
[318] K. Noack, U. Schaerer, and F. Calderazzo, *J. Organometallic Chem.*, 1967, **8**, 517.
[319] M. L. H. Green, M. Ishaq, and R. N. Whiteley, *J. Chem. Soc.* (*A*), 1967, 1508.
[320] R. A. Firth, H. A. O. Hill, B. E. Mann, J. M. Pratt, and R. G. Thorp, *Chem. Comm.*, 1967, 1013.
[321] G. N. Schrauzer and R. J. Windgassen, *J. Amer. Chem. Soc.*, 1967, **89**, 1999.
[322] H. O. Hill, K. G. Morallee, and R. E. Collis, *Chem. Comm.*, 1967, 888.
[323] E. G. Vassian and R. K. Murman, *Inorg. Chem.*, 1967, **6**, 2043.

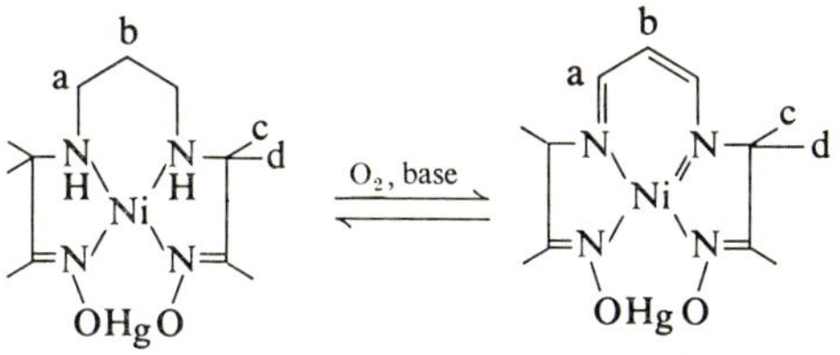

(75a) (75b)

shifts are transferred in molecules is of considerable theoretical interest. In Ni^{II} complexes of the ligand tribenzo[*b,f,j*][1,5,9]triazacycloduodecine (tri) (76) it was found that contact shifts depended only on the configuration of the ligand and not on the nature of any other small ligands.[324] Electrons can be delocalised into a π system across a Ni—N σ bond in $[Ni(PhCH_2NH_2)_6]^{2+}$. The phenyl proton shifts indicate that the shift is not pseudo-contact.[325] Similar delocalisation in a Ni—N σ bond has also been noted in aniline complexes. In the nickel and cobalt complexes L_2MX_2 the contact shifts of L increase in the order X = Cl < Br < I. It is suggested that changes in bond angles around the nickel, rather than changes in bond covalency, are the main cause of the increase. This is established by comparing complexes where L is a 3- or a 4-substituted pyridine. The former affects the contact shifts whereas the latter does not, so that a steric effect is important.[326]

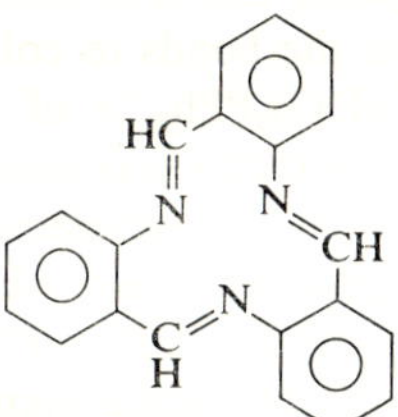

(76)

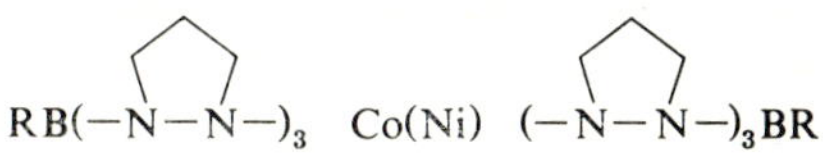

(77)

In the complexes (77) the isotropic and anisotropic contact shifts for M = Co have been separated and it has been demonstrated that the scalar contact shifts are very similar for both cobalt(II) and nickel(II).[327] Dipolar and scalar shifts exist in bis(benzoylacetonato)-complexes of Co^{II} and Ni^{II}.[328] Complexes of 4-vinylpyridine (vpy), vpy_4NiCl_2, vpy_4CoCl_2, vpy_2CoCl_2 have been studied. In the first two delocalisation occurs via σ bonds whereas for the latter, since there are three unpaired electrons in

[324] G. N. La Mar, *Inorg. Chem.*, 1967, **6**, 1921.
[325] R. J. Fitzgerald and R. S. Drago, *J. Amer. Chem. Soc.*, 1967, **89**, 2879.
[326] G. N. La Mar, *Inorg. Chem.*, 1967, **6**, 1941.
[327] J. P. Jesson, *J. Chem. Phys.*, 1967, **47**, pp. 579, 582.
[328] D. W. Kluiber and W. D. de W. Horrocks, jun., *Inorg. Chem.*, 1967, **6**, 166.

π-symmetry orbitals in the tetrahedral complex, π interaction predominates.[329] The interaction of ammonia and metal ions in hexammine complexes shows that σ effects predominate for Ni^{II} and Cu^{II} (large upfield shifts of NH_3 protons), whereas π effects predominate for Mn^{II} (large downfield shift); the two effects oppose for Co^{II} (small upfield shift).[330] Changes with temperature in the isotropic shifts of [$Cu(CF_3\cdot CO\cdot CH\cdot CO\cdot CF_3)_2$($\gamma$-picoline-*N*-oxide)] suggest an equilibrium between two paramagnetic species; a symmetrical low-temperature form with the spin on the copper and a less symmetrical high-temperature form with the spin delocalised onto the *N*-oxide,[331]

There is no pseudo-contact interaction present between the iron and cyclopentadienyl rings in the ferricinium cation.[332] The ^{13}C shifts for paramagnetic cyanide complexes of manganese and iron have been measured[333] and the activation energy of electron transfer between the $[Fe(CN)_6]^{4-}$ and $[Fe(CN)_6]^{3-}$ ions has been determined from ^{14}N measurements to be $E = 4{\cdot}2 \pm 0{\cdot}6$ kcal. mole^{-1}.[334]

A series of lanthanide complexes of the ligand (78) have been studied. The complexes are weak and exchange occurs. Contact shifts of the paramagnetic species indicate that the metal–ligand interaction is minimally covalent.[335] The ligand is *cis* in the complex whereas the free ligand is *trans*. The proton and ^{31}P n.m.r. of the lanthanide adducts of triaryl phosphate indicate a major difference in adduct structure between the light and heavy rare earths; exchange processes also occur in this system.[336]

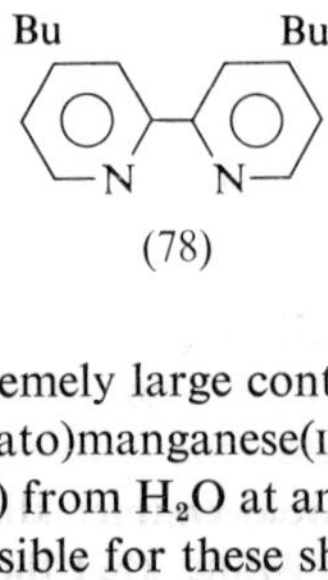

(78)

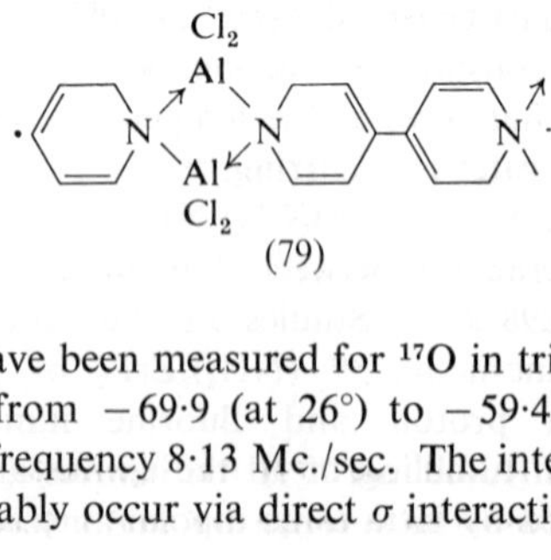

(79)

Extremely large contact-shifts have been measured for ^{17}O in tris(acetylacetonato)manganese(III) varying from $-69{\cdot}9$ (at 26°) to $-59{\cdot}4$ Kc./sec. (at 83°) from H_2O at an operating frequency 8·13 Mc./sec. The interactions responsible for these shifts presumably occur via direct σ interaction.[337]

329 N. H. Agnew, L. F. Larkworthy, and G. A. Webb, *Inorg. Nuclear Chem. Letters*, 1967, **3**, 303.
330 B. B. Wayland and W. L. Rice, *Inorg. Chem.*, 1967, **6**, 2270.
331 R. W. Kluiber and W. de W. Horrocks, *Inorg. Chem.*, 1967, **6**, 1427.
332 H. P. Fritz, H. J. Keller, and K. E. Schwarzhans, *J. Organometallic Chem.*, 1967, **7**, 105.
333 D. G. Davis and R. J. Kurland, *J. Chem. Phys.*, 1967, **46**, 388.
334 A. Loewenstein and G. Ron, *Inorg. Chem.*, 1967, **6**, 1604.
335 F. A. Hart, J. E. Newberry, and D. Shaw, *Chem. Comm.*, 1967, 45.
336 T. H. Siddall, W. E. Stuart, and D. G. Karraker, *Inorg. Nuclear Chem. Letters*, 1967, **3**, 479,
337 Z. Luz, B. L. Silver, and D. Fiat, *J. Chem. Phys.*, 1967, **46**, 469.

Contact shifts in the radical (79) have been used to confirm its structure,[338] and ^7Li shifts as a function of temperature in the system LiBr–THF–(fluorenone)$^-$ have been measured in an initial investigation of ionic interactions in these solutions.[339]

N.m.r. techniques have also been used to determine the magnetic susceptibility of paramagnetic dithiolate complexes of iron and cobalt,[340] and cyclopentadienylnitrosylchromium complexes of the type $(\pi\text{-}C_5H_5)Cr(NO)ClL$ (L = py, γ-picoline, R_3P, R_3As, or RCN).[341]

6 Solid State Nuclear Magnetic Resonance

While n.m.r. spectroscopy of solid-state systems is sometimes considered to be of more interest to the physicist than the chemist, a considerable amount of chemically significant information has been obtained during the year both on motions within solid materials and on the structures of certain crystals.

The motions of the guest molecules in a number of clathrate hydrates have been investigated in this way in both H_2O and D_2O host lattices. Thus the fluorine magnetic resonance of CF_4 and SF_6 clathrate hydrates shows that the guest molecules are very little restricted by the walls of the host cavities between 77 and 252°K and re-orient easily around chosen axes of symmetry at low temperature and randomly at higher temperature. Above 150°K the results for SF_6 are consistent with a rattling motion in the cavity.[342] For the guest molecules SO_2, ethylene oxide, propylene oxide, THF, dihydrofuran, Cl_2, Br_2, and Me_3N, it is found that the guest is rotating fast (τ *ca.* 10^{-9} sec.) at 208°K. In D_2O the guest proton linewidth remains constant as the temperature varies whereas in water the line becomes narrower as the temperature increases, indicating that motion of the host water is occurring.[343]

Proton n.m.r. of $Ni(CN)_2NH_3{\cdot}C_6H_6$ type clathrates shows that there is spin interaction between the Ni and the lattice protons at 100° but not at 298°K.[344] Studies of the ionic crystals NH_4BF_4 and ND_4BF_4 by fluorine n.m.r.,[345] $(NH_4)_3ZrF_7$ by proton n.m.r.,[346] and $(NH_4)_2SiF_6$ by both proton and fluorine n.m.r.[347] indicate the presence of rotational tumbling of all the anions and cations. Relaxation of BF_4^- is dominated by intra-ionic dipolar interactions above 318°K whereas below this temperature the protons give a contribution which can be removed by

[338] E. Ziegler, G. Fuchs, and H. Lehmkuhl, *Z. anorg. Chem.*, 1967, 355, 145.
[339] Gee. W. Canters, H. van Willigen, and E. de Boer, *Chem. Comm.*, 1967, 566.
[340] A. C. Balch, *Inorg. Chem.*, 1967, **6**, 2158.
[341] E. O. Fischer and H. Strametz, *J. Organometallic Chem.*, 1967, **10**, 323.
[342] C. A. McDowell and P. Raghunathan, *Mol. Phys.*, 1967, **13**, 331.
[343] S. Brownstein, D. W. Davidson, and D. Fiat, *J. Chem. Phys.*, 1967, **46**, 1454.
[344] Kimiko Umemoto and S. S. Danyluk, *J. Phys. Chem.*, 1967, **71**, 450.
[345] A. P. Caron, D. J. Huettner, J. L. Ragle, L. Sherk, and T. R. Stengle, *J. Chem. Phys.*, 1967, **47**, 2577.
[346] M. Pintar, G. Lahajnar, and J. Slivnik, *Mol. Phys.*, 1967, **12**, 117.
[347] R. Blinc and G. Lahajnar, *J. Chem. Phys.*, 1967, **47**, 4146.

deuteriation.[345] The results for the zirconate support the presence of the ZrF_7^{3-} ion.[346] The re-orientation of NH_4^+ in $(NH_4)_2SiF_6$ occurs by a process in which the cation is excited above the energy barrier to rotation, rotates rapidly, and then falls back with a different orientation. Hindered rotation of SiF_6^{2-} occurs above room temperature.[347] The λ transition at 242·8°K in NH_4Cl and the similar transition in ND_4Cl have been detected by proton and deuterium n.m.r., presumably because of some indirect relation between the nuclear correlation time and the degree of order in the crystal lattice.[348] The deuterio-hydrates of potassium oxalate[349] and cupric sulphate[350] have also been studied by deuterium n.m.r. The onset of D_2O flipping occurs at 25–50° for the former and is present in the latter at 0°. The D_2O resonance in the cupric salt is chemically shifted from liquid heavy water due to dipolar interaction with the Cu^{2+}.

Fluorine n.m.r. has shown that, for SF_6 on faujasite, relaxation is due to the diffusion of SF_6 in the magnetic field of paramagnetic iron impurities. The time between diffusional jumps is given by τ $1{\cdot}14 \times 10^{-12}e^{4000/RT}$.[351]

Information can also be gleaned about the structure of solids, including the positions of protons, which can otherwise be obtained only by using neutron scattering. Two sorts of NH_4 or ND_4 groups are shown to be present in ferroelectric ammonium sulphate.[352] Convincing evidence has also been obtained that the ion ND_4^+ in ND_4Cl is a perfect tetrahaedron.[353] The proton n.m.r. of polycrystalline polyborates has shown that the hydrogen present is bound as a hydroxy-group and not as H_2O, the results being consistent with the formulations $K[B_{12}O_{15}(OH)_7]$ and $K[B_3O_3(OH)_4]$.[354] Hydroxy-groups have also been detected in the products of thermal decomposition of $Sn(OH)_4$; in SnO_3H_2 the OH protons are in clusters of three whereas in $Sn_2O_5H_2$ they are isolated.[355]

The Co—H distance in $HCo(CO)_4$ has been measured as $1{\cdot}2 \pm 0{\cdot}1$ Å from the spin interaction with ^{59}Co ($I = \frac{7}{2}$) or $1{\cdot}4 \pm 0{\cdot}05$ Å from the measurement of the second moment.[356] An independent determination, which includes quadrupole effects,[357] gives the value as $1{\cdot}59 \pm 0{\cdot}4$ Å and also finds H—Mn distance in $HMn(CO)_5$ to be $1{\cdot}44 \pm 0{\cdot}03$ Å.

Knight shifts in the lanthanum hydrides have been discussed but cannot easily distinguish between a hydride and protonic structure.[358]

[348] D. E. Woessner and B. S. Snowden, *J. Phys. Chem.*, 1967, **71**, 952.
[349] J. W. McGrath and G. W. Ossman, *J. Chem. Phys.*, 1967, **46**, 1824.
[350] J. O. Clifford and J. A. S. Smith, *Mol. Phys.*, 1967, **13**, 297.
[351] H. A. Resing and J. K. Thompson, *J. Chem. Phys.*, 1967, **46**, 2876.
[352] D. E. O'Reilly and Tung Tsang, *J. Chem. Phys.*, 1967, **46**, 1291.
[353] M. Linzer and R. A. Forman, *J. Chem. Phys.*, 1967, **46**, 4690.
[354] K. Wegener, *J. Inorg. Nuclear Chem.*, 1967, **29**, 1847.
[355] E. W. Giesekke, H. S. Gutowsky, P. Kirkov, and H. A. Laitinen, *Inorg. Chem.*, 1967, **6**, 1294.
[356] T. C. Farrar, F. E. Brinckman, T. D. Coyle, A. Davison, and J. W. Faller, *Inorg. Chem.*, 1967, **6**, 161.
[357] G. M. Sheldrick, *Chem. Comm.*, 1967, 751.
[358] W. G. Bos and H. S. Gutowsky, *Inorg. Chem.*, 1967, **6**, 552.

A cubic modification of the crystal of barium dicalcium propionate has been shown to have this structure because of motion of the ethyl groups.[359]

Fluorine n.m.r. has been used to measure the average M—F distance in spinels containing a minor proportion of paramagnetic fluorides, MF_2, and the fluoride was found probably to have entered the lattice at random.[360] In the rare-earth trifluorides the fluorine resonance shifts and linewidths depend upon the magnetic moment of the metal ion and there is an interesting reversal of the sign of the shift between the two halves of the rare-earth series.[361] The clathrate salt $[Ag_7O_8]^+HF_2^-$ has also been examined by fluorine n.m.r. The fluorine spectrum is a broad four-line pattern consistent only with the presence of HF_2^- ions and not F^- ions which, in this situation, should give a narrow line.[362]

The ^7Li n.m.r. of $LiNbO_3$ has allowed an electric-field-gradient calculation to be made which shows that the charge on Nb is $+1{\cdot}59$ units and on O is $-0{\cdot}86$ units, indicating that the niobate ion is predominantly covalently bonded.[363]

Knight-shifted lines have been detected in the ^{51}V spectra of vanadium carbides, VC_x; the lines seen in a particular sample depend on the value of x and correspond to a varying co-ordination of vanadium by 3, 4, 5, or 6 carbon atoms. For six neighbouring carbons the ^{51}V line is narrower due to the high symmetry.[364]

Solid-state n.m.r. has been used in addition for analysis of ^1H, ^3H, ^3He, and ^7Li in irradiated lithium hydride,[365] and to determine fluoride in aluminium fluoride ores.[366] It has been used to measure anisotropies in chemical shifts in $^{51}V_2O_5$ [367] and an ENDOR measurement involving ^{23}Na has been made on *X*-irradiated sodium formate.[368]

7 Boron and Other Group III Elements

Proton, ^{11}B, and ^{19}F n.m.r. have been used extensively to obtain information about Group III compounds and work on these compounds comprises about 13% of the papers reviewed.

Polyhedral Boranes and Carboranes.—Boron n.m.r. indicates that the anion $B_7H_7^{2-}$ may be a pentagonal bipyramid (two doublets with intensity ratio 5 : 2), whereas in $B_8H_8^{2-}$ all the borons are in very similar environments.[369] Boron n.m.r. spectra are reported for 1, 12-, and 1,7-$B_{12}H_{10}L_2$

[359] J. M. Chezeau, J. Dufourq, and B. Lemarceau, *J. Chim. Phys.*, 1967, **64**, 412.
[360] A. Sobel, *J. Phys. Chem. Solids*, 1967, **28**, 185.
[361] V. Saraswati and R. Vijayaraghavan, *J. Phys. Chem. Solids*, 1967, **28**, 2111.
[362] A. C. Gossard, D. K. Hindermann, M. B. Robin, N. A. Kuebler, and T. H. Geballe, *J. Amer. Chem. Soc.*, 1967, **89**, 7121.
[363] G. E. Peterson, P. M. Bridenbaugh, and P. Green, *J. Chem. Phys.*, 1967, **46**, 4009.
[364] C. Froidevaux and D. Rossier, *J. Phys. Chem. Solids*, 1967, **28**, 1197.
[365] P. C. Souers, T. A. Jolly, and C. F. Cline, *J. Phys. Chem. Solids*, 1967, **28**, 1717.
[366] W. P. Ferren and N. Shane, *Analyt. Chem.*, 1967, **39**, 117.
[367] S. D. Gornostansky and C. V. Stager, *J. Chem. Phys.*, 1967, **46**, 4959.
[368] R. J. Cook and D. H. Whiffen, *J. Phys. Chem.*, 1967, **71**, 93.
[369] F. Klanberg, D. R. Eaton, L. J. Guggenberger, and E. L. Muetterties, *Inorg. Chem.*, 1967, **6**, 1271.

(L = CO or CO_2H), a singlet being seen for the B—L boron.[370] The boron spectrum for $[(B_{10}H_{12})_2Hg]^{2-}$ consists of a large unsymmetrical doublet and a small symmetrical high-field doublet.[371] The ^{11}B spectrum for normal and deuteriated $B_{10}H_{12}Y_2$ (Y = Rb, SEt, or EtNC) have also been reported and show considerable assignable structure.[372]

Triethylamine abstracts the ligands from $B_{10}H_{12}L_2$ species to give $B_{10}H_{10}^{2-}$ which was identified by its boron n.m.r. spectrum.[373] Proton resonance has been reported for $CdB_{10}H_{10}2Et_2O$.[374] It has also been used to follow deuterium exchange induced by HI in $\mu\mu'\mu''\mu'''$-5,6,7,8,9,10-decadeuteriodecaborane-14 when it was shown that the compound originally formulated as 5-iododecaborane is in fact 1-iododecaborane.[375]

Pentaborane(9) derivatives have been characterised by proton and boron

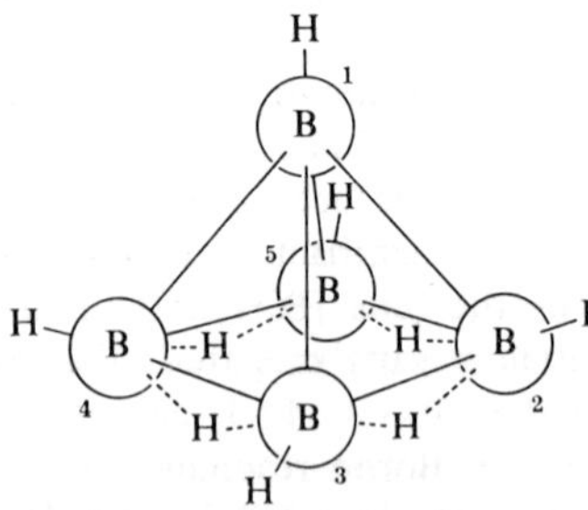

(80)

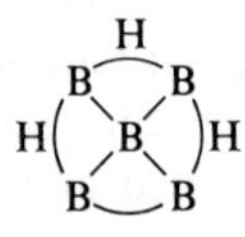

(81)

resonances and $^1H[^{11}B]$ double resonance experiments have been used to distinguish the two kinds of bridge hydrogen in (80).[376] One of the bridge hydrogens can be replaced by a $SiMe_3$ group as in the reaction:

$$LiB_5H_8 + Me_3SiCl \longrightarrow B_5H_8SiMe_3 + LiCl$$

The basal boron atoms adjacent to the Si have different chemical shifts from the other two basal borons;[377] the $SiMe_3$ group can migrate to a terminal position. These two papers have been criticised and a fresh interpretation of the $B_5H_8^-$ ion spectrum given.[378] The lithium salt is stable but the potassium and sodium salts are not. The ^{11}B spectrum consists of two doublets due to apical and base borons, the base boron doublet chemical shift and coupling constants being temperature-dependent due to bridge-proton exchange as in (81). Reaction of the lithium salt with diborane led

[370] W. H. Knoth, J. C. Sauer, J. H. Balthis, M. C. Miller, and E. L. Muetterties, *J. Amer. Chem. Soc.*, 1967, **89**, 4842.
[371] N. N. Greenwood and N. F. Travers, *Chem. Comm.*, 1967, 216.
[372] D. E. Hyatt, F. R. Scholer, and L. J. Todd, *Inorg. Chem.*, 1967, **6**, 630.
[373] M. F. Hawthorne, R. L. Pilling, and R. N. Grimes, *J. Amer. Chem. Soc.*, 1967, **89**, 1067.
[374] N. N. Greenwood and N. F. Travers, *J. Chem. Soc.* (*A*), 1967, 880.
[375] M. G. H. Wallbridge, J. Williams, and R. L. Williams, *J. Chem. Soc.* (*A*), 1967, 132.
[376] T. Onak, G. B. Dunks, I. W. Searcy, and J. Spielman, *Inorg. Chem.*, 1967, **6**, 1465.
[377] D. F. Gaines and T. V. Iorns, *J. Amer. Chem. Soc.*, 1967, **89**, 4249.
[378] R. A. Geanangel and S. G. Shore, *J. Amer. Chem. Soc.*, 1967, **89**, 6771.

to decaborane:

$$B_2H_6 + LiB_5H_8 \longrightarrow LiBH_4 + B_6H_{10} \xrightarrow[\text{in glyme}]{\text{reflux}} B_{10}H_{14}$$

It is known that ligand exchange and hydrogen tautomerism can both occur in certain borane adducts, *e.g.* in B_3H_7,Et_2O, since all three borons have the same chemical shift and are equally coupled to all seven protons. Ligand exchange would not be expected when the ligand is a much stronger donor than the solvent, and it is now found that no exchange occurs with B_3H_7,NMe_3 in benzene or in diethyl ether: two ^{11}B octets are observed in the ratio 2 : 1 indicating that ligand exchange is absent but that hydrogen tautomerism is still occurring. With the adduct of the weaker ligand THF, exchange occurred both in THF and in benzene and a single unresolved octet was observed.[379]

The carboranes continue to receive considerable attention. Here, proton n.m.r. is of greater use because well-resolved lines can be obtained from the protons on the carbon. Proton n.m.r. has been used to help in identifying the 1-methyl-2-tropenyliumyl-1,2-dicarbaclovodecaborane(2) cation, $[C_{10}H_{19}B_{10}]^+$.[380] The existence of the ions $[B_9H_9CH]^-$ and $[B_{11}H_{11}CH]$ has been predicted and their caesium salts prepared; the ^{11}B resonance of the first is three doublets in the ratio 1 : 4 : 4, and of the second is three doublets in the ratio 1 : 5 : 5.[381] Boron resonance has been used to assign structures to $[B_{10}H_{12}CH]^-$, $[B_{10}H_{12}CNR_3]^-$, $[B_{10}H_{10}CH]^-$, and $[B_{10}H_{10}CNR_3]^-$. The first two ions seem to be icosahedral with one site vacant and two bridge hydrogen atoms on nonadjacent edges of the open face, whereas the last two are closed structures with the CH or CNR_3 group sited symmetrically over the six boron atoms, which in the parent decaborane structure carried bridge hydrogen atoms.[382]

Boron n.m.r. has not, however, been able to resolve the structure of the carborane $B_6H_6C_2Me_2$ which may be intermediate between two regular forms.[383] The boron doublets in the ^{11}B spectrum of $C_2B_5H_7$ collapse to singlets in $C_2H_2B_5Me_5$, indicating that the Me groups are all on boron.[384] The carbon atoms are in the unusual adjacent position in 1,2-dicarbaclovopentaborane(5). Attempts were made to follow by proton resonance the pyrolytic re-arrangement to the more common 1,5-isomer but this was not formed.[385]

Structural assignments of carbaphosphaboranes and phosphaboranes have been made on the basis of proton and boron n.m.r.: 1,2-$B_{10}H_{10}CHP$,

[379] M. A. Ring, E. F. Witucki, and R. C. Greenough, *Inorg. Chem.*, 1967, **6**, 395.
[380] K. M. Harmon, A. B. Harmon, and B. C. Thompson, *J. Amer. Chem. Soc.*, 1967, **89**, 5309.
[381] W. H. Knoth, *J. Amer. Chem. Soc.*, 1967, **89**, 1274.
[382] D. E. Hyatt, F. R. Scholer, L. J. Todd, and J. L. Warner, *Inorg. Chem.*, 1967, **6**, 2229.
[383] H. V. Hart and W. N. Lipscomb, *J. Amer. Chem. Soc.*, 1967, **89**, 4220.
[384] H. V. Seklemian and R. E. Williams, *Inorg. Nuclear Chem. Letters*, 1967, **3**, 289.
[385] R. N. Grimes, *J. Organometallic Chem.*, 1967, **8**, 45.

in which the carbon and phosphorus atoms are adjacent, shows H—P coupling whereas 1,7-$B_{10}H_{10}$CHP shows no P—H coupling since P and C are well separated; $B_{11}H_{11}$PPh shows a 1 : 5 : 5 ^{11}B pattern and is thus an icosahedral structure and 1,2-$B_{10}Br_3H_7$CHP shows H—P coupling suggesting that all three bromine atoms are attached to boron.[386]

The spectra of a series of thiaboranes and azaboranes are also reported: $B_9H_{12}S^-$ has the S in the 6-position of the isoelectronic $B_{10}H_{14}^{2-}$ structure, the 5,10- and 7,8-positions are proton bridged and the presence of the BH_2 group in the 9-position is shown by an ^{11}B triplet. MeCN,$B_9H_{11}S^-$ is similar though two acetonitrile signals are seen: $B_{10}H_{12}S$ probably contains a BH_2 group in the open pentagonal face and $B_{10}H_{11}S$ contains no SH.[387] The boron spectrum of $[B_{10}H_{10}S)_2Co]^-$ is consistent with a bisicosehedral cobalt sandwich with S adjacent to Co, the unsymmetrical ($B_{10}H_{10}$S)Co(π-C_5H_5) has also been made.[387] $[Me_2N\text{-}NB_9H_{12}]^-$ gives two methyl proton signals and probably has two isomers; data are given for $B_9H_{11}S$ and $B_9H_{12}NH$.[387]

The transition-metal complexes of $[B_9C_2H_{11}]^{2-}$, which we represent as (cb), have been prepared: $[Fe\,cb_2]^{2-}$, $[Co\,cb_2]^-$, $[Ni\,cb_2]$, $[Ni\,cb_2]^-$, and $[Ni\,cb_2]^{2-}$. The boron n.m.r. of the first three complexes were similar; they are isoelectronic and contain the metal in the formal oxidation states Fe^{II}, Co^{III} and Ni^{IV}.[388] $[Fe(B_{10}H_{10}CH)_2]^{3-}$ and $[Ni(B_{10}H_{10}NHMePr)_2]$ have been prepared; the first has a very broad boron spectrum while the latter, a Ni^{IV} complex, has a structure similar to the unsubstituted carborane.[389] Two sandwich compounds of cobalt $[(\pi\text{-}C_5H_5)(B_7C_2H_4)Co]$ and $[(B_7C_2H_9)_2Co]^-$ probably have structures in which the cobalt completes the polyhedron; the proton spectrum of the latter shows two different CH groups and all the borons are different, so that the symmetry is low.[390] In the Mn^I complex $[(B_6C_2H_8)Mn(CO)_3]^-$ the CH groups are identical while the boron resonance provides four doublets in ratio 1 : 2 : 2 : 1. The structure may be as in (82).[391]

Derivatives of Monoborane and Diborane.—Complexes with phosphorus donor-compounds have received considerable attention during the year. Tetrafluorodiphosphine has basic properties and forms a borane adduct P_2F_4,BH_3. No boron–phosphorus coupling is observed so that the boron must be exchanging between phosphorus atoms.[392] Diborane reacts with phosphorus trioxide in chloroform solution to give successive co-ordination of one, two and three BH_3 groups and crystals of $P_4O_6,2BH_3$ and $P_4O_6,3BH_3$ were obtained. In solution an equilibria exists between the various complexes

386 J. L. Little, J. T. Moran, and L. J. Todd, *J. Amer. Chem. Soc.*, 1967, **89**, 5495.
387 W. R. Hertler, F. Klanberg, and E. L. Muetterties, *Inorg. Chem.*, 1967, **6**, 1696.
388 L. F. Warren, jun., and M. F. Hawthorne, *J. Amer. Chem. Soc.*, 1967, **89**, 470.
389 D. E. Hyatt, J. L. Little, J. T. Moran, F. R. Scholer, and L. J. Todd, *J. Amer. Chem. Soc.*, 1967, **89**, 3342.
390 M. F. Hawthorne and T. A. George, *J. Amer. Chem. Soc.*, 1967, **89**, 7114.
391 M. F. Hawthorne and A. D. Pitts, *J. Amer. Chem. Soc.*, 1967, **89**, 7115.
392 K. W. Morse and R. W. Parry, *J. Amer. Chem. Soc.*, 1967, **89**, 172.

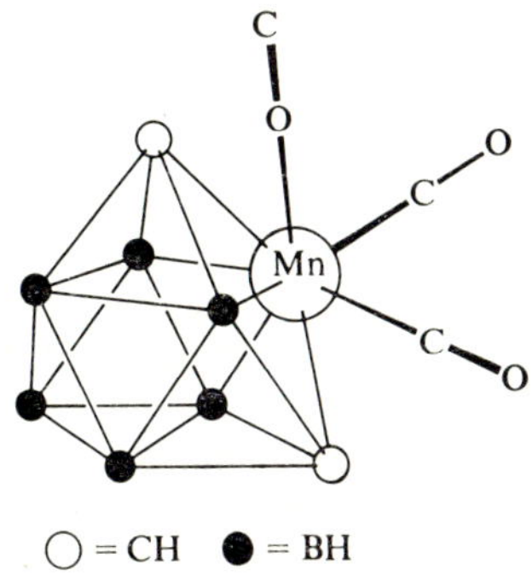

(82)

and the equilibrium constants are worked out. The substituent electronegativity affects the ^{31}P shifts and a theory of this effect is given.[393] The phosphine ligands PH_3, PF_3, and HPF_2 have been shown to give 1 : 1 borane adducts in which the stability varies in the order $HPF_2,BH_3 > PF_3,BH_3 \approx PH_3,BH_3$. The strongest complex is given by HPF_2 because of F—H—F hydrogen bonding in the HPF_2, which allows close P—B approach without steric hindrance from the fluorines. Proton, boron, and fluorine resonances were studied and the proton spectra were complex since all the nuclei couple; thus, for HPF_2,BH_3 each proton resonance of the borane quartet is split by further coupling with $J(P—H_B)$ 17·5, $J(F—H_B)$ 26, and $J(H_P—H_B)$ 4 c./sec.[394] Monobromodiborane can be made to form complexes to react with phosphine to give PH_3,BH_2Br and PH_3,BH_3. The proton data for the PH_3 groups are PH_3,BH_2Br: J(P—H) 405, J(H—H) 6 c./sec., δ 5·1 p.p.m., downfield from TMS; PH_3,BH_3: J(P—H) 375, J(H—H) 8 c./sec.; δ 4·4 p.p.m. downfield from TMS. It is suggested that the differing chemical shifts, δ, are due to the electronegativity of the bromine while the changes in the coupling constants indicate changes in the *s*-character of the P—H bond.[395] Ammonia and alkylamines will add to PH_3,BH_3 to give, in the case of ammonia, $[NH_4]^+[H_2P(BH_3)_2]^-$ which is fully identified using proton, boron, and phosphorus n.m.r. The proton resonance shows $J(H_P—H_B)$ 7·1 c./sec. in the anion and this coupling is lost if the preparation starts from PH_3,BD_3. The boron couples both with its hydrogen and with the phosphorus.[396] Monosilylphosphine forms adducts in the same way with BH_3 and with BCl_3.[397, 398]

Some boranocarbonates have been made, $M_2[H_3B{\cdot}CO_2]$ which are isoelectronic with carbonates: proton and boron n.m.r. show the boron to

[393] J. G. Riess and J. R. van Wazer, *J. Amer. Chem. Soc.*, 1967, **89**, 851.
[394] R. W. Rudolf and R. W. Parry, *J. Amer. Chem. Soc.*, 1967, **89**, 1621.
[395] J. E. Drake and J. Simpson, *Inorg. Nuclear Chem. Letters*, 1967, **3**, 87.
[396] J. W. Gilje, K. W. Morse, and R. W. Parry, *Inorg. Chem.*, 1967, **6**, 1761.
[397] J. E. Drake and J. Simpson, *Inorg. Chem.*, 1967, **6**, 1984.
[398] J. E. Drake and J. Simpson, *Chem. Comm.*, 1967, 249.

be in a highly symmetrical environment similar to that in BH_4^-. Thus the proton spectrum of $Na_2[H_3B{\cdot}CO_2]$ shows a 1 : 1 : 1 : 1 quartet ($^{11}B-H$ coupling) and a septet ($^{10}B-H$ coupling) and the ^{11}B spectrum is a 1 : 3 : 3 : 1 quartet.[399]

There has been some controversy about the structure of the amine adducts of diborane. The ^{11}B spectrum of B_2H_6 is a triplet of triplets and there was some question as to whether B_2H_6,NR_3 gives an identical pattern or a septet; the 19·2 Mc./sec. spectra suggest the latter, in which case the structure is presumably a dynamic form of R_3N,BH_2-H-BH_3.[400, 401] The hydrogens in diborane can be successively replaced by methyl groups using tetramethyl-lead to give eventually Me_3B. The following reactions were observed:

$$B_2H_6 + Me_4Pb \longrightarrow Me_3PbH + B_2H_5Me$$

$$2Me_3PbH \longrightarrow Me_6Pb_2 + H_2\uparrow$$

$$Me_6Pb_2 + B_2H_6 \longrightarrow Me_4Pb + Pb + B_2H_4Me_2 + H_2$$

Proton n.m.r. detected the presence of Me_3PbH and showed that the terminal hydrogens of B_2H_6 were replaced first, the bridge protons suffering a chemical shift (τ 10·53 $\rightarrow$ τ 9·65) as the terminal proton signal diminished.[402]

The i.r. spectra of the urea adducts $PhNH{\cdot}CO{\cdot}NHBu^t,BH_3$ suggest that the BH_3 is not attached to oxygen; the resonance of the amide protons is a single line so that the BH_3 must be exchanging between the nitrogen atoms.[403] BH_3 adducts of the ligand $P(NMeCH_2)_3CMe$ have been prepared with BH_3 on nitrogen or phosphorus or both.[404] Mono- and di-chloro-borane will add across olefinic double bonds in THF in the same way as will borane. The relative Lewis acidities of these boranes, of phenyl thio-borane, and of alkylboron halides, in THF, have been compared (by measuring the THF α-proton shifts) and further compared with the calculated electron densities on the boron.[405]

A correlation has been found between the i.r. B—H stretching frequency and J(B—H) from measurements carried out on some twenty boron hydrogen compounds. However, the results for $Al(BH_4)_3$ fall wide of the correlation, the value of J(B—H) being much lower for its stretching frequency than it apparently should be. This is rationalised in terms of the different time scales of the i.r. and n.m.r. experiments: the stretching frequency is that of the terminal B—H bonds only, whereas J is the average

399 L. J. Malone and R. W. Parry, *Inorg. Chem.*, 1967, **6**, 817.
400 J. F. Eastham, *J. Amer. Chem. Soc.*, 1967, **89**, 2237.
401 S. G. Shore and C. L. Hall, *J. Amer. Chem. Soc.*, 1967, **89**, 3947.
402 A. K. Holliday and G. N. Jessop, *J. Organometallic Chem.*, 1967, **10**, 291.
403 R. H. Cragg and N. N. Greenwood, *J. Chem. Soc.* (*A*), 1967, 961.
404 B. L. Laube, R. D. Bertrand, G. A. Casedy, R. D. Compton, and J. G. Verkade, *Inorg. Chem.*, 1967, **6**, 173.
405 D. J. Pasto and P. Balasubramaniyan, *J. Amer. Chem. Soc.*, 1967, **89**, 295.

of the terminal and bridging bonds.[406] The proton n.m.r. spectrum of $Al(BH_4)_3$ has been shown to be dependent on the heat treatment given to the material. Heating at 110° results in the irreversible development of the 1 : 1 : 1 : 1 quartet structure from the structureless room-temperature spectrum. It is suggested that there are two structures, octahedral and prismatic, with different ^{27}Al relaxation times.[407] Tetra-alkylammonium compounds of $[Al(BH_4)_4]^-$, $[Be(BH_4)_3]^-$, and $[Be_2(BH_4)_5]^-$ are reported. Boron n.m.r. shows that all BH_4 groups and all protons are equivalent in each compound. The latter compound may be formulated as in (83).[408]

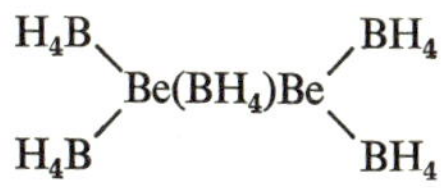

(83)

The alkali-metal borodeuterides, MBD_4, have been prepared and proton n.m.r. used to test for completion of deuteriation; 1–2% of hydrogen could be detected.[409]

The reactions under irradiation of sodium tetraphenylboronate have been studied and proton n.m.r. used to help in following the course of the reaction; in the absence of air the main product is the new compound 1-phenylcyclohexa-1,4-diene.[410] The reaction

$$R_2BH + I_2 + 2(\text{amine}) \longrightarrow [R_2B(\text{amine})_2]^+$$

has been studied using very bulky R groups. The reaction could only be made to go with difficulty and it is suggested that $[(\text{amine})_2I]^+$ is an intermediate in the reaction and that the total bulk interferes.[411]

Other Boron Adducts.—Studies of the adducts BF_3,H_2O and $BF_3,MeOH$ are reported.[412, 413] Solutions of the former in acetone at −80° show a quartet in the proton spectrum due to coupling with the three fluorines and two isotope-shifted triplets in the ^{19}F spectrum ($J(^{19}F{-}H)$ 2·93 ± 0·1 c./sec.). This proved the existence of the molecular species $H_2O \rightarrow BF_3$ in solution. If more than the stoicheiometric amount of water required for the 1 : 1 composition is present, then ligand exchange removes the fine structure. The assignments were confirmed by using D_2O- and ^{10}B-enriched BF_3.[412] The reactions occurring in low-temperature, sulphur dioxide solutions of

[406] Haruyuki Watanabe and Koichiro Nagasawa, *Inorg. Chem.*, 1967, **6**, 1069.
[407] P. C. Maybury and J. E. Ahnell, *Inorg. Chem.*, 1967, **6**, 1286.
[408] H. Nöth and M. Ehemann, *Chem. Comm.*, 1967, 685.
[409] J. G. Atkinson, D. W. McDonald, R. S. Stuart, and P. H. Tremaine, *Canad. J. Chem.*, 1967, **45**, 2583.
[410] J. L. R. Williams, J. C. Doty, P. J. Grisdale, T. H. Regan, and D. G. Borden, *Chem. Comm.*, 1967, 109.
[411] J. E. Douglass, G. R. Roehrig, and On-Hou Ma, *J. Organometallic Chem.*, 1967, **8**, 421.
[412] R. J. Gillespie and J. S. Hartman, *Canad. J. Chem.*, 1967, **45**, 859.
[413] R. J. Gillespie and J. S. Hartman, *Canad. J. Chem.*, 1967, **45**, 2243.

the methanol adducts in which [MeOH] > [BF_3] are:

$$\text{MeOH} + \text{BF}_3 \longrightarrow \text{MeOH,BF}_3$$

$$\text{MeOH,BF}_3 + \text{MeOH} \longrightarrow \text{Me{-}O(H)}\cdots\text{H{-}O(Me),BF}_3$$

$$\rightleftharpoons [\text{MeOH}_2]^+ + [\text{MeOBF}_3]^-$$

This explains the presence of only two proton resonances at low temperatures and is consistent with the known electrical conductivity of molten BF_3,2MeOH.[413]

It has been shown that ureas and thioureas co-ordinate boron halides via the nitrogen rather than the oxygen atom. Proton, boron, and fluorine n.m.r. spectra show that the boron is four-co-ordinate, that there is no barrier to rotation around the C—N bond, and that there is fast intramolecular exchange of BF_3 but slow intermolecular exchange leading to dissociated and associated BF_3 species.[414]

On the basis of their ^{11}B chemical shifts substituted pyridines are considered to show donor strengths toward boron trihalides in the order:[415] 2-EtC_5H_4N > 2-MeC_5H_4N > 4-EtC_5H_4N > 4-MeC_5H_4N, though it is elsewhere reported that 2-substitution causes some steric hindrance to complex formation.[416] Thus the proton shifts of substituted pyridine, on complexing to BF_3, are smallest for the α-substituted bases. BF_3 gives stable 1 : 1 adducts which can exist in the presence of an excess of pyridine. $AlBr_3$, however, was found to give complexes capable of exchange. Proton n.m.r. and i.r. spectra are reported for BF_3,$MeNH_2$, BF_3,Me_2NH, and BF_3,Me_3N.[417] Diphenylketimine, $Ph_2C{:}NH$, is a weaker donor than trimethylamine towards trimethylboron and this is paralleled by the fact that the methyl proton resonance of the acceptor is shifted less on co-ordination with $Ph_2C{:}NH$ than on co-ordination with Me_3N;[418] on heating the ketimine adduct to 160° this decomposes to give methane and the azomethine $Ph_2C{:}N{\cdot}BMe_2$. The reactions of triethylboron are also reported; triphenylboron does not react.[418] Azomethine derivatives of aluminium and gallium are reported and in this case the compounds are dimers.

Acetoxime, $Me_2C{:}N{\cdot}OH$, reacts with trimethylboron to give a six-membered BONBON ring (84) in which the Me_2C groups are equivalent, unlike those in the parent acetoxime. Other Group III metals and lithium also react.[419]

[414] N. N. Greenwood and B. H. Robinson, *J. Chem. Soc.* (*A*), 1967, 511.
[415] E. J. McLauchlan and E. F. Mooney, *Spectrochim. Acta*, 1967, **23**, *A*, 1227.
[416] H. H. Percampus and U. Krüger, *Chem. Ber.*, 1967, 71, 439.
[417] A. Derek, H. Clague, and A. Danti, *Spectrochim. Acta*, 1967, **23**, *A*, 2359.
[418] I. Pattinson and K. Wade, *J. Chem. Soc.* (*A*), 1967, 1098.
[419] K. Wade and J. R. Jennings, *J. Chem. Soc.* (*A*), 1967, 1333.

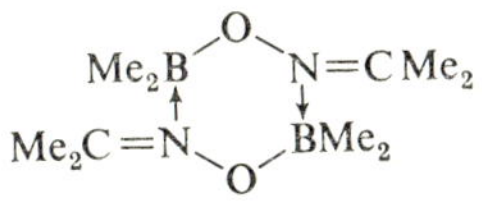

(84)

The monomer–dimer equilibrium:

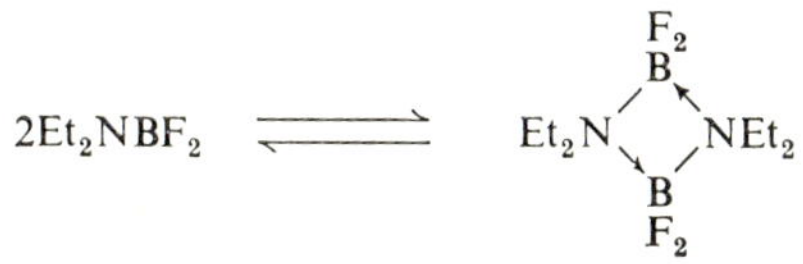

has been followed by fluorine n.m.r. spectra of the melt, which show two quartets separated by 4·35 p.p.m. The melt contains all monomer at 100° and all dimer at 30° (supercooled) and the dimer is the major component of the crystal and of solutions. Boron and proton n.m.r. are consistent with B—CH_2 coupling but not B—C—CH_3 coupling.[420]

A 1 : 1 adduct has been made of B_2Cl_4 and acetylene. Double-irradiation of ^{11}B sharpened the proton line and made visible the ^{13}C satellites which showed that J(C—C—H) was 17·5 c./sec. and that the *cis* form of the product $Cl_2BCH{:}CHBCl_2$ has been formed. Irradiation with u.v. light gave the *trans* isomer, J(C—C—H) 19·6 c./sec. The proton chemical shifts are: *cis*, $-6{\cdot}57$; and *trans*, $-7{\cdot}07$ p.p.m.[421]

Proton n.m.r. has been used to characterise the intermediates formed during the preparation of trimeric *N*-methylaminoborane, $(MeNHBH_2)_3$, from methylamine–borane, $MeNH_2,BH_2$; the reaction proceeds according to the sequence:

$$2(MeNH_2,BH_3) \longrightarrow [(MeNH_2)_2BH_2]^+BH_4^- \xrightarrow[-H_2]{+MeNH_2,BH_3}$$

$$\text{(A)}\quad [(MeNH_2)BH_2\cdot MeNH\cdot BH_2(MeNH_2)]^+BH_4^- \xrightarrow{-H_2}$$

$$(MeNH_2)BH_2\cdot MeNH\cdot BH_2(MeNH)\cdot BH_3 \longrightarrow (MeNHBH_2)_3$$

The chloride derivative of the intermediate (A) was isolated and characterised by its proton n.m.r. spectrum.[422]

Boron trifluoride reacts with boron at 2000° to give boron monofluoride which condenses as $(BF)_n$; on warming this gives several higher fluorides including B_2F_4 and B_3F_5. The latter has structure $BF_2\cdot BF\cdot BF_2$ with two ^{19}F resonances 87·2 p.p.m. apart; it melts and decomposes at $-50°$ to give B_8F_{12} and B_2F_4 and is very reactive. Co-condensation of BF with CO or

[420] N. N. Greenwood and J. Walker, *J. Chem. Soc.* (*A*), 1967, 959.
[421] T. D. Coyle and J. J. Ritter, *J. Amer. Chem. Soc.*, 1967, **89**, 5739.
[422] O. T. Beachley, *Inorg. Chem.*, 1967, **6**, 870.

PF_3 gives $(BF_2)_3B{\cdot}CO$ and $(BF_2)_3B{\cdot}PF_3$ which are stable crystalline compounds with equivalent BF_2 groups.[423]

Boron tribromide has been used to prepare anhydrous metal bromides of a large number of metals[424] namely Al, Sn, As, Sb, Bi, Ti, Zr, Nb, W, Fe, Pt, Cu, and Hg; $PhPBr_2$ and $SOBr_2$ were prepared similarly. It is suggested that reaction involves the formation of halogenoborate anions, *e.g.*

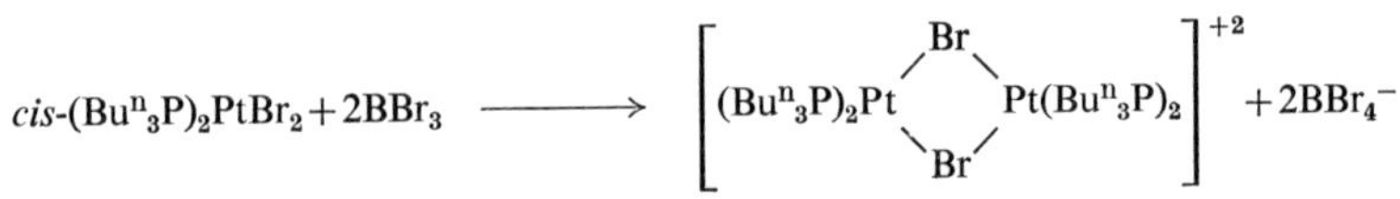

B^{11} n.m.r. was used to identify BX_4^- and ^{31}P n.m.r. to detect changes in the phosphine ligands.[424]

Boron–Nitrogen Compounds.—The ^{11}B shifts (from BF_3,Me_2O) have been correlated with the ^{13}C shifts (from benzene) in a series of seventeen alkanes and their analogous B—N compounds.[425] A good fit was obtained to the linear relation: $\delta(^{13}C) = 1{\cdot}44\ \delta(^{11}B) + 86{\cdot}0$ (p.p.m.). A perfluorovinylborazine has been prepared (^1H, ^{11}B, and ^{19}F n.m.r. reported) in which all vinyl groups are equivalent $(CF_2{:}CF{\cdot}BNMe)_3$.[426] The lithium derivative of the pyrrole analogue (85) gives a ferrocene type-compound with iron(II) chloride and forms the compound (86) with Ph_2BCl.[427] Poly-pyrazolyl and -imidazole borates (87) and (88) have also been prepared and identified by proton n.m.r.[428–430]

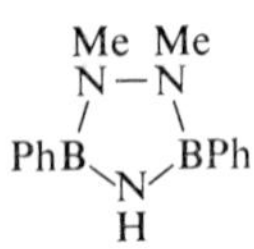

(85)

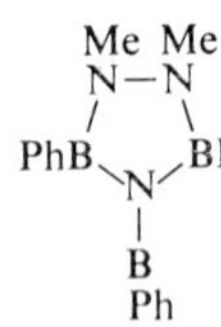

(86)

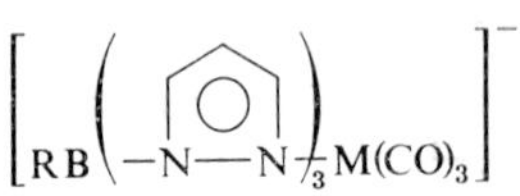

(87)

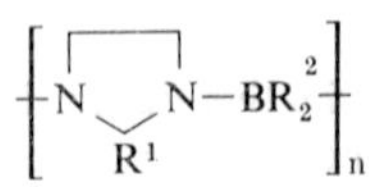

(88)

[423] P. L. Timms, *J. Amer. Chem. Soc.*, 1967, **89**, 1629.
[424] P. M. Druce, M. F. Lappert, and P. N. K. Riley, *Chem. Comm.*, 1967, 486.
[425] B. F. Spielvogel and J. M. Purser, *J. Amer. Chem. Soc.*, 1967, **89**, 5294.
[426] A. J. Klanica, J. P. Faust, and C. S. King, *Inorg. Chem.*, 1967, **6**, 840.
[427] H. Nöth and W. Regnet, *Z. anorg. Chem.*, 1967, **352**, 1.
[428] S. Trofimenko, *J. Amer. Chem. Soc.*, 1967, **89**, 3904.
[429] S. Trofimenko, *J. Amer. Chem. Soc.*, 1967, **89**, 6288.
[430] S. Trofimenko, *J. Amer. Chem. Soc.*, 1967, **89**, 3903.

Optically active forms of the cation (89) were separated manually and characterised by proton and boron n.m.r.[431] The cations $[Me_3N \cdot BH_2S(Me)]_2BH_2^+$ and $[Me_3N \cdot BH_2S(Me)BH_2 \cdot NMe_3]^+$ have been investigated by proton n.m.r. which supports a structure with free rotation as written.[432] The fluorine resonance of $(Me_2N)_2B(n\text{-}C_3F_7)$ is reported.[433]

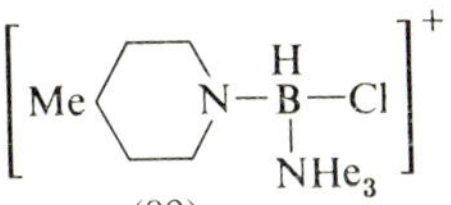

(89)

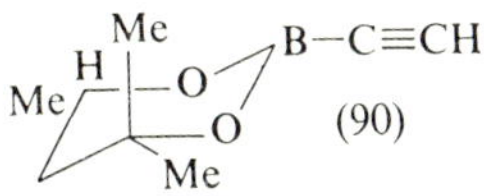

(90)

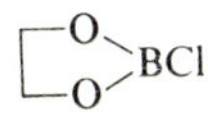

(91)

Boron–Oxygen Compounds.—The ring compound (90) has been prepared. Its ^{11}B signal is in a similar position to that of dimethylacetyleneboronate $(MeO)_2BC{:}CH$. Diels–Alder adducts of (90) have also been prepared and characterised by n.m.r. and i.r. spectroscopy.[434] The compound (91) can form B···O bonded dimers in CH_2Cl_2 and two ^{11}B peaks can be seen; addition of a donor such as R_2O, py, or R_3N gives three ^{11}B peaks.[435] The ^{11}B resonance is reported in the same paper for compounds analogous to (91) in which Cl is replaced by R or OR and in which one of the ring-oxygen atoms is replaced by a CH_2 group. Dibenzylphenylboronate $PhB(OCH_2Ph)_2$ is also associated in the neat liquid and concentrated solution: the ^{11}B chemical shifts are characteristic of tetrahedral boron in concentrated solution but change rapidly to the trigonal boron value at and below about 30% w/v of ester in carbon tetrachloride.[436]

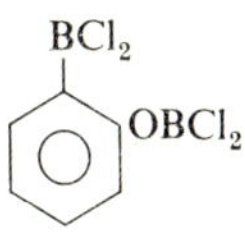

(92)

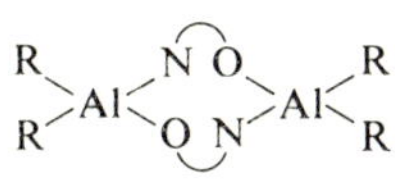

(93)

The compound (92) has been prepared and gives two ^{11}B peaks. On hydrolysis the chlorines are replaced by OH groups.[437] Sodium polyborate solutions (NaB_5O_8) show two ^{11}B peaks whose intensities are concentration dependent. It was found that the equilibria present are:

$$OH^- + B(OH)_3 \rightleftharpoons [B(OH)_4]^-; \quad pK = -5$$
$$2B(OH)_3 + [B(OH)_4]^- \rightleftharpoons [B_3O_3(OH)_4]^- + 3H_2O; \quad pK = -1{\cdot}9$$
$$4B(OH)_3 + [B(OH)_4]^- \rightleftharpoons [B_5O_6(OH)_4]^- + 6H_2O; \quad pK = -2$$

[431] G. E. Ryschkewitsch and J. M. Garrett, *J. Amer. Chem. Soc.*, 1967, **89**, 4240.
[432] R. J. Rowatt and N. E. Miller, *J. Amer. Chem. Soc.*, 1967, **89**, 5509.
[433] T. Chivers, *Chem. Comm.*, 1967, 157.
[434] W. G. Woods and P. L. Strong, *J. Organometallic Chem.*, 1967, **7**, 371.
[435] J. P. Laurent and J. P. Bonnet, *Bull. Soc. chim. France*, 1967, 2702.
[436] E. F. Mooney and P. H. Winson, *Chem. Comm.*, 1967, 341.
[437] C. A. Eggers and S. F. A. Kettle, *Inorg. Chem.*, 1967, **6**, 160.

Solutions of the salts $Na_2B_4O_7$ and $K_2B_{11}O_7$ were shown by ^{11}B n.m.r. to contain $B(OH)_3$ and $[B(OH)_4]^-$ at low concentrations, but mixed polyborates at high concentrations.[438]

Other Group III Elements.—Coupling has been detected in 1,2-dimethoxyethane solutions between aluminium and the protons in $LiAl(CH_3)_4$ ($J(^{27}Al—H)$ 6·34 c./sec.) and it is concluded that the Li^+ cation is bound to the solvent and is not exchanging to any great extent with the anion. In other solvents some ion pairing occurs, *e.g.* a chemical shift is observed between solutions in diethyl ether and 1,2-dimethoxyethane. The observation of Al—H coupling confirms the presence of symmetrical $AlMe_4^-$.[439]

The α-hydrogen atoms of pyridine in the adducts of $4\text{-}EtC_5H_4N,AlX_3$ become progressively broadened in the sequence X = Cl < Br < I and it is suggested that this is due to hydrogen bonding between the α-hydrogen and the halogen which is most effective for the largest halogen.[440]

The unusual chelates of aluminium with 2-methylquinolin-8-ol have been made in chloroform–methylamine solution and have been characterised by proton resonance.[441] The reaction of trialkyl- and triaryl-aluminium with bidentate ligands of type $R_2N(CH_2)_nCH_2O$— has been studied where $n = 1$, 2, and 8. The complexes are dimers but proton n.m.r. down to $-80°$ shows only one type of alkyl group which, together with i.r. data, shows them to have the structure (93).[442] The compound $Al(OCH_2\cdot CCl_3)_3$ is shown to be a dimer in methylene chloride and benzene but a monomer in dioxane. It is further demonstrated from the chemical shifts of the two sorts of CH_2 group that oxygen forms bridges in the dimer.[443]

Monomethylthallium(III) species have been prepared using the equilibrium:[444]

$$Me_2TlOAc + Hg(OAc)_2 \rightleftharpoons MeTl(OAc)_2 + MeHgOAc$$

The presence of all components was demonstrated by proton resonance.[445] Monoalkyl derivatives have also been made. The coupling constant $J(^{205}Tl—CH)$ is very solvent dependent and increases in the series $R_3Tl < R_2Tl^+ < RTlX_2$ (R = alkyl, X = alkoxy) owing to the increased *s* character of the Tl—C bond as the number of alkyl groups is decreased and to the increase in the effective nuclear charge on the thallium ion.[446] Typical values of the Tl—H coupling constant in chloroform solution are: $Me_2Tl(O_2CPr^i)$, 377; $MeTl(OAc)_2$, 892; $MeTl(O_2CPr^i)_2$, 902 c./sec. Changes

[438] R. K. Momii and N. H. Nachtrieb, *Inorg. Chem.*, 1967, **6**, 1189.
[439] J. P. Oliver and C. A. Wilkie, *J. Amer. Chem. Soc.*, 1967, **89**, 163.
[440] T. N. Huckerby, J. W. Wilson, and I. J. Worrall, *Chem. Comm.*, 1967, 1190.
[441] P. R. Scherer and Q. Fernando, *Chem. Comm.*, 1967, 1107.
[442] T. J. Hurley, M. A. Robinson, J. A. Scruggs, and S. I. Trotz, *Inorg. Chem.*, 1967, **6**, 1310.
[443] Takeo Saegusa and Takashi Ueshima, *Inorg. Chem.*, 1967, **6**, 1679.
[444] Hideo Kurosawa and Rokuro Okawara, *Inorg. Nuclear Chem. Letters*, 1967, **3**, 93.
[445] H. Kurosawa and R. Okawara, *Inorg. Nuclear Chem. Letters*, 1967, **3**, 21.
[446] Hideo Kurosawa and Rokuro Okawara, *J. Organometallic Chem.*, 1967, **10**, 211.

in hybridisation are also used to explain changes in coupling constants in dimethylthallium compounds between non-polar and polar solvents, J being smaller in the latter because the equilibrium:

$$Me_2TlX \rightleftharpoons Me_2Tl^+ + X^-$$

allows the formation of significant quantities of the sp hybridised cation.[447]

8 Compounds of Fluorine and Phosphorus

After proton and boron n.m.r., fluorine and phosphorus resonances have been the most extensively studied nuclei in inorganic systems. Compounds in which fluorine is bonded to oxygen or carbon are considered first, then miscellaneous fluorine compounds. As considerable overlap exists between fluorine and phosphorus chemistry a linking section dealing with fluorophosphorus compounds has been included before the other compounds of phosphorus are considered in the final section.

p-Difluorobenzene is suggested as a standard for ^{19}F variable temperature measurements. The temperature dependence of its shift with respect to methane has been found by a double-resonance method to be only 0·303 p.p.m. over the temperature range 0–100°.[448]

Compounds containing mainly oxygen and fluorine have caused considerable interest during the year and lend themselves readily to fluorine n.m.r. investigations. The ^{19}F chemical shifts of OF_2 and O_2F_2 at low temperatures are -249 and -865 p.p.m. downfield from $CFCl_3$. The O—F bond is long in O_2F_2 and it is suggested that this is a one-electron bond which introduces anti-shielding of the fluorine and is responsible for the very low-field shift.[449] The ^{19}F chemical shift of 'O_3F_2' is very temperature dependent and varies from -1900 at 85° to -865 p.p.m. at 145°K.[449] The ^{17}O n.m.r. spectrum of 'O_3F_2' shows three lines at -647, -971, and -1512 p.p.m. from H_2O; the line at -647 p.p.m. showed splitting, J 424 c./sec., and is due to O_2F_2. Increasing the temperature increased the proportion of O_2F_2, as determined by fluorine n.m.r. It is suggested that O_3F_2 consists of a mixture of O_2F_2 with $(O_2F)_n$.[450]

Considerable effort has gone into the preparation of bis(fluoroxy)-difluoromethane $CF_2(OF)_2$. This is a stable compound which has been made in low yield by fluorination of sodium trifluoroacetate or sodium oxalate,[451] or by the reaction:[452]

$$(FCO)_2O_2 + F_2 \xrightarrow{h\nu} FC(O)OF \xrightarrow{CsF-F_2} F_2C(OF)_2$$

[447] Hideo Kurosawa, Kiyoshi Yasuda, and Rokuro Okawara, *Bull. Chem. Soc. Japan*, 1967, **40**, 861.
[448] R. F. Spanier and E. R. Malinowski, *J. Chem. Phys.*, 1967, **47**, 1560.
[449] J. W. Nebgen, F. I. Metz, and W. B. Rose, *J. Amer. Chem. Soc.*, 1967, **89**, 3118.
[450] I. J. Solomon, J. K. Raney, A. J. Kacmarek, R. G. Maguire, and G. A. Noble, *J. Amer. Chem. Soc.*, 1967, **89**, 2015.
[451] P. G. Thompson, *J. Amer. Chem. Soc.*, 1967, **89**, 1811.
[452] R. L. Cauble and G. H. Cady, *J. Amer. Chem. Soc.*, 1967, **89**, 5161.

It is made in good yield by the fluorination of carbon dioxide.[453, 454] Its fluorine n.m.r. consists of two triplets (J 38·6 c./sec., ϕ −155·7 and +81·7 p.p.m.) which, together with i.r. data, confirm the formulation. The spectra of $SF_5{\cdot}OF$, SO_3F_2, and FC(O)OF are also reported.[452] The related compounds $CF_3{\cdot}CF(OF)_2$ and $(CF_3)_2C(OF)_2$ have also been prepared. In the former the coupling constants are $J(CF_3{-}CF)$, 1; $J(CF_3{-}OF)$ 10·3; and J(CF—OF) 26·2 c./sec.[455]

Bis(trifluoromethyl)trioxide $CF_3{\cdot}OOOCF_3$ has been prepared by addition of oxygen difluoride to carbonyl fluoride and gives a single ^{19}F line.[456] The parent and a higher fluorocarbon homologue have also been prepared by fractionation of the products obtained from the fluorination of acetates: fluorine n.m.r. data are presented for $CF_3{\cdot}OOOCF_3$, $CF_3{\cdot}OOOCF_2{\cdot}CF_3$, $CF_3{\cdot}OOCF_2{\cdot}CF$, and $CF_3{\cdot}OCF_2{\cdot}CF_3$. The reaction

$$CF_3OOOCF_3 \longrightarrow CF_3OOCF_3$$

proceeds with a half life of 65 weeks at 25°. The coupling constant, $J(CF_2{-}CF_3)$ in $-CF_2CF_3$ groups is small (1·5–2·2 c./sec.).[457]

The compound $CH_3{\cdot}CF_2{\cdot}NF{\cdot}Cl$ has been prepared and characterised by proton and fluorine n.m.r. (J(H—CF), 16·0; J(H—NF), 3 c./sec.).[458] The structure (94) is assigned to $(CF_3SNCO)_2$. The ^{13}C satellites of the ^{19}F singlet main line contain none of the fine quartet structure to be expected from a $(CF_3S)_2N$ group.[459]

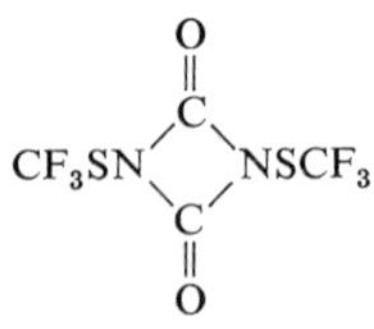

(94)

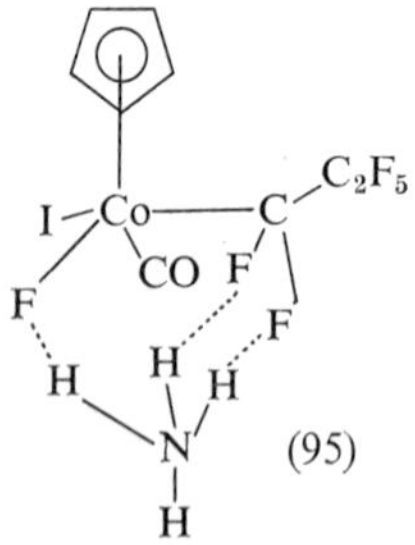

(95)

A series of complexes $(\pi\text{-}C_5H_5)Co(CO)C_3F_7X$ have been made. If X = SCN, a trimer is formed with one of the C_3F_7 groups differing from the others. The fluorine n.m.r. of the ammonium salt of the complex when X = IF^- was not fully understood but the presence of a strongly hydrogen-bonded ammonium group is suggested as in (95).[460]

453 R. L. Cauble and G. H. Cady, *J. Amer. Chem. Soc.*, 1967, **89**, 1962.
454 F. A. Hohorst and J. M. Shreeve, *J. Amer. Chem. Soc.*, 1967, **89**, 1809.
455 P. G. Thompson and J. H. Prager, *J. Amer. Chem. Soc.*, 1967, **89**, 2263.
456 C. R. Anderson and W. B. Fox, *J. Amer. Chem. Soc.*, 1967, **89**, 4313.
457 P. G. Thompson, *J. Amer. Chem. Soc.*, 1967, **89**, 4316.
458 M. Lustig, *Inorg. Chem.*, 1967, **6**, 1064.
459 A. J. Downs and A. Haas, *Spectrochim. Acta*, 1967, **23**, *A*, 1023.
460 P. M. Treichel and G. P. Werber, *J. Organometallic Chem.*, 1967, **7**, 157.

Certain manganese carbonyl species, *e.g.* $[Mn(CO)_4PPh_3]^-$, will replace F* in (96) and (97) forming an Mn—C bond. The position of substitution was determined from fluorine n.m.r. with the aid of the ^{14}N broadening pattern which is greatest for *o*-fluorine atoms.[461]

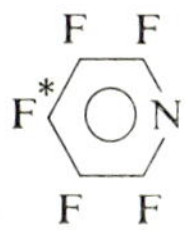
(96)

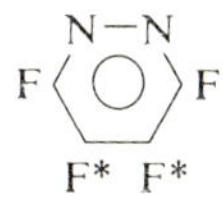
(97)

Miscellaneous Fluoro-compounds.—Two groups report the preparation of NF_4^+ from the reaction:

$$NF_3 + F_2 + MF_5 \longrightarrow [NF_4]^+[MF_6]^- \quad (M = As \text{ or } Sb).$$

The ion hydrolyses to give NF_3 and contains four equivalent fluorines exhibiting a 1 : 1 : 1 ^{19}F triplet with $J(^{14}N—^{19}F)$ 234 c./sec. and with very narrow lines; the ion must thus be tetrahedral.[462, 463]

The HF_2^- ion has been investigated by measuring the concentration dependence of the ^{19}F chemical shift in solutions of its salts; it was shown that the ion is in equilibrium with HF and F^-.[464] Evidence was also presented for hydrogen bonding between F^- and NH_4^+.

The preparation and fluorine n.m.r. data for $(NF_2)_2CO$ and $NF_2{\cdot}CO{\cdot}Cl$, and new routes to $NF_2{\cdot}OCF_3$ and $NF_2{\cdot}Cl$, are reported.[465, 466] The preparation and n.m.r. data of $NCN{:}S(F_2){:}O$ and $F_3CN{:}S(F_2){:}O$,[467] and of $SF_2{:}NCN$ and $F_2S{:}NCF_2N{:}SF_2$[468] are given. The compound $FSeO_2{\cdot}OR$ is found to be associated in solution in contrast to the sulphur analogues. If R = Et then two configurations are found. Fluorine n.m.r. shows that an equilibrium exists between $FSeO_2{\cdot}OR$ and small amounts of SeO_2F_2 and $(RO)_2SeO_2$; the major component exhibits a long-range coupling for R = Et, with J[F(SeOCC)H] 0·82 c./sec.[469]

The fluorine n.m.r. of aqueous titanium tetrafluoride has enabled the hydrolysis products to be identified as $[TiF_5,H_2O]^-$ giving a doublet-quintet pattern, $TiF_4,Ti(OH)_4(H_2O)_2$ giving two triplets, and $[TiF_6]^{2-}$ giving a singlet. $[TiF_6]^{2-}$ is stable in neutral solution. In aqueous ethanol only $[TiF_5(H_2O)]^-$ and $[TiF_5(EtOH)]^-$ are formed. If $TiF_4(DMF)_2$ is hydrolysed

[461] J. Cooke, M. Green, and F. G. A. Stone, *Inorg. Nuclear Chem. Letters*, 1967, **3**, 47.
[462] K. O. Christe, J. P. Guertin, A. E. Pavlath, and W. Sawodny, *Inorg. Chem.*, 1967, **6**, 533.
[463] W. E. Tolberg, R. T. Rewick, R. S. Stringham, and M. E. Hill, *Inorg. Chem.*, 1967, **6**, 1156.
[464] R. Haque and L. W. Reeves, *J. Amer. Chem., Soc.*, 1967, **89**, 250.
[465] G. W. Fraser and J. M. Shreeve, *Inorg. Chem.*, 1967, **6**, 1711.
[466] R. L. Cauble and G. H. Cady, *Inorg. Chem.*, 1967, **6**, 2117.
[467] M. Lustig and J. K. Ruff, *Inorg. Nuclear Chem. Letters*, 1967, **3**, 531.
[468] L. Glemser and U. Biermann, *Inorg. Nuclear Chem. Letters*, 1967, **3**, 223.
[469] R. Paetzold, R. Kurze, and G. Englehardt, *Z. anorg. Chem.*, 1967, **353**, 62.

then $[TiF_5(H_2O)]^-$, $[TiF_5(DMF)]^-$, and $[TiF_6]^{2-}$ are formed. In aqueous hydrofluoric acid there is an exchanging system containing $[TiF_6]^{2-}$ and $[TiF_4,2H_2O]$.[470]

The compound F_5TeNMe_2 gives an AX_4 ^{19}F spectrum whereas $F_4Te(NMe_2)_2$ gives an A_2X_2 spectrum and is therefore the *cis*-isomer.[471]

The Group IV metals, except possibly lead, form the hexafluoro-anions MF_6^{2-} in aqueous solutions. Fluorine exchange is rapid if M = Zr, Hf, Pb, and broadened lines are obtained, but is slow if M = Si, Ge, Sn, or Ti.[472] In the former case the reaction occurring is:

$$HfF_7^{3-} \rightleftharpoons F^- + HfF_6^{2-}$$

Amongst the elements for which the exchange is slow, solvent interaction occurs, *e.g.* for Sn where the coupling constant $J(^{119}Sn—^{19}F)$ in SnF_6^{2-} is solvent dependent; satellites due to the other tin isotopes are also visible. In GeF_6^{2-} there is a singlet flanked by ten satellites caused by the ^{73}Ge isotope (abundance 7·7%, $I = \frac{9}{2}$, $J(^{73}Ge—^{19}F)$ 98 c./sec.) which confirms the symmetrical nature of the ion; similarly satellites are seen for TiF_6^{2-} and the fluorine resonance consists of a single sharp line flanked by a satellite pattern which is a combination of six lines caused by the ^{47}Ti isotope (abundance 7·28%, $I = \frac{5}{2}$) and eight lines caused by the ^{49}Ti isotope (abundance 5·51%, $I = \frac{7}{2}$); $J(^{49}Ti—^{19}F)$ is 33 c./sec.[472]

The tungsten(VI) chloride fluorides WF_5Cl, WF_2Cl_4, $WFCl_5$, and possibly WF_4Cl_2 have been prepared. WF_5Cl gives a doublet-quintet ^{19}F pattern at $-30°$ with a coupling constant of 73 c./sec. between the fluorine atoms which are *cis* and *trans* to the chlorine.[473] Chlorine pentafluoride has a similar spectrum, with J 130 c./sec., and must therefore have C_{4V} symmetry; it was prepared according to the reaction:[474]

$$MClF_4 + F_2 = MF + ClF_5 \quad (M = K, Rb, Cs).$$

Fluorophosphorus Compounds.—The compounds HPF_4 and H_2PF_3 have been prepared and characterised by proton, fluorine, and phosphorus n.m.r. spectroscopy in the gas and liquid phases. In HPF_4 the fluorine atoms are equivalent and are exchanging thus giving rise to a ^{19}F spectrum of two doublets, a proton spectrum of two quintets, and a ^{31}P spectrum also of two quintets. In H_2PF_3 both the fluorine atoms and the protons are equivalent, leading to a sextet ^{31}P spectrum with $J(P—F)$ *ca.* $J(P—H)$ *ca.* 866 c./sec.; the proton spectrum comprises two quartets and the fluorine spectrum comprises two triplets in the gas phase but shows a spectrum which varies with temperature in the liquid phase.[475]

[470] Yu. A. Buslaev, D. S. Dyer, and R. O. Ragsdale, *Inorg. Chem.*, 1967, **6**, 2208.
[471] G. W. Fraser, R. D. Peacock, and P. M. Watkins, *Chem. Comm.*, 1967, 1248.
[472] P. A. W. Dean and D. F. Evans, *J. Chem. Soc.* (*A*), 1967, 698.
[473] G. W. Fraser, M. Mercer, and R. D. Peacock, *J. Chem. Soc.* (*A*), 1967, 1091.
[474] D. Pilipovitch, W. Maya, E. A. Lawton, H. S. Bauer, D. F. Sheehan, N. N. Ogimachi, R. D. Wilson, F. C. Gunderloy, jun., and V. E. Bedwell, *Inorg. Chem.*, 1967, **6**, 1918.
[475] P. M. Treichel, R. A. Goodrich, and S. B. Pierce, *J. Amer. Chem. Soc.*, 1967, **89**, 2017.

Auto-ionisation has been sought in compounds Me_2N,MF_5 (M = P, As, or Sb). Fluorine n.m.r. shows only the original molecular compound to be present and the spectra are temperature independent so that no exchange takes place. Extra species are only found in the presence of impurities or upon hydrolysis and it is felt that the conductimetric work which first indicated auto-ionisation is wrong.[476] PF_6^- was detected by fluorine n.m.r. in the products of the reaction of tetrabutylammonium halides with PF_5.[477]

The compounds OPF_2H and SPF_2H have been prepared and characterised. The coupling constants to phosphorus are large and typical of pentavalent phosphorus so that the structure is probably (98). Hydrolysis of SPF_2H gave SF_4, H_2S, and phosphorous acid containing F^- and HF_2^-, all identified by n.m.r.[478] The compounds $[SF_5O]^-$ $[PF_2O_2]^-$ and $[PF_2S_2]^-$

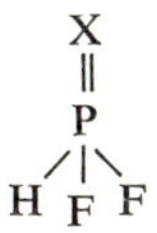

(X = O or S)

(98)

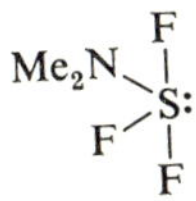

(99)

have been made by reactions involving alkali-metal fluorides with *e.g.*, SPF_3, or $SPCl_3$; the fluorine resonance of $[SF_5O]^-$ is an AB_4 pattern, J(P—F) in $[PF_2S_2]^-$ is greater than in $[PF_2O_2]^-$ (1164 and 952 c./sec.) and the electron density in its P—F bonds is greater, consistent with the stability of the $[PF_2S_2]^-$ ion to hydrolysis.[479]

Amine derivatives of SF_4 and PF_5 have been made by Si—N bond-cleavage reactions to give Me_2NSF_3, Me_2NPF_4, and $(MeNPF_3)_2$ for which ^1H, ^{19}F, and ^{31}P data are reported; Me_2NSF_3 gives a broad fluorine resonance at room temperature but two signals with intensity ratio 2 : 1 at $-100°$, consistent with that expected for structure (99).[480] Preparation and spectra of $H_2NP(O)F_2$ are also reported; it shows the expected proton, fluorine, and phosphorus patterns.[481] Phosphoryl fluoride amine derivatives are prepared according to the reaction:

$$Me_2NH + POF_3 \longrightarrow Me_2NP(O)F_2 + \text{solid}$$

Solid-state fluorine n.m.r. was used to identify PF_6^- and $PO_2F_2^-$. The ^{19}F spectrum of $Me_2NP(O)F_2$ is a doublet of septets, J(P—F) 997·5 c./sec., and the proton spectrum a doublet of triplets as expected.[482]

The spectra of the compounds $(CF_3)_2C(OPF_2)Br$, $(CF_3)_2C(OPF_2)I$, and $(CF_3)_2C(OPF_2)H$ are described; all magnetic nuclei are coupled.[483] It has

[476] F. N. Tebbe and E. L. Muetterties, *Inorg. Chem.*, 1967, **6**, 129.
[477] S. Brownstein, *Canad. J. Chem.*, 1967, **45**, 2403.
[478] T. L. Charlton and R. G. Cavell, *Inorg. Chem.*, 1967, **6**, 2204.
[479] M. Lustig and J. K. Ruff, *Inorg. Chem.*, 1967, **6**, 2115.
[480] G. C. Demitras and A. G. MacDiarmid, *Inorg. Chem.*, 1967, **6**, 1903.
[481] S. Kongpricha and W. C. Preusse, *Inorg. Chem.*, 1967, **6**, 1915.
[482] R. G. Cavell, *Canad. J. Chem.*, 1967, **45**, 1309.
[483] M. Lustig and W. E. Hill, *Inorg. Chem.*, 1967, **6**, 1448.

been found that $(CF_3P)_4$ can be cleaved by primary and secondary phosphines to give $(MeP)_5$, CF_3PH_2, Me_4P_2, and $CF_3P[PMe_2]_2$, this last substance giving a pair of triplets in the fluorine spectrum with J(P—CF) 40·2 and J(PP—CF) 6·4 c./sec.[484a]

Phosphorus Compounds.—A comprehensive account of ^{31}P n.m.r. has been published during the year; this includes a full treatment of the measurement and interpretation of high-resolution ^{31}P n.m.r. spectra and a compilation of data on some 3250 compounds for which ^{31}P chemical shift and coupling constant data are available.[484b]

Dynamic nuclear polarisation studies of phosphorus compounds have been carried out by irradiating the electron resonance of added tris-(t-butyl)-phenoxyl radicals while observing the effect on the ^{31}P resonance. In the case of phosphorus(III) systems large positive enhancements of the ^{31}P resonance are obtained, ranging from 10 times to, in the case of $(MeO)_3P$, 250 times. In the case of most phosphorus(V) compounds the enhancement is small or in the opposite (negative) sense with the signal inverted and it is suggested that this is due to steric interference caused by radical diffusion near phosphorus. This view is supported by an enhancement of +5 being obtained for the compound $Et_2P(O)H$ where the steric interference is minimal owing to the hydrogenic substituent.[485]

In a series of trialkyl- and cyclic-alkyl-phosphite–metal complexes $[CoL_5]^+$, NiL_4, $[NiL_5]^+$, $[CuL_4]^+$, and $[AgL_4]^+$ (L = $(MeO)_3P$, $RC(CH_2O)_3P$, and $(CH_2)_3(CHO)_3P$), it has been found that the ^{31}P shifts can be correlated with the metal oxidation-state but only if the complexes are both isostructural and isoelectronic, *e.g.* the ^{31}P resonance moves upfield in the sequence $NiL_4 \rightarrow [AgL_4]^+$ and $[CoL_5]^+ \rightarrow [NiL_5]^+$.[486]

Several compounds have been made based on the P_3N_3 hexacyclic ring with various substituents on the phosphorus atoms. $P_3N_3Cl_3(NMe_2)_2(OPh)$ has structure (100) since it gives two methyl doublets with J(P—H) 17·5 c./sec.; if the methylamine groups were geminal the coupling constant would be smaller.[487] Three isomers of $P_3N_3Cl_4(NHMe)_2$ and their ^{31}P n.m.r. spectra are reported: *gem*, *vic–trans* and *vic–cis*.[488] The nitrogen substituents in $P_3N_3Cl_4(NH_2)_2$ and $P_3N_3Cl_4(NPCl_3)_2$ are probably *gem*. The ^{31}P spectra are compared with those of the monosubstituted rings and could all be interpreted as AB_2 or ABC.[489] Multiple-substituted rings are also discussed.[490]

484a A. H. Cowley, *J. Amer. Chem. Soc.*, 1967, **89**, 5990.
484b M. M. Crutchfield, C. H. Dungan, J. H. Letcher, V. Mark, and J. R. van Wazer, *Topics Phosphorus Chem.*, 1967, **5**.
485 P. W. Atkins, R. A. Dwek, J. B. Reid, and R. E. Richards, *Mol. Phys.*, 1967, **13**, 175.
486 K. J. Coskran, R. D. Bertrand, and V. G. Verkade, *J. Amer. Chem. Soc.*, 1967, **89**, 4535.
487 C. T. Ford, F. E. Dickson, and I. I. Bezman, *Inorg. Chem.*, 1967, **6**, 1594.
488 W. Lehr, *Z. anorg. Chem.*, 1967, **352**, 27.
489 G. R. Feistel and T. Moeller, *J. Inorg. Nuclear Chem.*, 1967, **29**, 2731.
490 B. W. Fitzsimmons, C. Hewlett, K. Hills, and R. A. Shaw, *J. Chem. Soc. (A)*, 1967, 679.

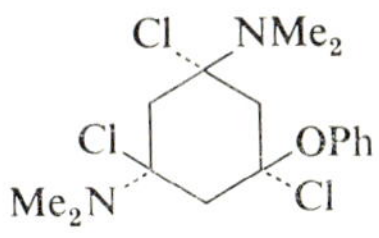

(100)

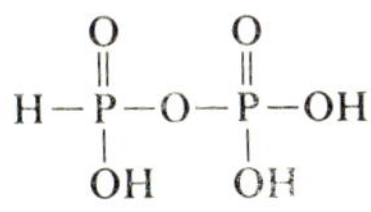

(101)

Proton n.m.r. shows that chloramination of dimethylaminophosphine with dimethylchloramine occurs on the phosphorus atom. The cations $[(Me_2N)_2PMe_2]^+$, $[(Me_2N)_3PMe]^+$, $[(Me_2N)_3PNH_2]^+$, $[(Me_2N)_3PCl]^+$, and $[(Me_2N)_3PBr]^+$ are obtained.[491] The ^{31}P spectra of the cations $[(H_2N)_3PN{:}P(NH_2)_3]^+$ and $[P_2(NH_2)_4(NHMe)_2]^+$, and the proton spectra of the compounds Ph(RNH$_2$)PX (X = O, S, and Se) are reported.[492, 493]

The ^{31}P n.m.r. shifts of the isohypophosphate ion (101), which contains two types of phosphorus atom, show that the most strongly acidic proton is associated with the phosphorus which carries two hydroxy-groups and it is this phosphorus atom which suffers the biggest chemical shift as the pH changes. Complexing with Li^+ affects both phosphorus atoms equally so that the acid is a bidentate chelating agent.[494] The pH-dependent ^{31}P chemical shifts and coupling constants in the oxyacid anions PO_4^{3-}, HPO_3^{2-}, $H_2PO_2^-$, $PhP(H)(O)^-$, $MePO_3^{2-}$ and $Me_2PO_2^-$ have been found to vary linearly with the degree of neutralisation. This is interpreted in terms of additive contributions from (*a*) variations in the effective electronegativity of the oxygen atoms bonded and non-bonded to protons, (*b*) occupation of phosphorus d_π orbitals, and (*c*) O—P—O bond angles. The interesting point is made that in an X—Y bonded pair it is the relative effective electronegativities of the two atoms which affect the paramagnetic shift of one of them, so that increasing the effective electronegativity of one substituent can cause either a decrease or an increase in the chemical shift of the other, depending on their relative electronegativities[495] (such a picture could for instance explain the anomalous sequences in the ^{11}B, ^{27}Al, and ^{71}Ga shifts of the Group III halides).

Proton n.m.r. has been used to determine the concentration of the components in mixed phosphorus and pyrophosphorus acids in dioxane by following the chemical shift of the single peak as it varies with the relative concentrations of the exchanging species.[496] I.r. and ^{31}P measurements on compounds of the type (102) suggest that the proton directly bonded to the phosphorus atom is capable of feeble hydrogen-bonding and it is found that the ^{31}P shifts correlates with $\Delta\nu$(P—H).[497] Data are also reported for

[491] S. R. Jain, L. K. Krannich, R. E. Highsmith, and H. H. Sisler, *Inorg. Chem.*, 1967, **6**, 1058.
[492] M. Becke-Goehring and B. Scharf, *Z. anorg. Chem.*, 1967, **353**, 320.
[493] A. P. Lane, D. A. Morton-Blake, and D. S. Payne, *J. Chem. Soc.* (*A*), 1967, 1492.
[494] R. L. Carroll and R. E. Mesmer, *Inorg. Chem.*, 1967, **6**, 1137.
[495] K. Moedritzer, *Inorg. Chem.*, 1967, **6**, 936.
[496] P. Bivel, F. Hossenlopp, and J. P. Ebel, *Bull. Soc. chim. France*, 1967, 1224.
[497] R. Wolf, D. Houalla, and F. Mathis, *Spectrochim. Acta*, 1967, **23**, *A*, 1641.

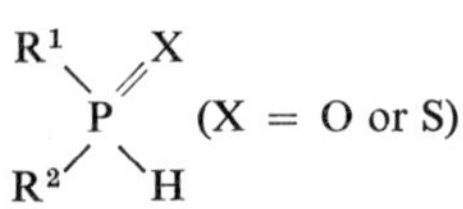

(102)

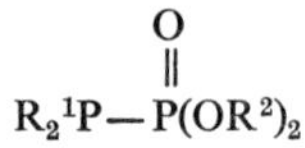

(103)

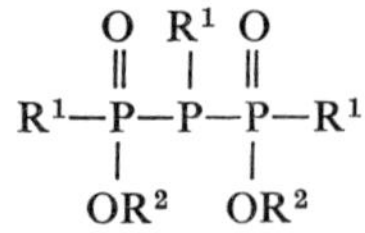

(104)

a range of pyrophosphoric and hypophosphoric acid esters[498] and of compounds of type (103) and (104).[499]

The cation PH_4^+ has been identified by ^{31}P and ^{1}H n.m.r. in the strongly acidic solvents H_2O,BF_3 and $MeOH,BF_3$ at 25°, and in concentrated sulphuric acid at lower temperatures; J(P—H) 547 c./sec. and the ^{31}P spectrum is a quintet.[500] In the series PH_2^- (J 140 c./sec., τ 11·4), PH_3 (J 190 c./sec., τ 8·1), and PH_4^+ (J 547 c./sec., τ 3·6) the shifts are in the order to be expected both from the relative charges and the magnetic anisotropy of the phosphorus atom. The coupling constants correlate with the changes in bond angle which leads to a decrease in the phosphorus 3*s* character in the order $PH_4^+ > PH_3 > PH_2^-$.[500] The cations $[(MeO)_4P]^+$, $[(MeO)_3PMe]^+$, and $[(MeO)_3PSMe]^+$, and probably $[(MeO)P(OEt)_3]^+$, have all been prepared and their proton and phosphorus resonances recorded.[501]

9 Less-common Resonances

Several papers have appeared describing unusual resonances. Chemical shifts and linewidths have been obtained for the ^{59}Co resonance of a series of organo-cobalt compounds and ions. The symmetrical ions gave narrow lines and spin–spin coupling could be observed: $Co(CN)_6^{3-}$, $J(^{59}Co—^{13}C)$ 127; $Co(CO)_4^-$, $J(^{59}Co—^{13}C)$ 287; and $Co(PF_3)_4^-$, $J(^{59}Co—^{31}P)$ 1222 and $J(^{59}Co—^{19}F)$ 57 c./sec. The last two ions are thus certainly tetrahedral. The complex $Co(CO)_3NO$ also gave a fairly narrow line. The carbonyl $Co_4(CO)_{12}$ gave two ^{59}Co lines showing that it contains one cobalt of one type and three of another, in agreement with the crystal structure. The shifts measured ranged up to 4220 p.p.m. between $Co(CN)_6^{3-}$ and $Co(PF_3)_4^-$. It was found possible to accurately predict the chemical shift of $(C_5H_5)_2CoCl$ from $K_3Co(CN)_6$ using ligand-field theory. Linewidths were obtained up to 23 gauss. Substances with broad lines showed complex solution behaviour which was not interpreted.[502] The use of ^{59}Co shifts in determining the stereochemistry of a complex is given in ref. 99.

Work has also been carried out on the ^{55}Mn resonance. In the complexes $XMn(CO)_5$ (X = halogen), it is found that the strongest metal–halogen bond (Mn—I, in this case) has least paramagnetic shielding (smallest $1/\Delta E$) though it is noted that this effect can be reversed by a change in oxidation

[498] R. K. Harris, A. R. Katritsky, S. Musierowicz, and B. Ternai, *J. Chem. Soc.* (*A*), 1967, 37.
[499] E. Fluck and H. Binder, *Inorg. Nuclear Chem. Letters*, 1967, **3**, 307.
[500] G. M. Sheldrick, *Trans. Faraday Soc.*, 1967, **63**, 1077.
[501] J. S. Cohen, *J. Amer. Chem. Soc.*, 1967, **89**, 2543.
[502] E. A. C. Lucken, K. Noack, and D. F. Williams, *J. Chem. Soc.* (*A*), 1967, 148.

state of the metal (cf. ref. 495). Shifts are calculated using ligand-field theory. The different linewidths are not understood, some being surprisingly broad while others [$Mn_2(CO)_{10}$ and $XMn(CO)_5$] are very narrow.[503] The ^9Be resonances of a series of beryllium compounds has also been observed; those of $BeCl_2,2Et_2O$ in Et_2O, Be acac, and beryllium 8-hydroxyquinaldate in $CHCl_3$ and aqueous $BeSO_4$ gave a single line in the same position as for beryllium acetate. The BeF_4^{2-} ion [$(NH_4)_2BeF_4$ in water] was shifted 2 p.p.m. upfield with $J(^9Be—^{19}F)$ 33 c./sec. The fluorine n.m.r. spectrum of this ion only showed coupling below 28° and gave a main quartet due to BeF_4^{2-}, and a minor one at higher field due perhaps to H_2O,BeF_3^-. The behaviour of BeF_4^{2-} is reminiscent of BF_4^-.[504]

Work on other less-usual resonances has also been reported earlier in the chapter: ^7Li refs. 161, 162, 339; ^{13}C ref. 33; ^{14}N refs. 33, 239, 252, 334; ^{17}O refs. 33, 211, 220–223, 228–231, 311, 337, 450; ^{27}Al refs. 210, 242; and ^{35}Cl ref. 204.

10 Appendix: Compounds not Referred to in Detail

This Appendix contains three Tables each of which lists compounds which have not been referred to in the main body of the chapter but for which n.m.r. data have been reported during the year. Unless some other nucleus is specifically mentioned it can be assumed that proton n.m.r. was used for characterisation. The Tables list typical compounds from each paper and in this sense are not exhaustive. In many cases, however, variations are due only to changes in substituents, R; in other cases 'etc.' indicates that a series of similar compounds is included in the reference.

The Tables are arranged according to vertical groups in the Periodic Table and contain references to types of compound in the order:

Organometallic compounds and complexes of the main-group elements; Compounds and complexes of the transition metals; π Complexes of transition metals.

It is hoped that the Tables will serve as a useful check list and as a source of references to particular compounds.

[503] F. Calderazzo, E. A. C. Lucken, and D. F. Williams, *J. Chem. Soc.* (*A*), 1967, 154.
[504] J. C. Kotz, R. Schaeffer, and A. Clouse, *Inorg. Chem.*, 1967, **6**, 620.

Table 1 *Organometallic compounds of Main-Group elements (comprising references to further papers containing n.m.r. data on compounds of the elements:* Al, Ga, In, *and* Tl; Si, Ge, Sn, *and* Pb; As *and* Sb; Se *and* Te)

	Compound	*Ref.*		*Compound*	*Ref.*
1	$[Me_2Al\text{-}NH_2]_3$	505	18	$H_3SiI,2py$	520
2	$(Et_3Al)_2SbEt_5$	506	19	$MeSiH_2I,2py$	520
3	$[Et_2GaN{:}CH\cdot R]_2$	507	20	$MeSiH_2I,Me_3N$	520
4	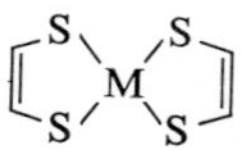 (M = Ga, In, Tl)	508	21	$[CH_2{:}CH(Me)Si—NPh]_3$	521
5	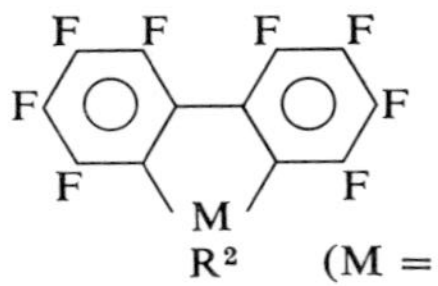 (^{19}F) (M = Si, Ge, Sn)	509	22	$R_3{}^1SiN{:}CR^2R^3$	522
			23	$(ClMe_2Si)_4Si$ (or C)	523
			24	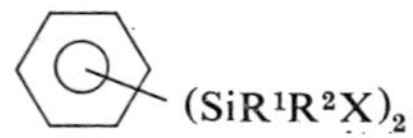	523
6	$RSiCl_3$ (fluorinated R) (^{19}F)	510	25	$(Me_3Si)_3SiSi(SiMe_3)_3$	524
7	$Me_2Si—SiMe_2$ and $[CH_2]_n$	511	26	$R_3SiOC(=CH_2)C_6H_4X$	525
8	$Me(SiMe_2)_nMe$	512			
9	$R_3SiCHCl_2$	513			
10	$R^1R^2R^3R^4Si$	514	27	$Ph_3SiOCH(CMe_3)$	526

13	Me_3SiR	517	28	RMe_2SiCH_2Li 	527
14	$MeOSiH_2Me$	518			
15	$H_nSi(OR)_{4-n}$ ($n = 2, 1$)	519			
16	$H_2Si(NR_2)_2$	519	29	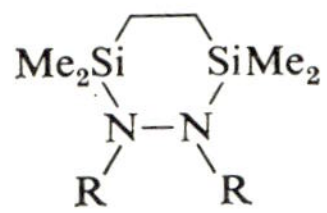	528
17	$HSiX_3$ (X = Br, I)	519			

[505] N. Wiberg, W. Ch. Joo and H. Henke, *Inorg. Nuclear Chem. Letters*, 1967, **3**, 267.
[506] Y. Takashi, *Organometallic Chem.*, 1967, **8**, 225.
[507] J. R. Jennings and K. Wade, *J. Chem. Soc.* (*A*), 1967, 1222.
[508] E. Hoyer, W. Dietzch, H. Muller, A. Zschunke, and W. Schroth, *Inorg. Nuclear Chem. Letters*, 1967, **3**, 457.
[509] S. C. Cohen and A. G. Massey, *J. Organometallic Chem.*, 1967, **10**, 471.
[510] D. Cooper, R. N. Haszeldine, and M. J. Newlands, *J. Chem. Soc.* (*A*), 1967, 2098.
[511] M. Kumado, K. Tamao, T. Takubo, and M. Ishikawa, *J. Organometallic Chem.*, 1967, **9**, 43.
[512] R. West, F. A. Kramer, E. Carberry, M. Kumado, and M. Ishikawa, *J. Organometallic Chem.*, 1967, **8**, 79.
[513] D. Seyferth, J. M. Burlitch, H. Dertonzos, and H. D. Simmons, *J. OrganometallicChem.*, 1967, **7**, 405.
[514] K. Yamamoto, K. Nakanishi, and M. Kumada, *J. Organometallic Chem.*, 1967, **7**, 197.
[515] F. W. G. Fearon and H. Gilman, *J. Organometallic Chem.*, 1967, **10**, 535.
[516] H. Burger and U. Goetze, *Inorg. Nuclear Chem. Letters*, 1967, **3**, 549.
[517] M. D. Curtis, *J. Amer. Chem. Soc.*, 1967, **89**, 4241.
[518] J. T. Wang and C. H. Van Dyke, *Inorg. Chem.*, 1967, **6**, 1741.
[519] H. J. Campbell-Ferguson, E. A. V. Ebsworth, A. G. McDiarmid, and T. Yoshioka, *J. Phys. Chem.*, 1967, **71**, 723.
[520] H. J. Campbell-Ferguson and E. A. V. Ebsworth, *J. Chem. Soc.* (*A*), 1967, 705.
[521] S. S. Dua and M. V. George, *J. Organometallic Chem.*, 1967, **10**, 219.
[522] Lui Heung Chan and E. G. Rochow, *J. Organometallic Chem.*, 1967, **9**, 231.
[523] H. Sakurai, T. Watanabe, and M. Kumada, *J. Organometallic Chem.*, 1967, **9**, P 11.
[524] H. Gilman and P. L. Harrell, *J. Organometallic Chem.*, 1967, **9**, 67.
[525] A. G. Brook and S. A. Fieldhouse, *J. Organometallic Chem.*, 1967, **10**, 235.
[526] G. J. D. Peddle, *J. Organometallic Chem.*, 1967, **10**, 377.
[527] D. J. Peterson, *J. Organometallic Chem.*, 1967, **9**, 373.
[528] C. G. Pitt and K. R. Skillern, *Inorg. Chem.*, 1967, **6**, 865.

Table 1 (*cont.*)

	Compound	*Ref.*
30	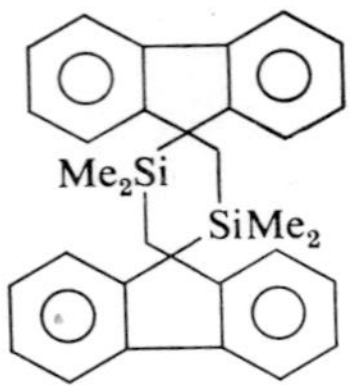 etc.	529
31	$Me_3SiC\colon CSnMe_3$	530
32	$Bu^n{}_3GeNPh\cdot CO_2R$	531
33	$Ph_3GeC(Ph)\colon C(R)H$	525
34	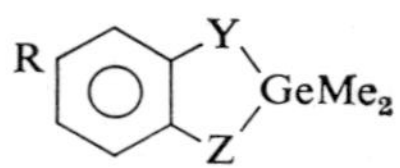 (Y, Z = O, O; S, S; O, NH)	532
35	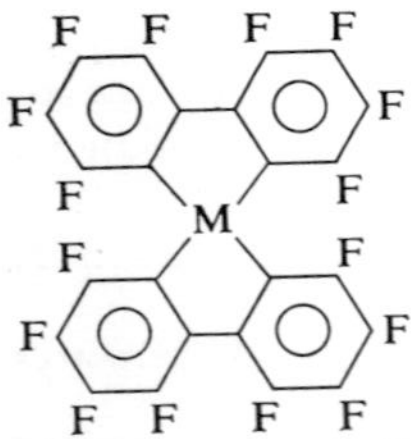 (M = Ge, Sn)	509
45	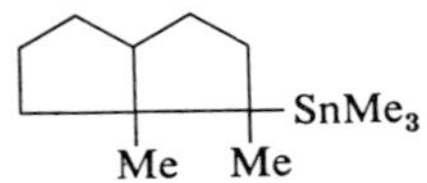 etc.	533
46	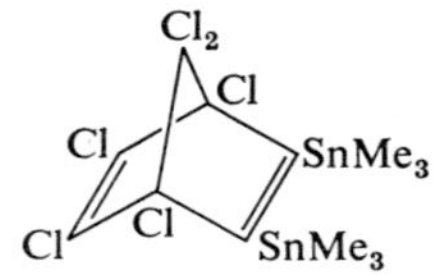 etc.	539
47	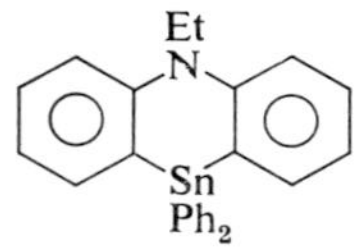 etc.	540
48	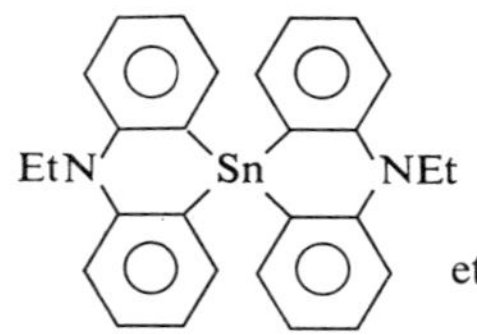 etc.	540
49	 M = Sn, Pb	508

38	$R_3{}^1SnC(R^2):CHR$, etc.	534	50	Ph_3PbN — N; $MeCO_2 \cdot C{=}CCO_2 \cdot Me$	541
39	$Me_3SnCXCl \cdot SnMe_3$	535	51	$(C_6F_5)_nAsCl_{3-n}$ (^{19}F)	542
40	$R_nSn(Et_2NCS_2)_{4-n}$	536	52	$(C_6F_5As)_4$ (^{19}F)	542
41	$R_2Sn(OMe)X$	537	53	$[(C_6F_5)_2As]_2$ (^{19}F)	542
42	$Me_2Sn(OAc)_2$	538	54	$(Me_2AsCH_2)_3CMe$	543
43	$[Me_2Sn(OAc)]_2O$	538	55	$(R_3SbCl)_2O$	544
44	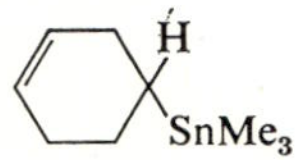 etc.	533	56	$Et_5Sb,2AlEt_3$	506
			57	RSeSH	545
			58	RSeH	545

[529] G. J. D. Peddle, *J. Organometallic Chem.*, 1967, **9**, 171.
[530] W. Findeiss, W. Davidsohn, and M. C. Henry, *J. Organometallic Chem.*, 1967, **9**, 435.
[531] Y. Ishii, K. Itoh, A. Nakamura, and S. Sakai, *Chem. Comm.*, 1967, 224.
[532] M. Wieber and C. D. Frohning, *J. Organometallic Chem.*, 1967, **8**, 459.
[533] R. H. Fish, H. G. Kuivik, and I. J. Tyminski, *J. Amer. Chem. Soc.*, 1967, **89**, 5861.
[534] A. J. Leusink, H. A. Budding, and J. W. Marsman, *J. Organometallic Chem.*, 1967, **9**, pp. 285, 295.
[535] D. Seyferth, and F. M. Armbrecht, jun., *J. Amer. Chem. Soc.*, 1967, **89**, 2790.
[536] F. Bonati and R. Ugo, *J. Organometallic Chem.*, 1967, **10**, 257.
[537] A. G. Davies, and P. G. Harrison, *J. chem. Soc.* (*C*), 1967, 298.
[538] Yutaka Maeda and Rokuru Okawara, *J. Organometallic Chem.*, 1967, **10**, 247.
[539] D. Seyferth and A. B. Evnin, *J. Amer. Chem. Soc.*, 1967, **89**, 1468.
[540] E. J. Kupchik and V. A. Perciaccante, *J. Organometallic Chem.*, 1967, **10**, 181.
[541] H. Gorth and M. C. Henry, *J. Organometallic Chem.*, 1967, **9**, 117.
[542] M. Green and D. Kirkpatric, *Chem. Comm.*, 1967, 57.
[543] R. D. Feltham, A. Kasenally, and R. S. Nyholm, *J. Organometallic Chem.*, 1967, **7**, 285.
[544] R. C. McKenney and H. H. Sisler, *Inorg. Chem.*, 1967, **6**, 1178.
[545] I. Lalezari and N. Sharghi, *Spectrochim. Acta*, 1967, **23**, *A*, 1948.

See also compounds 10, 13, 15 (Si); 10, 15, 34 (Ge); and 7, 10, 12, 15 (Sn) and 5 (B) in Table 2, and compounds 2, 6, 17 (B); 12, 22, 30 (Si); 29, 64 (Sn) in Table 3.

Table 2 *Transition-metal compounds and complexes (comprising references to further papers containing n.m.r. data on compounds of the elements:* Sc, Y, La, *and* Ln; Ti; Cr *and* Mo; Mn *and* Re; Fe *and* Ru; Co, Rh, *and* Ir; Ni, Pd, *and* Pt; Zn, Cd, *and* Hg)

	Compound	*Ref.*
1	$M(C_3F_7\cdot CO\cdot CH\cdot CO\cdot Bu^t)_3$ (M = Sc, Y, La, Ln)	546
2	$Ti(SR^1)_4(R^2SH)_x(R^3NH)_y$ (polymer)	547
3	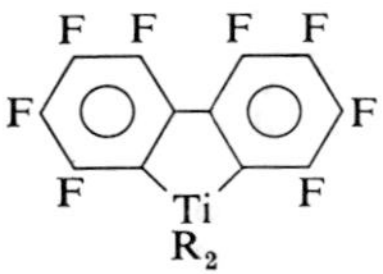 (^{19}F)	509
4	$LM(CO)_4$ (M = Cr, Mo; L = $Me_2AsC(R){:}C(R)AsMe_2$)	548
	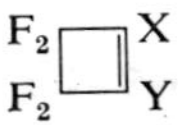 (X = Y = PPh or $AsMe_2$)	
5	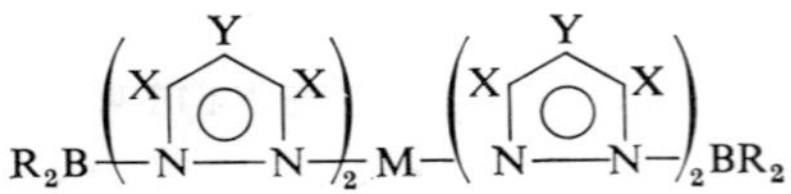 (M = Cr, Mn, Fe, Co, Ni, or Zn)	549
12	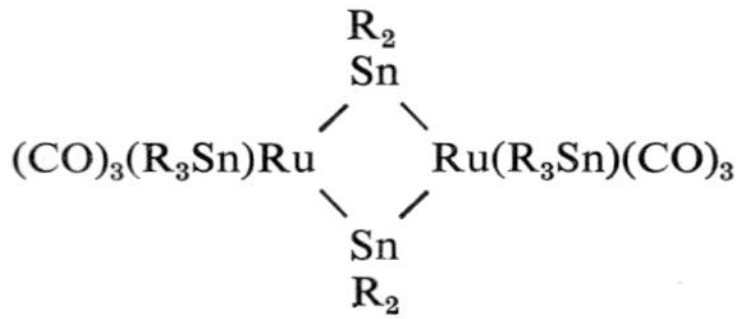	555
13	$(CO)_4(Me_3Si)Ru\text{-}Ru(Me_3Si)(CO)_4$	555
14	$Co_2(CO)_6MeC{:}CMe$, etc.	556
15	$[Co(CO)_3]_4R_2{}^1M(C{:}CR^2)_2$ (M = Si, Ge, Sn)	553
16	$(CF_3C)Co_3(CO)_9$ (^{19}F)	557
17	$(CO)_4Co(CF_2)_2Co(CO)_4$ (^{19}F)	557
18	$(CF_3CF)Co_2(CO)_7$ (^{19}F)	557
19	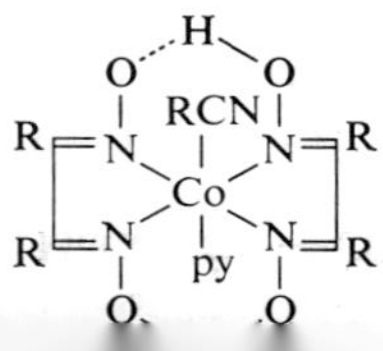	558

7	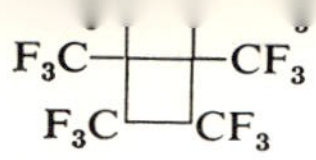 (^{19}F)	551	20	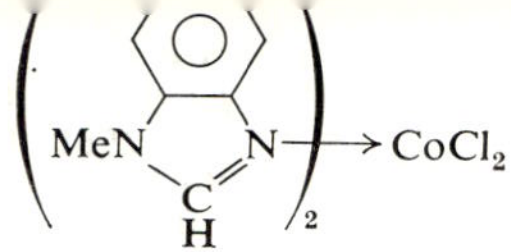	559
8	$ReH_3[(Ph_2P)CH_2\cdot CH_2(PPh_2)]_2$	552			
9	$ReH_3(Ph_3P)_2[Ph_2PCH_2\cdot CH_2\cdot PPh_2]$, etc.	552	21	$L^1L^2Co(R^7CO\cdot CH\cdot CO\cdot R^2)$ ($L^1L^2 = NO_3, NO_3; NO_3$, py; en)	560
10	$[R_2MFe(CO)_{4-n}(Ph_3P)]_{n-2}$ ($n = 0, 1$; M = Si, Ge, Sn)	553	22	$RhCl_3,L$ (L as in 4)	548
11		554	23	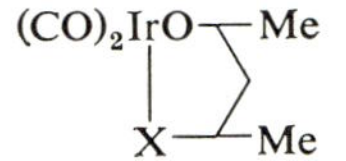 (X = O, NPh, NR)	561

[546] C. S. Springer, D. W. Meek, and R. E. Sievers, *Inorg. Chem.*, 1967, **6**, 1105.
[547] D. C. Bradley and P. A. Hammersley, *J. Chem. Soc.* (*A*), 1967, 1894.
[548] W. R. Cullen, P. S. Dhaliwal, and C. J. Stewart, *Inorg. Chem.*, 1967, **6**, 2256.
[549] S. Trofimenko, *J. Amer. Chem. Soc.*, 1967, **89**, 6288.
[550] H. Singer, *J. Organometallic Chem.*, 1967, **9**, 135.
[551] R. E. J. Bichler, M. R. Booth, and M. C. Clark, *Inorg. Nuclear Chem. Letters*, 1967, **3**, 71.
[552] H. Freni, R. Demichelis, and D. Giusto, *Inorg. Nuclear Chem.*, 1967, **29**, 1433.
[553] S. D. Ibekwe and M. J. Newlands, *J. Chem. Soc.* (*A*), 1967, 1783.
[554] M. M. Bagga, P. E. Baikie, O. S. Mills, and P. L. Pauson, *Chem. Comm.*, 1967, 1106.
[555] J. D. Cotton, S. A. R. Knox, and F. G. A. Stone, *Chem. Comm.*, 1967, 965.
[556] R. S. Dickson and D. B. W. Yawney, *Inorg. Nuclear Chem. Letters*, 1967, **3**, 209.
[557] B. L. Booth, R. N. Haszeldine, P. R. Mitchell, and J. J. Cox, *Chem. Comm.*, 1967, 529.
[558] Akira Misona, Yasuzo Uchida, Masanobu Hidai, and Hideo Knai, *Bull. Chem. Soc. Japan*, 1967, **40**, 2089.
[559] R. Price, *J. Chem. Soc.* (*A*), 1967, 521.
[560] C. J. Boucher, *Inorg. Chem.*, 1967, **6**, 2162.
[561] F. Bonati and R. Ugo, *J. Organometallic Chem.*, 1967, **7**, 167.

Table 2 (*cont.*)

	Compound	*Ref.*
24	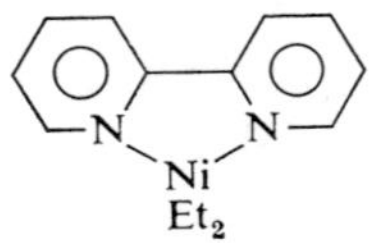	562
25	$Ni(CO)_2L$ (L as in 4)	548
26	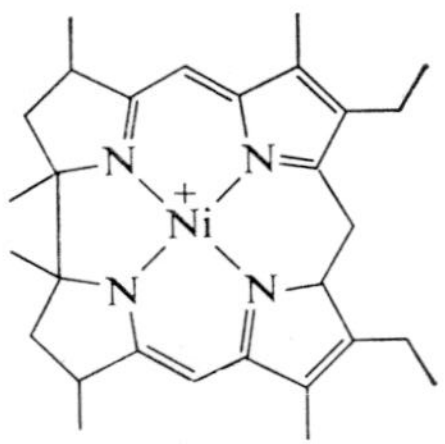	563
27	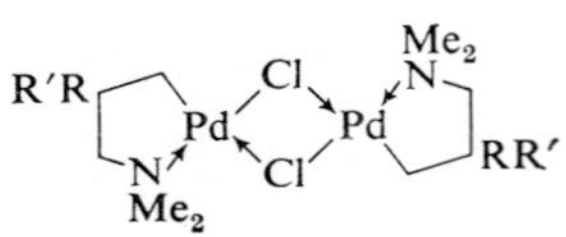	564
28	MCl_2,L (M = Pd, Pt; L as in 4)	548
29	$(Ph_3P)_2PtCO_3$	565
30	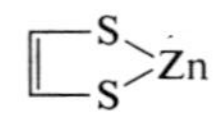	508
31	$Zn(Et_2NCS_2)_2$	536
32	$PhCH_2CdCl$	566
33	$HgCl_2,L$ (L as in 4)	548
34	$(Me_3Ge)_2Hg$	567
35	$ClHgC(=CH_2)—CMe_2(OMe)$ etc.	568
36	$XHgCH_2CO_2^-$	569
37	$Hg(CH_2CO_2^-)_2$	569

[562] T. Saito, M. Akai, Y. Uchida, and A. Misons, *J. Phys. Chem.*, 1967, **71**, 2370.
[563] R. Grigg, A. W. Johnson, and P. van der Broek, *Chem. Comm.*, 1967, 502.
[564] A. C. Cope, J. M. Kliegman, and E. C. Friedrich, *J. Amer. Chem. Soc.*, 1967, **89**, 287.
[565] C. J. Nyman, C. E. Wymore, and G. Wilkinson, *Chem. Comm.*, 1967, 407.
[566] P. R. Jones, P. D. Sherman, and K. Schwarzenberg, *J. Organometallic Chem.*, 1967, **10**, 521.
[567] C. Eaborn, W. A. Dutton, F. Glockling, and K. A. Hooton, *J. Organometallic Chem.*, 1967, **9**, 175.
[568] ... *Chem. Soc.*, 1967, **89**, 6261.

and other π complexes of the elements: Ti *and* Zr; Mo *and* W; Re; Fe *and* Ru; Co, Rh, *and* Ir; Ni, Pd, *and* Pt)

	Compound	*Ref.*		*Compound*	*Ref.*
1	$(\pi\text{-}C_5H_5)_2Ti(C⋮CPh)_2$	570	12	$(\pi\text{-}C_5H_5)Mo(CO)_3 \cdot SiMe_3$	576
2	$(\pi\text{-}C_5H_5)_2TiBH_4$ (^{11}B)	571	13	$(\pi\text{-}C_5H_5)Mo(CO)_2(Ph_3P)I$	577
3	$(\pi\text{-}C_5H_4R)_2TiCl_2$	572	14	$(\pi\text{-}C_5Me_5)Mo(CO)_3Me$	573
4	$(\pi\text{-}C_5H_4 \cdot CMe_2^-)_2Ti$	572	15	$[(\pi\text{-}C_5Me_5)Mo(CO)_2]_2$	573
5	$(\pi\text{-}C_5Me_5)TiCl_3$	573	16	$(\pi\text{-}C_3H_4R)$ bipy $Mo(CO)_2NCS$	578
6	$(\pi\text{-}C_5H_5)_2Zr(H)BH_4$ (^{11}B)	571	17	$(\pi\text{-}C_3H_4R)$ bipy py $Mo(CO)_2BF_4$	578
7	$[(\pi\text{-}C_5H_5)_2ZrH_2]_n$	571	18	$C_8H_8Mo(CO)_3$	579
9	$(\pi\text{-}C_5H_5)_2MoX_2$	574	19	$(MeCO \cdot CH{:}CH_2)_3Mo$ (^{19}F)	579
10	$[(\pi\text{-}C_5H_5)_2MoX_2]^-$	574	20	$(CF_3C⋮C)_3MoNCMe$	579
11	$(\pi\text{-}C_5H_5)Mo(CO)_3 \cdot (CH_2)_3Br$	575	21	$[(\pi\text{-}C_5H_5)_2WX_2]^{n+}$ ($n = 0, 1$)	574

[570] H. Kopf and M. Schmidt, *J. Organometallic Chem.*, 1967, **10**, 383.
[571] B. D. James, R. K. Nanda, and M. G. H. Wallbridge, *Inorg. Chem.*, 1967, **6**, 1979.
[572] M. F. Sullivan and W. F. Little, *J. Organometallic Chem.*, 1967, **8**, 277.
[573] R. B. King and M. B. Bisnette, *J. Organometallic Chem.*, 1967, **8**, 287.
[574] R. L. Cooper and M. L. H. Green, *J. Chem. Soc.* (*A*), 1967, 1155.
[575] R. B. King and M. B. Bisnette, *J. Organometallic Chem.*, 1967, **7**, 311.
[576] D. J. Cardin and M. F. Lappert, B. M. Kingston, and S. A. Keppie, *Chem. Comm.*, 1967, 1035.
[577] R. J. Haines, R. S. Nyholm, and M. H. B. Stiddard, *J. Chem. Soc.* (*A*), 1967, 94.
[578] C. G. Hull and M. H. B. Stiddard, *J. Organometallic Chem.*, 1967, **9**, 519.
[579] R. B. King, *J. Organometallic Chem.*, 1967, **8**, 139.

Table 3 (*cont.*)

	Compound	*Ref.*
22	$(\pi\text{-}C_5H_5)W(CO)_3\cdot SiMe_3$	576
23	$(\pi\text{-}C_5Me_5)Re(CO)_3$	573
24	$(\pi\text{-}C_5H_5)Fe(\pi\text{-}C_5H_4R)$	580
25	$(\pi\text{-}C_5H_5)Fe\{(PhO)_3P\}_2X$	581
26	$[(\pi\text{-}C_5H_5)Fe\{(PhO)_3P\}_2]_2$	581
27	$(\pi\text{-}C_5H_5)Fe(R)(CO)(Ph_3P)$	582
28	$(\pi\text{-}C_5H_5)(CO)_2FeC_6F_4\cdot CH_2\cdot Fe(CO)_2(\pi\text{-}C_5H_5)$ (^{19}F)	583
29	$(\pi\text{-}C_5H_5)Fe(CO)_2CF_3\cdot C:CCF_3\cdot SnMe_3$ (^{19}F)	551
30	$(\pi\text{-}C_5H_5)Fe(CO)_2Si(Cl)_2Me$	584
31	$(\pi\text{-}C_5H_5)Fe(CO)_2CH_2\cdot C_6F_5$ (^{19}F)	583
32	$(\pi\text{-}C_5H_4R^1)Fe(\pi\text{-}C_5H_4R^2)$	585
33	$[(\pi\text{-}C_5H_4)Fe(\pi\text{-}C_5H_4)]_n$	586
34	$(\pi\text{-}C_5H_4)Fe(\pi\text{-}C_5H_4)$ / $O:C\cdot[CH_2]_2\cdot CHMe$	585
35	$(\pi\text{-}C_5H_4)Fe(\pi\text{-}C_5H_4)$	587
42	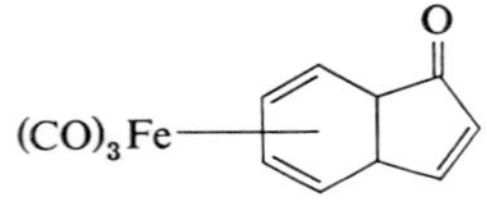	588
43	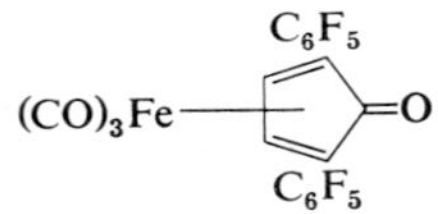	556
44	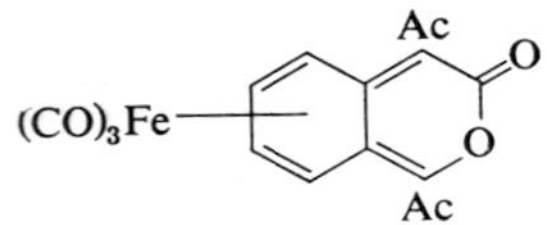	591
45	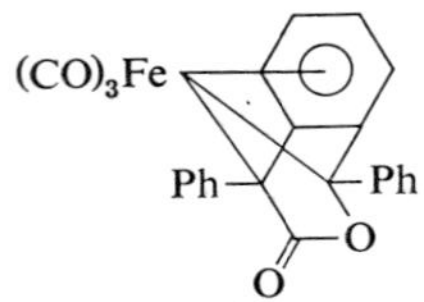	591
46	$[Ru(C_6H_6)Cl_2]_x$	592
47	$[Ru(C_7H_8)Cl_2]_x$	592
48	$(\pi\text{-}C_5H_5)Co(CO)_2(HgX_2)$	593
49	$[(\pi\text{-}C_5H_5)Co(C_8H_{11})]^+$	594

37	$[(\pi\text{-}C_5Me_5)Fe(CO)_2]_2$	573
38	$(\pi\text{-}C_4Me_4)Fe(CO)_3$	588
39	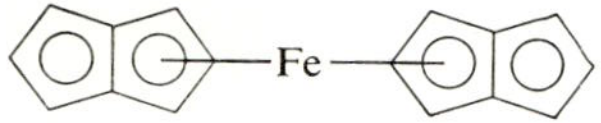	589
40	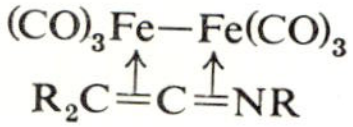	590
41	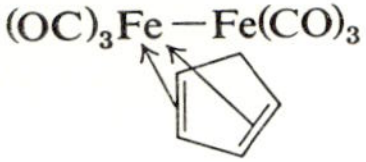	588
52	$(\pi\text{-}C_5Me_5)Co(CO)_2$	573
53	$(\pi\text{-}C_4Me_4)Co(CO)_2I$	596
54	$(\pi\text{-}C_4Me_4)Co(CO)_2Co(CO)_2R$	596
55	$(\pi\text{-}C_4Ph_4)Co(CO)_2X$	597
56	(PhC⋮CPh)Co	598
57	$(CF_3C⋮CCF_3)Co_2(CO)_6$ (^{19}F)	556
58	$[(\pi\text{-}Ph)RhCl_2Rh(\pi\text{-}Ph)]^{4+}$	599
59	$[(\pi\text{-}Ph)RhCl\ py]^{2+}$	599

[580] R. E. Bozak and J. L. Lakner, *Canad. J. Chem.*, 1967, **45**, 773.
[581] A. N. Nesmeyanov, Yu. A. Chapovsky, and Yu. A. Ustynyuk, *J. Organometallic Chem.*, 1967, **9**, 345.
[582] M. L. H. Green and C. R. Hurley, *J. Organometallic Chem.*, 1967, **10**, 188.
[583] M. I. Bruce, *J. Organometallic Chem.*, 1967, **10**, 495.
[584] W. Jetz and W. A. G. Graham, *J. Amer. Chem. Soc.*, 1967, **89**, 2773.
[585] W. M. Horspool and R. G. Sutherland, *Chem. Comm.*, 1967, 240.
[586] E. W. Neuse and R. K. Crossland, *J. Organometallic Chem.*, 1967, **7**, 347.
[587] T. H. Barr and W. E. Watts, *J. Organometallic Chem.*, 1967, **9**, P 3.
[588] R. Bruce, K. Moseley, and P. M. Maitlis, *Canad. J. Chem.*, 1967, **45**, 2011.
[589] T. J. Katz and J. J. Mrowca, *J. Amer. Chem. Soc.*, 1967, **89**, 1105.
[590] S. Otsuka, A. Nakamura, and T. Yoshida, *J. Organometallic Chem.*, 1967, **7**, 339.
[591] J. M. Holland and D. W. Jones, *Chem. Comm.*, 1967, 946.
[592] G. Winkhaus and H. Singer, *J. Organometallic Chem.*, 1967, **7**, 487.
[593] D. J. Cook, J. L. Dawes, and R. D. W. Kemmitt, *J. Chem. Soc. (A)*, 1967, 1547.
[594] J. Lewis and A. W. Parkins, *J. Chem. Soc. (A)*, 1967, 1150.
[595] J. F. Helling, S. C. Rennison, and A. Merijan, *J. Amer. Chem. Soc.*, 1967, **89**, 7141.
[596] R. Bruce and P. M. Maitlis, *Canad. J. Chem.*, 1967, **45**, 2017.
[597] A. Efraty and P. M. Maitlis, *J. Amer. Chem. Soc.*, 1967, **89**, 3744.
[598] H. Yamazaki and N. Hagihara, *J. Organometallic Chem.*, 1967, **7**, P 23.
[599] B. L. Booth, R. N. Haszeldine, and M. Hill, *Chem. Comm.*, 1967, 1118.

Table 3 (*cont.*)

	Compound	*Ref.*
60	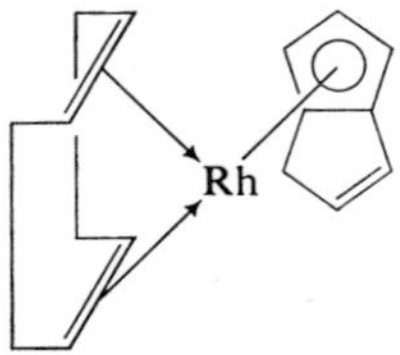	589
61	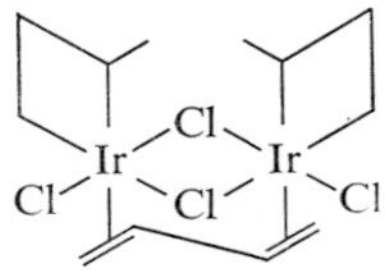	600
62	$(\pi\text{-}C_5H_5)Ni(Ph_3P)SnCl_3$	601
63	Pd–olefin complexes	602
64	$[(MeCH\cdot CH\cdot CHEt)PdCl]_2$	603
65	$(MeCH\cdot CH\cdot CH\cdot Et)Pd\ py$	603
66	$[(MeCH\cdot CR^1\cdot CHR^2)PdX]_2$	604
67	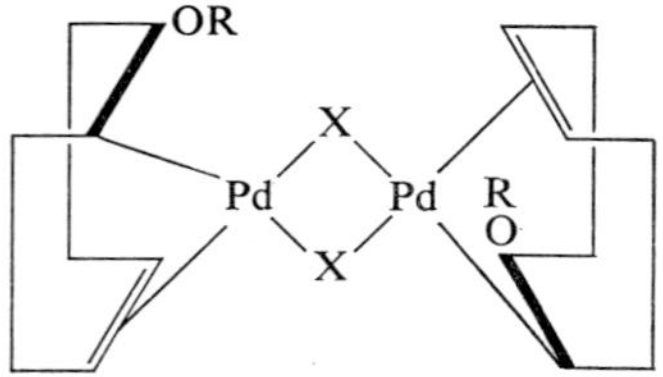	605
68	Pt–olefin complexes	602
69	$py_2Pt(C_3H_5)_2(I_3)_2$	606
70	$[PtH(SnCl_3)(Ph_3P)_2]_2C_8H_{12}$	607
71	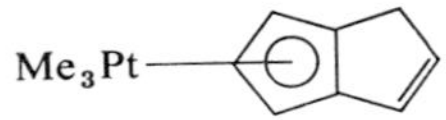	589
72	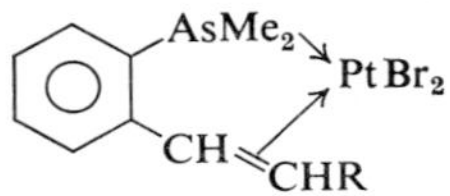	608

[600] B. L. Shaw and E. Singleton, *J. Chem. Soc.* (*A*), 1967, 1972.
[601] M. Van Der Akker and F. Jellinek, *J. Organometallic Chem.*, 1967, **10**, P 38.
[602] H. A. Tayim and J. C. Bailar, *J. Amer. Chem. Soc.*, 1967, **89**, 4330.
[603] I. I. Moiseev, A. P. Belov, and G. Yu. Pek, *Russ. J. Inorg. Chem.*, 1965, **10**, 180 (336).
[604] J. M. Rowe and D. A. White, *J. Chem. Soc.* (*A*), 1967, 1451.
[605] C. B. Anderson and B. J. Burreson, *J. Organometallic Chem.*, 1967, **7**, 181.
[606] C. R. Kistner, D. A. Drew, J. R. Doyle, and G. W. Rausch, *Inorg. Chem.*, 1967, **6**, 2036.
[607] H. A. Tayim and J. C. Bailar, *J. Amer. Chem. Soc.*, 1967, **89**, 3420.

2

Nuclear Quadrupole Resonance Spectra

Although nuclear quadrupole resonance spectroscopy, n.q.r., has been applied to studies of chemical compounds for a number of years, it has still not acquired any great prominence in the array of spectroscopic techniques now available to the modern structural chemist. This is borne out by the relatively few references which have been found to be appropriate to this Review. Even if all the organic molecules studied during the year were also included, the total number of papers would still be small by comparison with those which report results obtained using other spectroscopic techniques. This lack of general interest in the technique presumably reflects its limited applicability and the reluctance of individual chemists to invest time in acquiring the knowledge and expertise required to build and operate their own equipment.

An elementary presentation of n.q.r. spectroscopy has been given in a short review by Mairinger;[1] recommended further reading is contained in the 21 references.

The instrumental problems of n.q.r. are highlighted by the large percentage of papers discussing spectrometer techniques. Several spectrometers and modifications have been described[2–4], including one specially adapted for ^{14}N nuclei studies in the 0·5–6 MHz range.[5] A goniometer arrangement for observation of Zeeman splitting of n.q.r. lines in single crystals has been reported,[6] as has a system for improving the signal-to-noise ratio of a spectrum.[7] Data accumulation is by a time-averaging computer. Any magnetic-field independent components in the signal which cause baseline slope and irregularities are removed by subtracting a similar accumulation run at low magnetic field.

Work has been published during the year on thirteen nuclides of nine different elements. In the discussion which follows these are arranged in order of increasing atomic number and mass: ^{14}N; ^{35}Cl, ^{37}Cl; ^{59}Co; ^{69}Ga,

[1] F. Mairinger, *Allg. prakt. Chem.*, 1967, **18**, 33.
[2] P. Kesselring and M. Gautschi, *J. Sci. Instr.*, 1967, **44**, 911.
[3] J. D. Graybeal and R. P. Croston, *Rev. Sci. Instr.*, 1967, **38**, 122.
[4] D. Gill, M. Hayek, Y. Alon, and A. Simievic, *Rev. Sci. Instr.*, 1967, **38**, 1588.
[5] A. Colligiani, *Rev. Sci. Instr.*, 1967, **38**, 1331.
[6] G. E. Peterson and P. M. Bridenbaugh, *Rev. Sci. Instr.*, 1967, **38**, 387.
[7] D. A. Tong, *J. Sci. Instr.*, 1967, **44**, 875.

^{71}Ga; ^{79}Br, ^{81}Br; ^{115}In; ^{121}Sb, ^{123}Sb; ^{127}I; and ^{197}Au. As in previous years, work on the isotopes of chlorine dominates the field.

Nitrogen-14.—The ^{14}N n.q.r. resonances in KSCN and KSeCN at 78°K are similar and show a pair of closely spaced lines indicating a finite asymmetry parameter.[8] The SCN^- ion is known to be linear in the solid, so that η must arise primarily from the lattice term. The relevant quadrupole coupling constants and asymmetry parameters are 2·431 (0·0281) and 2·845 MHz (0·0497) for KSCN and KSeCN respectively. The eQq/h values are less than for organic thiocyanates (typically 3·5–3·6 MHz) because the covalent bond to carbon is no longer present.

Chlorine-35 and Chlorine-37.—A detailed ^{35}Cl and ^{37}Cl n.q.r. study of molecular motion and intermolecular forces in solid chlorine has been published.[9] Data were obtained between 20 and 172°K. Some covalent character was found in the intermolecular forces.

An interesting paper, not strictly covered by our primary classification, analyses the differences in the coupling constants measured from the ^{35}Cl, ^{37}Cl, ^{79}Br, ^{81}Br, and ^{127}I quadrupole coupling in the microwave spectra of hydrogen, deuterium, and tritium halides (see Table 1) by considering the

Table 1 *Quadrupole coupling constants* (MHz)

	H	D	T
^{35}Cl	67·51	67·41	67·21
^{37}Cl	—	53·14	52·96
^{79}Br	535·44	530·58	528·75
^{81}Br	447·76	443·39	441·26
^{127}I	1831·07	1823·38	1822·61

effects of vibration and rotation on the electric-field gradient.[10] The variations can be explained in terms of the change of size of the molecule arising from substitution of hydrogen by deuterium or tritium. The ^{35}Cl resonance at 12 MHz observed for tetraethylammonium hydrogen dichloride[11] indicates a symmetric HCl_2^- anion, in contrast to the 20 MHz asymmetric resonance of the methyl complex previously reported. Change in the molecular environment produces a remarkably large frequency shift. The charge redistribution which occurs on formation of HCl_2^- from HCl is mainly from the chloride ion to the vicinity of the other chlorine.

Precise measurements[12] of the temperature dependence of the ^{35}Cl n.q.r. frequency in $KClO_3$ are interpreted in terms of two lattice vibration modes

[8] R. Ikeda, D. Nakamura, and M. Kubo, *Bull. Chem. Soc. Japan*, 1967, **40**, 701.
[9] N. Nakamura and H. Chihara, *J. Phys. Soc. Japan*, 1967, **22**, 201.
[10] T. Tokuhiro, *J. Chem. Phys.*, 1967, **47**, 109.
[11] J. C. Evans and G. Y. Lo, *J. Phys. Chem.*, 1967, **71**, 3697.
[12] D. B. Utton, *J. Chem. Phys.*, 1967, **47**, 371.

at 60 and 126 cm.$^{-1}$. These values show some correlation with the known Raman spectrum.

New measurements have been made on the trimeric, and stable and metastable tetrameric phosphonitrilic chlorides, $(PNCl_2)_3$ and $(PNCl_2)_4$.[13, 14] Previous data are at variance. Differences in the number of lines are correlated with the limited crystallographic evidence. The stable $(PNCl_2)_4$ is still somewhat anomalous.

The ^{35}Cl resonance has been followed with changing temperature in both the covalent (PCl_5) and ionic ($PCl_4^+PCl_6^-$) modifications of phosphorus pentachloride.[15] PCl_5 shows a phase transition at 183°K, which has two resonance lines above, and three below. $PCl_4^+PCl_6^-$ has a transition at 102·3°K giving four lines above and seven below. Data are also given for $PCl_4^+SbCl_6^-$. Covalent PCl_5 is metastable and the change to the ionic form was followed. The estimated ionic characters of the P—Cl bonds were correlated with other PCl_nF_{5-n} halides.

^{35}Cl and ^{37}Cl resonances have been observed[16] in $PCl_4^+PCl_6^-$, $PCl_4^+PF_6^-$, $PCl_4^+SbCl_6^-$, $Et_4N^+PCl_6^-$, $NO^+SbCl_6^-$, $Et_4N^+SbCl_6^-$, $AsCl_4^+AlCl_4^-$, and $PbCl_4$, and the assignment of the lines made for the PCl_4^+, $AsCl_4^+$, PCl_6^-, and $SbCl_6^-$ species. The data in the two papers taken together are, however, inconsistent in some cases. The ^{35}Cl spectrum of $PClF_4$ at 77°K is consistent with an equatorial rather than an axial chlorine in a trigonal bipyramid[17] as already suggested from other evidence.

An earlier prediction, that the two ^{35}Cl resonances observed near 34 MHz in I_2Cl_6 came from terminal chlorines only, has been verified by finding a third frequency due to the bridging chlorines at 13·740 MHz.[18] The detailed calculations show considerable electron transfer from the iodine to the chlorine atoms. The related complex $IAlCl_6$ gives a spectrum appropriate to the structure $ICl_2^+AlCl_4^-$.

The ^{35}Cl, ^{37}Cl, ^{121}Sb, and ^{123}Sb resonances have been reported over a range of temperature in the complexes $SbCl_5$,MeCN, and $SbCl_5$,$POCl_3$, and $SbCl_5$,[19] and representative data for antimony are given in Table 2. $SbCl_5$,$POCl_3$ gives a ^{35}Cl spectrum entirely consistent with the known *X*-ray structure, there being two pairs of equivalent chlorine atoms. The parameters are discussed in detail with reference to the six-co-ordination about the antimony. Only two chlorine resonances are observed in $SbCl_5$,MeCN and the antimony has axial symmetry. A covalent six-co-ordinate axial structure is favoured rather than $SbCl_4(MeCN)_2^+SbCl_6^-$. Above 210°K $SbCl_5$ has a trigonal bipyramidal structure, but at lower

[13] M. Kaplansky and M. A. Whitehead, *Canad. J. Chem.*, 1967, **45**, 1669.
[14] M. Dixon, H. D. B. Jenkins, J. A. S. Smith, and D. A. Tong, *Trans. Faraday Soc.*, 1967, **63**, 2852.
[15] H. Chihara, N. Nakamura, and S. Seki, *Bull. Chem. Soc. Japan*, 1967, **40**, 50.
[16] J. V. DiLorenzo and R. F. Schneider, *Inorg. Chem.*, 1967, **6**, 766.
[17] R. R. Holmes, *J. Chem. Phys.*, 1967, **46**, 3718.
[18] J. C. Evans and G. Y.-S. Lo, *Inorg. Chem.*, 1967, **6**, 836.
[19] R. F. Schneider and J. V. DiLorenzo, *J. Chem. Phys.*, 1967, **47**, 2343.

Table 2 *Antimony n.q.r. parameters*

		Coupling constant (MHz)	*Temp.* (°K)	*Asymmetry parameter*, η
$SbCl_5$	^{121}Sb	$84{\cdot}63 \pm 0{\cdot}06$	249	$0{\cdot}00 {}^{+0{\cdot}017}_{-0{\cdot}00}$
	^{123}Sb	$107{\cdot}88 \pm 0{\cdot}09$	249	
$SbCl_5,POCl_3$	^{121}Sb	$204{\cdot}49 \pm 0{\cdot}14$	300	$0{\cdot}3068 \pm 0{\cdot}0024$
	^{123}Sb	$260{\cdot}66 \pm 0{\cdot}18$	300	
$SbCl_5,Me_3CN$	^{121}Sb	$213{\cdot}16 \pm 0{\cdot}17$	300	$0{\cdot}0360 {}^{+0{\cdot}0113}_{-0{\cdot}0165}$
	^{123}Sb	$271{\cdot}71 \pm 0{\cdot}22$	300	

temperatures it may be a chlorine-bridged dimer in a complex structure related to that of $NbCl_5$.

The ^{35}Cl n.q.r. resonance in Ph_3SnCl has been briefly mentioned and correlated with known values for the other members of the R_xSnCl_{4-x} series.[20]

The normal and deuteriated acid salts $KH(CCl_3CO_2)_2$ and $NH_4H(CCl_3CO_2)_2$ contain dimeric units with a possibly symmetric hydrogen bond. If it is symmetric there will be two crystallographically equivalent CCl_3 groups and only three ^{35}Cl or ^{37}Cl resonances. If the bond is not symmetric, so that the system tends towards CCl_3CO_2H and $CCl_3CO_2^-$, six frequencies are expected. All four systems show only three lines,[21] which means that both the proton and the deuteron are centrally located in the hydrogen bonds on the observation time-scale.

^{35}Cl n.q.r. frequencies have been given for $GeCl_4$ and $SnCl_4$, and for thirteen organo-chlorogermanes (see Table 3).[22] In all but three cases, the number of resonances equals the number of Ge—Cl bonds in the compound. $Cl_3Si{\cdot}CH_2{\cdot}CH_2{\cdot}GeCl_3$ shows only a very broad singlet which presumably contains unresolved fine structure, and $MeGeCl_3$ shows two lines in a 2 : 1 intensity ratio. The compound $\overline{CH_2{\cdot}CH{:}CH{\cdot}CH_2{\cdot}GeCl_2}$ gives only one sharp signal. The rule applied earlier to similar carbon and silicon compounds, namely that the mean n.q.r. frequency is linearly related to the sum of the Taft induction constants of the organic substituents, is found to apply equally well to the germanium analogues.

Eleven organo-tin chlorides (mainly of the type R_2SnCl_2) have been studied at three different temperatures.[23] There is a larger decrease in the coupling constant relative to the tetrahalide with organic substitution than with carbon and silicon compounds (see Table 4). The effects of variation in the aliphatic chain-length are much greater for tin than for carbon compounds. Discussion of the bonding suggests that there is an appreciable change of either ionic or π character of the Sn—Cl bond with organic

[20] T. S. Srivastava, *J. Organometallic Chem.*, 1967, **10**, 374.
[21] R. Blinc, M. Mali, and Z. Trontelj, *Phys. Letters*, 1967, **25**, *A*, 289.
[22] I. P. Biriukov, M. G. Voronkov, V. F. Mironov, and I. A. Safin, *Doklady Akad. Nauk U.S.S.R.*, 1967, **173**, 381.
[23] P. J. Green and J. D. Graybeal, *J. Amer. Chem. Soc.*, 1967, **89**, 4305.

Table 3 *N.q.r. frequencies for organo-chlorogermanes*

Compound	*Resonant frequencies* (MHz)
$GeCl_4$	25·745, 25·736, 25·717, 25·449
$MeCO_2 \cdot CH_2 \cdot GeCl_3$	23·819, 23·556, 22·993
$HGeCl_3$	23·631, 23·332, 23·332
$F_3C \cdot [CH_2]_2 \cdot GeCl_3$	23·395, 23·285, 23·260
$Br[CH_2]_2 \cdot GeCl_3$	23·627, 22·924, 22·924
$MeGeCl_3$	22·941(2) 22·707
$Cl_3Si \cdot CH_2 \cdot CH_2 \cdot GeCl_3$	22·85
$Cl[CH_2]_2 \cdot GeCl_3$	23·277, 22·587, 22·333
$ClCO \cdot [CH_2]_2 \cdot GeCl_3$	22·75, 22·67, 22·67
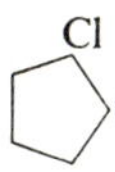 $GeCl_3$	23·4, 22·1, 22·1
 $GeCl_3$	22·624, 22·497, 22·329
$EtGeCl_3$	22·795, 23·300, 22·160
 $GeCl_2$	21·62
Me_2GeCl_2	20·322, 20·180
$SnCl_4$	24·294, 24·225, 24·137, 23·720

Table 4 *N.q.r. coupling constants for organo-tin chlorides*

	eqQ/h (MHz)	*Temp.* (°K)
Me_2SnCl_2	31·428	303
Et_2SnCl_2	31·014	303
$Pr^n{}_2SnCl_2$	30·618	303
$Bu^n{}_2SnCl_2$	31·974	303
$(n\text{-}C_8H_{17})_2SnCl_2$	30·272	303
Ph_2SnCl_2	35·687	303
Ph_3SnCl	33·736	303
$PhSnCl_3$	40·836	200
Bu^nSnCl_3	40·654	77

substituent. Intermolecular bonding in the solid is indicated in three of the compounds by a positive temperature coefficient of the coupling constant.

Cobalt-59.—^{59}Co resonances have been briefly reported and discussed [24] in a number of complexes of the type *trans*-$[CoCl_2\,en_2]^-$ (en = ethylenediamine) (see Table 5).

Table 5 *N.q.r. parameters for* ^{59}Co *resonances*

	eqQ/h (MHz)	η
trans-$[CoCl_2(NH_3)_4]Cl$	59·23	0·136
trans-$[CoCl_2\,en_2]Cl$	60·63	0·272
trans-$[CoCl_2\,en_2]NO_3$	62·78	0·132
trans-$[CoBr_2\,en_2]Br$	60·36	0·235

Gallium-69 and -71.—An earlier report of a discontinuity in the temperature dependence of the n.q.r. frequency of Ga in GaI_3 has been refuted.[25] The ^{69}Ga resonance occurred at 21·467 at 25°C and at 21·586 MHz at 77·3°K. The ^{71}Ga resonance was at 13·603 MHz at 77·3°K.

Bromine-79 and -81.—The ^{79}Br and ^{81}Br data for Br_3^- ions in PBr_7, $CsBr_3$, and NH_4Br_3 show differences in charge distribution with differing environment in this nearly linear but asymmetric ion.[26]

Indium-115.—The ^{35}Cl and ^{115}In resonances have been studied in $(NH_4)_2[InCl_5(H_2O)]$ as a function of temperature.[27] The ^{115}In resonances at 300°K occur at 7·453, 10·205, 10·346, and 14·552 MHz. Detailed analysis of the data on a point-charge model shows that the In—H_2O bond is more covalent than the In—Cl bond, although the complex is predominantly ionic. The ^{35}Cl frequencies were assigned to specific chlorine sites in the lattice, but the considerable temperature dependence of the data could not be explained adequately.

Antimony-121 and -123.—For data pertinent to $SbCl_5$, $SbCl_5$,CH_3CN, and $SbCl_5$,$POCl_3$, see under chlorine.[19]

Iodine-127.—The tri-iodide ion, in CsI_3, RbI_3, and NH_4I_3, shows three non-equivalent iodines.[28] Small anomalies in the ammonium salt are attributed to hydrogen bonding to terminal iodine atoms. The coupling constants are very sensitive to environment as shown by the data obtained

24 I. Watanabe and Y. Yamagata, *J. Chem. Phys.*, 1967, **46**, 407.
25 H. R. Brooker, *J. Chem. Phys.*, 1967, **47**, 1864.
26 G. L. Breneman and R. D. Willett, *J. Phys. Chem.*, 1967, **71**, 3684.
27 S. L. Carr, B. B. Garrett, and W. G. Moulton, *J. Chem. Phys.*, 1967, **47**, 1170.
28 A. Sasane, D. Nakamura, and M. Kubo, *J. Phys. Chem.*, 1967, **71**, 3249.

at *ca.* 20°C: NH_4I_3, 575·4, 1587·5, and 2441·4; RbI_3, 757·5, 1442·6, and 2450·7; CsI_3, 796·0, 1429·5, and 2454·7 MHz. The η values are all small.

Gold-197.—The first report of an n.q.r. signal for ^{197}Au ($I = \frac{3}{2}$) has appeared.[29] A triplet was observed at 300°K in AuCl, in the range 200–300 MHz, at 257·075, 257·173, and 257·377 MHz. The crystal structure of AuCl is not known, but it appears that there are three inequivalent sites in the unit cell. The value of eqQ/d is calculated to be 514 MHz which may be compared with the earlier, less accurate, value of 540 MHz from Mössbauer spectroscopy.

[29] P. Machmer, M. Read, and P. Cornil, *Inorg. Nuclear Chem. Letters*, 1967, **3**, 215.

3
Electron Spin Resonance Spectroscopy

1 Introduction

The foremost impression gained from the twelve-month survey of the literature embodied in this section is that electron spin resonance spectroscopy, e.s.r., does not belong to any one formal discipline. This reflects the great importance of the unpaired electron in all branches of physics and chemistry. It is also apparent that its practical application is restricted mainly to the specialist, because papers containing e.s.r. spectra are often concerned solely with that technique. There is little tendency as yet to use an e.s.r. spectrum merely as a routine diagnostic test as one would use for example an i.r. spectrum. However, in the hands of the expert it is a very powerful way of probing the structures of materials.

Despite the broad area of application of e.s.r. spectroscopy it was decided to limit this particular survey to inorganic and organometallic compounds consistent with the title of this volume. All references to organic compounds and organic free radicals have been omitted. Similarly, the area of solid-state physics which is concerned primarily with doped impurity cations in crystalline lattices, such as oxides, has been ignored. Pseudo-chemical compounds such as cations in alkali-metal halide lattices have also been omitted since these are so closely related to the wider oxide field that it would be difficult to separate the two. The allied problem of impurities and free radicals generated in solids by nuclear irradiation processes is likewise outside the scope of the present review. However, occasional references on these fields have been included in the text where a specific point seemed to be of particular interest or relevance to the compounds considered.

The sections which are included are:

Simple inorganic molecules and free radicals; complex radicals of main-group elements; and transition-metal complexes. This third section is by far the most extensive and reflects the great emphasis at present being placed on such complexes, particularly those of cobalt, nickel, and copper.

A few texts of general interest have been published during the year. The proceedings of a symposium on e.s.r. spectroscopy held in 1966 in Michigan, U.S.A., are now available in full.[1] The 'Altas of E.s.r. Spectra,' is mainly

[1] Symposium on E.s.r. Spectroscopy, Michigan, U.S.A., August 1966, *J. Phys. Chem.*, 1967, **71**, 1–214.

concerned with free radicals in solutions, polymers, and glasses, and is a compilation of 1200 spectra with illustrations and tabulated data to be used as a reference collection.[2] 'Introduction to Electron Spin Resonance,' is an elementary text[3] and 'Electron Spin Resonance in Semiconductors,' is a specialist work[4] dealing with the immensely important field of impurities in silicon and germanium. 'Electron Spin Resonance in Chemistry,' provides an excellent introduction to both theoretical and practical aspects of e.s.r.[5] Transition-metal ions, free radicals, *F*-centres, etc. are all discussed, and ample references to recent literature are provided, making it a useful source work.

A number of papers concerned with complete spectrometers or various aspects of their instrumentation have appeared. These describe spectrometers designed to operate with modulation frequencies of 37·5[6] and 9 GHz,[7] 900[8] and 280 MHz,[9] and 100 kHz.[10, 11] A new type of instrument using a novel detector in the form of a helical resonator has been described.[12] This has an internal radio-frequency field distribution of sufficient volume and homogeneity to be used for magnetic resonance experiments, and the high Q factors of these resonators make them very suitable detectors of low-frequency resonance absorption. Two types are described operating at 55 and 85 MHz with the option of higher harmonic frequencies.

2 Simple Molecules and Free Radicals

Of theoretical interest is a paper[13] which proposes a new expression, U^x, to facilitate comparison of spin polarisations in π radicals; it is defined as the ratio of the more familiar Q^x value to the isotropic hyperfine coupling constants, A^x. Formulae have been given for approximate determination of line-shape data for free-radical spectra.[14]

Most of the free radicals appropriate to this section contain oxygen. An exception is the BH_3^- radical anion of BH_3 which has been reported[15] in the e.s.r. spectrum of γ-irradiated polycrystalline KBH_4 at 77°K.

Controversy has arisen over a paper published in 1966 in which it was claimed that the ^{17}O–^{18}O species had been studied in gaseous oxygen, an

[2] B. H. J. Bielski and J. M. Gebicki, 'Atlas of E.s.r. Spectra,' Academic Press, New York, 1967.
[3] H. M. Assenheim, 'Introduction to Electron Spin Resonance,' Plenum Press, New York, 1967.
[4] G. Lancaster, 'Electron spin resonance in semiconductors,' Plenum Press, New York, 1967.
[5] P. B. Ayscough, 'Electron Spin Resonance in Chemistry,' Methuen, London, 1967.
[6] A. V. Patankar, *J. Sci. Instr.*, 1967, **44**, 354.
[7] H. A. Buckmaster and J. C. Dering, *Canad. J. Phys.*, 1967, **45**, 107.
[8] W. Duncan, *J. Sci. Instr.*, 1967, **44**, 437.
[9] M. J. Hill and S. J. Wyard, *J. Sci. Instr.*, 1967, **44**, 433.
[10] D. A. Giardino and L. Petrakis, *Rev. Sci. Instr.*, 1967, **38**, 1180.
[11] T. H. Wilmshurst, *J. Sci. Instr.*, 1967, **44**, 403.
[12] J. C. Collingwood and J. W. White, *J. Sci. Instr.*, 1967, **44**, 509.
[13] T. F. Hunter and M. C. R. Symons, *J. Chem. Soc.* (*A*), 1967, 1770.
[14] D. Sames, *Z. Physik*, 1967, **198**, 125.
[15] M. C. R. Symons and H. W. Wardale, *Chem. Comm.*, 1967, 758.

observation now queried.[16] The gas-phase spectrum[17] of the O_2F radical complements earlier data in the liquid and solid phases. Normally forbidden transitions, involving a nuclear spin flip, were detected, the principal axes of the *g*-tensors were assigned, and the derived parameters were used in molecular orbital calculations and gave results in good agreement with data from i.r. studies. The ClO radical in the gas phase has been described in detail.[18, 19] Hyperfine coupling from ^{35}Cl and ^{37}Cl was detected, and the electronic ground-state is now established as ^{2} $^2\Pi_{\frac{3}{2}}$; BrO in its ground state $^2\Pi_{\frac{3}{2}}$ is also described.[19] A species ClO_2 has been detected in irradiated $KClO_4$,[20] and has been partially characterised as having the peroxide structure Cl·O·O.[21]

The e.s.r. spectrum of gaseous NO_2 has been analysed to test existing electron spin rotation interaction theories for polyatomic free radicals.[22] Second-order field effects and rotational relaxation have to be considered to interpret the results fully.

Four of the five predicted fine-structure transitions are reported for the vibrationally excited $^{33}S^{16}O$ radical in the gas phase, and are used to discuss its electronic structure.[23] Hyperfine effects are seen from the ^{33}S ($I = \frac{3}{2}$). Ground- and excited-state SO radicals are reported elsewhere.[19] Other radicals detailed include NS ($^2\Pi_{\frac{3}{2}}$),[19] SeO in ground- ($^3\Sigma$) and excited-states ($^1\Delta$),[19, 24] SF and SeF ($^2\Pi_{\frac{3}{2}}$)[24] and IO ($^2\Pi_{\frac{3}{2}}$).[24] The CS_2^- radical anion spectrum[25] shows hyperfine effects from the metal surface which stabilises it and is probably not 'free' in the strict sense.

3 Complex Radicals

E.s.r. spectroscopy has now proved to be of value in the study of polyhedral boranes. A red, paramagnetic unstable species has been described[26] which, although not isolated in a pure compound, is believed to be the $B_8H_8^{\cdot-}$ radical anion. The e.s.r. spectrum shows over 300 resolved lines (Figure 1) because both boron and hydrogen produce hyperfine interactions, complicated even further by isotopic effects from ^{10}B ($I = 3$) and ^{11}B ($I = \frac{3}{2}$). The attempted simulation of the spectrum by assuming all the borons to be in equivalent positions, and similarly for the hydrogens, gives reasonable

16 A. Carrington, D. H. Levy, and T. A. Miller, *J. Phys. Chem.*, 1967, **71**, 2372; L. K. Keys, *J. Phys. Chem.*, 1967, **71**, 2373.
17 F. J. Adrian, *J. Chem. Phys.*, 1967, **46**, 1543.
18 A. Carrington, P. N. Dyer, and D. H. Levy, *J. Chem. Phys.*, 1967, **47**, 1756.
19 A. Carrington and D. H. Levy, *J. Phys. Chem.*, 1967, **71**, 2.
20 J. R. Byberg, S. J. K. Jensen, and L. T. Muus, *J. Chem. Phys.*, 1967, **46**, 131.
21 R. S. Eachus. P. R. Edwards, S. Subramanian, and M. C. R. Symons, *Chem. Comm.*, 1967, 1036.
22 T. J. Schaafsma, *Chem. Phys. Letters*, 1967, **1**, 16.
23 A. Carrington, D. H. Levy, and T. A. Miller, *Mol. Phys.*, 1967, **13**, 401.
24 A. Carrington, G. N. Currie, P. N. Dyer, D. H. Levy, and T. A. Miller, *Chem. Comm.*, 1967, 641.
25 J. E. Bennett, B. Mile, and A. Thomas, *Trans. Faraday Soc.*, 1967, **63**, 262.
26 F. Klanberg, D. R. Eaton, L. J. Guggenberger, and E. L. Muetterties, *Inorg. Chem.*, 1967, **6**, 1271.

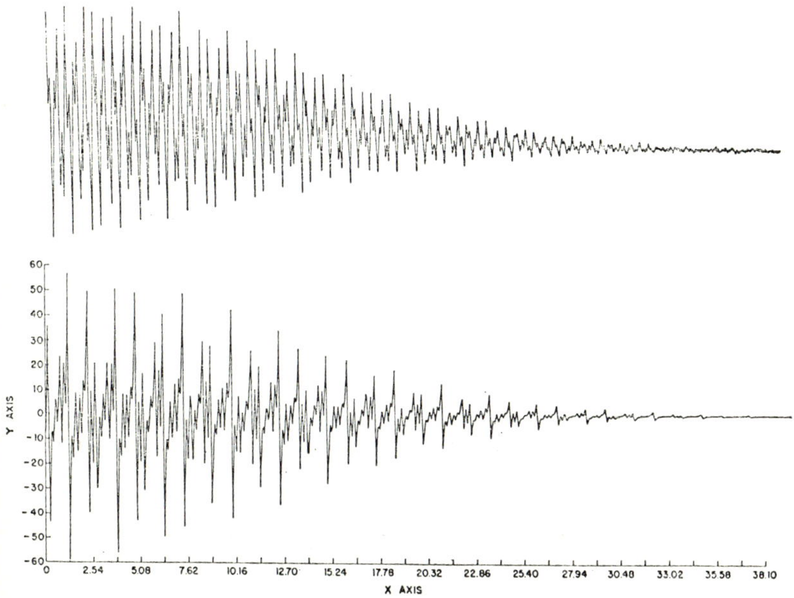

Figure 1 *E.s.r. spectrum ascribed to the* $B_8H_8^-$ *ion in tetrahydrofuran: upper, observed; lower, calculated;* x-*axis scale in gauss*
(Reproduced by permission from *Inorg. Chem.*, 1967, **6**, 1271)

agreement and definitely eliminates the corresponding B_7 and B_9 systems. It is concluded that the red species is a $B_8H_8^{\cdot -}$ ion with a probable D_{2d} dodecahedral geometry analogous to $B_8H_8^{2-}$. The spectrum is particularly interesting because it provides experimental verification of the concept of an 'aromatic' structure in boron-cage hydrides.

A purple coloration observed in a $B_{10}H_{14}$–$B_{10}H_{14}^{2-}$ reaction mixture was shown by e.s.r. to be due to a radical species, and other evidence established it to be $B_{10}H_{13}^{\cdot -}$.[27]

Kinetic measurements of the dissociation of the dithionate ion in aqueous solution ($S_2O_4^{2-} \rightleftharpoons 2SO_2^-$) have been made using an e.s.r. method.[28]

Sulphuric acid has long been known to give strongly coloured solutions with many materials, and studies on these systems are still in progress. Solutions of sulphur and S_4N_4 in sulphuric acid, oleum, and SbF_5 have been examined.[29] The sulphur radical produced is believed to have the form $X_2S{-}SX_2^+$ where X is derived from the medium. The sulphur nitride species is the unusual cyclic radical $S_2N_2^+$. The metals Nb, Ta, U, Pb, Na, K, Ca, In, Tl, and Sn all give green solutions with fluorosulphuric

[27] E. B. Rupp, D. E. Smith, and D. F. Shriver, *J. Amer. Chem. Soc.*, 1967, **89**, 5562.
[28] L. Burlamacchi, G.-C. Casini, O. Fagioli, and E. Tiezzi, *Ricerca sci.*, 1967, **37**, 97.
[29] D. A. C. McNeil, M. Murray, and M. C. R. Symons, *J. Chem. Soc.* (*A*), 1967, 1019.

acid. The e.s.r. spectra [30] demonstrate the presence of an identical paramagnetic species in all these solutions, showing that reduction to elemental sulphur has taken place.

γ-Irradiated KH_2AsO_4 contains an AsO_4^{4-} radical which appears to occupy a normal AsO_4^{3-} site hydrogen-bonded to other tetrahedra.[31] However, there is proton exchange among the various possible configurations around the tetrahedron, and the exchange frequency of this can be determined.

4 Transition-metal Complexes

A theoretical treatment has been given of e.s.r. linewidths for ions of the first transition series with quenched orbital angular momentum using a modified relaxation–matrix theory including third- and fourth-order perturbation terms.[32]

A variety of paramagnetic one-electron reduction species of metal phthalocyanines in tetrahydrofuran solutions have been investigated.[33]

Scandium.—Sc^{2+} impurity ions have been detected for the first time in CaF_2 and SrF_2 matrices.[34] Hyperfine effects from ^{45}Sc ($I = \frac{7}{2}$) and superhyperfine effects from eight equivalent ^{19}F nuclei show that the Sc^{2+} cation resides on a normal cation site. It has the $3d^1$ configuration and undergoes a Jahn–Teller distortion.

Titanium.—The ligand-field splitting diagrams of trichlorotrisethylenediaminetitanium(III), [Ti en_3]Cl_3, and the similar propylene diamine complex, [Ti pn_3]Cl_3, have been established from their e.s.r. spectra.[35] Assuming the $^2T_{2g}$ ground-state to be split by a trigonal field, the g values measured on the powders at room temperature indicate that the singlet $^2A_1$ level lies below the ^2E term, and that the trigonal-field splitting is large.

Studies [36] on the sodium titanium bronzes Na_xTiO_2 ($x = 0{\cdot}20$–$0{\cdot}25$) combined with magnetic data show no evidence for Ti^{3+} cations and suggest that the excess of electrons from the sodium are forming a conduction band.

Vanadium and Niobium.—The complex $(\pi\text{-}C_5H_5)_2VCS_2$ has been shown by a combination of i.r. and e.s.r. data to be correctly formulated as a π-CS_2 complex with the vanadium bonding to the carbon and one of the sulphur atoms in a three-membered ring structure.[37]

The practical observation in 1964 by Borsherts and Kikuchi of a five-line superhyperfine structure in the e.s.r. spectrum of vanadyl-doped zinc

[30] J. N. Brazier and A. A. Wolf, *J. Chem. Soc.* (*A*), 1967, 99.
[31] R. Blinc, P. Cevc, and M. Schara, *Phys. Letters*, 1967, **24A**, 214.
[32] D. Sames, *Z. Physik*, 1967, **198**, 71.
[33] D. W. Clack, N. S. Hush, and J. R. Yandle, *Chem. Phys. Letters*, 1967, **1**, 157.
[34] U. T. Höchli and T. L. Estle, *Phys. Rev. Letters*, 1967, **18**, 128.
[35] R. J. H. Clark and M. L. Greenfield, *J. Chem. Soc.* (*A*), 1967, 409.
[36] A. F. Reid and M. J. Sienko, *Inorg. Chem.*, 1967, **6**, 321.
[37] M. C. Baird, G. Hartwell, jun., and G. Wilkinson, *J. Chem. Soc.* (*A*), 1967, 2037.

Tutton's salt $(NH_4)_2Zn(SO_4)_2,6H_2O$ has now been interpreted.[38] The new interpretation is based on a $VO(H_2O)_4{}^{2+}$ species with proton hyperfine splitting superimposed on the normal eight ^{51}V ($I = \frac{7}{2}$) lines, and theoretical calculations simulate the observed spectrum quite well. The calculated spectrum for the $VO(H_2O)_5{}^{2+}$ cation is quite dissimilar, implying that a fifth water molecule is probably absent.

Data on vanadocene, $(\pi\text{-}C_5H_5)_2V$, in glassy solutions of 2-methyltetrahydrofuran complements earlier work on solutions in benzene and solid ferrocene.[39]

Interesting evidence for metal–metal coupling in anionic species has been found in some oxovanadium(IV) tartrates.[40] The e.s.r. spectra for the 1 : 1 complex anions of VO^{2+} and tartaric acid are different for the [{(+)-tartaric acid} VO^{2+}]$_2$ and [{(+)-tartaric acid} $(VO^{2+})_2$ {(−)-tartaric acid}] complexes. The latter shows hyperfine splitting from two equivalent vanadium nuclei.

Spin–lattice and spin–spin relaxation times for VO^{2+} in aqueous solutions have been measured using a dynamic nuclear polarisation spectrometer operating near 3000 MHz.[41]

Solution spectra of $[(Et_4N)_3(Nb_6Cl_{12})Cl_6]$ show an intrinsically isotropic hyperfine splitting.[42] The 49 detectable hyperfine components (see Figure 2)

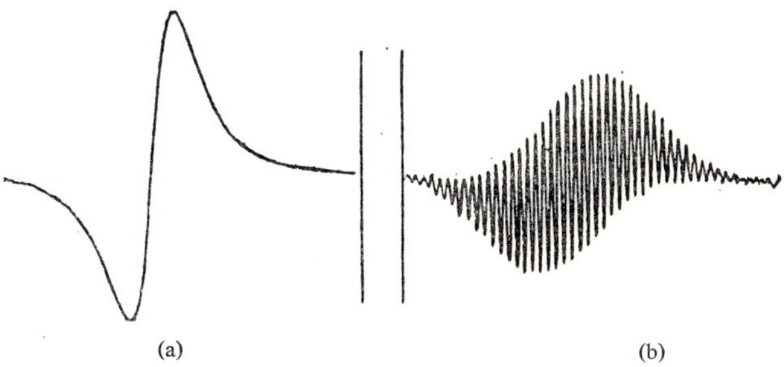

Figure 2 *E.s.r. spectra of* $(Et_4N)_3(Nb_6Cl_{12})Cl_6$: (a), *polycrystalline solid;* (b), *nitromethane solution*
(Reproduced by permission from *Inorg. Chem.*, 1967, **6**, 549)

are believed to be derived from the set of 55 such lines which a single electron delocalised overall six niobium atoms (^{93}Nb, $I = \frac{9}{2}$) in the cluster would be expected to show.

Chromium, Molybdenum, and Tungsten.—Studies of two nitrosylchromium complexes are reported. The $[Cr(CN)_5NO]^{3-}$ anion was doped into two crystalline forms of $K_3[Mn(CN)_5NO],2H_2O$. Least-squares analysis of the

[38] W. Burton Lewis, *Inorg. Chem.*, 1967, **6**, 1737.
[39] R. Prins, P. Biloen, and J. D. W. van Voorst, *J. Chem. Phys.*, 1967, **46**, 1216.
[40] R. E. Tapscott and R. L. Belford, *Inorg. Chem.*, 1967, **6**, 735.
[41] D. C. McCain and R. J. Myers, *J. Phys. Chem.*, 1967, **71**, 192.
[42] R. A. Mackay and R. F. Schneider, *Inorg. Chem.*, 1967, **6**, 549.

data [43] shows that the ^{53}Cr hyperfine tensor is collinear with the *g* tensor, but that the ^{14}N tensor is inclined at about 9° to it. It thus appears that the Cr—N—O bonds are not collinear. The similarity of the spectra with those of the same anion in a KCl matrix shows that the bent bonds are a feature of the complex ion itself and not induced by a particular lost lattice. The room-temperature spectrum [44] of $[Cr(NH_3)_5NO]^{2+}$ in dimethylformamide solutions shows ^{14}N ($I = 1$) hyperfine splittings from the NH_3 ligands which were not previously detected in aqueous solutions. Hyperfine parameters were measured for ^{53}Cr ($I = \frac{3}{2}$), ^{14}N (in NO), and ^{14}N (axial and equatorial NH_3) and the energy-level scheme is proposed to be

$$e(xz, yz) < b_2(xy) < e(\pi^*NO) < b_1(x^2 - y^2) \simeq a_1(z^2).$$

An attempt has been made to interpret the ligand hyperfine structure observed in the e.s.r. spectra of MOX_5^{2-} ions (M = Cr, Mo, or W; X = F or Br).[45] The ions have tetragonal symmetry and show both metal and ligand hyperfine interactions. A simple model was proposed for the $CrOF_5^{2-}$ ion in frozen aqueous solution which simulates the observed complex spectrum. Further work has since appeared [46] on $CrOF_5^{2-}$ (enriched ^{53}Cr, $I = \frac{3}{2}$) and $MoOF_5^{2-}$ (enriched ^{98}Mo, $I = 0$) where the reduction in fine structure is an aid to interpretation.

The cyanide $K_4[Mo(CN)_5NO],2H_2O$ has previously been formulated as containing $[Mo(CN)_5NO]^{4-}$ or $[Mo(CN)_5NO(OH)_2]^{4-}$. The one-electron oxidation product has been doped into $K_3Co(CN)_6$ and KBr for e.s.r. measurements and proves to be $[Mo(CN)_5NO]^{3-}$.[47] Glassy solution spectra correspond to a species with axial symmetry and showing ^{95}Mo ($I = \frac{5}{2}$) and ^{97}Mo ($I = \frac{5}{2}$) hyperfine effects. The species is six-co-ordinate in all phases, and the possibility of eight-co-ordination in the parent compound is probably ruled out. Ligand hyperfine structure was observed in the solids at 77°K, and interactions from ^{14}N ($I = 1$) in both the axial NO^+ and equatorial CN^-, and ^{13}C ($I = \frac{1}{2}$) in the equatorial CN^- ligand were identified. The ^{14}NO hyperfine tensor has axial symmetry about the structure axis, and the bond to the central metal is strongly covalent in character.

Solutions of $[MoOBr_5]^{2-}$ complexes dissolved in HBr contain at least three paramagnetic species, two of these being detected in the e.s.r. spectrum.[48] At high HBr concentrations, an axially symmetric complex which is believed to be $[MoOBr_5]^{2-}$ or $[Mo(OH)Br_5]^-$ is dominant. The inferred undetected species may be dimeric. A series of solid complexes of the type M_2MoOX_5 (M is *e.g.* NH_4^+, K^+, Rb^+, Me_3NH^+, or piperidinium; X = Cl or Br) and Et_4NMoCl_6 show three basically different types of

[43] B. R. McGarvey and J. Pearlman, *J. Chem. Phys.*, 1967, **46**, 4992.
[44] P. T. Manoharan, H. A. Kuska, and M. T. Rogers, *J. Amer. Chem. Soc.*, 1967, **89**, 4564.
[45] J. L. Verbeek and P. F. Cornaz, *Rec. Trav. chim.*, 1967, **86**, 209.
[46] J. L. Verbeek and A. T. Vink, *Rec. Trav. chim.*, 1967, **86**, 913.
[47] R. G. Hayes, *J. Chem. Phys.*, 1967, **47**, 1692.
[48] R. D. Dowsing and J. F. Gibson, *J. Chem. Soc. (A)*, 1967, 655.

spectrum.[48] Many of the compounds with large cations give the spectrum of an axially symmetric anion. Some of the others have an isotropic g tensor indicating an exchange interaction. Neither of these classes shows any dependence of the line shape with temperature. However, several of the compounds with smaller cations show anomalous spectra which change with temperature and seem to indicate a weak ferromagnetic exchange interaction between a number of molybdenum atoms.

Spectra of complexes containing the ion $[(\pi\text{-}C_5H_5)_2MX_2]^+$ (M = Mo or W; X = Cl or Br)[49] confirm the metal oxidation state to be +5. Hyperfine effects from two equivalent halogens and from ^{95}Mo ($I = \frac{5}{2}$) and ^{97}Mo ($I = \frac{5}{2}$) were detected. A weak ^{183}W ($I = \frac{1}{2}$) interaction may be present. The bromides show g values greater than that for the free electron as a result of spin–orbit coupling and covalency in a transition metal with less than a half-filled *d*-shell.

Solution spectra of molybdenum(v) in solutions of H_3AsO_4 are attributed to the complex $[MoO(H_2AsO_4)_4]^-$.[50] Hyperfine effects from ^{95}Mo ($I = \frac{5}{2}$), ^{97}Mo ($I = \frac{5}{2}$), and ^{75}As ($I = \frac{3}{2}$) were recorded. The structure is stated to contain four monodentate ligands in a square planar arrangement. Solutions containing HCl give the complexes $MoO(H_2AsO_4)_3Cl$, $MoO(H_2AsO_4)_2Cl_2$, $MoO(H_2AsO_4)Cl_3$, and $MoOCl_4$ in quantities dependent on the molar proportions. Data for enriched ^{98}Mo are mentioned.

Spectra[51] of a series of one-electron reduction products of the silicomolybdic acid $[SiMo_{12}O_{40}]^{4-}$ in solution show them all to contain Mo^V. ^{95}Mo and ^{97}Mo hyperfine splittings were recorded in only two of the four complexes studied.

Manganese.—E.s.r. spectra of manganous Mn^{2+} species are not usually observed unless the local symmetry is cubic because of the gross line-broadening expected in this situation. Data already available for the $[Mn(H_2O)_6]^{2-}$ and MnF_6^{4-} ions in aqueous solutions have now been augmented by observations of $[Mn(MeCN)_6]^{2+}$, $MnCl_4^{2-}$, and $MnBr_4^{2-}$ in non-aqueous solvent systems.[52] Fast ligand-exchange with the solvent was found in the $MnBr_4^{2-}$ systems, but any noncubic species were not observed.

Preliminary e.s.r. data[53] on the new heteropoly-12-niobomanganate(IV) anion indicate a structural similarity with the 9-molybdomanganate(IV) anion which is known to contain a trigonally distorted MnO ,octahedron.

Iron.—Electron irradiation of single crystals of sodium nitroprusside, $Na_2[Fe(CN)_5NO],2H_2O$, forms some $[Fe(CN)_5NO]^{3-}$ anions with the additional electron in an antibonding molecular orbital.[54] Studies of ^{15}N

49 R. L. Cooper and M. L. H. Green, *J. Chem. Soc.* (*A*), 1967, 1155.
50 I. N. Marov, Iu. I. Dubrov, V. K. Beliaeva, and A. N. Ermakov, *Doklady Akad. Nauk S.S.S.R.*, 1967, **176**, 590.
51 P. Rabette, C. Ropars and J.-P. Grivet, *Compt. rend.*, 1967, **265**, *C*, 153.
52 S. I. Chan, B. M. Fung, and H. Lütje, *J. Chem. Phys.*, 1967, **47**, 2121.
53 B. W. Dale and M. T. Pope, *Chem. Comm.*, 1967, 792.
54 J. Danon, R. P. A. Muniz, and A. O. Caride, *J. Chem. Phys.*, 1967, **46**, 1210.

and ^{13}C hyperfine effects indicate that the electron is localised mainly on the NO group.

E.s.r. spectra[55,56] of bis-(*NN*-di-isopropyldithiocarbamato)iron(III) chloride as the solid and in acetone solutions show *g* values consistent with the unusual $S = \frac{3}{2}$ configuration; a detailed analysis has been made.

Preliminary data on some five-co-ordinate iron and cobalt mononitrosyl-1,2-dithiolene complexes show a threefold anisotropy of the *g* tensor plus ^{14}N and ^{57}Co hyperfine interactions.[57]

The observed variation in e.s.r. lineshape with single-crystal orientation in high-spin Fe^{3+} in iron(III) met-myoglobin and low-spin Fe^{III} in myoglobin azide can be attributed largely to random variations of about 2° in the orientation of the local symmetry axes.[58]

Cobalt, Rhodium, and Iridium.—A blue low-spin cobalt(II) complex, formulated in 1954 as $Co(PhNC)_5(ClO_4)_2$ has been re-investigated.[59] It is now shown to be $Co(PhNC)_5(ClO_4)_2,1{\cdot}5H_2O$, and it is dehydrated by P_2O_5 to give a yellow compound. E.s.r., i.r., and visible spectra show the change to be reversible, and both complexes are low-spin, probably of C_{4v} symmetry. The blue complex is concluded to contain octahedral $[Co(PhNC)_5H_2O]^{2+}$ and the yellow to be square pyramidal $[Co(PhNC)_5]^{2+}$.

The complexes $Co(CNR)_4X_2$ (CNR = alkyl or aryl isonitrile; X = Cl, Br, or I) have been stated to be four-co-ordinate with ionic halides. There is one unpaired electron. The e.s.r. spectra in various solvents[60] show $^{35-37}Cl$ ($I = \frac{3}{2}$), $^{79-81}Br$ ($I = \frac{3}{2}$), and ^{127}I ($I = \frac{5}{2}$) hyperfine interactions with a weak ^{59}Co ($I = \frac{7}{2}$) isotropic hyperfine effect also sometimes detectable. This indicates some degree of covalency between the complex and the halides. The *g* values were used to determine a tentative order of the levels as $d_{xy} < d_{xz}, d_{yz} < d_{z^2} < d_{x^2-y^2}$. The unpaired electron is in a d_{z^2} orbital thereby allowing σ interaction with unfilled halogen *s* orbitals and thence a hyperfine effect.

$K_2Co_2(CN)_{10},4H_2O$ dissociates in ethylene glycol–water mixtures to give mainly the paramagnetic ($S = \frac{1}{2}$) monomeric $[Co(CN)_5]^{3-}$ species. At 300°K the e.s.r. spectrum[61] (see Figure 3) is only a broad featureless line, but at 77°K it shows two *g* values with hyperfine splitting from ^{59}Co ($I = \frac{7}{2}$) verifying that the ion has axial symmetry. The detailed parameters are consistent with a square pyramidal structure with the unpaired electron in a d_{z^2} orbital. The trigonal bipyramidal configuration is eliminated as a possibility.

55 R. L. Martin and A. H. White, *Inorg. Chem.*, 1967, **6**, 712.
56 H. H. Wickman and F. R. Merritt, *Chem. Phys. Letters*, 1967, **1**, 117.
57 J. A. McCleverty, N. M. Atherton, J. Locke, E. J. Wharton, and C. J. Winscom, *J. Amer. Chem. Soc.*, 1967, **89**, 6082.
58 P. Eisenberger and P. S. Pershan, *J. Chem. Phys.*, 1967, **47**, 3327.
59 J. M. Pratt and P. R. Silverman, *Chem. Comm.*, 1967, 117.
60 J. P. Maher, *Chem. Comm.*, 1967, 632.
61 J. J. Alexander and H. B. Gray, *J. Amer. Chem. Soc.*, 1967, **89**, 3356.

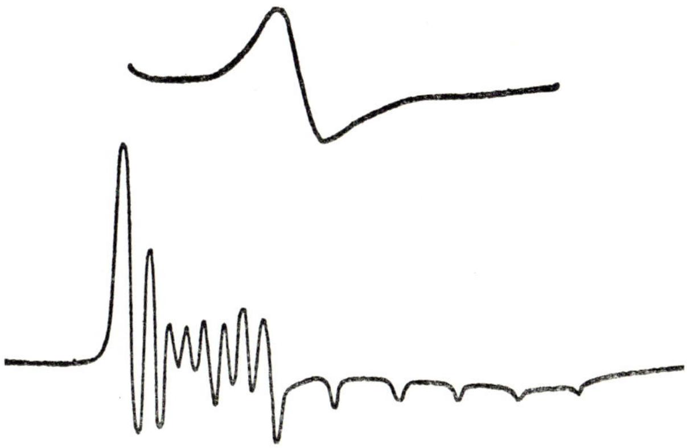

Figure 3 *E.s.r. spectra for* $Co(CN)_5^{3-}$ *in* 2 : 1 *ethylene glycol–water: upper curve,* 300°K; *lower curve,* 77°K *in frozen solution. Spectra in pure ethylene glycol are not significantly different*
(Reproduced by permission from *J. Amer. Chem. Soc.*, 1967, **89**, 3356)

Spectra have been obtained for ten cobalt(II) complexes which are magnetically anomalous,[62] including $K_2Ba[Co(NO_2)_6],H_2O$, and a number of complexes with tri- and bi-dentate ligands typified by [Co terpy$_2$]Cl_2,$5H_2O$ (terpy = terpyridyl) and [Co *o*-phen$_3$]$(ClO_4)_2$ (*o*-phen = *o*-phenylinediamine). Only the last mentioned is found to be pure high spin, the others being mixtures of high- and low-spin states. The magnetic moments calculated from the e.s.r. data are generally much lower than the corresponding conventionally determined values, showing that only the low-spin signal is appearing in the spectrum. Presumably the high-spin signal is broadened by relaxation. The e.s.r. intensity decreases with increasing temperature as the low-spin content decreases. Solutions were also studied and some hyperfine effects from ^{59}Co noted.

E.s.r. spectra of a number of peroxodicobalt anions of the form $[(CN)_5CoO_2Co(CN)_5]^{5-}$ and $[(NH_3)_5CoO_2Co(NH_3)_5]^{5+}$ show the two cobalt atoms to be equivalent.[63] The cyano- and ammine-complexes are so similar as to indicate the same square planar, staggered bridge structure for the former. The mixed anion–cation complex shows no signs of electron transfer. Similar compounds based on

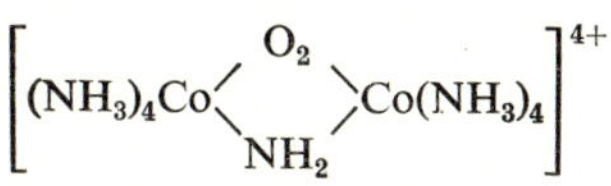

but with some of the NH_3 groups replaced by CN^- are all found to be very similar to each other. The hyperfine coupling constants show that generally there is a higher spin density at the cobalt in the ammines, consistent with

[62] J. G. Schmidt, W. S. Brey, jun., and R. C. Stoufer, *Inorg. Chem.*, 1967, **6**, 268.
[63] M. Mori, J. A. Weil, and J. K. Kinnaird, *J. Phys. Chem.*, 1967, **71**, 103.

the greater donor power of the CN^- forcing the unpaired electron towards the peroxo-group. Enrichment of 12% with ^{17}O in the peroxo-group gives a large ^{17}O ($I = \frac{5}{2}$) hyperfine splitting.[64] This is the first direct evidence that the unpaired electron is shared by the cobalt atoms via the oxygen bridge rather than by direct overlap of cobalt orbitals or via the amido-bridge.

Solution e.s.r. spectra of the five-co-ordinate cobalt dithiolate complexes of the type $LCoS_4C_4(CF_3)_4$ (L = PhO_3P, Ph_3P, Ph_3As, or Ph_3Sb), are consistent with an $S = \frac{1}{2}$ ground state.[65] Both the ^{59}Co splitting and the anisotropy of the g tensor are unusually low for a paramagnetic cobalt compound, indicating extensive delocalisation of the unpaired electron.

The powder e.s.r. spectra[66] of the complexes $Co^{II}[Ph_2P(CH_2)_2PPh_2]_2X_2$ (X = Cl, Br, or I) show them to have C_{2v} symmetry and three g values. In conjunction with spectral and magnetic measurements, the data have been used to deduce the likely electronic ground-state. The kinetics of a ceric oxidation reaction of a μ-amido-μ-peroxo-bis-(bisethylenediamine-cobalt) species have been studied by e.s.r.[67]

Green solutions of $RCoX(PhEt_2P)_2$ (R = Ph or Me) show a symmetric g factor and the complexes are presumed to be tetrahedral.[68] By contrast, yellow solutions of $R_2Co(PhEt_2P)_2$ (R = mesityl, pentachlorophenyl, etc.) show a broad anisotropic spectrum and they are presumed to be *trans*-square planar.[68] New complexes with pyridine derivatives of the form Me_2CoR_2 also appear to be square planar. The g factor decreases, owing to increased covalency towards the ligands, as the basicity of the ligand increases.

The powder spectrum of $(\pi\text{-}C_5H_5)_2Ir$ at 80°K shows hyperfine interaction with ^{191}Ir ($I = \frac{3}{2}$) and ^{193}Ir ($I = \frac{3}{2}$), but $(\pi\text{-}C_5H_5)_2Rh$ does not show effects from ^{103}Rh ($I = \frac{1}{2}$). Both compounds show axial symmetry.[69] Brief details have been given of the minority paramagnetic constituents in rhodium and iridium porphyrins.[70]

Nickel and Palladium.—The e.s.r. spectrum of $[Ph_4As][NiSe_4C_4(CF_3)_4]$ in a DMF–$CHCl_3$ glass at −170° (DMF = dimethylformamide) shows the three g values expected from such an isotropic system.[71] Hyperfine effects from ^{77}Se ($I = \frac{1}{2}$) were observed. Data are also available for the square planar complexes $Ni[o\text{-}C_6H_4(NH)S]_2$ and $Ni[(glyoxal)(o\text{-mercaptoanil})_2]$.[72] Several one-electron reduction species are formed in non-aqueous solutions.

64 J. A. Weil and J. K. Kinnaird, *J. Phys. Chem.*, 1967, **71**, 3341.
65 A. L. Balch, *Inorg. Chem.*, 1967, **6**, 2158.
66 W. deW. Horrocks, jun., G. R. van Hecke, and D. deW. Hall, *Inorg. Chem.*, 1967, **6**, 694.
67 M. Mori and J. A. Weil, *J. Amer. Chem. Soc.*, 1967, **89**, 3732.
68 K. Matsuzaki and T. Yasukawa, *J. Phys. Chem.*, 1967, **71**, 1160.
69 H. J. Keller and H. Wawersik, *J. Organometallic Chem.*, 1967, **8**, 185.
70 E. B. Fleischer and N. Sadasivan, *Chem. Comm.*, 1967, 159.
71 A. Davison and E. T. Shawl, *Chem. Comm.*, 1967, 670.
72 R. H. Holm, A. L. Balch, A. Davison, A. H. Maki, and T. E. Berry, *J. Amer. Chem. Soc.*, 1967, **89** 2866.

Spectra of complexes of bis(diphenylglyoximato)-nickel(II) and -palladium(II) with bromine and iodine are not fully understood but contribute towards a general picture of the complexes being inclusion-like and stabilised by charge-transfer interaction.[73] The complexes bis-(α-benzyldioxime)nickel(III) bromide and the equivalent palladium(III) compound have axial symmetry in solid and solution phases, confirming their square pyramidal structure.[74] ^{14}N hyperfine interactions are seen from the ligands. The data are fitted by an $S = \frac{1}{2}$ Hamiltonian showing the compounds to be low-spin trivalent complexes. Analysis of the g values shows that the unpaired electron is in a d_{z^2} orbital.

Copper.—A considerable volume of data has been published during the year on the e.s.r. spectra of copper complexes. The effects of the overlapping of components produced by copper(II) hyperfine interactions and oxygen- and nitrogen-ligand superhyperfine interactions on the intensities and widths observed in practice have been discussed in detail.[75]

Previously reported data for bis-(2-ammonioethyl)ammonium monochloride tetrachlorocuprate(II), $[(NH_3\cdot CH_2\cdot CH_2)_2NH_2]Cl[CuCl_4]$, had been interpreted on the basis of the $CuCl_5^{3-}$ ion, but an *X*-ray analysis [76] has now proved conclusively that the compound actually contains the $CuCl_4^{2-}$ ion, making re-interpretation of the e.s.r. spectrum necessary. Room-temperature spectra for the square planar $CuCl_4^{2-}$ ion in the complexes $(MeNH_3)_2CuCl_4$, $(EtNH_3)_2CuCl_4$, and $[Me_2CHNH_3]_2CuCl_4$ show only single broad lines as a result of dipolar broadening and exchange narrowing.[77] The g-values are consistent with a considerable degree of covalent bonding. Semi-empirical LCAO–MO calculations have been applied to the $CuCl_6^{4-}$ complex ion to consider the static and dynamic aspects of the Jahn–Teller effect in the system.[78] The results are compared with earlier e.s.r. data of Cu^{II} in NaCl and $CdCl_2$. The room-temperature e.s.r. spectrum has been briefly mentioned for the planar $Cu_2Cl_6^{2-}$ ion in $KCuCl_3$.[79]

The compounds $CuZrF_6,4H_2O$, $CuSiF_6,4H_2O$, and $CuNbOF_5,4H_2O$ have basically a chain structure of MF_6^{2-} or MOF_5^{2-} octahedra bonding to the axial positions of planar $[Cu(H_2O)_4]^{2+}$ units.[80,81] Broadening of the spectra was shown to be mainly due to a spin–spin interaction between neighbouring copper atoms. Although two different tetragonal Cu-site axes exist in the unit cell, these were not observed because of exchange coupling.

[73] A. S. Foust and R. H. Soderberg, *J. Amer. Chem. Soc.*, 1967, **89**, 5507.
[74] N. S. Garifianov, I. I. Kalinichenko, I. V. Ovchinnikov, and Z. F. Martianova, *Doklady Akad. Nauk S.S.S.R.*, 1967, **176**, 328.
[75] A. T. Nikitaev and K. I. Zamaraev, *Zhur. strukt. Khim.*, 1967, **8**, 429.
[76] B. Zaslow and G. L. Ferguson, *Chem. Comm.*, 1967, 822.
[77] R. D. Willett, O. L. Liles, jun., and C. Michelson, *Inorg. Chem.*, 1967, **6**, 1885.
[78] L. L. Lohr, jun., *Inorg. Chem.*, 1967, **6**, 1890.
[79] G. J. Maass, B. C. Gerstein, and R. D. Willett, *J. Chem. Phys.*, 1967, **46**, 401.
[80] R. Clad, *Compt. rend.*, 1967, **264**, *B*, 577.
[81] R. Clad, G. Heinrich, and J. Wucher, *Compt. rend.*, 1967, **264**, *B*, 510.

The e.s.r. spectrum of anhydrous copper sulphate[82] shows a very broad line which broadens even further below the Néel point of 35°K to the extent of being undetectable. This line is attributed to the anhydrous $CuSO_4$. A narrow line in the spectrum which does not show any change below the Néel point is considered to be due to surface hydration of the crystal.

The anisotropic metal hyperfine parameters of some copper acetylacetonates have been used to calculate the relative covalency with different substituents.[83] These calculations agree with the anticipated trends from general chemical considerations and contrast with earlier similar calculations from the isotropic parameters.

Room-temperature e.s.r. spectra of dichlorobis(pyridine-*N*-oxide)-copper(II), $[(C_5H_5NO)_2CuCl_2]$, and dichloroaquopyridine-*N*-oxide-copper(II), $[C_5H_5NOCuCl_2,H_2O]$ can be fitted by an $S = 1$ Hamiltonian, showing the presence of weak metal–metal interaction within the dimeric units.[84] Dilution with zinc to create chiefly Cu—Zn pairs gives data best fitted by the usual $S = \frac{1}{2}$ Hamiltonian for a single copper atom.[85] The complex similarly doped with nickel is the first to be reported containing dimeric units with exchange-coupled dissimilar transition metal ions. The system can still be described basically by an $S = \frac{1}{2}$ Hamiltonian but not as simply as in the zinc case.

Hydrated copper monochloroacetate, $Cu(ClCH_2\cdot CO_2)_2,2\cdot 5H_2O$, gives a hyperfine spectrum characteristic of an $S = 1$ state.[86] A 25% zinc-doped specimen shows an additional spectrum with $S = \frac{1}{2}$, having destroyed the weakly coupled interaction between pairs of copper atoms. The *g* values are identical, but differences are seen in the hyperfine coupling constants. Exchange coupling has also been investigated in a range of chloro-substituted cupric acetates.[87] The four compounds $Cu(CCl_3\cdot CO_2)_2,3H_2O$, $Cu(CCl_3\cdot CO_2)_2$, $Cu(CHCl_2\cdot CO_2),4H_2O$, and $Cu(CHCl_2\cdot CO_2)_2$ show the spectrum of an isolated cupric ion, the compounds $Cu(CH_2Cl\cdot CO_2)_2,2\cdot 5H_2O$ and $Cu(CH_3\cdot CO_2)_2,H_2O$ contain only exchange-coupled ion pairs, whereas $Cu(CH_3\cdot CO_2)_2$ and $Cu(CH_2Cl\cdot CO_2)_2$ give spectral evidence of both types of behaviour simultaneously, the relative proportions being temperature dependent.

A detailed analysis[88] of the e.s.r. and optical spectra of a purple unstable species found in solutions of CN^- and Cu^{II} in various solvents gives unequivocal evidence for the existence of the ion $[Cu(CN)_4]^{2-}$. The electron 'hole' appears to be considerably delocalised, and this may have some bearing on the instability of the ion. Some evidence was also

82 J. S. Wells, L. M. Matarrese, and D. J. Sukle, *J. Chem. Phys.*, 1967, **47**, 2259.
83 M. T. Rogers and R. E. Drullinger, *J. Phys. Chem.*, 1967, **71**, 109.
84 G. F. Kokoszka, H. C. Allen, jun., and G. Gordon, *J. Chem. Phys.*, 1967, **46**, 3013.
85 G. F. Kokoszka, H. C. Allen, jun., and G. Gordon, *J. Chem. Phys.*, 1967, **46**, 3020.
86 G. F. Kokoszka and H. C. Allen, jun., *J. Chem. Phys.*, 1967, **47**, 10.
87 A. Dall'Olio, G. Dascola, and V. Varacca, *Physica Status solidi*, 1967, **22**, 365.
88 A. Longo and T. Buch, *Inorg. Chem.*, 1967, **6**, 556.

found for a second species of higher symmetry in equilibrium with the first.

Data have been given [72] for the complex Cu[(glyoxal)(*o*-mercaptoanil)$_2$].

In single crystals of ethylenediaminecopper(II) sulphate, Cu en SO_4, the copper atom is found to be in a site of tetragonal symmetry with the principal axis along the *c* axis of the crystal.[89] No temperature dependence of the spectrum was found and a high degree of exchange narrowing was observed. The complexes with diphenylamine, $CuCl_2,nPh_2NH$ (n = 1–6) and $2CuCl_2,3Ph_2NH$ have been examined as polycrystalline powders.[90] $CuCl_2,Ph_2NH$ and $2CuCl_2,Ph_2NH$ show similar, very sharp spectra without hyperfine effects and with two *g* values. Additional small signal is more intense the greater the number of ligands. The formation of a charge-transfer type of molecular compound is postulated. The complex $Cu(NH_3)_4SO_4,H_2O$ shows anomalous line-broadening in all three lines at low temperatures as a result of a strong exchange interaction within the linear chains of copper atoms along the *c* axis.[91]

The nearly isotropic room-temperature e.s.r. spectrum of copper(II)-doped tris(phenanthroline)zinc(II) nitrate dihydrate becomes anisotropic on cooling and show hyperfine effects due to four equivalent nitrogen atoms.[92] The data can be explained in terms of a dynamical Jahn–Teller effect with a relatively high energy barrier between configurations. However, the observation of only two equivalent distortions instead of three is contrary to expectation and prevents full analysis by existing theoretical models.

The first hyperfine splittings seen for the nearest ligand atom of a near tetrahedral Cu^{II} complex at 77°K have been reported.[93] They are ^{14}N interactions in a 3% copper-doped complex of bis(*N*-isopropylsalicylaldiminato)zinc. The measured parameters are used to investigate covalency in the Cu—N bond.

In the complexes $[N(Ph_2PO)_2]_2Cu$ and $[N(Ph_2PNH)_2]Cu$ there is little interaction between copper atoms, but the addition product $[N(Ph_2PO)_2]_2Cupy_2$ differs considerably.[94] The imino-derivative shows much stronger covalency than the phosphineoxide complex.

The e.s.r. spectra and electronic structure [95] of bis(dipivaloylmethanido)-copper(II), $Cu(DPM)_2$, doped into the isostructural diamagnetic $Ni(DPM)_2$ are found to be very similar to those of $Cu(acac)_2$ (acac = acetyl acetone) as predicted.

89 S. Subramanian, J. Witters, A. van Itterbeek, J. Moreau, and Z. Rahman, *Physica*, 1967, **34**, 456.
90 V. P. Spacu, O. Constantinescu, I. Pascaru, M. Brezeanu, and F. Zalaru, *Z. anorg. Chem.*, 1967, **349**, 107.
91 S. Saito, *Phys. Letters*, 1967, **24**, *A*, 442.
92 G. F. Kokoszka, C. W. Reimann, H. C. Allen, jun., and G. Gordon, *Inorg. Chem.*, 1967, **6**, 1657.
93 H. P. Fritz, B. M. Golla, and H. J. Keller, *Z. Naturforsch.*, 1966, **21**, *B*, 1015.
94 H. J. Keller and A. Schmidpeter, *Z. Naturforsch.*, 1967, **226**, 231.
95 F. A. Cotton and J. J. Wise, *Inorg. Chem.*, 1967, **6**, 915.

Spectra of single crystals of the complex $(Cl.\cdot C_6H_4.\cdot SO_2NH)_2CuNH_3$ have been obtained between 1·6° and 300°K.[96] They consist of a single symmetrical absorption line anisotropic in width and *g* factor, and the system was found to be tetragonal. Detailed analysis of the *g* values indicates a high degree of covalency in the copper–ligand bonds which is consistent with the low solubility in water, alcohol, and acetone. E.s.r. and ^{14}N n.m.r. data on solutions of Cu^{II} in ethylenediamine–water mixtures give evidence for rapid ligand-exchange processes.[97] The solution species appear to be tetragonal with one or two monodentate ethylenediamine groups in axial positions. The relaxation processes are discussed in detail.

Spectra of copper-tetraphenylporphyrin and -tetrachlorotetraphenylporphyrin have been measured in dilute and frozen solutions.[98] They prove to be similar and contradict the earlier suggestion that the unpaired electron in copper porphyrins is delocalised over the whole conjugated bond system. The region of its delocalisation is presumably limited to the four central co-ordinating nitrogens.

Silver.—The e.s.r. and diffuse-reflectance spectra [99] of $[Ag(C_5H_5N)_4]S_2O_8$ are consistent with an assumed model of square planar bonding around the silver with a small distortion. The estimated molecular orbital parameters indicate strong covalent bonding in the complex.

Zinc and Cadmium.—One-electron reduction products of Zn and Cd[(glyoxal)(*o*-mercaptoanil)$_2$] in non-aqueous solutions are seen to be cation-stabilised free-radical complexes.[72] Cadmium oxide is a non-stoicheiometric *n*-type semiconductor with a sodium chloride crystal structure. The e.s.r. spectrum shows a minority species which can be assigned to the Cd^+ ion.[100] At low temperatures there is a hyperfine splitting of 54·2 Oe produced by the spin $\frac{1}{2}$ isotopes ^{111}Cd and ^{113}Cd. More interesting is the superposition of a super-hyperfine splitting of 6·75 Oe from a contact term with the twelve nearest-neighbour cadmium ions. The resonance is not to be confused with the separate conduction electron band also reported, and is clear evidence that the unpaired electron reaches far out into the crystal.

We thank B. A. Goodman for assistance, and J. D. Cooper for some translations from the Russian publications.

[96] Yu. A. Bratashevskii, N. N. Dykhanov, V. A. Moiseev, V. N. Topchii, and V. R. Shilov, *Zhur. fiz. Khim.*, 1967, **41**, 611.
[97] M. Alei, jun., W. B. Lewis, A. B. Dennison, and L. O. Morgan, *J. Chem. Phys.*, 1967, **47**, 1062.
[98] Yu. V. Glazkov and K. N. Solov'ev, *Zhur. priklad. Spektroskopii*, 1967, **6**, 262.
[99] H. G. Hecht and J. P. Frazier, jun., *J. Inorg. Nuclear Chem.*, 1967, **29**, 613.
[100] B. Elschner and M. Schlaak, *Phys. Letters*, 1967, **24**, *A*, 10.

4

Microwave Spectroscopy

This chapter has been ordered in terms of the number of atoms present in the molecule and then according to the vertical Groups of the Periodic Table.

A new edition of 'Spectroscopy at Microwave Frequencies' has been published[1] and this provides a thorough and comprehensive introduction to the subject.

The method of high-power microwave double-resonance has provided a new and useful means of studying collision-induced transitions between rotational levels.[2,3] In this technique, a strong microwave signal saturates a rotational transition and causes the enhancement or diminution of a different microwave transition under simultaneous observation. This effect was attributed to the transfer of some of the excess or deficiency of molecules by collision from the first to the second set of levels, and it has been observed for a variety of molecules. The effect has been confirmed using a modulated microwave double-resonance technique.[4]

Di- and Tri-atomic Molecules.—The microwave spectrum of the SO radical has been measured[5] and the equilibrium S—O distance is $1{\cdot}48108 \pm 0{\cdot}00005$ Å

Centrifugal distortion in the rotational spectrum of NSF has been analysed.[6,7] Values were obtained for the rotational and centrifugal distortion constants, and by combining i.r. and microwave data, a set of force constants were obtained. The interatomic distances are $r_{(S-N)}$ $1{\cdot}448 \pm 0{\cdot}002$ and $r_{(S-F)}$ $1{\cdot}643 \pm 0{\cdot}002$ Å, the bond angle $\widehat{NSF}$ is $116°55' \pm 2'$, and dipole moment value is $1{\cdot}902 \pm 0{\cdot}012$ D.

A gas-phase microwave parallel-plate absorption cell capable of very accurate Stark-effect measurements at high fields has been constructed.[8] With this cell the polarisability anistropy of OCS has been determined to

[1] D. J. E. Ingram, 'Spectroscopy at Microwave Frequencies', 2nd edn., Butterworth, London, 1967.
[2] T. Oka, *J. Chem. Phys.*, 1967, **47**, 13.
[3] T. Oka, *J. Chem. Phys.*, 1967, **47**, 4852.
[4] A. M. Ronn and E. Bright Wilson, jun., *J. Chem. Phys.*, 1967, **46**, 3262.
[5] T. Amano, E. Hirota, and Y. Morino, *J. Phys. Soc. Japan*, 1967, **22**, 399.
[6] A. M. Mirri and A. Guarnieri, *Spectrochim. Acta*, 1967, **23**, *A*, 2159.
[7] R. L. Cook and W. H. Kirchhoff, *J. Chem. Phys.*, 1967, **47**, 4521.
[8] L. H. Scharpen, J. S. Muenter, and V. W. Laurie, *J. Chem. Phys.*, 1967, **46**, 2431.

be $\alpha_{zz}-\alpha_{xx} = (5{\cdot}34 \pm 0{\cdot}06) \times 10^{-24}$ cm.^{3}, where z denotes the molecular axis.

Spectra of the carbonyl selenide molecule have also been determined for the ground- and vibrationally excited-states of a number of isotopic species, under the natural abundance ratio [9] (r(O—C) 1·157, r(C—Se) 1·708 Å).

Measurements on the vibrational satellite lines of FCN have been reported.[10] When these are combined with improved i.r. measurements they permit corrections to be made for the Fermi resonance existing between states with the quantum numbers ν_1, ν_2, ν_3, 1, and ν_1-1, ν_2+2, ν_3, 1. The correction has been neglected previously.

The spectrum of gaseous CsOH [11] in a high-temperature spectrometer indicates a linear or near-linear molecule with a large-amplitude, low-frequency bending mode. A large number of excited states involving the bending and Cs—O stretching modes have been identified. The Cs—O bond length is 2·40 ± 0·01 Å, the O—H bond length is approximately 0·97 Å and the value for the electric dipole moment is 7·1 ± 0·5 D. Relative intensity measurements indicate a Cs—O stretching mode at 400 ± 80 cm.$^{-1}$ and a bending mode in the neighbourhood of 300 cm.$^{-1}$ All the evidence supports a highly ionic Cs—O bond.

Tetra- and Penta-atomic Molecules.—The millimeter wave spectrum of NF_3 has been re-examined [12] because, in an attempt to determine the quadratic general force-field by a combination of i.r. and microwave data, the authors found that no reasonable force constants would explain the published values of D_J and D_{JK}. The new values are D_J 0·014534 ± 0·000068 and $D_{JK} = -0{\cdot}022694 \pm 0{\cdot}000096$ Mc./sec.

The rotational constants which fit the spectrum of $^{80}SeOF_2$ are $A = 6570{\cdot}35$, $B = 6282{\cdot}45$, and $C = 4156{\cdot}51$ Mc./sec. and the corresponding values for $^{78}SeOF_2$ are $A = 6585{\cdot}06$, $B = 6298{\cdot}94$, and $C = 4157{\cdot}73$ Mc./sec.[13] The structural parameters obtained from these constants are r(Se—O) 1·580 ± 0·002, r(Se—F) 1·727 ± 0·002 Å, $\widehat{FSeF}$ 92°21′ ± 5′, and $\widehat{OSeF}$ 104°49′ ± 2′.

Stark shift-measurements give a dipole moment of 3·18 ± 0·02 D along an axis at an angle of 49°9′ to the SeO bond and in the plane that contains this bond and bisects the FSeF angle. The geometrical and dipole-moment results suggest that the SeO bond has considerable double-bond character.

The observed Si—I bond length for trifluoroiodosilane is 2·387 ± 0·02 Å. This shortening has been found in other molecules and is attributed to ionic and double-bond character.[14]

9 Y. Morino and C. Matsuhura, *Bull. Chem. Soc. Japan*, 1967, **40**, 1101.
10 W. J. Lafferty and D. R. Lide, jun., *J. Mol. Spectroscopy*, 1967, **23**, 94.
11 D. R. Lide and R. L. Kuczkowski, *J. Chem. Phys.*, 1967, **46**, 4768.
12 A. M. Mirri and G. Cazzoli, *J. Chem. Phys.*, 1967, **47**, 1197.
13 I. C. Bowater, R. D. Brown, and F. R. Burden, *J. Mol. Spectroscopy*, 1967, **23**, 272.
14 L. C. Sams, jun., and A. W. Jache, *J. Chem. Phys.*, 1967, **47**, 1314.

Hexa- and Hepta-atomic Molecules—The spectra of ten isotopic species of germyl cyanide have been measured.[15] Assuming r(Ge—H) 1·529 Å and a tetrahedral distribution around the germanium, then r(C—N) is 1·155 ± 0·001 Å and r(Ge—C) 1·919 ± 0·001 Å. From an analysis of the splittings observed in the $1 \rightarrow 2$ transitions of $GeH_3 \cdot CN$, a quadrupole coupling constant for the nitrogen nucleus of −5·0 ± 0·1 Mc./sec. has been obtained. Analysis of the Stark effect gives a dipole moment value of 3·99 ± 0·05 D.

The molecular structure of gaseous $H_2B_2O_3$ has been determined by measuring the microwave spectra of six isotopic species.[16] The molecule (1) possesses C_{2v} symmetry and the B and O atoms form a five-membered planar ring. The dipole moment of the molecule is 0·95 ± 0·01 D.

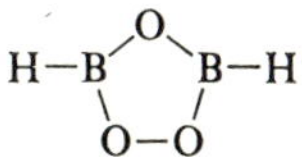

(1)

The barrier to internal rotation of the methyl group of methyl thiocyanate has been determined from the Q-branch splittings in the $v = 0$ torsional state and the value is 1600 ± 80 cal./mole.[17] The corresponding barriers for methyl isocyanate and methyl isothiocyanate are 83 ± 15 and 304 ± 50 cal./mole respectively. A normal-co-ordinate analysis of the three molecules has been carried out using a valence force-field. The lowest frequency mode in MeSCN was found to contain a large contribution from the S·C⋮N bend in addition to the CSC bend. The lowest modes in MeNCO and MeNCS are likewise not pure CNC bends.

Woods[18] has reported the first value of a threefold barrier of a CF_3 group to be obtained from a microwave spectrum. The final averaged barrier to internal rotation of fluoral ($CF_3 \cdot CHO$) is 885 ± 75 cal./mole. This may be compared with the value of 1150 cal./mole for acetaldehyde. The dipole moment components are estimated to be μ 0·15, and μ_b 1·64 D.

Two distinct sets of pure rotational transitions have been observed in the microwave spectrum (8–36 kMc./sec.) of fluoroacetyl fluoride.[19] They arise from two rotamers, one with two fluorine atoms *trans* to one another and the other where they are *cis*. Several sets of satellite lines were observed and assigned to torsionally excited *trans* molecules, to a skeletal bending mode and to a combination of these;

$$\Delta E = E_{cis} - E_{trans} = 910 \pm 100 \text{ cal./mole.}$$

[15] R. Varma and K. S. Buckton, *J. Chem. Phys.*, 1967, **46**, 1565.
[16] W. V. F. Brooks, C. C. Costain, and R. F. Porter, *J. Chem. Phys.*, 1967, **47**, 4186.
[17] R. G. Lett and W. H. Flygare, *J. Chem. Phys.*, 1967, **47**, 4730.
[18] R. C. Woods, *J. Chem. Phys.*, 1967, **46**, 4789.
[19] E. Saegebarth and E. Bright Wilson, jun., *J. Chem. Phys.*, 1967, **46**, 3088.

The molecular dipole moment of IOF_5 (C_{4v} symmetry) is $1{\cdot}08 \pm 0{\cdot}10$ D. Rotational constants and nuclear quadrupole coupling constants have also been calculated.[20]

Molecules with Eight, Nine, or Ten Atoms.—The microwave spectra of several isotopic species of phosphorus trifluoride–borane (PF_3,BH_3) have been measured and the bond lengths and angles have been determined (P—B $1{\cdot}835 \pm 0{\cdot}006$ Å).[21] The dipole moment from Stark splittings is $1{\cdot}64 \pm 0{\cdot}02$ D. Each of the observed ground-state transitions was accompanied by a number of weaker satellite lines from excited vibrational states. A previous Raman study of the liquid led to the assignment of eleven of the twelve fundamental modes. The remaining torsional mode $\nu_6(A_2)$ is inactive in the i.r. and Raman. The vibrational satellite spectrum, which is associated with the excited $\nu_{12}(E)$ and $\nu_6(A_2)$ states, was found to be strongly perturbed and it was fitted to the theory of a first-order Coriolis interaction. The analysis of this perturbation and the measurement of ν_{12} in the i.r. permitted the inactive mode ν_6 to be determined as 197 ± 5 cm.$^{-1}$ This is equivalent to a barrier to internal rotation of $3{\cdot}24 \pm 0{\cdot}15$ kcal. mole^{-1}. The enthalpy of dissociation for the reaction $2F_3P,BH_3 \rightleftharpoons 2PF_3 + B_2H_6$ (298°) was determined as 10·99 kcal. mole.

The microwave spectrum of germylsilane, H_3GeSiH_3, has been measured and the centrifugal distortion constant $D_J = 1{\cdot}8 \pm 0{\cdot}1$ kc./sec. was used to estimate the Ge—Si stretching frequency.[22] It gave a value of 350 ± 50 cm.$^{-1}$ Making reasonable assumptions for the parameters of the $—SiH_3$ and $—GeH_3$ groups, the Ge—Si bond distance is $2{\cdot}357 \pm 0{\cdot}004$ Å. Since the sum of the covalent radii gives a value of 2·39 Å, the slight shortening may indicate the presence of some multiple bonding. It was impossible to determine the barrier to internal rotation, because the lines were too weak, but the authors give an estimated value of 1100 cal./mole.

The dipole moment of $Co(CO)_3NO$ has been calculated from the non-resonant microwave absorption of the vapour.[23] The electric dipole moment is small (*ca.* 0·7 D or less) on the basis of dielectric-constant measurements of dilute benzene solutions. The true value, however, cannot be obtained unambiguously from solution data, partly because of solvent effects and more importantly because of uncertainties regarding the magnitude of atomic (vibrational) contribution to polarisation. The conventional microwave spectrum has not been observed. The authors report the accurate determination of the dipole moment in the gas phase from observations of the non-resonant microwave absorption (dielectric relaxation spectrum). The method is applicable to symmetric top molecules and may be applied either to pure gases or to mixtures

[20] S. B. Pierce and C. D. Cornwell, *J. Chem. Phys.*, 1967, **47**, 1731.
[21] R. L. Kuczkowski and D. R. Lide, jun., *J. Chem. Phys.*, 1967, **46**, 357.
[22] A. P. Cox and R. Varma, *J. Chem. Phys.*, 1967, **46**, 2007.
[23] R. H. Mann and A. A. Maryott, *J. Chem. Phys.*, 1967, **47**, 4275.

diluted with a non-dipolar foreign gas (μ 0·362 D, the uncertainty is *ca.* 3%).

The rotational constants and values of the second moment are given for 1,3,2-dioxaborolane, (2), and for the ^{10}B-substituted species, the 4-D substituted species and the ^{13}C-enriched species.[24]

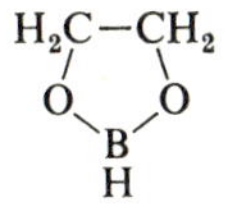

(2)

The B—O bond is approximately 33% ionic with about 10% π-bonding character. A ring-warping vibration was found which maintains a twofold axis of symmetry, but which shows a small hump in the planar configuration. The hump is probably as low or lower than the ground-state energy, perhaps of the order of 40 cm.$^{-1}$ above the potential minimum. The ring-warping vibration can be described as a torsion about the CC bond with simultaneous motion of one carbon atom above the OBO plane and the other carbon atom below it.

More Complicated Species.—Static dielectric constants and specific volumes have been measured for benzene solutions of aluminium acetylacetonate at 25, 35, 45, and 55°. The dielectric constants and losses at frequencies 9133 and 24,560 Mc./sec. were also determined and the total molar polarisations and dielectric relaxation times have been calculated from these data.[25] The static measurements confirm that the aluminium chelate is non-polar. Dielectric constants and losses of dilute benzene solutions of aluminium, chromic, and zirconium acetylacetonate complexes have been measured.[26] The data have been used to calculate small dielectric relaxation times, very large atomic polarisations and apparently small dipole-moment values. The most probable relaxation times are so small as to indicate that they are associated with intramolecular motion and not with the orientation of a permanent dipole molecular rotation. It is concluded, therefore, that the small apparent dipole moments arise from the atomic polarisation and not from permanent molecular moments.

The dipole moment, atomic polarisation, and dielectric relaxation have been measured for some polymethylpolysiloxane molecules, $Me_3Si[OSiMe_2]_xOSiMe_3$ (x = 1, 2, 3, or 5). The squares of the dipole moment, atomic polarisations, and relaxation times are found to be linear functions of the size of the molecule as measured by x.[27]

[24] J. H. Hand and R. H. Schwendeman, *Trans. Amer. Cryst. Assoc.*, 1966, **2**, 178; *Chem. Abs.*, 1967, **67**, 16435q.
[25] E. N. Dicarlo and R. E. Stronski, *Nature*, 1967, **216**, 679.
[26] S. Dasgupta and C. P. Smyth, *J. Amer. Chem. Soc.*, 1967, **89**, 5532.
[27] S. Dasgupta and C. P. Smyth, *J. Chem. Phys.*, 1967, **47**, 2911.

Microwave absorption phenomena have been observed in single crystals of the rare-earth metals terbium, dysprosium, holmium, and erbium.[28] All these metals show absorption at the critical-field transitions. In terbium and dysprosium additional absorption lines have been observed which are associated with the ferromagnetic spin configuration. The microwave results are discussed in terms of the normal-mode frequencies derived on the basis of the spin-wave approximation.

[28] D. M. S. Bagguley and J. Liesengang, *Proc. Roy. Soc.*, 1967, *A*, **300**, 497.

5
Vibrational Spectra

General Introduction.—The applications of vibrational spectroscopy to inorganic chemistry are continually increasing and the present collection of some 1200 references published during 1967 has presented formidable problems of organisation and presentation of the data.

The report is divided into four parts each comprising a separate sub-chapter:

Part I. The definitive spectra of small molecules and ions.

Part II. Characteristic frequencies of inorganic complexes.

Part III. Vibrational spectra of co-ordinated ligands.

Part IV. Appendices containing references to further compounds for which partial vibrational data are available.

There is also an extensive tabulation of compounds for which spectroscopic information has been obtained for characterisation but which do not justify fuller reporting. Not all the data have necessarily been obtained for the first time; some of the compounds have been studied previously but the information published in 1967 is given. It is hoped that these Tables will be a convenient source for tracing references to specific compounds back to the literature.

It should be emphasised that the assignments listed in Part II do not imply that the vibrations are always 'pure'. Mixed vibrations are frequently encountered and many authors have added qualifying statements to their assignments. Large Tables of data for individual compounds have been listed in the Appendix.

The main emphasis in gathering the references has been placed on the chemical and structural applications of the techniques, and papers concerned primarily with either instrumental methods or theoretical aspects have been omitted.

Three books concerned with important aspects of vibrational spectroscopy have been published. Adams [1] has given a critical survey of the i.r. and Raman spectra of metallic and organometallic compounds; Clark [2] has reviewed metal–halogen vibrational frequencies; and Wollrab [3] has reviewed

[1] D. M. Adams, 'Metal–Ligand and Related Vibrations,' Arnold, London, 1967.
[2] R. J. H. Clark, 'Halogen Chemistry, ed. V. Gutmann, vol. 3, Academic Press, 1967, p. 85.
[3] J. E. Wollrab, 'Rotational Spectra and Molecular Structure,' Academic Press, 1967.

recent theoretical and experimental advances in the study of rotational spectra together with a development of the theory used to treat rotational energies.

Several useful Reviews on vibrational spectroscopy have been published during 1967, covering both instrumental advances and specific areas of chemistry. Laser Raman spectroscopy has been discussed by Brandmüller [4] and a more general and elementary Review of the Raman effect and its applications has been given by Tobias.[5] Hester [6] has reviewed the Raman spectra of co-ordination compounds published in the 1960's and current applications of spectroscopy to the problems of high-temperature chemistry have been reported by Brewer.[7]

The vibrational spectra of the following classes of compound have been the subject of Review papers: boron–nitrogen compounds,[8, 9] carboranes,[10] organometallic acetylenes of the main Groups III–V,[11] halogen azides,[12] siloxane compounds of the transition metals,[13] the nitro-prusside ion,[14] and co-ordination compounds containing metal–metal bonds.[15]

Calderazzo [16] has reviewed halogeno–metal carbonyls and has included a small section on carbonyl frequencies. The collection of papers based on lectures presented at the 1966 Infrared Spectroscopy Institute [17] is of more interest to organic chemists.

A major problem in the interpretation of solid-state low-frequency i.r. spectra is the presence of lattice modes, and the study of solid samples under high pressure appears to offer a useful method of detecting lattice vibrations so that internal vibrational modes can be distinguished and assigned.[17a] The symmetric and antisymmetric metal–halogen stretching vibrations for a number of co-ordination compounds have also been studied; in all compounds examined, the peak positions are relatively insensitive to pressure, but the intensity of ν_{sym} is decreased to a much greater extent than ν_{asym}. The authors conclude that the technique can be used to distinguish between ν_{sym} and ν_{asym}.

4 J. Brandmüller, *Naturwiss.*, 1967, **54**, 293.
5 R. S. Tobias, *J. Chem. Educ.*, 1967, **44**, 1, 70.
6 R. Hester, *Co-ordination Chem. Rev.*, 1967, **2**, 319.
7 Leo Brewer, *Nat. Acad. Sci. Nat. Res. Council*, 1967, **1470**, 3; see also *Chem. Abs.*, 1967, **67**, 27, 394*b*.
8 A. Meller, *Organometallic Chem. Rev.*, 1967, **2**, 1.
9 V. I. Spitsyn, I. D. Kolli, R. A. Rodionov, A. I. Grigor'ev, T. G. Sevast'yanova, and V. M. Perfi'lev. *Zhur. neorg. Khim.*, 1967, **12**, 1158; see also *Chem. Abs.*, 1967, **67**, 69,038*u*.
10 R. Köster and M. A. Grassberger, *Angew. Chem. Internat. Edn.*, 1967, **6**, 218.
11 W. E. Davidsohn and M. C. Henry, *Chem. Rev.*, 1967, **67**, 73.
12 K. Dehnicke, *Angew. Chem. Internat. Edn.*, 1967, **6**, 240.
13 F. Schindler and H. Schmidbaur, *Angew. Chem. Internat. Edn.*, 1967, **6**, 683.
14 J. H. Swinehart, *Co-ordination Chem. Rev.*, 1967, **2**, 385.
15 B. J. Bulkin and C. A. Rundell, *Co-ordination Chem. Rev.*, 1967, **2**, 371.
16 F. Calderazzo, 'Halogen Chemistry,' ed. V. Gutmann, vol. 3, Academic Press, 1967, p. 383.
17 *Progr. I.R. Spectroscopy*, 1967, **3**.
17a C. Postmus, K. Nakamoto, and J. R. Ferraro, *Inorg. Chem.*, 1967, **6**, 2194.

It has not generally been appreciated that many common inorganic compounds, when dissolved in an aqueous medium at the appropriate pH, show absorption bands in the 950–1550 cm.$^{-1}$ region, where their spectra can be easily examined. The well-known difficulties of obtaining satisfactory spectra from powders are obviated by the use of aqueous solutions. Goulden and Manning [18] have studied the i.r. absorption spectra of the common inorganic ions in aqueous solution, using calcium fluoride cell windows, and they have constructed some structural correlation charts which permit unequivocal identification of most of the ions.

Savoie and Tremblay [19] have found that the intensities of Raman lines of small crystals can be appreciably enhanced by immersing them in a liquid with an equal, or nearly equal, refractive index. This usually gives better results than the hollow-cone method especially with optically isotropic samples, provided small crystals (2–4 mm.) can be obtained.

The fundamental Raman-active frequencies of octahedral or tetrahedral metal cluster compounds are predicted [20] to lie in the ratios 2 : 2 : 1 for the A_1, F_2, and E modes respectively. This approximation has been found adequate to interpret the low-frequency Raman spectra of $Bi_6(OH)_{12}^{6+}$, $Pb_4(OH)_4^{4+}$, and $Tl_4(OEt)_4$. For the molecule $Ir_4(CO)_{12}$, three strong shifts are observed at 208, 164, and 105 cm.$^{-1}$ and these are assigned respectively to the A_1, F_2, and E modes of the tetrahedral Ir_4 cluster; the frequency ratios are 2 : 1·57 : 1·01.

A computational scheme for the calculation of Raman intensities and depolarisation ratios has been developed [21] and a method of predicting the *PR* separations of band envelopes produced by prolate and oblate top molecules has been presented.[22] The method is simpler than those of Gerhard and Dennison, and Badger and Zumwalt, and the agreement with observed values is quite satisfactory.

A number of hydrates such as $CuSO_4,H_2O$, $CuSO_4,5H_2O$, $Co(NO_3)_2,2H_2O$, etc., have been examined by the inelastic scattering of 'cold' neutrons.[23] This overcomes the difficulties frequently met with in the use of i.r. techniques to observe or assign the low-frequency motions of water molecules in these and other complex solids owing to weak or diffuse absorptions, or to the presence of a number of other normal or lattice modes in the same frequency region. The neutron scattering experiment involves the transfer of vibrational or rotational quanta from the sample to the incident neutrons, so that the intensity from the various modes depends on the Boltzmann factor of populated states. Since the proton-scattering cross-section is quite large and almost entirely incoherent,

[18] J. D. S. Goulden and D. J. Manning, *Spectrochim. Acta*, 1967, **23**, *A*, 2249.
[19] R. Savoie and J. Tremblay, *J. Opt. Soc. Amer.*, 1967, **57**, 329, *Chem. Abs.*, 1967, **66**, 99,978*x*.
[20] C. O. Quicksall and T. G. Spiro, *Chem. Comm.*, 1967, 839.
[21] B. S. Averbukh, *Optics and Spectroscopy*, 1967, **23**, 28.
[22] W. A. Seth Paul and G. Dijkstra, *Spectrochim. Acta.*, 1967, **23**, *A*, 2861.
[23] J. J. Rush, J. R. Ferraro, and A. Walker, *Inorg. Chem.*, 1967, **6**, 346.

the spectra of inelastically scattered neutrons from the crystal hydrates will primarily reflect the motions of the water molecules. In addition, the neutron technique has no selection rules involving dipole moments or polarisabilities. Broad bands observed in the spectra at neutron energy gains of 500–800 cm.$^{-1}$ are assigned to the wagging and rocking modes of the co-ordinated and hydrogen-bonded water molecules. Bands around 400–500 cm.$^{-1}$ are attributed both to $M—OH_2$ stretching modes and to the H_2O torsional vibrations around the bisectrix. Maxima are also observed at energy transfers below 300 cm.$^{-1}$, which are tentatively assigned to hydrogen-bond stretching vibrations and possibly to $H_2O—M—OH_2$ deformation modes. Comparison of the various spectra appears to indicate that the average strength of binding of the water molecules does not change significantly in proceeding from the higher to the lower hydrates. The technique has also been used to study the low-frequency modes of urea, thiourea,[24] and solid sulphur dioxide hydrate.[25]

Finally, there is a growing number of reports which refer to the vibrational frequencies of co-ordinated molecular oxygen and molecular nitrogen. The $Cr(O_2)$ group absorbs in the range 860–885, $W(O_2)$ at 899 and 948, $Pd(O_2)$ at about 875, and $Pt(O_2)$ at 830 cm.$^{-1}$; fuller details and references are given in Part III. Likewise, molecular nitrogen complexes, for which fuller details are given on page 186, absorb as follows: $Co(N_2)$, 2080–2088; $Ru(N_2)$, 2105–2167; and $Os(N_2)$, 2010–2064 cm.$^{-1}$.

PART I: Definitive Spectra of Small Molecules and Ions

1 Diatomic Molecules and Ions

The i.r. spectrum of HF gas contains an extraneous line at 390 cm.$^{-1}$ which is very pressure dependent and has been assigned to the HF trimer.[26] As the pressure is lowered, however, the *R* and *Q* branches of this band become considerably weaker than the *P* branch. HF dimer, which has an absorption at 384·5 cm.$^{-1}$, is probably responsible for this peculiar feature. The vibrational spectra of crystalline HF and DF and of the dilute isotopic mixed crystals have been studied;[27] unlike the results of previous workers, four bands are now found in the stretching region of pure DF crystals as well as the previously reported four bands in the corresponding region of the HF spectrum.

Brunel and Peyron[28] examined HCl at low temperatures and found three fundamentals at 2756, 2718, and 2707 cm.$^{-1}$. This observation confirms a chain structure and shows that the chains are not planar.

Since the early work of Becker and of Pimentel, it has been tacitly accepted that small molecules do not have rotational freedom when trapped

[24] J. J. Rush, *J. Chem. Phys.*, 1967, **47**, 4278.
[25] A. W. Naumann and G. J. Safford, *J. Chem. Phys.*, 1967, **47**, 867.
[26] J. Leland Hollenberg, *J. Chem. Phys.*, 1967, **46**, 3271.
[27] J. S. Kittelberger and D. F. Hornig, *J. Chem. Phys.*, 1967, **46**, 3099.
[28] L. C. Brunel and M. Peyron, *Compt. rend.*, 1967, **264**, *C*, 821.

in solid nitrogen. However, multiplets have now been observed for HCl and HBr at liquid helium temperatures in the region of the fundamental of each molecule,[29, 30] and an interpretation is given based on rotation of the HCl and HBr molecules. A similar interpretation has been advanced for HCl and DCl in SF_6 matrices at low temperatures.[31] However, no structure attributable to rotation has been observed for carbon monoxide.[31]

The i.r. spectrum of the vapour species above lithium fluoride has been examined.[32] The spectrum is interpreted in terms of monomeric, dimeric, and trimeric molecules. For the monomer $\nu(^7Li-F) = 867$, and $\nu(^6Li-F) = 917$ cm.$^{-1}$. Isotope shifts are consistent with a cyclic configuration for the dimer. The ^7Li dimer has the following fundamentals $\nu(B_{3u})$ 655, $\nu(B_{2u})$ 570, and $\nu(B_{1u})$ 300 cm.$^{-1}$, and the remaining i.r.-inactive frequencies are estimated to be at $\nu(A_g)$ 560, $\nu(A_g)$ 280, and $\nu(B_{1g})$ 490 cm.$^{-1}$. The following absorptions are tentatively assigned to i.r.-active fundamentals of the ^7Li trimer: 230, 267, 509, and 736 cm.$^{-1}$.

The spectra of ammonium and alkali metal salts in dimethylsulphoxide solutions have been examined.[33] The absorption band is essentially independent of the nature and mass of the anion and it has been assigned to the interaction of the cation with solvent molecules: LiX (X = Cl, Br, I, NO_3, and ClO_4) 429; NH_4X (X = Cl, Br, I, NO_3, ClO_4, and SCN) 214; and NaX (X = Cl, Br, I, NO_3, SCN, and BPh_4) *ca.* 200 cm.$^{-1}$

The i.r. spectrum of vaporised magnesium fluoride isolated at liquid hydrogen temperatures contains a band at 740 cm.$^{-1}$ which has been assigned to MgF.[34] Other bands relating to the species MgF_2 are discussed in the section on triatomic molecules and ions.

I.r. lattice vibrations have been measured for AgCl, AgBr, and AgI,[35] and also for solid N_2, CO_2, and CO.[36]

The BrO^- anion in aqueous solution gives rise to a polarised Raman line at 620 cm.$^{-1}$.[37]

Finally, the Raman spectra for I_2, Br_2, and ICl have been determined in thirty-three solvents.[38] The observed Raman frequencies for the halogens and interhalogen were considerably lower than the corresponding gas-phase values, and the data have been interpreted on the basis of charge-transfer interactions. The Raman band at 163 cm.$^{-1}$ in the pyridine–iodine system was interpreted as $\nu(N-I)$.

[29] K. B. Harvey and H. F. Shurvell, *Chem. Comm.*, 1967, 490.
[30] K. B. Harvey and H. F. Shurvell, *Canad. J. Chem.*, 1967, **45**, 2689.
[31] R. Ranganath, T. E. Whyte, jun., T. Theophanides, and G. C. Turrell, *Spectrochim. Acta*, 1967, **23**, *A*, 807.
[32] A. Snelson, *J. Chem. Phys.*, 1967, **46**, 3652.
[33] B. W. Maxey and A. I. Popov, *J. Amer. Chem. Soc.*, 1967, **89**, 2230.
[34] D. E. Mann, G. V. Calder, K. S. Seshadri, D. White, and M. J. Linevsky, *J. Chem. Phys.*, 1967, **46**, 1138.
[35] G. L. Bottger and A. L. Geddes, *J. Chem. Phys.*, 1967, **46**, 3000.
[36] A. Ron and O. Schnepp, *J. Chem. Phys.*, 1967, **46**, 3991.
[37] J. C. Evans and G. Y.-S. Lo, *Inorg. Chem.*, 1967, **6**, 1483.
[38] P. Klaboe, *J. Amer. Chem. Soc.*, 1967, **89**, 3667.

2 Triatomic Molecules and Ions

The data which have appeared during 1967 are summarized in Table 1.

Table 1 *Vibrational frequencies* (cm.$^{-1}$) *of some triatomic molecules, ions, and free radicals*

Species	ν_1	ν_2	ν_3	*Conditions*	*Ref.*
HOCl	3581	1239	729	Solid matrix	39
HOBr	3590	1164	626		
$^{16}OF_2$[a]	925, 915	461	821	Argon matrix	40
$^{18}OF_2$	898, 889	457	794		
$Na^+HF_2^-$[b]	675	1200	1550	Crystalline	41
$Na^+DF_2^-$		860	1140		
Cs^+ClHCl^-[c]		631	1670	Crystalline (20°K)	42
Cs^+ClDCl^-[c]		449	1320		
Cs^+ClHBr^-[c]		508	1705		
Cs^+ClHI^-[c]		485	2200		
Cs^+ClDI^-[c]		350	1640		
$Me_4N^+BrHBr^-$[d]	126	1038	1420	Solid	43
$Me_4N^+BrDBr^-$[d]	123	752	1070		
$Cs^+ICl_2^-$[e]	268	129	218	Solid	44, 45
$Me_4N^+IBr_2$[e]	160	98	171		
$Cs^+I_2Br^-$[e]	117	84	168		
$Cs^+I_3^-$[e]	103	69	149		
Br_3^-	170		*ca.* 210	Aqu. sln.	37
ClF_2^-	475		635	Solid	46
$ClF_2^+(BF_4^-)$	798	537	813	Solid	47
$ClF_2^+(AsF_6^-)$	810	558, 520	818		

[a] The ν_1 doublet of OF_2 is probably due to a Fermi resonance between ν_1 and $2\nu_2$.

[b] Characteristic bands have also been observed in these regions for CaF_2,2HF, SrF_2,HF, SrF_2,2·5HF, BaF_2,HF, BaF_2,3HF, and BaF_2,4·5HF; these are indicative of HF_2^- anions.[55] The i.r. spectra of $K^+H_2F_3^-$ and $K^+D_2F_3^-$ have been recorded in the solid state.[56] The observed bands are assigned and a normal-co-ordinate analysis was carried out to determine the Urey–Bradley force constants. The results suggest that the hydrogen atoms are not exactly halfway between fluorine atoms, contrary to the case of the HF_2^- anion.

[c] The ν_2 values of these ions occur at approximately half the frequency previously assigned to ν_2 for $R_4N^+[XHY]^-$ salts at room temperature. This requires a re-interpretation of the earlier bihalide spectra and the earlier values are re-assigned as $2\nu_2$.

[d] The data indicate that an ion such as HBr_2^- may exist either in a linear symmetrical form or in a linear unsymmetrical form, the environment being the determining factor. I.r. structure, which was observed in the broad region 500–1300 cm.$^{-1}$, shown by salts of the linear symmetrical anion, was assigned to $\nu_3 \pm n\nu_1$. This interpretation may be extended to a large class of compounds which contain very strong hydrogen bonds, O—H—O, and which show a similar very broad absorption region, the origin of which has not previously been satisfactorily explained.

[e] Differences in the spectra of the polyhalide anions as the cation is changed have been attributed to differences in the anion structures and/or differences in the crystal lattice. Thus, although the compound of the formula NMe_4IBr_2 seems to contain a linear anion, the spectra suggest that the caesium analogue does not. In agreement with previous observations, the stretching force-constants of these anions have values which are approximately half those of the corresponding halogen or interhalogen molecule.

Table 1 (*cont.*)

Species	ν_1	ν_2	ν_3	*Conditions*	*Ref.*
ICN	~ 485	~ 304	~ 2188	Vapour	48
$H^{14}NF$		1432	1000	Argon matrix	49
$H^{15}NF$		1429	981		
$D^{14}NF$		1069	1000		
HCO	2482	1083	1861	Gas	50
HNO	3450	1110	1563		
HO_2	3414	1101	1385		
ClO_2^-	790	400	840	Solid	51
BrO_2^-	775	400	800		
BrO_2^-	709	324	680	Aqu. soln.	37
HCCl		1201	815	Matrix	52
SeCTe			1179	Solid	53
$SCBr_2$	1097		685		
SCI_2	1062		602		
NSCl	1325	414	273	Gas	54
MgF_2[f]	477	240	840	Vapour isolated in argon matrix	34

[f] The laser Raman spectra of crystalline MgF_2, ZnF_2, FeF_2, MnF_2, and TiO_2 have also been reported.[57] The crystals have D_{4h} symmetry and the four Raman-active vibrations predicted by group theory have been assigned for each crystal.

The matrix isolation technique has also been used to characterise CCl_2,[58] ClO_2,[59] Cl_2O,[60] and $(ClO)_2$.[60] Bands at 982 and 945 cm.$^{-1}$ are attributed to the latter in which pairs of ClO molecules are weakly linked as in the NO dimer.

[39] I. Schwager and A. Arkell, *J. Amer. Chem. Soc.*, 1967, **89**, 6006.
[40] J. S. Ogden and J. J. Turner, *J. Chem. Soc.* (*A*), 1967, 1483.
[41] A. Ažman and A. Ocvirk, *Spectrochim. Acta*, 1967, **23**, *A*, 1597.
[42] J. W. Nibler and G. C. Pimentel, *J. Chem. Phys.*, 1967, **47**, 710.
[43] J. C. Evans and G.Y.-S. Lo, *J. Phys. Chem.*, 1967, **71**, 3942.
[44] A. G. Maki and R. Forneris, *Spectrochim. Acta*, 1967, **23**, *A*, 867.
[45] G. C. Hayward and P. J. Hendra, *Spectrochim. Acta*, 1967, **23**, *A*, 2309.
[46] K. O. Christe, W. Sawodny, and J. P. Guertin, *Inorg. Chem.*, 1967, **6**, 1159.
[47] K. O. Christe and W. Sawodny, *Inorg. Chem.*, 1967, **6**, 313.
[48] S. Hemple and E. R. Nixon *J. Chem. Phys.*, 1967, **47**, 4273.
[49] M. E. Jacox and D. E. Milligan, *J. Chem. Phys.*, 1967, **46**, 184.
[50] J. F. Ogilvie, *Spectrochim. Acta*, 1967, **23**, *A*, 737.
[51] B. Tanguy, B. Frit, G. Turrell, and P. Hagenmuller, *Compt. rend.*, 1967, **264**, 301.
[52] M. E. Jacox and D. E. Milligan, *J. Chem. Phys.*, 1967, **47**, 1626.
[53] R. Stendel, *Angew. Chem. Internat. Edn.*, 1967, **6**, 635.
[54] A. Müller, G. Nagarajan, O. Glemser, S. F. Cyvin, and J. Wegener, *Spectrochim. Acta*, 1967, 23, *A*, 2683.
[55] D. D. Ikrahi, E. D. Ruchkin, and N. S. Nikolaev, *Zhur. strukt. Khim.*, 1967, **8**, 354; see also *Chem. Abs.*, 1967, **67**, 58,936*j*.
[56] A. Ažman, A. Ocvirk, D. Hadži, P. A. Giguère, and M. Schneider, *Canad. J. Chem.* 1967, **45**, 1347.
[57] S. P. S. Porto, P. A. Fleury, and T. C. Damen, *Phys. Rev.*, 1967, **154**, 522.
[58] D. E. Milligan and M. E. Jacox, *J. Chem. Phys.*, 1967, **47**, 703.
[59] A. Arkell and I. Schwager, *J. Amer. Chem. Soc.*, 1967, **89**, 5999.
[60] M. M. Rochkind and G. C. Pimentel, *J. Chem. Phys.*, 1967, **46**, 4481.

The first spectroscopic detection of any of the X_3 radicals has been reported.[61] Cl_3 gives rise to a series of i.r. bands near 374 cm.$^{-1}$ which are assigned to the ν_3 fundamental of a linear or slightly asymmetric radical.

Xenon dichloride has been detected in the i.r. as a feature with fine structure at 313 cm.$^{-1}$.[62]

Finally, the spectra of the following triatomic molecules have also been reported in the current literature: SO_2,[63, 64] N_2O,[65, 66] CO_2,[65] CS_2,[64] COS,[64] and FNO.[57]

3 Tetra-atomic Molecules and Ions

Force constants [68] and mean amplitudes of vibration [69] have been calculated for some mixed boron halides, BXY_2, and i.r. intensities and bond moments have been obtained for BCl_3.[70]

The Raman spectrum of gaseous nitrogen trifluoride at 8 atm. pressure has been examined [71] and earlier assignments are confirmed. The molecular constants of NF_3 have been obtained from i.r. rotational fine-structure [72] and a general force-field has been calculated [73] using vibrational frequencies, centrifugal stretching, and Coriolis coupling constants.

Data for some other tetra-atomic species containing nitrogen are in Table 2, which indicates the point group symmetry and assignments. I.r. data on isotopic hyponitrous acids have also been obtained; [76] the acid appears to have a C_{2h} symmetry in the solid state.

Table 2 *Vibrational frequencies* (cm.$^{-1}$) *of some tetra-atomic nitrogen species*

Molecule	*Assignments*						*State*	*Ref.*
C_{2v}	$\nu_1(a_1)$	$\nu_2(a_1)$	$\nu_3(a_1)$	$\nu_4(a_2)$	$\nu_5(b_1)$	$\nu_6(b_1)$		
cis-N_2F_2	1525	896	341	(550)	952	737	Liquid	74
C_{2h}	$\nu_1(a_g)$	$\nu_2(a_g)$	$\nu_3(a_g)$	$\nu_4(b_u)$	$\nu_5(b_u)$	$\nu_6(b_u)$		
trans-N_2F_2[a]	1522	1010	600	363·5	990	423	Liquid	74

[61] L. Y. Nelson and G. C. Pimentel, *J. Chem. Phys.*, 1967, **47**, 3671.
[62] L. Y. Nelson and G. C. Pimentel, *Inorg. Chem.*, 1967, **6**, 1758.
[63] H. Gerding and J. W. Ypenburg, *Rec. Trav. chim.*, 1967, **86**, 458.
[64] J. G. David and H. E. Hallam, *Spectrochim. Acta*, 1967, **23**, *A*, 593.
[65] J. E. Cahill, K. L. Treuil, R. E. Miller, and G. E. Leroi, *J. Chem. Phys.*, 1967, **47**, 3678.
[66] A. LeRoy and P. Jouve, *Compt. rend.*, 1967, **264**, *B*, 1656.
[67] L. H. Jones, L. B. Asprey, and R. R. Ryan, *J. Chem. Phys.*, 1967, **47**, 3371.
[68] T. A. Ford and W. J. Orville-Thomas, *Spectrochim. Acta*, 1967, **23**, *A*, 579.
[69] G. Nagarajan and A. Mueller, *Monatsh.*, 1967, **98**, 73.
[70] O. Brieux de Mandirola, *Spectrochim. Acta*, 1967, **23**, *A*, 767.
[71] J. Shamir and H. H. Hyman, *Spectrochim. Acta*, 1967, **23**, *A*, 1899.
[72] R. J. L. Popplewell, F. N. Masri, and H. W. Thompson, *Spectrochim. Acta*, 1967, 23, *A*, 2797.
[73] A. M. Mirri, *J. Chem. Phys.*, 1967, **47**, 2823.
[74] S. T. King and J. Overend, *Spectrochim. Acta*, 1967, **23**, *A*, 61.
[75] M. N. Hughes, *J. Inorg. Nuclear Chem.*, 1967, **29**, 1376.
[76] G. E. McGraw, D. L. Bernitt, and I. C. Hisatsune, *Spectrochim. Acta*, 1967, **23**, *A*, 25.

Table 2 (*cont.*)

Molecule			*Assignments*				*State*	*Ref.*
C_{2h}	$\nu_1(a_g)$	$\nu_2(a_g)$	$\nu_3(a_g)$	$\nu_4(a_u)$	$\nu_5(b_u)$	$\nu_6(b_u)$		
$Na_2N_2O_2$	1383	1115	692	365	1025	485	Solid	75
$^{14}N_2O_2{}^{2-}$	1419	1121	696	492	1031	371	Solid	76
$^{14}N^{15}NO_2{}^{2-}$	1398	1113	696	487	1022	366	Solid	76
$^{15}N_2O_2{}^{2-}$	1377	(1098)	(695)	482	1015	362	Solid	76
C_{2v}	$\nu_1(a_1)$	$\nu_2(a_1)$	$\nu_3(a_1)$	$\nu_4(b_1)$	$\nu_5(b_1)$	$\nu_6(b_2)$		
$F^{14}NO_2$	1309·6	822·4	567·8	1791·5	559·6	742·0	Gas	77
$Cl^{14}NO_2$	1267·1	792·6	369·6	1684·6	408·1	652·3	Gas	77
C_{3v}	$\nu_1(a_1)$	$\nu_2(a_1)$	$\nu_3(a_1)$	$\nu_4(e)$	$\nu_5(e)$	$\nu_6(e)$		
NF_3O	1691	743	528	883	558	400	Gas	78

[a] Two Type *A* bands of *trans*-N_2F_2 have been resolved and analysed to yield estimates of rotational constants.[79] The values of the latter are consistent with the geometry previously obtained by Bauer from electron diffraction measurements.

The i.r. spectra of gaseous and solid HNCS and DNCS are reported.[80] The spectra of both molecules isolated in an argon matrix are also discussed. The shift of ν(N—H) and ν(N—D) on solidification is 440 and 293 cm.$^{-1}$ respectively, which suggests a very short and strong hydrogen bond.

The vibration–rotation spectra of HN_3 and DN_3 have been studied in the 400–1000 cm.$^{-1}$ region:[81]

HN_3 ν_3, 1263·7; ν_5, 534·2; ν_6, 607·0 cm.$^{-1}$
DN_3 ν_4, 953·8; ν_5, 492·2; ν_6, 588·4 cm.$^{-1}$

The i.r. spectra of gaseous $PFCl_2$, PF_2Br, and $PFBr_2$ are assigned.[82]

Campbell and Turner[83] discuss the i.r. spectra of some ^{18}O isotopically substituted bromate anions such as $Br^{16}O_2{}^{18}O^-$, $Br^{16}O^{18}O_2{}^-$, and $Br^{18}O_3{}^-$, and they indicate the changes that take place with increasing ^{18}O concentration. The anions are prepared by an unusual surface reaction. The effect of low temperature on the reflection spectrum of monocrystalline BrO_3Na has also been examined.[84]

The Raman spectrum of molten stannous chloride and of molten mixtures of $SnCl_2$ and KCl indicate that chain polymers of the type $[SnCl_2]_n$ containing three-co-ordinate Sn^{II} predominate in pure molten $SnCl_2$ over a wide temperature range.[85] As the mole fraction of KCl is increased, depolymerisation takes place with the ultimate formation of pyramidal $SnCl_3{}^-$ ions.

[77] D. L. Bernitt, R. H. Miller, and I. C. Hisatsune, *Spectrochim. Acta*, 1967, **23**, *A*, 237.
[78] E. C. Curtis, D. Pilipovich, and W. H. Maberly, *J. Chem. Phys.*, 1967, **46**, 2904.
[79] S.-T. King and J. Overend, *Spectrochim. Acta*, 1967, **23**, *A*, 2875.
[80] J. R. Durig and D. W. Wertz, *J. Chem. Phys.*, 1967, **46**, 3069.
[81] D. M. Levine and D. A. Dows, *J. Chem. Phys.*, 1967, **46**, 1168.
[82] A. Müller, E. Niecke, and O. Glemser, *Z. anorg. Chem.*, 1967, **350**, 256.
[83] C. Campbell and J. J. Turner, *J. Chem. Soc.* (*A*), 1967, 1241.
[84] M. Galtier, J. Barcelo, and C. Deloupy, *Compt. rend.*, 1967, **265**, *B*, 1322.
[85] J. H. R. Clarke and C. Solomons, *J. Chem. Phys.*, 1967, **47**, 1823.

Two planar D_{3h} ions have been characterised (frequencies in cm.$^{-1}$): [86, 87]

	ν_1	ν_2	ν_3	ν_4
AlF_3	—	300	965	270
$BaCl_3$	—	425	597	372 (NaCl–$BaCl_2$ quenched melt)
	—	421	600	377 (KCl–$BaCl_2$ quenched melt)

The spectra of halogen derivatives of selenium and tellurium in the far-i.r. region continue to attract attention. Data are summarized in Table 3. The explanation of the MX_4 spectra in terms of MX_3^+ ions having approximately C_{3v} symmetry appears to be acceptable but it is not necessarily the correct model. Thus Hayward and Hendra [88] have examined the Raman and i.r. spectra of solid $SeCl_4$, $TeCl_4$, and $TeBr_4$. They observe four bands in the ν(M—X) stretching region which suggests a tendency towards a C_{2v} covalent structure for these compounds.

Table 3 *Halogeno-cations of sulphur, selenium, and tellurium*

	ν_1	ν_2	ν_3	ν_4	*Lattice mode*	*Ref.*
SCl_3^{+a}	525		502			89
$SCl_3^+(AsF_6^-)$	519	284	543	214		90
$SeCl_3^+(Cl^-)$	371	145, 165, 190, 205	348	145, 165, 190, 205	80	91
$SeCl_3^{+a}$	412		395			89
$SeCl_3^+(AsF_6^-)$	437	200	390	168		90
$SeBr_3^+(Br^-)$	265	127	227, 247	107		91
$TeCl_3^+(Cl^-)$	358	191	347	150	57	91
$TeCl_3^+(Cl^-)$	364	186	353	151		92
$TeCl_3^{+a}$	383		365			89
$TeCl_3^+(AsF_6^-)$	412	170	385	150		90
$TeBr_3^+(Br^-)$	242	111	222	95, 85		92
$TeBr_3^+(Br^-)$	240	125, 110, 92	222	125, 110, 92	46	91

[a] The values are made up of a mixture of solid-state and solution data.

The Raman spectra of LaF_3, FeF_3, PrF_3, and NdF_3 have been obtained using a laser source at room temperature and at low temperatures.[93] Assignments are made on the basis of polarisation and intensities of the

86 A. Snelson, *J. Phys. Chem.*, 1967, **71**, 3202.
87 (The late) J. R. Chadwick, P. J. Cranmer, and H. C. Marsh, *J. Inorg. Nuclear Chem.*, 1967, **29**, 1532.
88 G. C. Hayward and P. J. Hendra, *J. Chem. Soc.* (*A*), 1967, 643.
89 I. R. Beattie and H. Chadzynska, *J. Chem. Soc.* (*A*), 1967, 984.
90 W. Sawodny and K. Dehnicke, *Z. anorg. Chem.*, 1967, **349**, 169.
91 J. W. George, N. Katsaros, and K. J. Wynne, *Inorg. Chem.*, 1967, **6**, 903.
92 D. M. Adams and P. J. Lock, *J. Chem. Soc.* (*A*), 1967, 145.
93 R. P. Bauman and S. P. S. Porto, *Phys. Rev.*, 1967, **161**, 842.

lines. The spectra are consistent with a hexamolecular unit cell of D_{3d}^4 symmetry. The absorption and dichroism in the far-i.r. of a single crystal of PrF_3 at low temperatures has been reported.[94]

LeMay and Bailar[95] have examined twelve iodate salts (IO_3^-) and find both ν_1 and ν_3 in the region 600–900 cm.$^{-1}$. They also assign the following modes for ten iodato-complexes: ν_1(O—I str.), 630–710; ν_{3a}(IO_2 asym. str.), 770–808; and ν_{3b}(IO_2sym.str.), 720–758 cm.$^{-1}$.

4 Penta-atomic Molecules and Ions

The spectrum of difluorocyanamide $F_2N—C{\equiv}N$ has been assigned[96] and the molecule is believed to be non-planar with C_{1v} symmetry.

Halogen nitrates containing ^{14}N and ^{15}N have been investigated in the gaseous and solid phases.[97] All nine fundamentals are assigned for $FONO_2$ and eight for $ClONO_2$. Thermodynamic properties, barriers to internal rotation, and valence force constants have been computed.

The i.r. spectra of dilute solid solutions of SiH_4, GeH_4, and SnH_4 in each other, and of SiH_4 and GeH_4 in SiF_4 and SF_6 have been examined.[98] The spectra of silane and germane matrices show striking changes in contour when the host crystal passes through a phase transition, and the relation of these two to the site symmetry in the crystal, and the possibility of free rotation in the high-temperature phase is discussed.

Precise values for the vibrational frequencies (cm.$^{-1}$) of GeD_3I (C_{3v}) have been obtained and assigned:[99]

$\nu_1(a_1)$	$\nu_2(a_1)$	$\nu_3(a_1)$	$\nu_4(e)$	$\nu_5(e)$	$\nu_6(e)$
1509·9	582·8	249·5	1529·5	614·5	405·6

Similar information for GeH_2I_2 (C_{2v}) is:[100]

$\nu_1(a_1)$	$\nu_2(a_1)$	$\nu_6(b_1)$	$\nu_7(b_1)$	$\nu_8(b_2)$	$\nu_9(b_2)$
2090	821	2110	451	706	294

The other modes (ν_3, ν_4, and ν_5) were not observed.

The absolute i.r. gas-phase intensities for the triply degenerate vibrations of SnH_4 and SnD_4 have been determined.[101]

Normal-co-ordinate analyses for MX_4-type molecules and ions with tetrahedral symmetry have been carried out and are reported in a series of papers by Müller and co-workers,[102–104] which include large tables of data for these molecules. The mean-square amplitudes of vibration of some

94 A. Hadni and P. Strimer, *Compt. rend.*, 1967, **265**, *B*, 811.
95 H. E. LeMay, jun., and J. C. Bailar, jun., *J. Amer. Chem. Soc.*, 1967, **89**, 5577.
96 N. B. Colthup, *Spectrochim. Acta*, 1967, 23, *A*, 2167.
97 R. H. Miller, D. L. Bernitt, and I. C. Hisatsune, *Spectrochim. Acta*, 1967, 23, *A*, 223.
98 D. C. McKean and A. A. Chalmers, *Spectrochim. Acta*, 1967, 23, *A*, 777.
99 J. E. Griffiths, *Canad. J. Chem.*, 1967, **45**, 2639.
100 S. Cradock and E. A. V. Ebsworth, *J. Chem. Soc.* (*A*), 1967, 12.
101 I. W. Levin, *J. Chem. Phys.*, **46**, 1176.
102 A. Muller and B. Krebs, *J. Mol. Spectroscopy*, 1967, **24**, 180.
103 A. Müller and A. Fadini, *Z. anorg. Chem.*, 1967, **349**, 164.
104 G. Nagarajan and A. Müller, *Z. anorg. Chem.*, 1967, **349**, 82.

twenty-five molecules of the type ZXY_3 (halides and hydrohalide compounds of Group IV elements) have also been calculated.[105] Finally, values for the Urey–Bradley force constants for a number of main group tetrahalides have been correlated with the ionic character of the M—X bond.[106]

The vibrational frequencies of numerous tetrahedral cations and anions in the solid state are collected in Table 4. Notable among these is the new

Table 4 *Vibrational frequencies* (cm.$^{-1}$) *of some tetrahedral ions in the solid state*

	$\nu_1(A_1)$	$\nu_2(E)$	$\nu_3(F_2)$	$\nu_4(F_2)$	*Lattice mode*	*Ref.*
$NF_4^+(AsF_6^-)$	813	488	1159	611		107
$AlCl_4^-$			493[a]	~ 183		108, 109
$(Ph_4As^+)TlCl_4^-$	312	60	296, 278	~ 88		110
$K^+TlBr_4^-$	182	58	196	58		110
$(C_4H_9)_4N^+TlI_4^-$	133		156	~ 56		110
$(Et_4N^+)_2ZnCl_4^{2-}$			274	130	80	111
$(Et_4N^+)_2ZnBr_4^{2-}$			199, 209	91	70	111
$(Et_4N^+)_2ZnI_4^{2-}$			167	71	54	111
PCl_3Br^+	490		643			112

[a] The currently accepted value of ν_3 = 580 cm.$^{-1}$ is wrong.

cation NF_4^+ and the force constants of this species have been calculated and compared with those of NF_3, BF_4^-, and CF_4.[107] The spectra of the tetrahalogenothallate anions are consistent with tetrahedral symmetry[110] even though the potassium, rubidium, and calcium salts of $TlBr_4^-$ have previously been described as square planar.

The data in Table 4 on the anions ZnX_4^{2-} have been extended to include the i.r. spectra of the tetraethylammonium salts of the mixed anions, and this information, together with the related Raman data, is summarized in Table 5.

The compound $(C_6H_{13}N_2^+)BiCl_4^-$ is reported[113] to have Raman bands at 290, 267, 220, 95, and 80 cm.$^{-1}$.

Brown *et al.*[114] have suggested the term 'semi-co-ordinated' for anions which are very weakly co-ordinated to a metal and for which the distortion

[105] G. Nagarajan and A. Mueller, *Monatsh. Chem.*, 1967, **98**, 68.
[106] W. A. Yeranos and J. D. Graham, *Spectrochim. Acta*, 1967, **23**, *A*, 732.
[107] K. O. Christe, J. P. Guertin, A. E. Pavlath, and W. Sawodny, *Inorg. Chem.*, 1967, **6**, 533.
[108] D. E. H. Jones and J. L. Wood, *J. Chem. Soc.* (*A*), 1967, 1140.
[109] D. E. H. Jones and J. L. Wood, *Spectrochim. Acta*, 1967, **23**, *A*, 2695.
[110] T. G. Spiro, *Inorg. Chem.*, 1967, **6**, 569.
[111] G. B. Deacon, J. H. S. Green, and F. B. Taylor, *Austral. J. Chem.*, 1967, **20**, 2069.
[112] J. A. Salthouse and T. C. Waddington, *J. Chem. Soc.* (*A*), 1967, 1096.
[113] R. P. Oertel and R. A. Plane, *Inorg. Chem.*, 1967, **6**, 1960.
[114] D. S. Brown, J. D. Lee, B. G. A. Melsom, B. J. Hathaway, I. M. Procter, and A. A. G. Tomlinson, *Chem. Comm.*, 1967, 369.

Table 5 *Vibrational frequencies* (cm.$^{-1}$) *of some mixed tetrahalogenozincates*[111]

	ν(Zn—Cl)		ν(Zn—Br)		ν(Zn—I)	
	I.r.	Raman[a]	I.r.	Raman[a]	I.r.	Raman[a]
$ZnCl_4^{2-}$	277	280				
$ZnCl_2I_2^{2-}$	283				163	158
					147	
$ZnClI_3^{2-}$	184				166	140
$ZnBr_4^{2-}$			204	172		
			199			
$ZnBr_3I^{2-}$			203		167	155
			183		148	
$ZnBr_2I_2^{2-}$			205		167	142
			184			
$ZnBrI_3^{2-}$			206		167	130
			199			
ZnI_4^{2-}					167	122

[a] Raman spectra values from M. L. Delvaulle, *Bull. Soc. chim. France*, 1955, 1294.

is insufficient to produce a full lowering of the symmetry of the anion. An example is the tetrafluoroborate ion in [Cu en_2]$(BF_4)_2$ (en = ethylenediamine). For the free BF_4^- anion, ν_1 and ν_2 are virtually absent from the i.r. spectrum but in the complex, ν_1 and ν_2 show a relative intensity when compared to ν_3 and ν_4 of 30–40% (ν_1/ν_3 and ν_2/ν_4); ν_3 shows signs of splitting (but not as much as in Me_3SnBF_4) and ν_4 remains unchanged from its position and intensity in the free ion. An *X*-ray crystal structure of the complex shows that the tetrafluoraborate groups are in sites of tetragonal symmetry and the anion is only slightly distorted from tetrahedral symmetry.

Far i.r. spectra indicate strong association between incompletely substituted alkylammonium cations and $GaCl_4^-$ in non-dissociating solvents such as *cis*-1,2-dichloroethylene.[115] Thus, $Pr_4^nN^+GaCl_4^-$ gives a single band for ν(Ga—Cl) at 374 cm.$^{-1}$ whereas $Et_3NH^+GaCl_4^-$ shows three bands (at 359, 383, and 390 cm.$^{-1}$). This is consistent with a lower symmetry (C_{3v}) due to hydrogen bonding to the chlorine.

The anion $[CuCl_4]^{2-}$ shows distortions from tetrahedral symmetry in the solid state.[116] Thus ν(Cu—Cl) is split in $Cs_2[CuCl_4]$ (257 and 292 cm.$^{-1}$) and in $(Me_4N)_2[CuCl_4]$ (236 and 278 cm.$^{-1}$) indicating a flattened tetrahedral structure (D_{2d}). A similar distortion is found in $(Me_4N)_2[CuBr_4]$ [ν(Cu—Br) at 177 and 218 cm.$^{-1}$] but the anion is even more distorted in $Cs[CuBr_4]$, the C_s symmetry giving rise to three bands at 172, 189, and 224 cm.$^{-1}$. The compound $(Et_4N)_2[CuCl_4]$ when dissolved in nitromethane shows ν_3 at 237 and 278 cm.$^{-1}$ indicating that $[CuCl_4]^{2-}$ retains a distorted structure in solution.[117]

[115] P. L. Goggin and T. G. Buick, *Chem. Comm.*, 1967, 290.
[116] D. M. Adams and P. J. Lock, *J. Chem. Soc. (A)*, 1967, 620.
[117] D. Forster, *Chem. Comm.*, 1967, 113.

The vibrational spectra of several MO_4 molecules and ions have been published during the year. For those in Table 6 the spectra can be adequately described in terms of a regular tetrahedral symmetry. For

Table 6 *Vibrational frequencies* (cm.$^{-1}$) *of some tetrahedral* XO_4 *and* XS_4 *molecules and ions*

Species	ν_1	ν_2	ν_3	ν_4	*Ref.*
VO_4^{3-}	824	340	790	340	118
$Mn^{16}O_4^-$	844	385	910	407	119
$Mn^{18}O_4^-$	806	368	873	386	119
$(AsPh_4^+)TcO_4^-$			895	336	120
$(AsPh_4^+)ReO_4^-$			912	323	120
OsO_4 gas	970	333^a	960·5	326^a	121
MoS_4^{2-}	460	(175)	481	195	122
WS_4^{2-}	487	(179)	466	186	122

^{a} Measured in solution; values in parentheses are calculated.

several other XO_4 molecules, however, the site symmetry of the anion is significantly lower than tetrahedral and the vibrational spectra are considerably more complex. Thus, the i.r. spectra of some chromate(V) and chromate(VI) complexes are interpreted in terms of the site symmetry of the anion [123] and the Raman spectra of $CaWO_4$, $SrWO_4$, $CaMoO_4$, and $SrMoO_4$ have been assigned according to the point group C_{4h}.[124] The i.r. spectrum of $KTcO_4$ is also assigned according to factor group C_{4h}.[120]

Orbital-valence force-field constants have been calculated for several transition-metal oxyanions.[118, 125]

Isomorphism and polymorphism of the compounds Li_3PO_4, Li_3AsO_4, and Li_3VO_4 have been investigated and vibrational interactions between some LiO_4 stretching frequencies and the PO_4, AsO_4, and VO_4 bending modes are revealed by abnormal ^6Li–^7Li isotopic shifts.[126]

Flint and Goodgame [127] give a useful table of the internal modes of the sulphate, selenate, and thiosulphate anions. Sulphate vibrations are also discussed by Harmelin [128] and Petrov *et al.*[129] Funck [130] has assigned

[118] B. Krebs and A. Müller, *J. Mol. Spectroscopy*, 1967, **22**, 290.
[119] S. Pinchas, D. Samuel, and E. Petreanu, *J. Inorg. Nuclear Chem.*, 1967, **29**, 335.
[120] A. Müller and W. Rittner, *Spectrochim. Acta*, 1967, 23, *A*, 1831.
[121] R. S. McDowell, *Inorg. Chem.*, 1967, **6**, 1759.
[122] A. Müller, B. Krebs, and G. Gattow, *Spectrochim. Acta*, 1967, **23**, *A*, 2809.
[123] W. P. Doyle and P. Eddy, *Spectrochim. Acta*, 1967, **23**, *A*, 1903.
[124] S. P. S. Porto and J. F. Scott, *Phys. Rev.*, 1967, **157**, 716.
[125] A. Müller and B. Krebs, *Spectrochim. Acta*, 1967, **23**, *A*, 1591.
[126] P. Tarte, *J. Inorg. Nuclear Chem.*, 1967, **29**, 915.
[127] C. D. Flint and M. Goodgame, *J. Chem. Soc.* (*A*), 1967, 1718.
[128] M. Harmelin, *Compt. rend.*, 1967, **264**, *B*, 794.
[129] K. I. Petrov, V. I. Ivanov, and V. G. Pervykh, *Zhur. strukt. Khim.*, 1967, **8**, 356.
[130] E. Funck, *Ber. Bunsengesellschaft Phys. Chem.*, 1967, **71**, 170.

$\nu_{asym}(BO_4)$ for some borate complexes using the isotope effect ($^{10}B/^{11}B$). The spectra suggest a mean position of 1030 cm.$^{-1}$ for this vibration. Data have also been published on YPO_4,[131] NH_4MnO_4,[132] $MgCrO_4,5H_2O$,[133] $V_2O_5,I_2O_5,4H_2O$,[134] $M_4{}^{I}P_2O_7$, and $M_2{}^{II}P_2O_7$.[135]

The vibrational frequencies and assignments for several substituted oxyanions of general formula MO_3X^{n-} are given in Table 7. The symmetry

Table 7 *Vibrational frequencies* (cm.$^{-1}$) *of some oxyanions*, MO_3X^{n-}

Species	ν_1	ν_2	ν_3	ν_4	ν_5	ν_6	*Ref.*
$K^+[SO_3F^-]^a$	1084	741	571	1285	587	405	136
$Ca^{2+}[SO_3F^-]_2{}^a$	1123	842	570	1290	600	419	136
$Cs^+[SeO_3F^-]$	885	685	495	922, 965, 985	555	365	137
$[SReO_3]$	938	504	330	891	310	240	138, 139
$[(K^+)_2NReO_3{}^{2-}]$	878	1022	315	830	273	380	140

^{a} Lattice modes occur at 138 cm.$^{-1}$ for $K^+[SO_3F^-]$ and at 250 cm.$^{-1}$ for $Ca^{2+}[SO_3F^-]_2$.

of the anions is approximately C_{3v} but if one or two atoms of the fluorosulphate ion form bonds with the cation, then the symmetry of the anion is reduced to C_s. Goubeau and Milne[136] find increasing cation–anion interactions in the series of fluorosulphates $K < Ca < Zn < Cu^{II} < Fe^{III}$ and suggest that covalent bonding occurs in the copper(II) and iron(II) compounds. Interactions between two anions and solid-state effects cause an appreciable lowering of the symmetry from C_{3v} for $NaSO_3Cl$.[141] The mean amplitudes of vibration have been calculated for sixteen molecules and ions of the type MO_3X^{n-} with C_{3v} symmetry,[142] *e.g.* $SPO_3{}^{3-}$, $FSO_3{}^-$, $SReO_3{}^-$, etc.

Molecular force-fields and average vibrational amplitudes have been calculated for thionyl and selenyl halides.[143, 144]

Further data have been published on thiophosphoryl halides and are summarised in Tables 8 and 9.

[131] R. W. Mooney and S. Z. Toma, *J. Chem. Phys.*, 1967, **46**, 3364.
[132] E. J. Baran and P. J. Aymonino, *Z. anorg. Chem.*, 1967, **354**, 85.
[133] R. G. Darrie, W. P. Doyle, and (in part) I. Kirkpatrick, *J. Inorg. Nuclear Chem.*, 1967, **29**, 979.
[134] G. Ladwig, *Z. anorg. Chem.*, 1967, **353**, 311.
[135] E. Steger and B. Kässner, *Z. anorg. Chem.*, 1967, **355**, 131.
[136] J. Goubean and J. B. Milne, *Canad. J. Chem.*, 1967, **45**, 2321.
[137] A. J. Edwards, M. A. Mouty, and R. D. Peacock, *J. Chem. Soc.* (*A*), 1967, 557.
[138] A. Müller, B. Krebs, and E. Diemann, *Z. anorg. Chem.*, 1967, **353**, 259.
[139] J. F. Leroy and G. Kaufmann, *Bull. Soc. chim. Fr.*, 1967, 1001.
[140] A. Müller, B. Krebs, and W. Höltje, *Spectrochim. Acta*, 1967, 23, *A*, 2753.
[141] E. Steger and I. C. Ciurea, *Z. anorg. Chem.*, 1967, **350**, 225.
[142] A. Mueller and G. Nagarajan, *Z. anorg. Chem.*, 1967, **349**, 87.
[143] A. Mueller and G. Nagarajan, *Z. Phys. Chem.*, 1967, **235**, 57.
[144] R. Pichai, M. G. Krishna Pillai, and K. Ramaswamy, *Austral. J. Chem.*, 1967, **20**, 1055.

Table 8 *Vibrational frequencies* (cm.$^{-1}$) *of some gaseous* (*or liquid*) *thiophosphoryl halides with* C_{3v} *symmetry*

	$\nu(P{=}S)$	$\nu_{sym}(PX_3)$	$\delta_{sym}(PX_3)$	$\nu_{asym}(PX_3)$	$\delta_{asym}(PX_3)$	$\delta(SPX_3)$	
Compound	$\nu_1(A_1)$	$\nu_2(A_1)$	$\nu_3(A_1)$	$\nu_4(E)$	$\nu_5(E)$	$\nu_6(E)$	*Ref.*
SPF_3 (gas)	694	980	445	944	409	276	145, 146, 147
$SPCl_3$ (gas)	770	431	250	547	250	174[a]	146
$SPBr_3$ (liquid)	726	299	165	431	133	175	145

[a] Liquid-phase value.

Table 9 *Mixed thiophosphoryl halides with* C_s *symmetry; frequencies in* cm.$^{-1}$

	SPF_2Cl [146]	SPF_2Br [145]	$SPCl_2F$ [146, 148]
PF sym. str.	946	938	912
PF asym. str.	920	911	
P=S str.	738	719	753
P—Cl asym. str.			575
P—Cl(Br) sym. str.	541	477	478
wagging	395	389	268
twisting	317	288	192
deformation	361	297	192
rock			
P=S out-of-plane bend	251[a]	231	337
P=S in-plane bend	198	175	325

[a] Liquid-phase value.

The vibrational assignment proposed for SPF_3 is in marked contrast to that currently accepted. The frequency of ν(P—S) is found to decrease as the number of fluorine atoms attached to the phosphorus atom increases. Several investigators have shown that ν(P=O) increases with increased electronegativity of the substituent atom. Similar suggestions have been proposed for ν(P=S) but this is incorrect. The effect is explained as a perturbation caused by $\nu_{sym}(PCl_3)$. The latter is considerably lower for $SPCl_3$ than for $OPCl_3$ and it interacts with ν(P=S). The two energy levels are pushed apart and consequently ν(P=S) is higher and $\nu(PCl_3)$ is lower than would be found if no perturbation took place. Cavell [147] has also assigned this molecule but he reverses the assignments for ν_1 and ν_2, *i.e.* ν(P=S) 983 cm.$^{-1}$.

145 A. Müller, B. Krebs, E. Niecke, and A. Ruoff, *Ber. Bunsengesellschaft Phys. Chem.*, 1967, **71**, 571.
146 J. R. Durig and J. W. Clark, *J. Chem. Phys.*, 1967, **46**, 3057.
147 R. G. Cavell, *Spectrochim. Acta*, 1967, **23**, *A*, 249.
148 R. A. Nyquist, *Spectrochim. Acta*, 1967, 23, *A*, 845.

The i.r. spectra of some $PF_2S_2^-$ salts have been examined.[149] For $CsPF_2S_2$, ν_{asym}(P—F) 818, ν_{sym}(P—F) 805, and $\nu(PS_2)$ 735sh, 710 cm.$^{-1}$ If the sulphur atoms are replaced successively by oxygen the positions of ν(P—F) shift to higher wavenumbers: $PF_2S_2^-$, ν_{asym}(P—F) 804, ν_{sym}(P—F) 781; PF_2SO^-, ν_{asym}(P—F) 814, ν_{sym}(P—F) 797; $PF_2O_2^-$, ν_{asym}(P—F) 850, ν_{sym}(P—F) 835 cm.$^{-1}$. The i.r. spectra of KMF_3 (M = Ni, Mg, and Zn) and $NaNiF_3$ have been measured at room temperature and at liquid nitrogen temperature.[150] Three absorption bands for KMF_3 were observed and these have been assigned to f_{1u} lattice modes of the regular cubic perovskite structure: $KNiF_3$, 150, 255, and 458; $KMgF_3$, 156, 300, and 498; $KZnF_3$, 142, 205, and 440 cm.$^{-1}$. The forms of the normal modes of the lattice vibrations have been fully described on the basis of a normal-co-ordinate analysis. The $NaNiF_3$ spectrum is much more complicated implying a structure having an appreciable deformation from the regular cubic perovskite. $KNiF_3$ has also been examined by Balkanski *et al.*[151] at 300 and 90°K. An extra side-band is observed near 500 cm.$^{-1}$ and its intensity and some features of the vibrational absorption spectrum in the visible region suggest that it may arise from a one-photon process.

The vibrational spectra and assignment of several square planar complex halides of palladium, platinum, and gold in the solid state are given in Table 10. When the complexes are dissolved in water or acid the stretching

Table 10 *Square planar* MX_4^{n-} *ions*[152,153]; *frequencies in* cm.$^{-1}$

Ion	ν_1	ν_2	ν_3	ν_4	ν_5	ν_6	ν_7	*Lattice modes*
K_2PdCl_4	310	275		198		336	193	95, 110
K_2PdBr_4	192	165	130	125		260	140	85, 100
K_2PtCl_4	333	306	168	196		321	191	89, 111
K_2PtBr_4	205	190	135	125		232	135	89, 104
K_2PtI_4	142	126	105			180	127	55, 78
$KAuCl_4$	347[a]	324[a]	151	171[a]		350	179	88
$KAuBr_4$	212[a]	196[a]		102[a]				
$KAuI_4$	148	110	113	75		192	113	56, 69·5

[a] Aqueous solution frequencies.

frequencies are relatively unaffected but the deformation modes may fall by as much as 25 cm.$^{-1}$ (ν_4 of solid K_2PtCl_4 196 cm.$^{-1}$; ν_4 of aqueous K_2PtCl_4 170 cm.$^{-1}$).

[149] M. Lustig and J. K. Ruff, *Inorg. Chem.*, 1967, **6**, 2115.
[150] I. Nakagawa, A. Tsuchida, and T. Shimanouchi, *J. Chem. Phys.*, 1967, **47**, 982.
[151] M. Balkanski, P. Moch, and M. K. Teng, *J. Chem. Phys.*, 1967, **46**, 1621.
[152] P. J. Hendra, *J. Chem. Soc* (*A*), 1967, 1298.
[153] P. J. Hendra, *Spectrochim. Acta*, 1967, **23**, *A*, 2871.

The salts M_2PdCl_4 and M_2PdBr_4 (M^+ is NH_4^+, K^+, Rb^+, or Cs^+) have also been examined in the far-i.r. region.[153a] The [$PdCl_4^{2-}$] complexes gave ν_3 160–175, ν_6 327–336, and ν_7 183–205 cm.$^{-1}$, and the [$PdBr_4^{2-}$] complexes gave ν_3 114–140 cm.$^{-1}$, ν_6 249–260 cm.$^{-1}$, and ν_7 130–169 cm.$^{-1}$.

A theoretical treatment has been given for the harmonic vibrations of a square planar MX_4 model, with application to XeF_4, and including a study of vibration–rotation interactions and centrifugal distortion.[154]

5 Hexa-atomic Molecules and Ions

Several complete assignments of trigonal bipyramidal molecules and ions of formula MX_5 have appeared and are summarised in Table 11. The data

Table 11 *Assignments for some trigonal bipyramidal species* MX_5; *frequencies in* cm.$^{-1}$

Species	ν_1	ν_2	ν_3	ν_4	ν_5	ν_6	ν_7	ν_8	*Ref.*
PF_5 (gas)	817	640	944	575	1026	532	300	514	155
AsF_5 (gas)	733	642	784	400	809	366	123	388	155
$NbCl_5$ (matrix)[a]	(348·5)	(292·5)	396	126	444	159	99	(139·2)	156
$Et_4N^+GeCl_5^-$	348	236	~ 310	~ 200	395	~ 200			157
$Bu_4N^+SnCl_5^-$	335	258	323	157	359	157			157

[a] Values in parentheses are calculated.

on PF_5 and AsF_5 call for little comment but $NbCl_5$ is only monomeric in matrices of N_2 (5°K) or cyclohexane (77°K) at high dilution.[156] In contrast to several earlier assignments, however, the molecule is dimeric (D_{2h}) in the solid phase and room temperature and in solutions in organic solvents. A full assignment of the thirty i.r. and Raman fundamental modes has been proposed.[156] $TaCl_5$ and WCl_5 are also dimeric solids.[158, 159] The dimeric structure of $TaCl_5$ is retained in solution but WCl_5 may be monomeric in solutions in non-polar solvents.[158]

The mean vibrational amplitudes and thermodynamic functions of VF_5 have been calculated[160] and some tentative assignments have been proposed[161] for $[N_2H_5]^+[UF_5]^-$ and $[NH_3OH]^+[UF_5]^-$. Data for $GeCl_5^-$ and $SnCl_5^-$ are in Table 11. The anions SiF_5^- and GeF_5^- have also been characterized in the i.r. region: $(Ph_4As^+)(SiF_5^-)$ shows absorptions at 925w,

[153a] C. H. Perry, D. P. Athans, E. F. Young, J. R. Durig, and B. R. Mitchell, *Spectrochim. Acta*, 1967, **23**, *A*, 1137.
[154] G. Hagen, *Acta Chem. Scand.*, 1967, **21**, 465.
[155] L. C. Hoskins and R. C. Lord, *J. Chem. Phys.*, 1967, **46**, 2402.
[156] R. D. Werder, R. A. Frey, and H. Günthard, *J. Chem. Phys.*, 1967, **47**, 4159.
[157] I. R. Beattie, T. Gilson, K. Livingston, and (in part) V. Fawcett and G. A. Ozin, *J. Chem. Soc.* (*A*), 1967, 712.
[158] R. A. Walton and B. J. Brisdon, *Spectrochim. Acta*, 1967, **23**, *A*, 2489.
[159] P. M. Boorman, N. N. Greenwood, M. A. Hildon, and H. J. Whitfield, *J. Chem. Soc.* (*A*), 1967, 2017.
[160] A. Mueller, *Z. Chem.*, 1967, **7**, 35; *Chem. Abs.*, 1967, **66**, 89,815*c*.
[161] B. Frlec, *J. Inorg. Nuclear Chem.*, 1967, **29**, 2112.

915vw, 875br,vs, 790vs, 770vs, 480, and 445s cm.$^{-1}$;[162] $[Ph_4As]^+[GeF_5]^-$ shows absorptions at 650s, 635s, 575s, 560s, and 325s cm.$^{-1}$.[162]

Assignments for two tetragonal–pyramidal MX_5 species having C_{4v} symmetry are shown in Table 12. It should be noted that IF_5 can be assigned

Table 12 *Tetragonal pyramidal* MX_5 *species; frequencies in* cm.$^{-1}$

Species	ν_1	ν_2	ν_3	ν_4	ν_5	ν_6	ν_7	ν_8	ν_9	*Ref.*
$SbCl_5^-$	445	285	180	420		117?	300	255	90	163
IF_5	694, 755	598	316	574		275	635	376	191	164

on the basis of C_{4v} only if one of the observed polarised Raman lines is not a fundamental. It is suggested that the strong polarised doublet at *ca.* 700 cm.$^{-1}$ is caused by Fermi resonance between ν_1 and $(2\nu_9+\nu_3)$.[164] The Raman spectrum of solid $[(EtNH_3)_2]^{2+}[BiCl_5]^{2-}$ has three bands at 276s(sh), 228m(sh), and *ca.* 100vw cm.$^{-1}$ which can be assigned to the anion.[113]

The vapour-phase i.r. spectra of $MoOF_4$ and WOF_4 have been reported.[165] The spectra are very similar and suggest a monomeric structure in this phase. In contrast, solid $MoOF_4$ is known to consist of distorted octahedral groups linked in endless chains by *cis*-bridging fluorine atoms.

Considerable interest has centred on assignments for the mixed phosphorus pentahalides during the year, though agreement has not yet been reached. One scheme, which refers to data on $MePF_4$, $ClPF_4$, and Cl_2PF_3, is given in Table 13. A revised assignment has been proposed for Cl_2PF_3.

Table 13 *Suggested assignments for* $MePF_4$, $ClPF_4$ *and* Cl_2PF_3;[166] *frequencies in* cm.$^{-1}$

	Assignment	$MePF_4$	$ClPF_4$	Cl_2PF_3	*Assignment*
a_1	PF_2 str.	932	895	893	PF str.
	PF_2' str.	596	691	665	PF_2' str.
	PF_2 in-plane bend	514	(510)	338	PCl_2 in-plane bend
	PX str.	725	434	407	PCl_2 str.
	PF_2' bend	179	144	124	PF_2' bend
a_2	PF_2F_2' twist	397	356	368	PCl_2F_2' twist
b_1	PF_2 str.	1009	921	427	PCl_2 str.
	PXF_2 in-plane str.	467	(490)	488	PCl_2F in-plane bend
	PF_2' bend	179	144	124	PF_2' bend
b_2	PF_2' str.	843	903	925	PF_2' str.
	PXF_2 out-of-plane bend	538	560	500	PCl_2F out-of-plane bend
	PXF_2F_2'	412	356	368	PCl_2FF_2' rock

[162] H. C. Clark and K. R. Dixon, *Chem. Comm.*, 1967, 717.
[163] H. A. Szymanski, R. Yelin, and L. Marabella, *J. Chem. Phys.*, 1967, **47**, 1877.
[164] R. J. Gillespie and H. J. Clase, *J. Chem. Phys.*, 1967, **47**, 1071.
[165] A. J. Edwards, G. R. Jones, and B. R. Steventon, *Chem. Comm.*, 1967, 462.
[166] R. R. Holmes, *J. Chem. Phys.*, 1967, **46**, 3730.

The force-constants calculated from these data show that the equatorial P—F bonds are considerably stronger than the axial P—F bonds in these molecules. An alternative assignment for Cl_2PF_3 together with the first analysis of Br_2PF_3 is given in Table 14.[167] $ClPF_4$ and CF_3PCl_4 have been

Table 14 *Assignments for* PF_3Cl_2 *and* PF_3Br_2;[167] *frequencies in* cm.$^{-1}$

		Assignment	PF_3Cl_2	PF_3Br_2
ν_1	a_1	P—F eq. str.	925	913
ν_2	a_1	P—F_2' sym. ax. str.	650	607
ν_3	a_1	P—F_2' ax. bend	429	420
ν_4	a_1	$PCl_2(Br_2)$ sym. str.	407	318
ν_5	a_1	$PCl_2(Br_2)$ bend	*a*	*a*
ν_6	a_2		*a*	*a*
ν_7	b_1	$PCl_2(Br_2)$ asym. str.	667	587
ν_8	b_1	P—F bend (*yz*)	562	510
ν_9	b_1	PF_2' wag	337	335
ν_{10}	b_2	PF_2' ax. sym. str.	895	878
ν_{11}	b_2	P—F bend (*xy*)	490	458
ν_{12}	b_2	P—F_2' rock	*a*	*a*

[a] Not observed.

fully assigned and a normal-co-ordinate analysis on the data shows that the symmetry co-ordinates adequately represent the normal modes except for ν_4, ν_6, and ν_7 where appreciable mixing is indicated.[168, 169]

Downs and Schmutzler [170] focus attention on some significant spectroscopic and stereochemical trends in the series PF_5, $MePF_4$, Me_2PF_3, and, Me_3PF_2. An apparent peculiarity of the trigonal bipyramidal systems is that the symmetric stretch of the axial unit has a depolarisation ratio of only slightly less than $\frac{6}{7}$. The mode in question really represents an out-of-phase motion of the axial and equatorial ligands of the trigonal bipyramid and unless the stretching force-constants of the two types of bonds differ considerably, the simple Wolkenstein theory of bond polarisability leads to a value close to $\frac{6}{7}$. A second conclusion is that for every degenerate mode of a D_{3h} system (PF_5 or Me_3PF_2) two distinct vibrations can usually be identified in the spectra of $MePF_4$ or Me_2PF_3 (C_{2v} symmetry). Finally, it would not be surprising if the as yet unknown compound Me_4PF were ionic, since the axial P—F bond suffers a marked weakening as the number of methyl groups increases.

167 J. A. Salthouse and T. C. Waddington, *Spectrochim. Acta*, 1967, **23**, *A*, 1069.
168 R. R. Holmes, *J. Chem. Phys.*, 1967, **46**, 3724.
169 R. R. Holmes, *J. Chem. Phys.*, 1967, **46**, 3718.
170 A. J. Downs and R. Schmutzler, *Spectrochim. Acta*, 1967, **23**, *A*, 681.

Treichel *et al.*[171] have characterised HPF_4, DPF_4, H_2PF_3, and D_2PF_3. The gas-phase i.r. spectra are very similar to those of $MePF_4$ and Me_2PF_3, which suggests that the molecules have equatorially substituted trigonal bipyramidal structures and assignments are made on the basis of this similarity.

Brunvoll[172] has calculated generalised mean-square amplitudes of vibration, mean amplitudes of vibration, and shrinkage effects for PF_5, PCl_3F_2, PCl_5, SbF_5, SbF_3Cl_2, $SbCl_5$, $NbCl_5$, and $TaCl_5$ using published frequencies.

I.r. and Raman data[173] on P_2I_4 indicate a *trans*-$I_2P—PI_2$ structure with C_{2h} symmetry in the solid state, and in CS_2 solution in which it had been reported previously to be gauche. P_2I_4 is thus similar in structure to P_2Cl_4 and unlike N_2H_4, P_2H_4, and N_2F_4. The data for the solid are in Table 15.

Table 15 *I.r. and Raman spectra of solid* P_2I_4[173]*; frequencies in* $cm.^{-1}$

Symmetry	*ν*	*Assignment*	*Symmetry*	*ν*	*Assignment*
a_g	319, 307	ν(P—P) and ν(P—I)	b_g	328	ν(P—I)
	117	PI_2 wag		92	PI_2 twist
	78	PI_2 scissors			
a_u	330	ν(P—I)	b_u	301	ν(P—I)
	89	PI_2 twist		112	PI_2 wag
	52	PI_2 rock (torsion)		62?	PI_2 scissors

A theoretical analysis of methyl mercaptan in the far-i.r. region shows that the absorption spectrum is associated with end-over-end rotations of the molecule as well as with internal rotational transitions.[174] Values for the rotational constant and the centrifugal stretching constant are deduced.

The normal vibrations have been analysed and the main bands have been assigned for trithiocarbonic acid $SC(SH)_2$ and its deuterio-derivative $SC(SD)_2$.[175] Using the Hückel MO Method, π-bond orders and π-electron densities have been calculated for $SC(SH)_2$.

The fundamental modes of S_6 and certain combination bands have been assigned and the results are compared with those of S_8.[176]

[171] P. M. Treichel, R. A. Goodrich, and S. B. Pierce, *J. Amer. Chem. Soc.*, 1967, **89**, 2017.
[172] J. Brunvoll, *Acta Chem. Scand.*, 1967, **21**, 473.
[173] S. G. Frankiss, F. A. Miller, H. Stammreich, and Th. T. Sans, *Spectrochim. Acta*, 1967, **23**, *A*, 543.
[174] E. D. Palik, D. G. Burkhard, and R. L. Wallis, *J. Mol. Spectroscopy*, 1967, **23**, 425.
[175] A. Müller, B. Krebs, and G. Gattow, *Z. anorg. Chem.*, 1967, **349**, 74.
[176] L. A. Nimon, V. D. Neff, R. E. Cantley, and R. O. Buttlar, *J. Mol. Spectroscopy*, 1967, **22**, 105.

6 Octahedral Molecules and Ions

Octahedral molecules and ions continue to be a fruitful source of vibrational data and the many species studied during 1967 have been collected together in Tables 16–19.

Table 16 *Vibrational frequencies* (cm.$^{-1}$) *of some* MF_6 *molecules*

Molecule	ν_1	ν_2	ν_3	ν_4	ν_5	*Conditions*	*Ref.*
SeF_6[a]			778·5	436		Gas	177
MoF_6	744	651			320	Anhydrous HF soln.	
	741	643				Gas	178
	741	645			322	Liquid	
WF_6[a]	775	678			325	Anhydrous HF soln.	
			712	258		Gas	177, 178
	772	672			316	Liquid	
ReF_6	756	578			249	Anhydrous HF soln.	
	755	(596)			246	Gas	178
	753	600				Liquid	
OsF_6	734					Anhydrous H.F. soln.	
						Gas	178
	733					Liquid	
UF_6	668	527				Anhydrous HF soln.	
	666·6	(535)		186·4 (184)		Gas	178, 179
	666	524				Liquid	
NpF_6				198·6 (198)		Gas	179
PuF_6				206·0 (203)		Gas	179

[a] Force-constants have been calculated. Values in parentheses are those predicted earlier from combination bands.

Table 17 *Vibrational frequencies* (cm.$^{-1}$) *of some* MX_6^- *ions*

Ion	ν_1	ν_2	ν_3	ν_4	ν_5	*Ref.*
$K^+PF_6^-$ [a]	751	580	830	558	477	180
$(Et_4N^+)PCl_6^-$	360	283	444	285	238	157
$Cs^+AsF_6^-$ [a]	685	576	699	392	372	180
$(NF_4^+)AsF_6^-$	687		709	406	378	107
$(IF_6^+)AsF_6^-$ [a]	683	583	695	404	377	181
$(Et_4N^+)AsCl_6^-$	337	289	333	220	202	157
$Li^+SbF_6^-$ [a]	668	558	669	350	294	180
$K^+SbCl_6^-$			349	181		182

[177] S. Abramowitz and I. W. Levin, *Inorg. Chem.*, 1967, **6**, 538.
[178] B. Frlec and H. H. Hyman, *Inorg. Chem.*, 1967, **6**, 1596.
[179] B. Frlec and H. H. Claassen, *J. Chem. Phys.*, 1967, **46**, 4603.
[180] G. M. Begun and A. C. Rutenberg, *Inorg. Chem.*, 1967, **6**, 2212.
[181] K. O. Christie and W. Sawodny, *Inorg. Chem.*, 1967, **6**, 1783.
[182] K. J. Wynne and W. L. Jolly, *Inorg. Chem.*, 1967, **6**, 107.

Table 17 (*cont.*)

Ion	ν_1	ν_2	ν_3	ν_4	ν_5	*Ref.*
$(AsCl_4^+)SbCl_6^-$	335	291	350	180	174	157
$(pyH^+)SbCl_6^-$ [b]	329	280	348	182	170	157
$K^+VF_6^-$	670	530?			310	183
$Cs^+VCl_6^-$			335			184
$(Me_4N^+)NbBr_6^-$			240			185
$(Me_4N^+)TaBr_6^-$			214, 234sh			185
$(N_2H_6^{2+})(MoF_6^-)_2$			620			186
$(N_2H_6^{2+})(OsF_6^-)_2$			640			186
$(Et_4N^+)MoCl_6^-$			330			184
$(Me_4N^+)PaBr_6^-$			215			185, 187
$(NH_3OH^+)UF_6^-$	628		526			188

[a] Force-constants have been calculated.
[b] In nitromethane solution.

Table 18 *Vibrational frequencies* (cm.$^{-1}$) *of some* MX_6^{2-} *ions*

Ion	ν_1	ν_2	ν_3	ν_4	ν_5	ν_6	*Lattice mode*	*Ref.*
K_2SiF_6 [a]			741	483				189
Na_2SiF_6 [b]	663	477	741	483	408			180
K_2GeF_6 [b]	624	471	603	339, 359	335			180, 189
$(Et_4N)_2GeCl_6$	318	213	310	213	191			157
Na_2SnF_6	592	477	559	300	252			180
K_2SnCl_6	311	235	325	172	165		79	190
$(Et_4N)_2SnCl_6$ [c]			300	165				157
K_2SnCl_6 [d]	323	249	314	174	184	[124]	83	191

Values in round brackets are calculated frequencies. Values in square brackets indicate unassigned low-frequency bands. py = pyridine.

[a] Most hexafluorides of K, Rb, and Cs crystallise with either cubic (O_h), trigonal (D_{3d}), or hexagonal (C_{3v}) symmetry.[189] The phases are temperature dependent and the contour of ν_4 is very characteristic of the structural type (singlet for O_h, doublet for D_{3d}, and C_{3v}). In theory the last two symmetries are indistinguishable using i.r. spectroscopy but ν_4 for hexagonal symmetry is very asymmetric and thus qualitatively recognisable. The origin of the asymmetry is not understood.

[b] Force-constants have been calculated.

[c] In nitrobenzene solution.

[d] A large range of cations are examined and force-constants have been calculated using a modified Urey–Bradley force-field. They show a general decrease with increase in cation size. Some cation lattice-modes are also observed and they decrease with increase in cation mass.

---

[183] H. Selig and B. Frlec, *J. Inorg. Nuclear Chem.*, 1967, **29**, 1887.
[184] R. A. Walton and B. J. Brisdon, *Spectrochim. Acta*, 1967, **23**, *A*, 2222.
[185] D. Brown and P. J. Jones, *J. Chem. Soc.* (*A*), 1967, 247.
[186] B. Frlec, H. Selig, and H. H. Hyman, *Inorg. Chem.*, 1967, **6**, 1775.
[187] D. Brown and P. J. Jones, *J. Chem. Soc.* (*A*), 1967, 243.
[188] B. Frlec and H. H. Hyman, *Inorg. Chem.*, 1967, **6**, 2233.
[189] D. H. Brown, K. R. Dixon, C. M. Livingston, R. H. Nuttall, and D. W. A. Sharp, *J. Chem. Soc.* (*A*), 1967, 100.
[190] M. Debeau and M. Krauzman, *Compt. rend.*, 1967, **264**, *B*, 1724.
[191] D. M. Adams and D. M. Morris, *J. Chem. Soc* (*A*), 1967, 1669.

Table 18 (*cont.*)

Ion	ν_1	ν_2	ν_3	ν_4	ν_5	ν_6	*Lattice mode*	*Ref.*
K_2SnBr_6	190	144	224	118	109		78	190
K_2PbCl_6	292	222	283	144	153		60–85	190
$(NH_4)_2PbCl_6$[d]	296	220	278	154	157	[84]	90, 166	191
K_2SeCl_6[e]	299	255	280	140–160	165			192, 89
K_2TeCl_6[e]	301	253	243	139	150		98, 108	193
K_2TeCl_6[e]		250	~ 250	120–130	145			89
$(NH_4)_2TeCl_6$[e]	299	249	~ 250	120–130	143			89
K_2TeBr_6[e]	174	153	198		75	[76, 100]		193
K_2TiF_6			585	278, 312				189
$(NH_4)_2TiF_6$[f]	598				230			194
$(Et_2NH_2)_2TiCl_6$			325					184
$(Et_2NH_2)TiBr_6$			238, 258					184
$(NH_4)_2ZrF_6$[f]	581				228			194
$(Et_2NH_2)_2ZrCl_6$			295					184
$(pyH)_2ZrCl_6$			297					195
$(Et_2NH_2)_2ZrBr_6$			~ 200					184
$(NH_4)_2HfF_6$[f]	613				275			194
$(pyH)_2HfCl_6$			273					195
$(Et_3NH)_2VCl_6$			305, 355					184
$(N_2H_6)MoF_6$			540, 600					186
$(Et_2NH_2)_2MoCl_6$[g]			~ 300					184
$(N_2H_6)TcF_6$			545					186
K_2ReCl_6	346	(275)	313	172	159			196
K_2ReBr_6	213	(174)	217	118	104			196
K_2RuCl_6			346	188				197
$(N_2H_6)OsF_6$			550					186
K_2OsCl_6	345	(274)	313	177	165			196
K_2IrCl_6[h]	352	(225)	335	168 (184)	190			196
K_2PdCl_6	317	292	358	175	164			196
K_2PdCl_6	324	294			172			153
K_2PdBr_6	198	176	253	130	100			196

[e] It is suggested that ν_3 for octahedral 'inert-pair ions' may be anomalously broad and band 'half-widths' should be measured more often as a further piece of spectral information.

[f] In aqueous solution. The symmetry of the hexafluorides is lowered to D_{3d} in the solid state.

[g] ν_3 for this complex is *ca.* 30 cm.$^{-1}$ lower than for analogous alkali-metal salts of this anion. This effect is probably not due to the size of the ion so much as to hydrogen bonding between the cationic protons and anionic halogen atoms which may result in a weakening of the Mo—Cl bond.

---

192 T. Barrowcliffe, I. R. Beattie, P. Day, and K. Livingston, *J. Chem. Soc.* (*A*), 1967, 1810.
193 D. M. Adams and D. M. Morris, *J. Chem. Soc.* (*A*), 1967, 2067.
194 P. A. W. Dean and D. F. Evans, *J. Chem. Soc.* (*A*), 1967, 698.
195 R. J. H. Clark and W. Errington, *J. Chem. Soc.* (*A*), 1967, 258.
196 P. J. Hendra and P. J. D. Park, *Spectrochim. Acta*, 1967, **23**, *A*, 1635.
197 J. M. Fletcher, W. E. Gardner, A. C. Fox, and G. Topping, *J. Chem. Soc.* (*A*), 1967, 1038.

Table 18 (*cont.*)

Ion	ν_1	ν_2	ν_3	ν_4	ν_5	ν_6	*Lattice mode*	*Ref.*
K_2PtCl_6	348	318	342	183	171	[88]		190
$K_2PtCl_6{}^h$	344	320	345	183 (163)	162			196
$K_2PtCl_6{}^i$	350	320	341	186	171	[82]	186	198
K_2PtCl_6	349	320			171			153
K_2PtBr_6	207	190	240	90	97			196
K_2PtBr_6	217	195	243	78	115	100		198
K_2PtBr_6	215	191			110			153
K_2PtBr_6	215	191	241		111	[77, 85]		190
$(Me_4N)_2ThCl_6$			258					187
$(Me_4N)_2ThBr_6$			179					187
$(Me_4N)_2PaCl_6$			257					187
$(Me_4N)_2PaBr_6$			182					187
$(Me_4N)_2UCl_6$			259					187
$(Me_4N)_2UBr_6$			181					187
$(Me_4N)_2NpCl_6$			265					187

^{h} The frequencies in parentheses were published previously by Adams and Gebbie.

^{i} The authors examine $M_2{}^IPtCl_6$ complexes for a series of cations. The Pt—Cl bond-stretching force-constant is found to decrease regularly with increase in cationic size. Lattice modes due to cations also show a regular mass dependence. ν_4 for these complexes is always intense and sharp ($\Delta\nu_{\frac{1}{2}}$ *ca.* 10) but ν_3 varies and is at least twice as wide as ν_4. It is always asymmetrical on the low-frequency side, probably due to ^{37}Cl. For the complex $(Ph_2I)_2PtCl_6$, ν_6 has been observed directly and this is ascribed to a change of structure.

Table 19 *Vibrational frequencies* (cm.$^{-1}$) *for some* $MX_6{}^{3-}$ *ions*

Ion	ν_1	ν_2	ν_3	ν_4	ν_5	ν_6	*Lattice mode*	*Ref.*
$[Co(NH_3)_6]^{3+}InCl_6{}^{3-}$	277	193	250	157	(149)			192
K_3TlCl_6,H_2O	280	262	294	222, 246	155	136		110
$[Co(NH_3)_6]^{3+}TlCl_6{}^{3-}$	264	192	230	146	(135)			192
$Rb_3TlBr_6,8/7H_2O$	161	153	190, 195	134, 156	95	80	57	110
$[Co(NH_3)_6]^{3+}SbCl_6{}^{3-}$	267	214	178		(111)			192
Cs_3BiCl_6	259	222	175		(108)			192
$(EtNH_3)_3BiCl_6$	256	213						113
$(N_2H_5)_3BiCl_6$	260	215			100			113
$(N_2H_5)_3BiBr_6$	156	133			63			113

Cs_4PbCl_6 has ν_1 at 202 cm.$^{-1}$ and ν_3 at 140 cm.$^{-1}$.

Raman spectra of XeF_6 [199] in all three phases have been examined at different temperatures. The spectra are complex and the authors conclude that either ground-state vapour molecules possess a symmetry which is *lower* than octahedral or they have some very unusual electronic properties which markedly influence the region of the spectrum usually considered to be the vibrational–rotational region.

[198] D. M. Adams and D. M. Morris, *J. Chem. Soc.* (*A*), 1967, 1666.
[199] E. L. Gasner and H. H. Claassen, *Inorg. Chem.*, 1967, **6**, 1937.

The complete gas-phase i.r. and liquid Raman spectra have been interpreted for $ClSF_5$(C_{4v} symmetry) and CF_3SF_5.[200] The results indicate that $\nu_6(b_1)$ occurs at 396 and $\nu_{11}(e)$ at 270 cm.$^{-1}$ for the chloride which is the opposite of a previous assignment made by Woodward *et al.*

Mean amplitudes of vibration, shrinkage effects, and Coriolis constants have also been calculated for $ClSF_5$.[201]

$WClF_5$ has been characterised as a yellow solid, liquid, and vapour.[202] The principal i.r. bands of the solid and vapour occur at 743vs, 703vs, 667vs, 400vs, 302m, 278m, 254s, and 228m cm.$^{-1}$. The first three peaks show unusual splittings.

The appearance of two bands of approximately equal intensity in the region of the ν(Mo—O) stretching vibrations in the spectra of salts containing $[MoO_2F_4]^{2-}$ ions indicates non-linearity of the MoO_2 group.[203] The frequencies of these bands are ν_1 944, 893, and 914, and ν_3 894, 852, and 857 cm.$^{-1}$ in the spectra of $Cs_2Mo^{16}O_2F_4$, $Cs_2Mo^{18}O_2F_4$, and $Cs_2Mo^{16}O^{18}OF_4$ respectively.

7 Larger Molecules and Ions

The far-i.r. spectrum of methyl thiocyanate vapour (MeSCN) has been re-assigned (70–650 cm.$^{-1}$).[204]

The vibrational spectrum of germyl isothiocyanate (gas and solid phases) and germyl isothiocyanate d_3 (gas phase) has been recorded[205] and it is provisionally concluded that the molecule has a bent skeleton with a GeNC angle of *ca.* 156°.

The mean-square amplitudes of vibration of S_8 have been calculated.[206]

The Ga_2Cl_6 dimeric unit persists in the melt, in solution, and in the gas phase. Vibrational assignments for this molecule have been made with the aid of normal-co-ordinate analysis.[207]

The far-i.r. spectra (60–4000 cm.$^{-1}$) of the ions $[M_2X_6]^{2-}$ (M = Pd or Pt; X = Cl, Br, or I) have also been assigned.[208] A normal-co-ordinate analysis has been made for the in-plane i.r.-active modes and the bond stretching force-constants for the bridge bonds vary from 75–100% of the values calculated for terminal bonds.

[200] J. E. Griffiths, *Spectrochim. Acta*, 1967, 23, *A*, 2145.
[201] K. Venkateswarlu and S. Mariam, *Bull. Soc. roy. Sci. Liège*, 1967, **36**, 178; see also *Chem. Abs.*, 1967, **67**, 58,972*t*.
[202] G. W. Fraser (Miss), M. Mercer, and R. D. Peacock, *J. Chem. Soc.* (*A*), 1967, 1091.
[203] Y. A. Buslaev, Y. Y. Kharitonov, and R. L. Davidovich, *Izv. Akad. Nauk. S.S.S.R., Neorg. Materialy*, 1967, **3**, 589.
[204] G. A. Crowder, *J. Mol. Spectroscopy*, 1967, **23**, 108.
[205] G. Davidson, L. A. Woodward, K. M. Mackay, and P. Robinson, *Spectrochim. Acta*, 1967, 23, *A*, 2383.
[206] K. Venkateswarlu and K. B. Joseph, *Bull. Soc. roy. Sci. Liège*, 1967, **36**, 173; see also *Chem. Abs.*, 1967, **67**, 58,973*u*.
[207] I. R. Beattie, T. Gilson, and (in part) P. Cocking, *J. Chem. Soc.* (*A*), 1967, 702.
[208] D. M. Adams, P. J. Chandler, and R. G. Churchill, *J. Chem. Soc.* (*A*), 1967, 1272.

The i.r. spectra of potassium and ammonium pentafluoroxyuranate, $M_3^IUO_2F_5$, have been determined and force-constant values have been calculated.[209]

The vibrational spectra of eleven divalent metal pyrophosphates have been recorded and the spectra are compared with predictions based on a normal-co-ordinate analysis of the ion, assuming a D_{3h} symmetry.[210] Discrepancies between observed and calculated values, due to departure of the structure from D_{3h}, are discussed. A theoretical discussion of the normal vibrations of the $P_2O_7^{4-}$ ion with D_3, D_{3h}, and D_{3d} symmetries has also been published.[210a] A study of some isomorphic complex hydroxides of the general formula (Na, K)$_2$M^{IV}(OH)$_6$ has been reported.[211] It is shown that co-ordinated water molecules are absent and that the M—O bond has appreciable covalent character.

Maroni and Spiro[212] have examined the Raman spectra of solutions and the i.r. and Raman spectra of crystals containing polynuclear hydroxy complexes of Pb.^{II} For an OH : Pb ratio of 1·00, the complex present is $Pb_4(OH)_4^{4+}$; the four lead atoms are tetrahedrally situated and the OH groups lie on the faces of the tetrahedron. The fundamental modes occur at 404 and 130 (a_1), 455 and 84 (e), and 505 ?, *ca.* 340 and *ca.* 60 cm.$^{-1}$ (f_2). When the OH : Pb ratio is 1·33, the complex is probably $Pb_6(OH)_8^{4+}$ and the spectra may be interpreted in terms of an octahedral structure. The Raman spectrum of the solid has bands at 68, *ca.* 90, 144, 386, and 455 cm.$^{-1}$, while the i.r. spectrum of the solid shows a further absorption at 365 cm.$^{-1}$. The structure is probably similar to $Mo_6Cl_8^{6+}$ but it is not possible to exclude other less symmetrical structures.

Two groups of workers have examined the vibrational spectra of molybdenum cluster compounds $[(Mo_6Cl_8)X_6]^{2-}$. Hartley and Ware[213] obtained the Raman spectra of $H_2[(Mo_6Cl_8)X_6],8H_2O$ in methanol solution and the i.r. spectra of the solids. They describe the results of a preliminary normal-co-ordinate analysis based on their data and point out some of the unusual vibrational properties of these rigid heavy-atom clusters. They avoid the concept of group frequencies with the exception of very low Mo—X deformation modes.

In contrast, Cotton and co-workers[214] give detailed group frequency assignments, *i.e.* ν(Mo—Cl) 290–350 cm.$^{-1}$ for $Mo_6Cl_8^{4+}$. For $Mo_6Cl_8Y_6^{2-}$ (Y = Cl, Br, or I), ν(Mo—Mo) is at 220, 232, and 233 cm.$^{-1}$ respectively. The authors also make some more tentative assignments for $Mo_6Br_8Y_6^{2-}$ (Y = Cl, Br, I) and for $W_6Cl_4^{2-}$ [ν(W—Cl) 290–320 cm.$^{-1}$] and they conclude that the force constants and hence the metal–metal bonding in

209 H. Brusset, H. Gillier-Pandraud, and N. Q. Dao, *Compt. rend.*, 1967, **265**, *C*, 1209.
210 A. Hezel and S. D. Ross, *Spectrochim. Acta*, 1967, **23**, *A*, 1583.
210a R. W. Mooney, S. Z. Toma, and J. Brunvoll, *Spectrochim. Acta*, 1967, **23**, *A*, 1541.
211 M. Maltese and W. J. Orville-Thomas, *J. Inorg. Nuclear Chem.*, 1967, **29**, 2533.
212 V. A. Maroni and T. G. Spiro, *J. Amer. Chem. Soc.*, 1967, **89**, 45.
213 D. Hartley and M. J. Ware, *Chem. Comm.*, 1967, 912.
214 F. A. Cotton, R. M. Wing, and R. A. Zimmerman, *Inorg. Chem.*, 1967, **6**, 11.

$Mo_6X_8Y_6^{2-}$ species are not appreciably influenced by changes in the nature of Y.

A vibrational analysis of the Nb_6Cl_{12} cluster has also been made [215] and assignments have been suggested in terms of terminal and bridging chlorine stretching modes and motions involving primarily the Nb_6 octahedron.

PART II: Characteristic Frequencies in Inorganic Complexes

Inorganic complexes frequently show vibrational absorption bands or Raman shifts which are characteristic of the groups in the complex. It is not suggested that these characteristic frequencies are group frequencies in the sense that only the group in question contributes to the vibration. In compounds of high symmetry the true vibrational modes often incorporate motions of several atoms besides those within the group itself. In this sense the frequencies which characterise the molecule are not pure group frequencies, but, as a matter of experience, typical groupings are often associated with absorptions in a characteristic region of the spectrum. They are, therefore, useful diagnostically, though, in the absence of a full normal-co-ordinate analysis, care should be exercised in any attempt to interpret variations in characteristic frequencies in terms of variations in the bonding, *e.g.* ionic–covalent, σ–π, etc.

In the sections which follow the data are arranged according to the sequence of elements within each vertical Group of the Periodic Table, the main groups being considered first and then the transition-element groups. Within each Group the metal–ligand vibrations are treated systematically according to the donor atom in the ligand, the sequence again following vertical grouping in the Periodic Table.

Characteristic vibrations of the ligands themselves, *e.g.* ν(C—O), ν(C—N), $\nu(N_2)$, etc., and the more complicated spectra of co-ordinated organic ligands, are discussed in Part III. Compounds for which partial or unassigned vibrational data have been published are listed in tabular form in Part IV.

1 Main-Group Elements

A. Group II Elements.—Coates and Tranch[216] have prepared phenylberyllium hydride–trimethylamine (1) and they report two bands in the i.r. spectrum of the mull (1342 and 1316 cm.$^{-1}$) which are characteristic of compounds containing the BeH_2Be bridge. The corresponding bands for the trimethylphosphine complex $[PhBeH,PMe_3]_2$ occur at 1356 and 1312 cm.$^{-1}$.

Data on beryllium–borohydride anions are given in the following section on compounds containing B—H bonds.

[215] R. A. Mackay and R. F. Schneider, *Inorg. Chem.*, 1967, **6**, 549.
[216] G. E. Coates and M. Tranah, *J. Chem. Soc.* (*A*), 1967, 615.

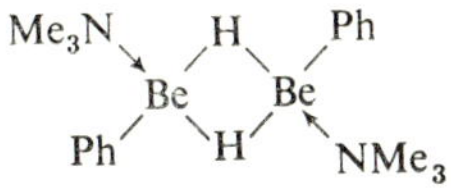

(1)

The vibrational spectra of di-t-butylberyllium are consistent with a monomeric structure having a linear C—Be—C skeleton.[217] This provides one of the rare examples of beryllium exhibiting two-co-ordination at room temperature; ν(Be—C) occur at 450 and 545 cm.$^{-1}$.

Both tetrakis(dimethylformamide)beryllium(II) and acetylacetonatobis-(dimethylformamide)beryllium(II) in dimethylformamide (DMF) exhibit an intense, highly polarised Raman line at 478 cm.$^{-1}$ which is assigned as ν_{sym}(B—O).[217a]

B. Group III Elements.—*Compounds Containing* B—H *Bonds.* Malone and Parry[218] have assigned the boranocarbonate anion (2) and compared their results with the spectra of K_2CO_3 and $Na[H_3CCO_2]$. For (2) ν(B—H) *ca.* 2200, ν_{asym}(C—O) *ca.* 1400, ν_{sym}(C—O) *ca.* 1218, and δ(B—H) *ca.* 1193, 1150 cm.$^{-1}$.

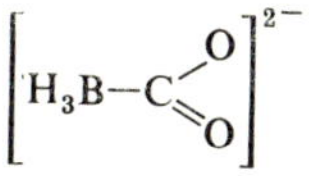

(2)

The following assignments have been made for tetrafluorodiphosphine–borane (P_2F_4,BH_3):[219]

ν (cm.$^{-1}$)		ν (cm.$^{-1}$)	
2432	ν(B—H)	727	$\rho(BH_3)$
1102	δ(B—H)	598	ν(P—B)
1042	δ(B—H)	442	δ(F—PF)
902	ν_{asym}(P—F)	395	τ(B—P)
850	ν_{sym}(P—F)	370	δ(F—P—F)

A band at 670–680 cm.$^{-1}$ was left unassigned. The i.r. and Raman spectra of an ionic phosphine borane, $Na^+(BH_3 \cdot PH_2 \cdot BH_3)^-$, have been measured in the solid state and in solution in a variety of solvents.[220] The frequencies are assigned on the basis of a C_{2v} model for the anion. A partial assignment of diborane(4)-bis(trifluorophosphine) has been carried out.[221] $\nu(BH_2)$ for B_2H_4,$(PF_3)_2$ occurs at 2403 and 2353 cm.$^{-1}$, BH_2 was 1120 and ν(B—P) 612 cm.$^{-1}$.

217 G. E. Coates, P. D. Roberts, and A. J. Downs, *J. Chem. Soc.* (*A*), 1967, 1085.
217a N. A. Matwiyoff and W. G. Movius, *J. Amer. Chem. Soc.*, 1967, **89**, 6077.
218 L. J. Malone and R. W. Parry, *Inorg. Chem.*, 1967, **6**, 817.
219 K. W. Morse and R. W. Parry, *J. Amer. Chem. Soc.*, 1967, **89**, 172.
220 R. E. Hester and E. Mayer, *Spectrochim. Acta*, 1967, 23, *A*, 2218.
221 W. R. Deever and D. M. Ritter, *J. Amer. Chem. Soc.*, 1967, **89**, 5073.

Lane and Burg[222] have assigned the unstable arsinoboron heterocycle $[(CF_3)_2AsBH_2]_3$ by comparison with $[(CF_3)_2PBH_2]_3$ and $[(CF_3)_2PDB_2]_3$.

The spectra of three triple borohydrides of aluminium and beryllium have been examined:[223]

	ν(B—H)term. (cm.$^{-1}$)	ν(M—H—B)bridging (cm.$^{-1}$)
$[(C_8H_{17})_3NC_3H_7][Al(BH_4)_4]$	2469, 2404	2151
$[(C_8H_{17})_3NC_3H_7][Be(BH_4)_3]$	2440, 2400	2238, 2162
$[(C_8H_{17})_3NC_3H_7][Be_2(BH_4)_5]$	2440, 2395	2240, 2180

ν(B—H) term. for $Ph_3P,Al(BH_4)_3$[224] occur at 2505 and 2440; ν(Al—H—B) bridging occurs at 2132 cm.$^{-1}$. The compounds $Zr(BH_4)_4$ and $Hf(BH_4)_4$ give similar results [(νB—H)term. 2400–2600, ν(B—H)bridging 2100–2200 cm.$^{-1}$].[225] The spectrum of $Al(BH_4)_3,6NH_3$, however, shows bands which are characteristic of the BH_4^- anion (2260 and 1100 cm.$^{-1}$) and the structure formulated as $[Al(NH_3)_6]^{3+}[BH_4]_3^-$.[224]

Octahydrotriboratetracarbonyl metallates of Cr, Mo, and W,[226] $[(CO)_4MB_3H_8]^-$, exhibit ν(B—H) absorptions at 2489, 2435, 2134, and 2109 cm.$^{-1}$; these are compatible with the known crystal structure of the ion (3):

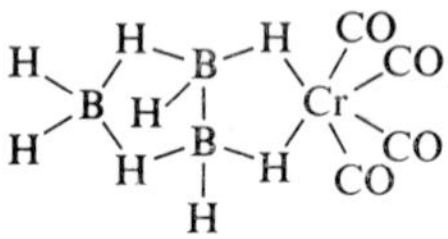

(3)

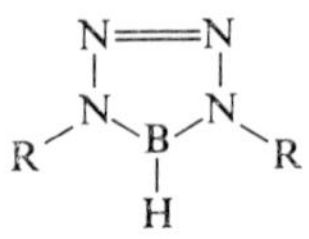

(4) (R = Me, Ph)

The effects of solvent on the B—H stretching frequency of some substituted cyclotetrazenoboranes, (4) have been studied.[227] ν(B—H) increases with increasing polarity of the solvent. No correlation was observed with the dielectric relationship $(\varepsilon\text{-}1)/(2\varepsilon+1)$ and non-linear relative-shift plots are obtained against ν(N—H), ν(C=O), etc. The results suggest that the solvent interactions occur with parts of the molecule remote from the B—H bond.

Burg and Sandhu[228] examined the i.r. effects of axial steric interference in $(Me_2NBH_2)_3$ by comparing its spectrum with that of the sterically free trimer $(Me_2PBH_2)_3$. Steric interference among the axial methyl groups of $(Me_2NBH_2)_3$ has an interesting effect on the vibrational modes. The sterically free trimer exhibits two ν(C—H) whereas the hindered trimer

[222] A. P. Lane and A. B. Burg, *J. Amer. Chem. Soc.*, 1967, **89**, 1040.
[223] H. Nöth and M. Ehemann, *Chem. Comm.*, 1967, 685.
[224] P. H. Bird and M. G. H. Wallbridge, *J. Chem. Soc.* (*A*), 1967, 664.
[225] V. V. Volkov. E. V. Sobolev, E. E. Zaev, and O. I. Shapiro, *Zhur. strukt. Khim.*, 1967, **8**, 461.
[226] F. Klanberg and L. J. Guggenberger, *Chem. Comm.*, 1967, 1293.
[227] J. B. Leach and J. H. Morris, *J. Chem. Soc.* (*A*), 1967, 1590.
[228] A. B. Burg and J. S. Sandhu, *J. Amer. Chem. Soc.*, 1967, **89**, 1626.

shows seven ν(C—H). Axial B—H groups are also affected since three bands are seen for $(Me_2NBH_2)_3$, cf. two ν(B—H) for $(Me_2PBH_2)_3$.

The octahydrotriboratobis(triphenylphosphine)copper(I) complex (5) exhibits characteristic terminal and bridging B—H stretching frequencies [ν(B—H)term. 2495, 2462, and 2418; ν(B—H)bridging 2035–2100 cm.$^{-1}$]; the structure (5) is proposed.[229]

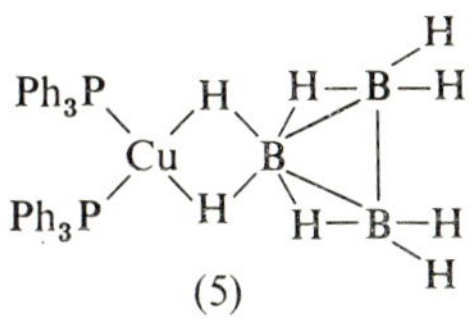

(5)

μ-Trimethylsilyl-pentaborane(9) is the first example of a compound containing a B—Si—B three-centre bond.[230] Gaseous μ-$SiMe_3B_5H_8$ shows major absorptions at 2960 and 2905 [ν(C—H)], 2600 and 2570 [ν(B—H) term.], and 1820 [ν(BHB) bridging]. A broad band at *ca.* 1410 cm.$^{-1}$ is present in B_5H_9 and all its derivatives. Watanabe *et al.*[231] have revealed a correlation between the boron–hydrogen stretching frequency and the boron–hydrogen n.m.r. coupling constant. They obtained a linear plot of ν(B—H) *vs.* $J(^{11}B—H)$ for twenty compounds.

Low *et al.*[232] have studied diborane–trimethylboron equilibria and have attempted to characterise the various compounds isolated from equilibrium mixtures by comparing their i.r. spectra with those reported previously. Discrepancies were noted and these authors have shown that large changes in the spectra occur over a period of time. For example, the band near 1600 cm.$^{-1}$, ascribed to the ν(B—H) bridging mode, decreases significantly in intensity, indicating a disruption of the diborane skeleton.

Trefimenko [233, 234] has assigned ν(B—H) for pyrazaboles and some poly-(1-pyrazolyl)borates(6) with a wide range of metals (M), and Grimes [235] has studied some methyl derivatives of 1,2-dicarbaclovopentaborane(5).

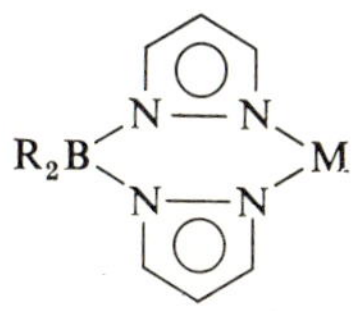

(6)

[229] S. J. Lippard and D. Ucks, *Chem. Comm.*, 1967, 983.
[230] D. F. Gaines and T. V. Iorns, *J. Amer. Chem. Soc.*, 1967, **89**, 4249.
[231] H. Watanabe and K. Nagasawa, *Inorg. Chem.*, 1967, **6**, 1068.
[232] M. J. D. Low, R. Epstein, and A. C. Bond, *Chem. Comm.*, 1967, 226.
[233] S. Trofimenko, *J. Amer. Chem. Soc.*, 1967, **89**, 3165.
[234] S. Trofimenko, *J. Amer. Chem. Soc.*, 1967, **89**, 3170.
[235] R. N. Grimes, *J. Organometallic Chem.*, 1967, **8**, 45.

Boron–hydrogen stretching frequencies (and other frequencies) have been assigned for the following additional compounds in Table 20.

Table 20 *References to further papers containing data on* ν(B—H)

Compound	*Ref.*	*Compound*	*Ref.*
$C_6H_{13}BH(NC_5H_5)_2$	236	$(Me_4N)_2B_{20}H_{24}Cd$	239
$PhNHCONHMe,BH_3$	237	$P_3Ph_6B_3H_{6-n}Br_n$ (n = 1–6)	240
$PhNHCONHEt,BH_3$	237	$P_3Ph_6B_3H_{6-n}I_n$ (n = 1–3)	240
$PhNHCONEt_2,BH_3$	237	$H_3N_3B_3H_2OMe$	241
$PhNHCONHBu^t,BH_3$	237	$H_3N_3B_3H_2OEt$	241
$EtNHCONHBu^t,BH_3$	237	$ClH_2B_3N_3H_3$	242
$[Zn(NH_3)_{2\cdot 76}(H_2O)_{1\cdot 24}]B_7H_7$	238	$Cl_2HB_3N_3H_3$	242
$Cs_2B_7H_7$	238	$Cl_2HB_3N_3Me_3$	242
$Rb_2B_8H_8$	238	$ClH_2B_3N_3Me_3$	242
$CdB_{10}H_{12},2Et_2O$	239	$Me_2NH_2B_3N_3H_3$	242
$CdB_{10}H_{12},2THF$	239	$(Me_2N)_2HB_3N_3H_3$	242

Compounds Containing Al—H *Bonds.* Halogen-substituted aluminium hydride adducts have been studied by Schmidt and Flagg[243] who have characterised ν(Al—H), ν(Al—D), δ(Al—H) and δ(Al—D) for the following tetrahydrofuran (THF) complexes: AlH_2Cl,2THF, AlH_2Br,2THF, AlH_2I,2THF, AlD_2Cl,2THF, AlD_2I,2THF, $AlDCl_2$,2THF, $AlHCl_2$,2THF, $AlHBr_2$,2THF, $AlHI_2$,2THF, and AlHClBr,2THF. Bands are observed in the general ranges: ν(Al—H) 1700–1900, ν(Al—D) 1200–1400, δ(Al—H) 700–850, and δ(Al—D) 500–650 cm.$^{-1}$.

Compounds Containing M—C *Bonds* (M = Al, Ga, In, *and* Tl). Four bands have been assigned to *X*-sensitive Al—C modes in the dimeric molecule triphenylaluminium:[244] 332, 420, 446, and 680 cm.$^{-1}$.

The ν(Ga—C) mode has been assigned to 538 cm.$^{-1}$ for $(Me_2GaF)_3$ and 525 for $(Et_2GaF)_3$.[245]

Clark and Pickard[246] have characterised In—C stretching modes for the Me_2In group in Me_2InX derivatives (X = halide, $C_5H_7O_2$, oxinate, etc.) ν_{sym}(In—C) = 490 cm.$^{-1}$ and ν_{asym}(In—C) = 500–550 cm.$^{-1}$.

The following bands in the range 450–550 cm.$^{-1}$ have been assigned to the ν(Tl—C) mode:[247] Me_2TlCl, 550; Et_2TlCl, 447, 511; MeEtTlCl, 462, 530; MePhTlCl, 513; and EtPhTlCl, 485 cm.$^{-1}$.

[236] J. E. Douglass, G. R. Roehrig, and O.-H. Ma, *J. Organometallic Chem.*, 1967, **8**, 421.
[237] R. H. Cragg and N. N. Greenwood, *J. Chem. Soc.* (*A*), 1967, 961.
[238] F. Klanberg, D. R. Eaton, L. J. Guggenberger, and E. L. Muetterties, *Inorg. Chem.*, 1967, **6**, 1271.
[239] N. N. Greenwood and N. F. Travers, *J. Chem. Soc.* (*A*), 1967, 880.
[240] W. Gee, J. B. Holden, R. A. Shaw, and B. C. Smith, *J. Chem. Soc.* (*A*), 1967, 1545.
[241] M. Nadler and R. F. Porter, *Inorg. Chem.*, 1967, **6**, 1739.
[242] R. Maruca, O. T. Beachley, jun., and A. W. Laubengayer, *Inorg. Chem.*, 1967, **6**, 575.
[243] D. L. Schmidt and E. E. Flagg, *Inorg. Chem.*, 1967, **6**, 1262.
[244] H. F. Shurvell, *Spectrochim. Acta*, 1967, **23**, *A*, 2925.
[245] H. Schmidbaur, H. F. Klein, and K. Eiglmeier, *Angew. Chem. Internat. Edn.*, 1967, **8**, 806.
[246] H. C. Clark and A. L. Pickard, *J. Organometallic Chem.*, 1967, **8**, 427.
[247] M. Tanaka, H. Kurosawa, and R. Okawara, *Inorg. Nuclear Chem. Letters*, 1967, **3**, 565.

Compounds Containing B—N *or* Al—N *Bonds.* Several monosubstituted triethylamine–boranes have been prepared and ν(B—N) assigned;[248] Et_3N,BH_2H, 645; Et_3N,BH_2Cl, 651; Et_3N,BH_2Br, 651; Et_3N,BH_2I, 650; Et_3N,BH_2Ph, 628 cm.$^{-1}$.

Greenwood and Robinson[249] have studied the site of co-ordination in boron trihalide adducts of several ureas and thioureas. Both spectroscopic and chemical evidence favour co-ordination to boron *via* the nitrogen atom rather than the oxygen.

Assignments for ν(Al—N) have been suggested for the following complexes:[250] Me_3N,Et_2AlCl, 303; Me_3N,Et_2AlBr, 306; Me_3N,Et_2AlI, 312; Me_3N,Et_3Al, 306; py,Et_2AlCl, 302; py,Et_2AlBr, 292; and py,Et_3Al, 309 cm.$^{-1}$.

Compounds Containing B—O, Al—O, *or* B—S *Bonds.* The ^{11}B—O ring mode for $(ROBX_2)_3$ (X = F or Cl) is independent of R but markedly dependent on X as shown in Table 21.[251]

Table 21 ν(^{11}B-O) (cm.$^{-1}$) *for several boroxoles*[251]

Compound	ν(^{11}B—O)	*Compound*	ν(^{11}B—O)
$(MeOBF_2)_3$	1253	$(MeOBCl_2)_3$	1339
$(EtOBF_2)_3$	1255	$(EtOBCl_2)_3$	1332
$(PrOBF_2)_3$	1256	$(PrOBCl_2)_3$	1330
$(BuOBF_2)_3$	1256	$(BuOBCl_2)_3$	1335

The i.r. spectra of inorganic aluminates show characteristic vibrational frequencies of 'AlO_4' tetrahedra and 'AlO_6' octahedra:[251a]

Condensed $A10_4$ tetrahedra	700–900 (cm.$^{-1}$)
Isolated $A10_4$ tetrahedra	650–800
Condensed $A10_6$ octahedra	500–680
Isolated $A10_6$ octahedra	400–530

Hurley *et al.*[252] confirm that ν(Al—O—Al) bridging vibrations occur as a strong absorption in the region 750–820 cm.$^{-1}$.

The spectra of twenty organic thioboranes have been examined,[252a] including members of the new classes $R_2NB(SR^1)_2$, $(R_2N)_2BSR^1$, $(R_2N)BrBSR^1$, and $(R_2N)ClBSR^1$. The recognition of a triplet absorption centred at 930 cm.$^{-1}$ is especially significant as diagnostic for their identification, *i.e.* ν(B—S) of three-co-ordinate boron compounds.

[248] J. N. G. Faulks, N. N. Greenwood, and J. H. Morris, *J. Inorg. Nuclear Chem.*, 1967, **29**, 329.
[249] N. N. Greenwood and B. H. Robinson, *J. Chem. Soc.* (*A*), 1967, 511.
[250] C. A. Smith and M. G. H. Wallbridge, *J. Chem. Soc.* (*A*), 1967, 7.
[251] W. Gerrard, E. F. Mooney, and W. G. Peterson, *J. Inorg. Nuclear Chem.*, 1967, **29**, 943.
[251a] P. Tarte, *Spectrochim. Acta*, 1967, **23**, *A*, 2127.
[252] T. J. Hurley, M. A. Robinson, J. A. Scruggs, and S. I. Trotz, *Inorg. Chem.*, 1967, **6**, 1310.
[252a] R. H. Cragg, M. F. Lappert, and B. P. Tilley, *J. Chem. Soc.* (*A*), 1967, 947.

Compounds Containing M—*Halogen Bonds* (M = B, Al, Ga, In, *and* Tl). Boron–halogen vibrations have been assigned for numerous compounds in the far-i.r. region (see Table 22) and it is concluded that ν_{sym}(B—X) occurs

Table 22 *Boron–halogen vibrations* (cm.$^{-1}$)

Compound	ν_{sym}	δ_{asym}	δ_{sym}	τ
$PhBCl_2$	[a]	330	230	127
$PhBBr_2$	[a]	282, 271	161	<80
p-$ClC_6H_4BCl_2$	[a]	342	209	125
$Pr_4{}^nN,BBr_4$	242(ν_1)	163(ν_4)	117	
Et_3N,BF_3	[a]	305	342	130
Et_3N,BCl_3	297	272, 222	252	<80
Et_3N,BBr_3	245	190, 128	176	<80
py,BF_3	[a]	300	334	188
py,BCl_3	330	218, 286, 208	246	<80
py,BBr_3	270	232, 125	181	<80

[a] >450 cm.$^{-1}$.

at considerably lower wave-numbers in compounds where the co-ordination numbers of boron is four rather than three.[253] The torsonal vibrations (τ) about A—BX_n bonds are observed above 80 cm.$^{-1}$ only when the mass of X is low enough (F), or where there is an appreciable high energy barrier to rotation, *i.e.* systems where the A—BX_n bond has considerable π-character as in arylboron dichlorides.

The boron–halogen stretching modes for alkoxyboron difluorides and dichlorides,[251] for Me_2NBCl_2,2py,[254] and for BCl_3,3($NH_2C_6H_4 \cdot CO_2H$) [255] have also been assigned.

Addition compounds of $AlCl_3$ and $AlBr_3$ with Me_2O and Et_2O have been examined.[256] Both 1 : 1 and 1 : 2 complexes result and the i.r. spectra at room temperature and at −180° show no change as the phase changes. The 1 : 1 complexes are assigned on the basis of an L → AlX_3 monomeric structure.

The spectra of the compounds $(Me_2GaF)_3$ and $(Et_2GaF)_3$ have been partially assigned [245] and ν(Ga—F) occurs at 420 and 425 cm.$^{-1}$ in the two compounds. The far-i.r. spectra of some Ga^{III} halide complexes with triarylphosphines and diphosphines have been investigated.[257] The chloride complexes show two strong bands in the region 340–400 cm.$^{-1}$. Charac-

253 P. G. Davies, M. Goldstein, and H. A. Willis, *Inorg. Nuclear Chem. Letters*, 1967, **3**, 249.
254 M. F. Lappert and G. Srivastava, *J. Chem. Soc.* (*A*), 1967, 602.
255 S. Prasad and N. P. Singh, *Z. anorg. Chem.*, 1967, **350**, 332.
256 J. Le Calvé, J. Deroualt, M.-T. Forel, and J. Lascombe, *Compt. rend.*, 1967, 264, *B*, 448.
257 A. J. Carty, *Canad. J. Chem.*, 1967, **45**, 3187.

teristic frequencies for these complexes are ν_{sym}(Ga—Cl) 348 ± 2 and ν_{asym}(Ga—Cl) 383 ± 3 cm.$^{-1}$. The corresponding values for the bromide and iodide complexes are: ν_{sym}(Ga—Br) 235 ± 1, ν_{asym}(Ga—Br) 286 ± 4, and ν_{asym}(Ga—I) 241 ± 4 cm.$^{-1}$. ν_{sym}(Ga—I) is expected to occur at *ca.* 160 cm.$^{-1}$ [see Greenwood *et al., J. Chem. Soc.* (*A*), 1966, 699] and this region was not investigated.

The Raman spectrum of $InCl_3,2Et_2O$, recorded just above its melting point, indicates that both molecules of ether are co-ordinated to indium and that the $InCl_3$ unit is planar as in (7).[258] Assignments were ν_{sym}(In—Cl)

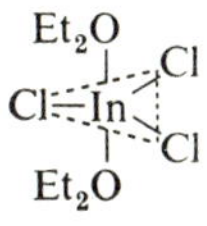

(7)

317, ν_{asym}(In—Cl) 351, and $\delta(InCl_3)$ 105 cm.$^{-1}$. Walton[259] has examined the far-i.r. spectra of $InCl_3$ with bidentate and tridentate nitrogen donors, where the structure may be *cis*- or *trans*-octahedral (C_{3v} or C_{2v}). ν(In—Cl) increases markedly on changing from *cis*- to *trans*-$InCl_3L_3$, *e.g.* for $InCl_3$,3py, ν(In—Cl) 274, 239 (*cis*-structure) and for $InCl_3$,terpy, ν(In—Cl) 314, 305, 270, and 247 cm.$^{-1}$ (*trans*-structure).

The far-i.r. spectra of halogenobis(pentafluorophenyl)thallium(III) compounds $(R_F)_2TlX$ (X = F, Cl, or Br) are consistent with dimeric (halogen bridged) structures in the solid state:[260]

	$(R_F)_2$Tl—F	—Cl	—Br
ν(Tl—X) (cm.$^{-1}$)	320	215	151
δ(Tl—X) (cm.$^{-1}$)	165	130	74

When X = Cl or Br the compounds are monomeric in acetone but dimeric in benzene. This is in contrast to $R_2Tl^+X^-$ complexes (R = alkyl or aryl) which are ionic and probably linear.

C. Group IV Elements.—The complex hexamethylcyclohexadienylrhenium tricarbonyl (8) exhibits an anomalously low C—H stretching frequency (2790 cm.$^{-1}$) and an *X*-ray crystal structure determination confirms that the vibration involves an *exo*-hydrogen atom.[261]

Compounds Containing M—H *Bonds* (M = Si, Ge, *and* Sn). It is not appropriate to discuss all organo-silicon compounds containing Si—H bonds, but data on some compounds of more inorganic interest are summarized in Table 23.

[258] M. J. Taylor, *J. Chem. Soc.* (*A*), 1967, 1462.
[259] R. A. Walton, *J. Chem. Soc.* (*A*), 1967, 1485.
[260] G. B. Deacon, J. H. S. Green, and W. Kynaston, *J. Chem. Soc.* (*A*), 1967, 158.
[261] P. H. Bird and M. R. Churchill, *Chem. Comm.*, 1967, 777.

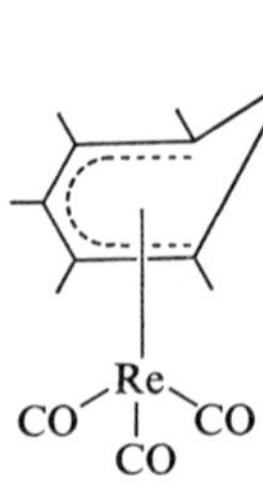

(8)

Table 23 *Some compounds containing* Si–H *bonds; frequencies in* cm.$^{-1}$

Compound	ν(Si—H)	*Ref.*	*Compound*	ν(Si—H)	*Ref.*
$SiH_3I,2py$	2206, 2218	262	$C_4H_8NSiH_3$	2175	263
$MeSiH_2I,2py$	2160, 2190	262	$C_5H_{10}NSiH_3$	2190	263
SiH_3Br,Me_3N	2185, 2220	262	$Ph_4C_4Si(H)Ph$	2120	264
SiH_3I,Me_3N	2181, 2225	262	$Ph_4C_4(H)Me$	2122	264
$MeSiH_2I,Me_3N$	2200, 2218	262			

The Ge—H stretching vibration has been assigned at 2056 cm.$^{-1}$ in $Ph_4C_4Ge(H)Ph$ and at 2060 cm.$^{-1}$ in $Ph_4C_4GeH_2$.[264] The compounds $HC\vdots CGeH_3$ and $HC\vdots CGeD_3$ have been fully assigned using C_{3v} symmetry.[265]

Kawakami *et al.*[266] have carried out some n.m.r. and i.r. spectral studies of halogenodialkyltin hydrides, R_2SnHX (R = Me, Et, Pr^n, Pr^i, and Bu^n; X = H, Cl, Br, and I). They assign ν(Sn—H) and compare the results with values for $J(^{117}Sn—H)$ and $J(^{119}Sn—H)$. Both coupling-constant values and ν(Sn—H) increase in the order X = H < I < Br < Cl.

Compounds Containing M—C *Bonds* (M = Ge, Sn, *and* Pb). Assignments for ν(C—C) and ν(Si—C) fall outside the scope of this review. Values for ν(Ge—C), ν(Sn—C), and ν(Pb—C) are summarized in Table 24 and call for no further comment.

Table 24 *Compounds containing* M—C *bonds* (M = Ge, Sn, *and* Pb)*; frequencies in* cm.$^{-1}$

Compound	ν(M—H)	*Ref.*
Me_4Ge	602	267
Me_6Ge_2	590, 552	267
Et_4Ge	570	267
Et_6Ge_2	565, 528	267
$Pr^n{}_4Ge$	639, 567, 553	267

262 H. J. Campbell-Ferguson and E. A. V. Ebsworth, *J. Chem. Soc.* (*A*), 1967, 705.
263 B. J. Aylett and J. Emsley, *J. Chem. Soc.* (*A*), 1967, 1918.
264 M. D. Curtis, *J. Amer. Chem. Soc.*, 1967, **89**, 4241.
265 R. W. Lovejoy and D. R. Baker, *J. Chem. Phys.*, 1967, **46**, 658.
266 K. Kawakami, T. Saits, and R. Okawara, *J. Organometallic Chem.*, 1967, **8**, 377.
267 F. Glockling and J. R. C. Light, *J. Chem. Soc* (*A*), 1967, 623.

Table 24 (*cont.*)

Compound	ν(M—H)		*Ref.*
Pr^i_4Ge	559, 549		267
$Pr^i_6Ge_2$	543, 536, 505		267
Bu^n_4Ge	641, 556		267
Bu^i_4Ge	647, 641		267
$Bu^i_6Ge_2$	639, 610		
$Bu_2Ge(SR)_2$ / $Bu_3Ge(SR)$	R = Et, Pr^n, Pr^i, Bu, Ph, etc.	690 (ν_{asym}Ge—C); 655 (ν_{sym}Ge—C)	268
$Me_2Sn\cdot SO_4$	600		269[a]
$Me_2Sn\cdot CO_3$	576, 523		269[a]
$Me_3Sn\cdot OCOH$	555, 513		270
$py\cdot Me_3Sn\cdot OCOH$	543		270
$(Me_3Pb)_2SO_4$ / $Me_3Pb\cdot(O_3SMe)$ / $Me_3Pb\cdot(O_2SMe)$	490–500 ($\nu_{asym}PbC_3$)		271

[a] The presence of a single ν(Sn—C) for Me_2SnSO_4 is consistent with a linear geometry for Me_2Sn^{2+}.

Compounds Containing M—M *or* M—M′ *Bonds* (M, M′ = Si, Ge, *and* Sn). Table 25 summarizes data concerning vibrations between heavy elements of Group IV.

Table 25 *Vibrations between heavy elements of Group* IV; *frequencies in* cm.$^{-1}$

Compound	ν	*Ref.*
$Me_3Si—SnMe_3$	ν(Si—Sn) 322	272
$Me_3Ge—SnMe_3$	ν(Ge—Sn) 240	273
$Bu_3Ge—SnPh_3$	ν(Ge—Sn) 235	273
$Ph_3Ge—SnMe_3$	ν(Ge—Sn) 225	273
$Ph_3Ge—SnEt_3$	ν(Ge—Sn) 230	273
$Me_3Sn—SnPh_3$	ν(Sn—Sn) 194	273
$Et_3Sn—SnPh_3$	ν(Sn—Sn) 208	273
$Et_3Sn—SnBu_3$	ν(Sn—Sn) 199	273

Compounds Containing M—N *Bonds* (M = Si, Ge, *and* Sn). Frequencies ascribed to Si—N vibrations are in Table 26, from which it can be seen that the values vary between 300–940 cm.$^{-1}$, depending markedly on the co-ordination number of silicon and the type of bonding involved.

[268] R. C. Mehrotra, V. D. Gupta, and (Miss) D. Sukhani, *J. Organometallic Chem.*, 1967, **9**, 263.
[269] H. C. Clark and R. G. Goel, *J. Organometallic Chem.*, 1967, **7**, 263.
[270] P. B. Simons and W. A. G. Graham, *J. Organometallic Chem.*, 1967, **8**, 479.
[271] U. Stahlberg, R. Gelius, and R. Müller, *Z. anorg. Chem.*, 1967, **355**, 230.
[272] H. Schumann and S. Ronecker, *Z. Naturforsch*, 1967, **22**, *B*, 452.
[273] N. A. D. Carey and H. C. Clark, *Chem. Comm.*, 1967, 292.

Table 26 *Silicon–nitrogen vibrations* (cm.$^{-1}$)

Compound	ν(Si—N)	*Ref.*
SiH_3Br,Me_3N	510	262
SiH_3I,Me_3N	542	262
SiD_3I,Me_3N	540	262
$MeSiH_2I,Me_3N$	581	262
Me_3SiI,Me_3N	573	262
H_2SiCl_2,2py	316 (νSi—py)	262
H_3SiI,2py	299, 308	262
$MeSiH_2I$,2py	270, 301	262
$Ar_2C{:}NSiR_3$ (for 28 compounds)	~ 900	274
H_3SiNMe_2	670	275
$H_3SiNC_4H_8$	695	263
$H_3SiNC_5H_{10}$	705	263
$Si(NMe_2)_4$	570 sym.; 710 asym.	276
$[Me_2SiNH]_3$	}	
$[Me_2SiNH]_4$	} 930–940 [ν(N—Si—N)]	277
$[MeEtSiNH]_3$	}	
$[MeEtSiNH]_4$	}	
$\overline{NMeSiMe_2NMeSiMe_2SiMe_2}$ (cyclic)	~ 930 [ν(N—Si—N)]	278

Comparative figures to those of $Si(NMe_2)_4$ in Table 26 and force-constant values are also available for the germanium and tin homologues:[276]

$Ge(NMe_2)_4$	$\nu_{sym}(GeN_4)$ 551;	$\nu_{asym}(GeN_4)$ 578 cm.$^{-1}$
$Sn(NMe)_4$	$\nu_{sym}(SnN_4)$ 516;	$\nu_{asym}(GeN_4)$ 535 cm.$^{-1}$

The following values of ν(Sn—N) have been given for a series of six-co-ordinate adducts of the tin(IV) halides:[279] $SnCl_4$,bipy 207, 254; $SnBr_4$,bipy 195, 260; SnI_4,bipy 198, 252 (bipy = 2,2′-bipyridyl); $SnCl_4$,2MeCN 207, 222; $SnCl_4$2PhCN 220, 248; $SnCl_4$,$2CH_2CHCN$ 192, 204 (cm.$^{-1}$). The presence of the doublet for the monodentate ligands implies that these are in the *cis*-configuration.

Compounds Containing Ge—P, Sn—P, *or* Sn—As *Bonds*. Data which have appeared during 1967 are summarized in Table 27.

[274] L.-H. Chan and E. G. Rochow, *J. Organometallic Chem.*, 1967, **9**, 231.
[275] B. J. Aylett and J. Emsley, *J. Chem. Soc.* (*A*), 1967, 652.
[276] H. Bürger and W. Sawodny, *Spectrochim. Acta*, 1967, **23**, *A*, 2841.
[277] K. A. Andrianov and G. V. Kotreler, *J. Organometallic Chem.*, 1967, **7**, 217.
[278] U. Wannagat, E. Bagusch, and F. Höfler, *J. Organometallic Chem.*, 1967, **7**, 211.
[279] M. F. Farona and J. G. Grasselli, *Inorg. Chem.*, 1967, **6**, 1675.

Table 27 *Vibrations between heavy atoms in Groups* IV *and* V(cm.$^{-1}$)

Compound	ν	*Ref.*
$(H_3Ge)_3P$	$\nu_{sym}(Ge_3P)$ 322, $\nu_{asym}(Ge_3P)$ 366	280
$(Me_3Ge)_3P$	$\nu_{sym}(Ge_3P)$ 320, $\nu_{asym}(Ge_3P)$ 397	281
$(Ph_3Ge)_3P$	$\nu_{sym}(Ge_3P)$ 308, $\nu_{asym}(Ge_3P)$ 375	281
$(Me_3Sn)_3P$	$\nu_{sym}(Sn_3P)$ 284, $\nu_{asym}(Sn_3P)$ 351	281
$(Ph_3Sn)_3P$	$\nu_{sym}(Sn_3P)$ 296, $\nu_{asym}(Sn_3P)$ 347	281
$(SnCl_4,Ph_3P)_2$	ν(Sn—P) 529	282
$SnCl_4,2Ph_3P$	ν(Sn—P) 513	282
$SnCl_4,2Ph_3As$	ν(Sn—As) 330	282
$SnBr_3,2Ph_3As$	ν(Sn—As) 334	282

Compounds Containing M—O *Bonds* (M = Si, Ge, *and* Sn). The Si—O—Si valency oscillation occurs at 1030–1110 cm.$^{-1}$ in $[(C_6H_{13})SiO(OH)]_8$, $[(C_7H_{15})SiO(OH)]_6$, $[(C_8H_{17})SiO(OH)]_6$, and $[(iso\text{-}C_9H_{19})SiO(OH)]_8$.[283] The Si—O—Sn vibration occurs at 1030 cm.$^{-1}$ in $ClMe_2SiOSnBu_2Cl$ and at 1020 cm.$^{-1}$ in $ClPhMeSiOSnoct_2Cl$.[284]

Clark *et al.*[285] have assigned ν(Sn—O) for a series of Ph_3PO and Ph_3AsO complexes with SnX_4 (X = F, Cl, Br, and I), and for $MeSnI_3$,2L and Me_2SnX_2,2L (X = Cl, Br, and I), and $(SnI_2,4L)I_2$. They find ν(Sn—O) occurs at 310–320 cm.$^{-1}$ for the Ph_3PO complexes and at 380–390 cm.$^{-1}$ for the Ph_3AsO complexes. For a series of alkoxide complexes, however, ν(Sn—O) is assigned to the region 500–580 cm.$^{-1}$.[286]

Several authors have assigned bands due to ν(Ge—O—Ge) and ν(Sn—O—Sn) vibrations.[284, 287–91] The stability of M—O—M bonds decreases in the order Si > C > Ge > Sn (Pb does not form such compounds); ν(Ge—O—Ge) occurs at *ca.* 850 cm.$^{-1}$ and ν(Sn—O—Sn) in the broad range 600–800 cm.$^{-1}$.

Compounds Containing Sn—S *Bonds.* The compound $Ph_3Sn\cdot S\cdot P(:S)Ph_2$ has ν(Sn—S) at 340 cm.$^{-1}$,[292] and the cyclic trimer $(Me_2SnS)_3$ has ν(Sn—S) at 290, 324, 344, and 357 cm.$^{-1}$.[293]

280 S. Cradock, E. A. V. Ebsworth, G. Davidson, and L. A. Woodward, *J. Chem. Soc.* (*A*), 1967, 1229.
281 G. Engelhardt, P. Reich, and H. Schumann, *Z. Naturforsch.*, 1967, **22**, *B*, 352.
282 R. Rivest, S. Singh, and C. Abraham, *Canad. J. Chem.*, 1967, **45**, 3137.
283 K. A. Andrianov and B. A. Izmaylov, *J. Organometallic Chem.*, 1967, **8**, 435.
284 A. G. Davies and P. G. Harrison, *J. Organometallic Chem.*, 1967, **10**, P31.
285 Miss J. P. Clark, Miss V. M. Langford, and C. J. Wilkins, *J. Chem. Soc.* (*A*), 1967, 792.
286 J. S. Morrison and H. M. Haendler, *J. Inorg. Nuclear Chem.*, 1967, **29**, 393.
287 Y. Maeda and R. Okawara, *J. Organometallic Chem.*, 1967, **10**, 247.
288 K. Sisido, S. Kozima, and T. Hanada, *J. Organometallic Chem.*, 1967, **9**, 99.
289 M. Frankel, D. Gertner, D. Wagner, and A. Zilkha, *J. Organometallic Chem.*, 1967, **9**, 83.
290 B. Jezowska-Trzebiatowska, J. Hanuza, and W. Wojciechowski, *Spectrochim. Acta*, 1967, **23**, *A*, 2631.
291 A. G. Davies and P. G. Harrison, *J. Organometallic Chem.*, 1967, **7**, 13.
292 H. Schumann, P. Jutzi, A. Roth, P. Schwabe, and E. Schauer, *J. Organometallic Chem.*, 1967, **10**, 71.
293 H. Schumann, *Z. anorg. Chem.*, 1967, **354**, 192.

Compounds Containing M-*Halogen Bonds* (M = C, Si, and Sn). The i.r. spectrum of $CF_3Mn(CO)_5$ has been investigated in the gas phase. ν(C—F) occur at 1045 and 1063 cm.$^{-1}$ and the shapes of the band envelopes indicate that these are in the *e* and a_1 modes respectively.[294] A calculation of the C—F force constants in $CF_3Mn(CO)_5$ and CF_3I yielded values of 4·6 and 5·9 mdyne Å^{-1}; this very substantial reduction in $CF_3Mn(CO)_5$ is attributed to C—F bond weakening by overlap of the filled metal $d\pi$ orbitals with C—F σ^* antibonding orbitals.

Some previously published results for *cis*-CFH:CFMn$(CO)_5$ have been re-examined[295] and it is concluded that the lengthening of the C—F distances, the low frequencies of ν(C—F), and the anomalous ^{19}F n.m.r. spectra provide the best evidence to date for the π-bonding properties of fluorinated organic groups.

Carbon–iodine stretching frequencies (cm.$^{-1}$) have been assigned for the following series of donor–acceptor complexes:[296]

	CF_3I	CF_3I,NMe_3	CF_3I,quinuclidine	CF_3I,SMe_2
ν(C—I) solid (−185°)	280	256	256	270
ν(C—I) gas (22°)	286	272		

For some solid charge-transfer complexes between iodoform and 1,4-dithian ($C_4H_8S_2$), 1,4-diselenan ($C_4H_8Se_2$), and S_8, the $\delta_{sym}(CI_3)$ which is absent in the i.r. spectrum of CHI_3 appears in the spectrum of the dithian and diselenan complexes and the degenerate $CI_3(\nu_6)$ deformation splits into a doublet.[297]

Silicon-chlorine vibrations have been investigated for some phenylchlorosilanes[298] and some dichloro-derivatives derived from octaphenylcyclotetrasilane and decaphenylcyclopentasilane.[299] The results are shown in Figure 1.

The spectrum of $(Ph_4As)SnCl_3$ (Nujol mull) shows $\nu_{sym}(SnCl_3)$ at 289 and $\nu_{asym}(SnCl_3)$ at 252 cm.$^{-1}$. The corresponding bands for free $SnCl_3^-$ in solution occur at 297 and 256 cm.$^{-1}$ whereas those for CH_3SnCl_3 lie at 366 and 384 cm.$^{-1}$.[300] This demonstrates that an increase in the oxidation state from Sn^{II} to Sn^{IV} leads to a significant stiffening of the Sn—Cl bond; the bonds in metal–$SnCl_3^-$ complexes are expected to lie between these two extremes and to show the highest ν(Sn—Cl) values when bonded to metals with the greatest electron-pair affinity, *e.g.* $(PPh_3)_3CuSnCl_3$. This premise is explored for twelve complexes and the generalisation is made that for a ligand L in a complex LX_n, the L—X force constant will increase upon co-ordination of L to an electron acceptor, if X is significantly more electronegative than L, and it will decrease for the converse. These effects will be

[294] F. A. Cotton and R. M. Wing, *J. Organometallic Chem.*, 1967, **9**, 511.
[295] H. C. Clark and J. H. Tsai, *J. Organometallic Chem.*, 1967, **7**, 515.
[296] N. F. Cheetham and A. D. Pullin, *Chem. Comm.*, 1967, 233.
[297] G. C. Hayward and P. J. Hendra, *Spectrochim. Acta*, 1967, **23**, *A*, 1937.
[298] A. L. Smith, *Spectrochim. Acta*, 1967, **23**, *A*, 1075.
[299] H. Gilman and D. R. Chapman, *J. Organometallic Chem.*, 1967, **8**, 451.
[300] D. F. Shriver and M. P. Johnson, *Inorg. Chem.*, 1967, **6**, 1265.

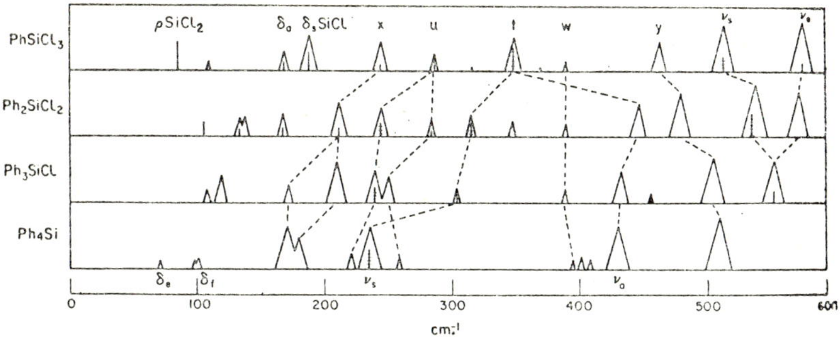

Figure 1 *Schematic representation of the low-frequency spectra for the phenyl chlorosilanes. Triangles represent i.r. absorption, lines represent Raman shifts, and cross-bars indicate polarised Raman lines. Heights are roughly proportional to intensities. The scale changes at about* 300 cm.$^{-1}$, *below which bands are much weaker than shown*
(Reproduced by permission from *Spectrochim. Acta*, 1967, **23**, *A*, 1075)

most pronounced with the strongest acceptors. Two conditions are important, however: co-ordination must not disrupt or alter the π-bonding within the ligand, and σ-donation from L to acceptor must predominate over any back π-bonding from the acceptor.

SnX_2L_2 complexes [X = Cl, Br, and I; L = acetylacetone (acac) or dibenzoylmethanate] show unusually high ν(Sn—X) stretching bands for six-co-ordinate tin, *e.g.* ($SnCl_2$ $acac_2$, ν(Sn—Cl) 332 cm.$^{-1}$, cf. Me_3SnCl, ν(Sn—Cl *ca.* 330 cm.$^{-1}$); the authors suggest that the Sn—O bands are primarily ionic in character.[301]

Co-ordination compounds of organo-tin halides with 4,4′-bipyridyl have been characterised by i.r. spectra.[302] It is concluded that the 1 : 1 adducts of R_2SnCl_2 are polymeric whereas the organo-tin monohalide complexes are monomeric. $SnCl_4$ alone forms a 2 : 1 adduct with (monodentate) 4,4′-bipyridyl which occupies *trans*-positions.

Farona and Grasselli [279] have studied the low-frequency i.r. spectra of $SnCl_4$,2L (L = MeCN, CH_2CHCN, and PhCN), $SnBr_4$,2MeCN, SnX_4,bipy (X = Cl, Br, and I), and SnX_4,2DMF (X = Cl and Br) and assign ν(Sn—Cl) in the range 275–370 cm.$^{-1}$ and ν(Sn—Br) in the range 195–250 cm.$^{-1}$. The assignments made in this investigation reverse those made recently by other workers for the ν(Sn—Cl) and ν(Sn—N) modes of the nitrile adducts and for the ν(Sn—Cl) and ligand vibrations of the DMF derivatives.

D. Group V Elements.—*Compounds Containing* P—H *Bonds.* Phosphacyclopropane, $\overline{CH_2\cdot CH_2\cdot PH}$, has ν(P—H) at 2287 cm.$^{-1}$.[303]

[301] W. H. Nelson, *Inorg. Chem.*, 1967, **6**, 1509.
[302] R. C. Poller and D. L. B. Toley, *J. Chem. Soc.* (*A*), 1967, 1578.
[303] R. I. Wagner, L. D. Freeman, H. Goldwhite, and D. G. Rowsell, *J. Amer. Chem. Soc.*, 1967, **89**, 1102.

Van den Akker and Jellinek[304] have prepared some addition compounds of Et_3P and Ph_3P with HF, HCl, HBr, and HI. The i.r. spectra of these adducts indicate two different types of compound: Et_3P,HCl, Et_3P,HBr, Et_3P,HI, Ph_3P,HF, Ph_3P,HBr, and Ph_3P,HI all contain strong PH—X bonds, whereas Ph_3P,2HF, Ph_3P,2HCl, and Et_3P,2HF are salts and show characteristic anionic absorptions (frequencies in cm.$^{-1}$):

	ν_1(calc.) (cm.$^{-1}$)	ν_2 (cm.$^{-1}$)	ν_3 (cm.$^{-1}$)
$(Ph_3PH)^+HF_2^-$	560	1200	1560
$(Et_3PH)^+HF_2^-$	560	1210	1590
$(Ph_3PH)^+HCl_2^-$	560	1180	1670

In these latter compounds ν(P—H) is at 2370 and 2380 cm.$^{-1}$ for $(Ph_3PH)^+HF_2^-$ and $Ph_3PH^+HCl_2^-$ respectively.

The ligand Ph_2PH shows ν(P—H) in the region 2270–2350 cm.$^{-1}$ in adducts with metal carbonyls:[305] $Cr(CO)_5$,Ph_2PH, 2270; $Mo(CO)_5$,Ph_2PH, 2290; $W(CO)_5$,Ph_2PH, 2345; $Fe(CO)_4$,Ph_2PH, 2350; $Fe(CO)_3$,$(Ph_2PH)_2$, 2336; *cis*-$Mo(CO)_4$,$(Ph_2PH)_2$, sym. 2320, asym. 2323; $Mo(CO)_3$,$(Ph_2PH)_3$, 2312 cm.$^{-1}$.

Compounds Containing P—C, As—C, *or* Sb—C *Bonds*. The acetylides of phosphorus, arsenic, and antimony show the following M—C frequencies, thirty-eight of the forty-two active fundamentals of these molecules having been assigned on the basis of C_{3v} symmetry (frequencies in cm.$^{-1}$):[306]

Compound	P(C⋮CH)$_3$	As(C⋮CH)$_3$	Sb(C⋮CH)$_3$
ν_{sym}(M—C)	615	526	477
ν_{asym}(M—C)	*ca.* 646	517	449·5
δ_{sym}(M—C_3)	120	89	72?
δ_{asym}(M—C_3)	100·5	100	94

For the compounds $(Me_3Sb^{2+})(ClO_4^-)_2$ and $(Me_3Sb^{2+})(NO_3^-)_2$ ν_{sym}(Sb—C) occurs at 536 cm.$^{-1}$ and ν_{asym}(Sb—C) at 582 cm.$^{-1}$.[307] The assignment of the symmetric SbC_3 mode enables the following comparisons to be made:[307]

Compound	Me_2Hg	Me_2Tl^+	Me_2Pb^{2+}	Me_3In	Me_3Sn^+	Me_3Sb^{2+}
ν_{sym}(M—C) (cm.$^{-1}$)	514	498	479	467	520	536
$10^5 \times k$(M—C) (dyne cm.$^{-1}$)	2·46	2·29	2·11	1·93	2·39	2·55

It will be noticed that the frequencies and force-constants diminish with increasing charge in the dimethyl series, but increase with increasing charge in the trimethyl series.

[304] M. Van den Akker and F. Jellinek, *Rec. Trav. chim.*, 1967, **86**, 275.
[305] J. G. Smith and D. T. Thompson, *J. Chem. Soc.* (*A*), 1967, 1694.
[306] F. A. Miller and D. H. Lemmon, *Spectrochim. Acta*, 1967, **23**, *A*, 1099.
[307] A. J. Downs and I. A. Steer, *J. Organometallic Chem.*, 1967, **8**, P.21.

For a large range of substituted stibine compounds of general formulae $[R_3Sb(Cl)]_2NH$ and $[R_3Sb(Cl)]_2O$ ν(Sb—alkyl) occurs in the range 500–591 and ν(Sb—phenyl) occurs at about 450 cm.$^{-1}$.[308]

Compounds Containing P=O, P=S, As=O, or Sb=S *Bonds.* Müller *et al.*[309, 310] have examined ν(P=O) for a series of phosphorus compounds and conclude that the electronegativity of the isothiocyanate group lies between the values for chlorine and bromine; values for ν(P=O) are in Table 28. Fild *et al.*[311] have used a similar relationship [λ(P=O)(μ) = 3·995/(39·96 $-$ ΣX_i)] to deduce the electronegativity, *X*, of the C_6F_5 group for a series of pentafluorophenylphosphorus compounds $R_n(C_6F_5)_{3-n}PO$ (R = Me, Et, and Ph). For $(C_6F_5)_3PO$ the electronegativity of the C_6F_5 group is 2·6.

Table 28 *Values of* ν(P=O) (cm.$^{-1}$) *for some phosphoryl halides and pseudohalides*

Compound	ν(P=O)	*Compound*	ν(P=O)	*Compound*	ν(P=O)
Cl_3PO	1290	Cl_2FPO	1331	ClF_2PO	1358
$(SCN-)_3PO$	1250	$(SCN-)_2FPO$	1308	$(SCN-)F_2PO$	1355
Br_3PO	1261	Br_2FPO	1303	BrF_2PO	1350

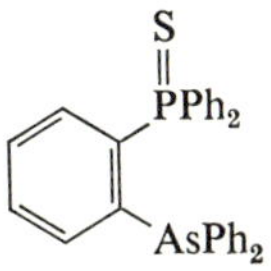

(9)

Diphenyl-(*o*-diphenylarsinophenyl)phosphine sulphide (9) exhibits ν(P=S) at 638 cm.$^{-1}$; the complexes of this ligand listed in Table 29 all show ν(P=S) in the range 593–609 cm.$^{-1}$ indicating that the ligand (9) is bidentate.[312] These results may be compared with those of the LAu^ICl complex [ν(P=S) 637 cm.$^{-1}$) where the ligand is monodentate via the arsenic atom.[312]

Table 29 *Values of* ν(P=S) (cm.$^{-1}$) *for some bidentate complexes* (*see text*)

Complex	ν(P=S)	*Complex*	ν(P=S)	*Complex*	ν(P=S)
$LPdCl_2$	603	$LPd(SCN)_2$	602	$LPtCl_2$	609
$LPdBr_2$	602	$LPd(SeCN)_2$	603	$LPtBr_2$	602
$LPdI_2$	603	$[L_2Pd](NO_3)_2$	593	$[L_2Cu]ClO_4$	609

308 R. L. McKenney and H. H. Sisler, *Inorg. Chem.*, 1967, **6**, 1178.
309 H. W. Roesky and A. Müller, *Z. anorg. Chem.*, 1967, **353**, 265.
310 A. Müller, E. Niecke, and O. Glemser, *Z. anorg. Chem.*, 1967, **350**, 246.
311 M. Fild, I. Hollenberg, and O. Glemser, *Z. Naturforsch.*, 1967, **22**, *B*, 248.
312 P. Nicpon and D. W. Meek, *Inorg. Chem.*, 1967, **6**, 145.

The following 1 : 1 complexes of the ligand Ph_3PS (L) with ν(P=S) 639 cm.$^{-1}$ show ν(P=S) in the range 592–604 cm.$^{-1}$: $LHgCl_2$, $LHgBr_2$, $LHgI_2$, LCuCl, and LCuBr.[313]

Durig *et al.*[314] have given a complete vibrational assignment for Me_2PSCl, Me_2PSBr, and Me_2POCl. An alternative explanation is presented for the origin of one of the two bands of the proposed doublet for ν(P=S) observed for the first two compounds. They suggest a coupling of ν(P=S) with $\nu_{sym}(PC_2)$: this shifts $\nu_{sym}(PC_2)$ to higher frequencies and it becomes coincident with $\nu_{asym}(PC_2)$; ν(P=S) is correspondingly perturbed to lower wavenumbers.

The trimer $(PhPS)_3$ has two strong bands at 650 and 710 cm.$^{-1}$ which may be assigned to ν(P=S); P—S—P bending modes are observed between 550–600 cm.$^{-1}$.[315] Values of ν(As=O) for a number of complexes of Ph_3AsO are shown in Table 30. Harris *et al.*[319] report evidence for a novel

Table 30 *Values of* ν(As=O) (cm.$^{-1}$) *for some complexes of* Ph_3AsO

Complex	ν(As=O)	*Ref.*
$CrCl_2,(Ph_3AsO)_2$	850, 879	316
$CrBr_2,(Ph_3AsO)_4$	860	316
$Cr(ClO_4)_2(Ph_3AsO)_4$	865	316
$UO_2(NO_3)_2,(Ph_3AsO)_2$	862–870	317
$Eu(NO_3)_3(Ph_3AsO)_4$	893	318[a]
$Eu(NO_3)_3(Ph_3AsO)_4,Me_2CO$	902	318
$Gd(NO_3)_3(Ph_3AsO)_3$	901, 930	318
$Gd(NO_3)_3(Ph_3AsO)_3,(Me_2CO)_2$	899, 910, 942	318
$Gd(NO_3)_3(Ph_3AsO)_2$,EtOH	903 921,	318

[a] The origin of the splitting of ν(As=O) is obscure.

triphenylarsine oxide derivative containing a short hydrogen bond. $[Ph_3As—O\cdots H\cdots O—AsPh_3]_2{}^+[Hg_2Br_6]^{2-}$. The complex $UCl_4,(diars.O_2)_2$ (diars = *o*-phenylenebisdimethylarsine) exhibits ν(As=O) at 832 and 851 cm.$^{-1}$.[317] Shindo and Okawara[320] have assigned ν(Sb=S) to a very strong band at 433 cm.$^{-1}$ in the new complex $[Me_3SbS]_2SnMe_2Cl_2$.

Compounds Containing M–*Halogen Bonds* (M = N, P, As, *and* Sb). Salts of the $N_2F_3{}^+$ cation (10) have been characterized;[321] they give strong bands

313 J. A. W. Dalziel, A. F. le C. Holding, and B. E. Watts, *J. Chem. Soc.* (*A*), 1967, 358.
314 J. R. Durig, D. W. Wertz, B. R. Mitchell, F. Block, and J. M. Greene, *J. Phys. Chem.*, 1967, **71**, 3815.
315 M. Baudler, K. Kipker, and H. W. Valpertz, *Naturwiss.*, 1967, **54**, 43.
316 L. F. Larkworthy and D. J. Phillips, *J. Inorg. Nuclear Chem.*, 1967, **29**, 2101.
317 J. Selbin and J. D. Ortego, *J. Inorg. Nuclear Chem.*, 1967, **29**, 1449.
318 D. R. Cousins and F. A. Hart, *J. Inorg. Nuclear Chem.*, 1967, **29**, 2965.
319 G. S. Harris, F. Inglis, K. J. McKechnie, K. K. Cheung, and G. Ferguson, *Chem. Comm.*, 1967, 442.
320 M. Shindo and R. Okawara, *Inorg. Nuclear Chem. Letters*, 1967, **3**, 75.
321 A. R. Young, jun., and D. Moy, *Inorg. Chem.*, 1967, **6**, 178.

associated with ν(N—F) at 922, 1100, and 1295 cm.$^{-1}$ and with ν(N=N) at 1500 cm.$^{-1}$. In the compound $(EtO)_2C{:}NCl$, ν(N—Cl) occurs at 698 cm.$^{-1}$.[322]

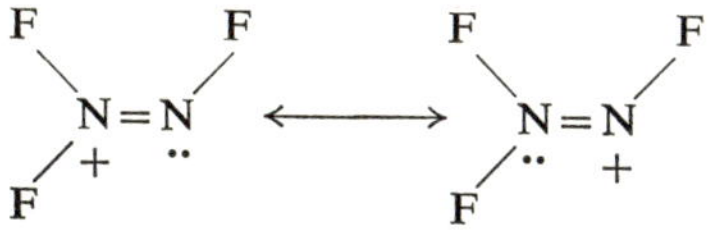

(10)

Phosphorus–fluorine stretching vibrations have been assigned for a series of tetrakis(fluorophosphine) derivatives of nickel(0) L_4Ni [L = PF_3, CF_3PF_2, $(CF_3)_2PF$, etc.];[323] P—F bridging-modes have also been assigned.[324]

For the compound (11), the ν(P—F) modes occur at 855, 858, 901, and 909 cm.$^{-1}$, *i.e.* both bridging and terminal P—F stretches occur in the region which has previously been regarded as characteristic of terminal PF_3 groups.

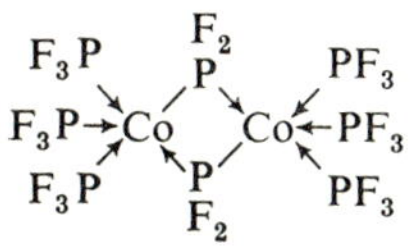

(11)

The presence of PF_2 bridges was confirmed by ^{19}F n.m.r. and mass spectra.

Bands at 388 and 409 cm.$^{-1}$ in a liquid-film spectrum of $PhPBr_2$ have been assigned tentatively to ν(P—Br).[325]

Compounds of 1,10-phenanthroline (*o*-phen) with some pentavalent phosphorus, arsenic, and antimony chlorides have been prepared and ionic structures [*o*-phen PCl_4^+]Cl^-, [*o*-phen $SbCl_4^+$]$SbCl_6^-$, and [*o*-phen $AsCl_4^+$]$SbCl_6^-$ are proposed on the basis of i.r. spectra and other information.[326]

O'Brien *et al.*[327] have investigated the spectra of some pentavalent trimethyl- and triphenyl-arsenic derivatives. The trimethylarsenic dihalide complexes Me_3AsX_2 have been assigned by comparing their spectra with those of the deuteriated analogues $(CD_3)_3AsCl_2$ and $(CD_3)_3AsBr_2$. The compounds have a pentacovalent trigonal-bipyramidal structure in the solid state when X = F or Cl, but an ionic structure with tetracovalent arsenic and a halide ion when X = Br or I.

E. Group VI Elements.—Some silyl and trimethylsilyl derivatives of selenium and tellurium have been reported and vibrational assignments are:[328]

322 E. Allenstein and R. Fuchs, *Chem. Ber.*, 1967, **100**, 2604.
323 J. F. Nixon, *J. Chem. Soc.* (*A*), 1967, 1136.
324 Th. Kruck and W. Lang, *Angew. Chem. Internat. Edn.*, 1967, **6**, 454.
325 A. Finch, P. J. Gardner, and K. K. Sen Gupta, *J. Chem. Soc.* (*B*), 1967, 1162.
326 M. J. Deveney and M. Webster, *Inorg. Nuclear Chem. Letters*, 1967, **3**, 195.
327 M. H. O'Brien, G. O. Doak, and G. G. Long, *Inorg. Chim. Acta*, 1967, **1**, 34.
328 H. Bürger and U. Goetze, *Inorg. Nuclear Chem. Letters*, 1967, **3**, 549.

Compound	(cm.$^{-1}$)	(cm.$^{-1}$)
$(Me_3Si)_2Se$	$\nu_{sym}(Si_2Se)$ 363	$\nu_{asym}(Si_2Se)$ 369
$(Me_3Si)_2Te$	$\nu_{sym}(Si_2Te)$ 330	$\nu_{asym}(Si_2Te)$ 323
$(H_3Si)_2Te$	$\nu_{sym}(Se_2Te)$ 334	$\nu_{asym}(Si_2Te)$ 335

Two groups have studied the relationship between the position of ν(S—N) in the range 800–830 cm.$^{-1}$ and the S—N bond length.[329, 330] They both conclude that, for a whole range of compounds, the position of the absorption band is linearly related to the S—N bond length and, in the case of compounds containing —NSO groups, the symmetric and antisymmetric frequencies are also linearly related.

The ν(S—S) vibrations for some polythionates, have been assigned as follows:[331] $S—SO_3^{2-}$, 435 cm.$^{-1}$; $S—S_xO_3^{2-}$, 430 cm.$^{-1}$; $S_x—SO_3^{2-}$, 423 cm.$^{-1}$. The ν(S—S) vibrations are notably weak in the i.r. region but for sulphur-rich dithiolate complexes of Ni^{II} a strong band at *ca.* 480 cm.$^{-1}$ is observed which is in keeping with the large dipole-moment change expected for S—S vibrations in cyclic structures.[332]

Some thiourea (tu) complexes with Te^{II} have been reported[333] (Te tu_4Cl_2, Te tu_2Cl_2, and Te tu_2Br_2); it appears that the Te—S bond is relatively weak and that the Te—S stretching vibrations are therefore at low frequencies (230–260 cm.$^{-1}$).

Di-2-naphthyl trisulphane-2-oxide, $(C_{10}H_7)_2S_3O$, the first example of a stable compound which contains the —S—SO—S— group has been prepared;[334] the strongest band in the i.r. spectrum occurs at 1117 cm.$^{-1}$ and has been assigned to ν(S=O).

The i.r. spectra of the trimethylsulphoxonium and trimethylsulphonium cations (Me_3SO^+ and Me_3S^+) have been reported;[335] assignments of the frequencies are made and the results of a normal-co-ordinate calculation for the trimethylsulphonium ion is reported.

Tellurium–fluorine stretching modes for $(Me_2N)_2TeF_4$ occur at 928, 643, 613, and 588, and for Me_2NTeF_5 vapour at 930, 698, and 629 cm.$^{-1}$.[336]

In some thiourea complexes of Te^{II}, ν(Te—Cl) occurs at *ca.* 270 and ν(Te—Br) at *ca.* 180 cm.$^{-1}$.[333]

F. Group VII Elements.—Pyridine–halogen complexes have been examined both in the solid state[337] and in solution.[338] In polar solvents the ionisation

329 A. J. Banister, L. F. Moore, and J. S. Padley, *Spectrochim. Acta*, 1967, **23**, *A*, 2705.
330 H. Garcia-Fernandez, *Compt. rend.*, 1967, **265**, *C*, 88.
331 U. Agarwala, C. E. Rees, and H. G. Thode, *Canad. J. Chem.*, 1967, **45**, 181.
332 D. Coucouvanis and J. P. Fackler, jun., *J. Amer. Chem. Soc.*, 1967, **89**, 1346.
333 P. J. Hendra and Z. Jović, *J. Chem. Soc.* (*A*), 1967, 735.
334 R. Steudel, P. W. Schenk, and J. Bilal, *Z. anorg. Chem.*, 1967, **353**, 250.
335 J. A. Creighton, J. H. S. Green, D. J. Harrison, and S. M. Waller, *Spectrochim. Acta*, 1967, 23, *A*, 2973.
336 G. W. Fraser, R. D. Peacock, and P. M. Watkins, *Chem. Comm.*, 1967, 1248.
337 F. Watari, *Spectrochim. Acta*, 1967, **23**, *A*, 1917.
338 I. Haque and J. L. Wood, *Spectrochim. Acta*, 1967, **23**, *A*, 959.

2py IX $\rightleftharpoons$ $py_2I^+ + IX_2^-$ takes place and a similar ionisation is postulated for the corresponding γ-picoline–halogen complexes.[339]

Yagi and Popov[340] have studied some mixed pentahalide anions and conclude that complexes containing $I_2Cl_3^-$ ions nearly always show strong bands at *ca.* 310 cm.$^{-1}$ whereas $I_2Cl_2Br^-$ ions usually show a band at *ca.* 294 cm.$^{-1}$.

2 Transition Elements

As with the main-group elements, information on compounds containing the structural unit M—X will be presented in the sequence of vertical groups of the transition elements M in the Periodic Table. Within each group the elements X will then also be arranged according to vertical groups in the Periodic Table. Characteristic vibrations of the ligands themselves, *e.g.* ν(C—O), ν(C—N), $\nu(N_2)$, etc., and the more complicated spectra of co-ordinated organic ligands are discussed in Part III. Compounds for which partial or unassigned vibrational data have been published are listed in tabular form in Part IV.

Vibrations involving bonds between heavy-metal atoms themselves bridge, to some extent, the main-group and transition elements. For this reason, data which have been published on these vibrations during 1967 are collected together in Table 31.

Table 31 *Vibrations* (cm.$^{-1}$) *involving heavy-metal atoms*

Complex	ν	*Ref.*
$(Et_3P)_2\cdot Pt(Cl)SiMe_3$	352, ν(Si—Pt)	341
$(Et_3P)_2\cdot Pt(GePh_3)SiMe_3$	337, ν(Si—Pt)	341
$Ph_3P\cdot Au\cdot SiPh_3$	305, ν(Si—Au)	342
$(Me_3As)_2Pd_2Cl_4$	~ 280, ν(As—Pd)	343
$(Et_3As)_2Pd_2Cl_4$	~ 339, ν(As—Pd)	343
$Me_3Sn\cdot Mn(CO)_5$	182, ν(Sn—Mn)	273
$Ph_3Sn\cdot Mn(CO)_5$	174, ν(Sn—Mn)	273
$Me_2ClSn\cdot Mn(CO)_5$	197, ν(Sn—Mn)	273
$Me_2BrSn\cdot Mn(CO)_5$	191, ν(Sn—Mn)	273
$Me_2ISn\cdot Mn(CO)_5$	179, ν(Sn—Mn)	273
$MeCl_2Sn\cdot Mn(CO)_5$	201, ν(Sn—Mn)	273
$Cl_3Sn\cdot Mn(CO)_5$	201, ν(Sn—Mn)	273
$Me_3SnFe(CO)_2(\pi\text{-}C_5H_5)$	185, ν(Sn—Fe)	273
$Ph_3Sn\cdot Fe(CO)_2(\pi\text{-}C_5H_5)$	174, ν(Sn—Fe)	273
$Me_3Sn\cdot Co(CO)_4$	176, ν(Sn—Co)	273
$Me_3Sn\cdot Mo(CO)_3(\pi\text{-}C_5H_5)$	172, ν(Sn—Mo)	273
$Ph_3Sn\cdot Mo(CO)_3(\pi\text{-}C_5H_5)$	169, ν(Sn—Mo)	273
$Mn_2(CO)_{10}$	157, ν_1(Raman)	344

[339] I. Haque and J. L. Wood, *Spectrochim. Acta.*, 1967, **23**, *A*, 2523.
[340] Y. Yagi and I. A. Popov, *J. Inorg. Nuclear Chem.*, 1967, **29**, 2223.
[341] F. Glockling and K. A. Hooton, *J. Chem. Soc.* (*A*), 1967, 1066.
[342] M. C. Baird, *J. Inorg. Nuclear Chem.*, 1967, **29**, 367.
[343] R. J. Goodfellow, P. L. Goggin, and L. M. Venanzi, *J. Chem. Soc.* (*A*), 1967, 1897.
[344] D. M. Adams, J. B. Cornell, J. L. Dawes, and R. D. W. Kemmitt, *Inorg. Nuclear Chem. Letters*, 1967, **3**, 437.

Table 31 (*cont.*)

Complex	ν	*Ref.*
Solution of Zn in fused $ZnCl_2$	175, ν(Zn—Zn)	345
$Ir_4(CO)_{12}$	208, (A_1); 164, (F_2); 105, (E)	20
$Hg[Cr(CO)_3(\pi\text{-}C_5H_5)]_2$	186, ν_3(i.r.)	344
$Hg[Mn(CO)_5]_2$	167, ν_1(Raman); 188, ν_3(i.r.)	346, 344
$ClHgMn(CO)_5$	184, ν(Mn—Hg)	346
$IHgMn(CO)_5$	176, ν(Mn—Hg)	346
$Hg[Fe(CO)_3NO]_2$	200, ν_{asym}(Fe—Hg—Fe)	347, 346
$[HgFe(CO)_4]_n$	200, ν_{asym}(Fe—Hg—Fe)	347, 346
$Hg[Fe(CO)_2(\pi\text{-}C_5H_5)]_2$	200, ν_3(i.r.)	344
$(ClHg)_2Fe(CO)_4$	218, 226, ν($Fe—Hg_2$)	346
$Hg[Co(CO)_4]_2$	161, ν_1(Raman); 196, ν_3(i.r.)	347, 346, 344
$Hg[Mo(CO)_3(\pi\text{-}C_5H_5)]_2$	178, ν_3(i.r.)	346, 344
$Hg[W(CO)_3(\pi\text{-}C_5H_5)]_2$	133, ν_1(Raman); 166, ν_3(i.r.)	346, 344

A. Titanium, Zirconium, and Hafnium.—Dithio-oxamide, $[C(S)NH_2]_2$, co-ordinates to $TiCl_4$ and $TiBr_4$ via the nitrogen atoms of the amide groups and ν(Ti—N) occurs at 457–490 cm.$^{-1}$.[348] Ti en_3Cl_3 and Ti pn_3Cl_3 (pn = propylenediamine) have been examined and the spectra indicate that all the chlorine atoms are ionic and that the complexes contain trisbidentate cations.[349] For the ethylenediamine complex ν(Ti—N) is tentatively assigned to the band at 516 cm.$^{-1}$ and for the propylenediamine complex it is at 528 cm.$^{-1}$.

I.r. measurements indicate that the solid complexes $TiX_3,6H_2O$ (X = Cl, Br) should be formulated as *trans*-$[Ti(H_2O)_4X_2]X,2H_2O$ and ν(Ti—O) occurs at 500 for the chloride complex and at 489 cm.$^{-1}$ for the bromide complex;[350] likewise ν(Ti—Cl) = 336 and ν(Ti—Br) = 294 cm.$^{-1}$. Several complexes of TiF_4 with *p*-substituted pyridine-*N*-oxide have been examined[351] and ν(Ti—O) assigned to the region 270–306 cm.$^{-1}$. Ti—O—Ti chain vibrations give characteristic bands at 825 and 938 cm.$^{-1}$ for $TiOSO_4,H_2O$[352] and at *ca.* 720 cm.$^{-1}$ for $(\pi\text{-}C_5H_5)_2Ti(X)—O—Ti(X)(\pi\text{-}C_5H_5)_2$.[353] Beattie *et al.*, however, could find no evidence for a Ti—O—Ti vibration in TiO $acac_2$ even though molecular weight measurements indicate a dimeric formulation.[354]

Bands at 414, 418, and 448 cm.$^{-1}$ have been tentatively assigned to ν(Ti—S) in $Ti(SEt)(NMe_2)_3$, $Ti(SEt)_2(NMe_2)_2$, and $Ti(SPr^i)_2(NMe_2)_2$ respectively.[355]

345 D. H. Kerridge and S. A. Tariq, *J. Chem. Soc.* (*A*), 1967, 1122.
346 P. N. Brier, A. A. Chalmers, J. Lewis, and S. B. Wild, *J. Chem. Soc.* (*A*), 1967, 1889.
347 W. Beck and K. Noack, *J. Organometallic Chem.*, 1967, **10**, 307.
348 S. C. Jain and R. Rivest, *J. Inorg. Nuclear Chem.*, 1967, **29**, 2787.
349 R. J. H. Clark and M. L. Greenfield, *J. Chem. Soc.* (*A*), 1967, 409.
350 H. L. Schläfer and H. P. Fritz, *Spectrochim. Acta*, 1967, **23**, *A*, 1409.
351 F. E. Dickson, E. W. Gowling, and F. F. Bentley, *Inorg. Chem.*, 1967, **6**, 1099.
352 M. L. Reynolds and T. J. Wiseman, *J. Inorg. Nuclear Chem.*, 1967, **29**, 1381.
353 R. Coutts and P. C. Wailes, *Inorg. Nuclear Chem. Letters*, 1967, **3**, 1.
354 I. R. Beattie and V. Fawcett, *J. Chem. Soc.* (*A*), 1967, 1583.
355 D. C. Bradley and P. A. Hammersley, *J. Chem. Soc.* (*A*), 1967, 1894.

$Ti^{IV}F_2$ $acac_2$ has been examined as a solid in the far-i.r.,[196, 356] and ν(Ti—F) occur at 633 and 618 cm.$^{-1}$. The presence of two bands is taken to indicate a *cis*-octahedral structure. TiF_4 complexes with $2Me_2SO$, $(MeOCH_2^-)_2$, phen, bipy, and 2py. All give Ti—F stretching frequencies in the range 550–670 cm.$^{-1}$ often with a weak band near 450 cm.$^{-1}$; the Ti—F deformations occur at 254–311 cm.$^{-1}$.[196]

Complexes of bivalent titanium dichloride L_2TiCl_2 (L = MeCN, C_4H_8O, $C_5H_{10}O$, py, bipy, and phen) give bands between 300–320 cm.$^{-1}$ which may be assigned to ν(Ti—Cl) (cf.Ti $^{II}Cl_2$ which has ν(Ti—Cl) *ca.* 290 cm.$^{-1}$).[357] The corresponding bromide complexes show broad bands at *ca.* 280 cm.$^{-1}$ but the spectra are less reliable.

$LTiX_2$ and L_2TiX_2 (X = Cl, Br, LH = 8-quinolinol (8-quin), salicylaldehyde, or acetylacetone) and TiX_4,2LH (X = F, Cl) have been examined and it has been suggested that $TiCl_4$,(8-quin), together with the analogous Sn complex, are monomeric five-co-ordinate species in the solid state (this conclusion, however, is at variance with the Mössbauer spectrum which indicates $SnCl_4$,(8-quin) to be six-co-ordinate[358]). For analogous compounds, Douek *et al.*[359] show that the relationship ν(Sn—X)/ν(Ti—X) $\simeq$0·90 is independent of the halogen and other ligand and applies to the compounds discussed in this paper as well as those previously reported in the literature. The authors also confirm that the relationship known for both four- and six-co-ordinate complexes of the transition metals: ν(M—Br)/ν(M—Cl)$\simeq$0·75 applies to the present compounds.

A similar relation concerning the relative position of ν(M—I) for tetrahedral species MX_4^{2-} (M = first-row transition metal): ν(M—I)/ν(M—Cl) $\simeq$0·65 is shown to be valid for six-co-ordinate compounds of Ti, Ge, and Sn.[359]

Complexes of zirconium(III) halides with several nitrogen-donor ligands have been studied in the 200–400 cm.$^{-1}$ range.[360] The i.r. spectrum of solid $ZrCl_3$,2py shows two Zr—Cl stretching modes (at 293 and 270sh) and these occur in the same region as the ν(Zr—Cl) modes of polymeric $ZrCl_3$. The latter contains six-co-ordinate zirconium and it is likely that the pyridine complex also contains similarly co-ordinated zirconium because of chlorine bridge bonds.

B. Vanadium, Niobium, and Tantalum.—Three bands have been observed at 386, 390, and 429 cm.$^{-1}$ in the far-i.r. spectrum of $[V(NH_3)_6]Cl_2$ and V—N stretching vibrations are expected to occur in this region.[361] For the complexes $[V\ en_3]Cl_3$ and $[V\ pn_3]Cl_3$, ν(V—N) has tentatively been assigned to a band at about 526 cm.$^{-1}$.[349]

356 R. C. Fay and R. N. Lowry, *Inorg. Nuclear Chem. Letters*, 1967, **3**, 117.
357 G. W. A. Fowles and T. E. Lester, *Chem. Comm.*, 1967, **1**, 47.
358 N. N. Greenwood and J. N. R. Ruddick, *J. Chem. Soc.* (*A*), 1967, 1679.
359 I. Douek, M. J. Frazer, Z. Goffer, M. Goldstein, B. Rimmer, and H. A. Willis, *Spectrochim. Acta*, 1967, **23**, *A*, 373.
360 G. W. A. Fowles, B. J. Russ, and G. R. Willey, *Chem. Comm.*, 1967, 646.
361 H. Behrens and K. Lutz, *Z. anorg. Chem.*, 1967, **354**, 184.

For $VCl_3,3C_4H_8O$, ν(V—O) occurs at 266 cm.$^{-1}$.[362] It has been found that $VX_3,2L$ (X = Cl or Br; L = Me_2S or C_4H_8S) and $VCl_3,2Et_2S$ are monomeric in solutions of nondonor solvents and that ν(V—S) occurs in the range 259–266 cm.$^{-1}$.[363]

Numerous compounds containing the V=O group have been investigated and values of ν(V=O) are summarized in Table 32. Some β-ketoenolate

Table 32 *Values for* ν(V=O) (cm.$^{-1}$) *in some vandium complexes*

Compound	ν(V=O)	*Ref.*
K_2VOCl_4	912	364
Rb_2VOCl_4	952	364
Cs_2VOCl_4	989	364
$[VO(S_2CNC_2H_6)_2]$	982	365
$[VO(S_2CNC_4H_{10})_2]$	984	365
$[VO(S_2CNC_4H_8)_2]$	992	365
$VOCl_2,2NMe_3$	990	366
$VOCl_2,2NHMe_2$	1000	366
$VOCl_2,4NH_2Me$	972	366
$VOCl_2,2SMe_2$	995	366
$VOCl_2,2SEt_2$	1010	366
$VOCl_2,2PhCN$	1030	367
$VOCl_2,2MeCN$	1005	367
$VOCl_2,2C_4H_8O_2$	995	367
$VOCl_2,C_4H_8O_2$	995	367
$VOCl_2,2MeCN\cdot O\cdot 5C_4H_8O_2$	997	367
$[VO\cdot HOC(CO_2)(CH_2CO_2)_2]^-$	972	368

complexes of oxovanadium(IV) also give ν(V=O) in the range 925–1006 cm.$^{-1}$.[369] Bovey and Clark[370] observe ν(V=O) at 1010 cm.$^{-1}$ for some $(VO)^{2+}$ complexes with aliphatic, straight-chain 2-hydroxy-carboxylic acids. They assign ν(V—O) to a band at 481 cm.$^{-1}$.

Twenty-one complexes of oxovanadium(IV) with various substituted pyridine-*N*-oxide ligands have been prepared and characterised.[371] Two types of behaviour have been observed: ν(V=O) *ca.* 950 cm.$^{-1}$ suggesting a strong interaction with the ligand *trans* to V=O, and ν(V=O) *ca.* 995 cm.$^{-1}$ suggesting no *trans* axial co-ordination. Kharitonov and Lipatova[372] find that the i.r. spectrum of $NbOCl_2$ has a broad band at 600–900 cm.$^{-1}$, which shows that the compound has a polymeric chain structure —NbONbO—.

[362] G. W. A. Fowles, P. T. Greene, and T. E. Lester, *J. Inorg. Nuclear Chem.*, 1967, **29**, 2365.
[363] M. W. Duckworth, G. W. A. Fowles, and P. T. Greene, *J. Chem. Soc.* (*A*), 1967, 1592.
[364] A. Feltz, *Z. anorg. Chem.*, 1967, **355**, 120.
[365] B. J. McCormick, *Inorg. Nuclear Chem. Letters*, 1967, **3**, 293.
[366] K. L. Baker, D. A. Edwards, G. W. A. Fowles, and R. G. Williams, *J. Inorg. Nuclear Chem.*, 1967, **29**, 1881.
[367] A. Feltz, *Z. anorg. Chem.*, 1967, **354**, 225.
[368] B. M. Nikolova and G. St. Nicolov, *J. Inorg. Nuclear Chem.*, 1967, **29**, 1013.
[369] J. Selbin, G. Maus, and D. L. Johnson, *J. Inorg. Nuclear Chem.*, 1967, **29**, 1735.
[370] D. M. H. Bovey and E. R. Clark, *J. Inorg. Nuclear Chem.*, 1967, **29**, 755.
[371] R. G. Garvey and R. O. Ragsdale, *J. Inorg. Nuclear Chem.*, 1967, **29**, 745.
[372] Y. Y. Kharitonov and N. P. Lipatova, *Izv. Akad. Nauk S.S.S.R., Neorg. Materialy*, 1967, **3**, 405; see also *Chem. Abs.*, 1967, **67**, 58,873*m*.

The appearance of a strong band at *ca.* 930 cm.$^{-1}$ in the spectra of $M_2^INbOCl_2$ and $M_2^INbOCl_4$, however, indicates the presence of an Nb=O group. Griffith and Wickens[373] have measured the Raman and i.r. spectra of solution of oxides of Groups VIA, VA, and IVA metals in aqueous hydrofluoric, hydrochloric, and perchloric acids. Evidence was obtained for the existence in such solutions of fluoro- and chloro-complexes of Mo and W with *cis*-dioxo-ligands, of oxopentahalogeno-species of Nb and V, and of unsubstituted halogeno-complexes of Ta, Ti, Zr, and Hf. Polynuclear complexes were formed in perchloric acid solutions.

Vanadium(III) chloride and bromide complexes, (12) and (13), with some nitrogen and oxygen donors have been studied as mulls and ν(M—X) assigned as in Table 33. In six-co-ordinate complexes ν(V—Cl) occurs in the 300–380 cm.$^{-1}$ region. The region is higher, however, for four- or five-co-ordinate complexes.[362, 373a]

Table 33 *Metal–halogen frequencies* (cm.$^{-1}$) *in some vanadium complexes*

Compound	ν(V—X)	*Configuration*
VCl_3,3py	352sh, 312, 283	*trans*-(C_{2v})
$VCl_3,3C_4H_8O$	366, 327, 300	*trans*-(C_{2v})
VBr_3,3py	289, 280sh	*cis*-(C_{3v})
$VBr_3,3C_4H_8O$ (in C_4H_8O)	282	Complex probably undergoes structural change in solution

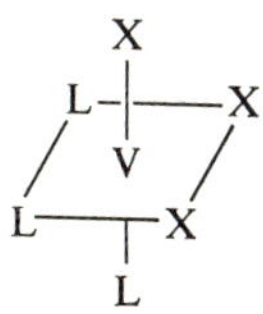

cis (C_{3v})
Two i.r.-active ν(M—X) modes
(12)

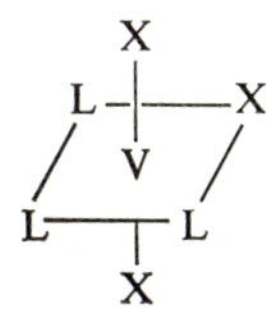

trans (C_{2v})
Three i.r.-active ν(M—X) modes
(13)

C Chromium, Molybdenum, and Tungsten.—Values for the metal–carbon stretching frequency in some π-cyclopentadienyltrihalogenometal dicarbonyls are:[374]

	ν(Mo—C) (cm.$^{-1}$)		ν(W—C) (cm.$^{-1}$)
(π-C_5H_5)$Mo(CO)_2Cl_3$	390,397	(π-C_5H_5)$W(CO)_2Cl_3$	374, 384
(π-C_5H_5)$Mo(CO)_2Br_3$	400	(π-C_5H_5)$W(CO)_2Br_3$	375, 392
(π-C_5H_5)$Mo(CO)_2I_3$	382, 409	(π-C_5H_5)$W(CO)_2I_3$	378, 407

The metal–carbon vibrations were found to lie in the range 300–500 cm.$^{-1}$ for a series of phosphine-substituted carbonyls of formulae $M(CO)_{6-x}(Ph_3P)_x$ and $M(CO)_{6-2y}(diphos)_y$; (M = Cr, Mo, and W; x = 1, 2, and 3; and y = 1 and 2; diphos = 1,2-bisdiphenylphosphinoethane).[375]

[373] W. P. Griffith and T. D. Wickins, *J. Chem. Soc.* (*A*), 1967, 675.
[373a] G. W. A. Fowles and P. T. Greene, *J. Chem. Soc.* (*A*), 1967, 1869.
[374] M. L. H. Green and W. E. Lindsell, *J. Chem. Soc.* (*A*), 1967, 686.
[375] A. A. Chalmers, J. Lewis, and R. Whyman, *J. Chem. Soc.* (*A*), 1967, 1817.

For $[Cr(NH_3)_6]Cl_3$,[376] ν(Cr—N) occurs at 468 and 457 cm.$^{-1}$ and δ(Cr—N) modes are assigned to a set of bands at 273, 283, and 291 cm.$^{-1}$. They are significantly sharpened and strengthened in going to $[Cr(NH_3)_6][CuCl_5]$ for which ν(Co—N) is 461 and δ(Cr—N) is 280 cm.$^{-1}$. For $M(CO)_3(MeCN)_3$ complexes, metal–nitrogen stretching frequencies were assigned as follows: ν(Cr—N)552 and 495; ν(Mo—N) 533 and 481; ν(W—N) 538 and 489 cm.$^{-1}$.[377] The complexes $NH_4[Cr(NCS)_4(NH_3)_2]$ and $ND_4[Cr(NCS)_4(ND_3)_2]$ have metal–nitrogen stretching frequencies at 501 and 487 cm.$^{-1}$ respectively.[378] ν(Cr—N) for $[Cr\,en_3]Cl_3$ occurs at 543 cm.$^{-1}$ and for $NH_4[CrF_4\,en]$ it has been assigned to bands at 550 and 526 cm.$^{-1}$.[379] Hughes and McWhinnie[380] have studied six *cis*- and five *trans*-bisethylenediamine complexes of Cr^{III} together with the *N*-deuteriated complexes. Earlier workers have assigned ν(Cr—N) and ethylenediamine ring deformation modes to the region 395–550 cm.$^{-1}$. The *cis*-complexes all show in this region four bands which move to lower wavenumbers in the *N*-deuteriated compounds. The authors therefore conclude that ν(Cr—N) must be strongly coupled with the ethylenediamine ring modes. The *trans*-complexes show three bands in the 395–550 cm.$^{-1}$ region only one of which moves (*ca.* 540 cm.$^{-1}$) in the *N*-deuterio-complexes. The remaining two bands (440 and 490 cm.$^{-1}$) are consequently assigned to ν(Cr—N) and the former band must be a ring-deformation mode. Several hydroxy-bridged complexes of Cr^{III} with bipy and phen, *e.g.* $[bipy_2Cr(OH)]_2(NO_3)_4,3H_2O$, have been examined in the far-i.r. region, and Cr—N vibrations occur in the range 343–378 cm.$^{-1}$.[381] Possible ligand-field effects in ν(M—O) of some first-row transition-metal alkoxides have been investigated.[382]

	ν(M—O) (cm.$^{-1}$)		ν(M—O) (cm.$^{-1}$)
$Cr(OMe)_2$	470 (?), 515	$Ni(OMe)_2$	375, 425
$Mn(OMe)_2$	307, 360	$Cu(OMe)_2$	435, 520
$Fe(OMe)_2$	330, 370, 420 (?)	$Zn(OMe)_2$	325, 465
$Co(OMe)_2$	491, 340, 412		

Although the spectra are not well resolved ($\Delta\nu_{\frac{1}{2}}$ *ca.* 100 cm.$^{-1}$) the relative positions of the band maxima appear to provide evidence for a well defined dependence on *d*-orbital population. The pair of ν(M—O) values for $t_{2g}^3e_g^1$ and $t_{2g}^6e_g^3$ are highest and therefore $t_{2g}^3\,e_g^2$ and $t_{2g}^6\,e_g^4$ are the lowest, and a steady increase occurs from Mn to Cu.

For $CrCl_3,3C_4H_8O$, ν(Cr—O) occurs at 275 cm.$^{-1}$.[362] Ferraro *et al.*[381] have assigned in the spectra of some hydroxy-bridged complexes of Cr^{III},

[376] G. C. Allen and N. S. Hush, *Inorg. Chem.*, 1967, **6**, 4.
[377] M. F. Farona, J. G. Grasselli, and B. L. Ross, *Spectrochim. Acta*, 1967, **23**, *A*, 1875.
[378] M. A. Bennett, R. J. H. Clark, and A. D. J. Goodwin, *Inorg. Chem.*, 1967, **6**, 1625.
[379] H. L. Schläfer, H. Gausmann, and H. U. Zander, *Inorg. Chem.*, 1967, **6**, 1528.
[380] M. N. Hughes and W. R. McWhinnie, *J. Chem. Soc.* (*A*), 1967, 592.
[381] J. R. Ferraro, R. Driver, W. R. Walker, and W. Wozniak, *Inorg. Chem.*, 1967, **6**, 1586.
[382] R. W. Adams, R. L. Martin, and G. Winter, *Austral. J. Chem.*, 1967, **20**, 773.

e.g. $[bipy_2Cr(OH)_2]_2^{4-}$, to the region 547–567 cm.$^{-1}$. For similar complexes with other transition elements, ν(M—O) decreases in the expected order $Cr^{III} > Fe^{III} > Cu^{III}$.

A large number of complexes containing the group Mo=O or W=O have been investigated and the values of ν(M=O) are summarized in Table 34. Values occur in the range 950–1005 cm.$^{-1}$ and are independent

Table 34 *Compounds containing* Mo=O *and* W=O *bonds; frequencies in* cm.$^{-1}$

Compound	ν(M=O)	*Ref.*
Cs_2MoOCl_5	955	383, 384
$(C_5H_6N)_2MoOCl_5$	952	384
Cs_2MoOBr_5	955	384
$(C_5H_6N)_2MoOBr_5$	955	384
$(Me_4N)MoOCl_4$	950–995	383
$(n\text{-}C_4H_9)_4N,MoOCl_4$	1000–1015	383
$(PhCH_2)Me_3N,MoOCl_4$	975–995	383
$(Ph_4As)MoOCl_4$	985–1000	383, 384
$(Ph_4As)MoOBr_4$	985	384
omp $MoOCl_3$	990	385
$Na_4[MoO_2(CN)_4],8H_2O$	790–850	386
$K_4[MoO_2(CN)_4],6H_2O$	800	386
$K_3[MoO(OH)(CN)_4],2H_2O$	921	386, 387
$K_3[MoO(OD)(CN)_4],2D_2O$	920	386
$[(Me_2(NCS)_2Mo{=}O]_2O$	~ 900	388
$MoO_2[Me_2(NCS)_2]_2$	~ 900	388
$K_2[MoO(O_2)_2\ ox]$	996	389
$K_2[MoO_2(O_2)\ ox,H_2O]$	912, 965	389
$K_2[Mo_2O_5\ ox_2(H_2O)_2]$	917, 962	389
$NO^+[MoO_2F_3]^-$	930, 980	390
Cs_2WOCl_5	958	384
Rb_2WOCl_5	966	384
Cs_2WOBr_5	980	384
$(C_9H_8N)WOCl_4$	972	384
$(Ph_4As)WOCl_4$	978	384
$(C_9H_8N)WOBr_4$	978	384
WOX_3,bipy	979	391
WO_2Cl_2L (L=bipy or phen)	910–920 and 950–970	392
$[WO(OH)(CN)_4]^{3-}$	~ 875	386, 387
$K_4[WO_2(CN)_4],6H_2O$	720	386
$K_2[WO(O_2)_2\ ox]$	972	389
$K_2[WO_2(O_2)\ ox,H_2O]$	907, 938	389

ox = Oxalate; omp = octamethylpyrophosphoramide.

383 O. Piovesana and C. Furlani, *Inorg. Nuclear Chem. Letters*, 1967, **3**, 535.
384 B. J. Brisden, D. A. Edwards, D. J. Machin, K. S. Murray, and R. A. Walton, *J. Chem. Soc.* (*A*), 1967, 1825.
385 M. D. Joesten, *Inorg. Chem.*, 1967, **6**, 1598.
386 S. J. Lippard and B. J. Russ, *Inorg. Chem.*, 1967, **6**, 1943.
387 S. J. Lippard, H. Nozaki, and B. J. Russ, *Chem. Comm.*, 1967, 118.
388 F. W. Moore and M. L. Larson, *Inorg. Chem.*, 1967, **6**, 998.
389 W. P. Griffith and T. D. Wickins, *J. Chem. Soc.* (*A*), 1967, 590.
390 O. Glemser, J. Wegener, and R. Mews, *Chem. Ber.*, 1967, **100**, 2474.
391 G. W. A. Fowles and J. L. Frost, *J. Chem. Soc.* (*A*), 1967, 671.
392 B. J. Brisdon, *Inorg. Chem.*, 1967, **6**, 1791.

of the metal when only one oxygen atom is attached, but values are frequently much lower when more than one oxygen atom is attached to the metal.

The following ν(Cr—P) assignments have been reported:[378] $(Me_3PH^+)[Cr(NCS)_4(Me_3P)_2^-]$, 275; and $(Et_3PH^+)[Cr(NCS)_4(Et_3P)_2^-]$, 305 cm.$^{-1}$.

Moore and Larson[388] have examined some dialkyldithiocarbamate complexes of Mo^V and Mo^{VI}, $[(R_2NCS_2)_2Mo{=}O]_2O$ and $MoO_2[R_2NCS_2]_2$, and they tentatively assign ν(Mo—S) to the range 460–515 cm.$^{-1}$.

The Cr—F stretching frequency occurs as a very strong band at 489 cm.$^{-1}$ in the complex $NH_4[CrF_4\,en]$.[379] Solid $CrCl_3,3C_4H_8O$ has a *trans* (C_{2v}) configuration and ν(Cr—Cl) occurs at 308, 344, and 366 cm.$^{-1}$.[362] The corresponding Cr—Cl vibration in a series of complexes (π-C_5H_5)Cr(NO)·L·Cl (where L contains N, P, or As donor atoms) is in the range *ca.* 290–315 cm.$^{-1}$.[393]

Glemser *et al.*[390] have assigned ν(Mo—F) for $NO^+[MoO_2F_3]^-$ to a band at 640 cm.$^{-1}$ and for $NO^+[WOF_5]^-$, ν(W—F) occurs at 625 and 450 cm.$^{-1}$. Green and co-workers have assigned ν(Mo—X) and ν(W—X) in the following complexes:[374, 394]

Compound	ν(Mo—X) (cm.$^{-1}$)	*Compound*	ν(W—X) (cm.$^{-1}$)
$(\pi\text{-}C_5H_5)_2MoCl_2$	262, 293	$(\pi\text{-}C_5H_5)_2WCl_2$	266, 283
$(\pi\text{-}C_5H_5)Mo(CO)_2Cl_3$	265sh, 273, 313	$(\pi\text{-}C_5H_5)W(CO)_2Cl_3$	274, 315
$(\pi\text{-}C_5H_5)Mo(CO)_2Br_3$	174, 194, 232	$(\pi\text{-}C_5H_5)W(CO)_2Br_3$	173, 187, 219
$(\pi\text{-}C_5H_5)Mo(CO)_2I_3$	126, 143, 153	$(\pi\text{-}C_5H_5)W(CO)_2I_3$	127, 143, 159

For the trihalogeno-complexes the configuration is as in (14)

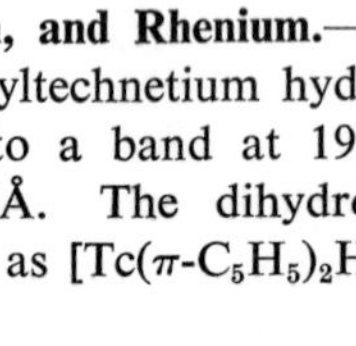

(14)

D. Manganese Technetium, and Rhenium.—Fischer and Schmidt[395] have prepared dicyclopentadienyltechnetium hydride, $(\pi\text{-}C_5H_5)_2TcH$, and they have assigned ν(Tc—H) to a band at 1930 (KBr). The Tc—H force constant is 2·20 mdyne/Å. The dihydrodicyclopentadienyltechnetium cation can be precipitated as $[Tc(\pi\text{-}C_5H_5)_2H_2]^+PF_6^-$ and ν(Tc—H) occurs at 1984 cm.$^{-1}$.

Davison and Faller[396] have attempted to establish the orientation of the hydrogen atom in $HMn(CO)_5$ and $HRe(CO)_5$. The spectra are consistent with C_{4v} symmetry for the molecules and ν(M—H) is very intense, indicating

393 E. O. Fischer and H. Strametz, *J. Organometallic Chem.*, 1967, **10**, 323.
394 R. L. Cooper and M. L. H. Green, *J. Chem. Soc.* (*A*), 1967, 1155.
395 E. O. Fischer and M. W. Schmidt, *Angew. Chem. Internat. Edn.*, 1967, **6**, 93.
396 A. Davison and J. W. Faller, *Inorg. Chem.*, 1967, **6**, 845.

a large polarisation change. This suggests a totally symmetric stretching mode and the hydrogen atom presumably lies on the fourfold axis; ν(M—H) occurs at 1780 and 1824 cm.$^{-1}$ for $HMn(CO)_5$ and $HRe(CO)_5$ respectively. Some other Re—H stretching frequencies are in Table 35.

Table 35 *Some* Re—H *stretching frequencies* (cm.$^{-1}$)[397]

Compound	ν(Re—H)	*Compound*	ν(Re—H)
$ReH_2diphos_2Cl$	2020–2040	$ReH_3diphos_2$	1860
$ReH_2diphos_2Br$	2010–2030	$ReH_3diphos\,(Ph_3P)_2$	1820, 1900, 1960
$ReH_2diphos_2I$	2050	$[ReH_4diphos_2]Cl$	1950
$ReH_2diphos\,(Ph_3P)_2I$	2000–2040	$[ReH_4diphos\,(Ph_3P)_2]Cl$	1970

Cotton *et al.*[398] have prepared $[ReO_2\,py_4]Cl,2H_2O$ and they suggest that the single strong band at 825 cm.$^{-1}$ is characteristic of a *trans*-dioxo, O=Re=O, grouping. The complex $Re_2O_3py_4Cl_4$ shows ν(Re=O) at 970 cm.$^{-1}$ but other bands, which may be assigned to ν(Re—O) of a bridging Re—O—Re group, occur in the range 675–710 cm.$^{-1}$. They propose the structure (15). ν(Re=O) occurs at 985 cm.$^{-1}$ for

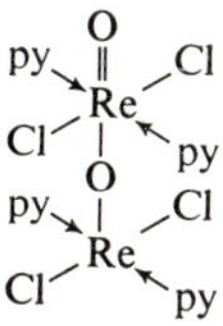

(15)

$(Ph_3P)_2ReOCl_3$,[399] and for a series of complexes such as $Cs_2[ReOCl_5]$ $Cs_2[ReOBr_5][ReOCl_3\,phen]$, $[ReO(OH)\,en_2]Cl_2$, etc., ν(Re=O) occurs in all cases as a strong band in the range 960–1000 cm.$^{-1}$.[400]

Despite the many halogen-containing derivatives of manganese, technetium, and rhenium, few have had metal–halogen frequencies assigned. An exception is the group of pentacarbonylmanganese halides:[401]

$Mn(CO)_5Cl$	$Mn(CO)_5Br$	$Mn(CO)_5I$
ν(Mn—Cl) 291 cm.$^{-1}$	ν(Mn—Br) 218 cm.$^{-1}$	ν(Mn—I) 185 cm.$^{-1}$

E. Iron, Ruthenium, and Osmium.—Mays and Simpson[402] have prepared $HFeCo_3(CO)_{12}$ and $HRuCo_3(CO)_{12}$ and these compounds are the first tetrahedral metal atom clusters in which transition metals from different periods are bonded together. The authors believe that i.r. and mass spectra

397 M. Freni, R. Demichelis, and D. Giusto, *J. Inorg. Nuclear Chem.*, 1967, **29**, 1433.
398 F. A. Cotton, W. R. Robinson, and R. A. Walton, *Inorg. Chem.*, 1967, **6**, 223.
399 G. Rouschias and G. Wilkinson, *J. Chem. Soc.* (*A*), 1967, 993.
400 D. K. Chakrabarti, V. I. Panesh, and B. N. Ivanov-Emin, *Zhur. neorg. Khim.*, 1967, **12**, 697; see also *Chem. Abs.*, 1967, **67**, 69,032*n*.
401 V. Valenti, F. Cariati, C. Forese, and G. Zerbi, *Inorg. Nuclear Chem. Letters*, 1967, **3**, 237.
402 M. J. Mays and R. N. F. Simpson, *Chem. Comm.*, 1967, 1024.

of the two complexes are consistent with structure (16) in which a hydrogen atom is located inside the tetrahedral cage. The spectrum of $DFeCo_3(CO)_{12}$ is identical with that of the hydride, between 800–2500 cm.$^{-1}$ and there is no peak which can be assigned to ν(M—H).

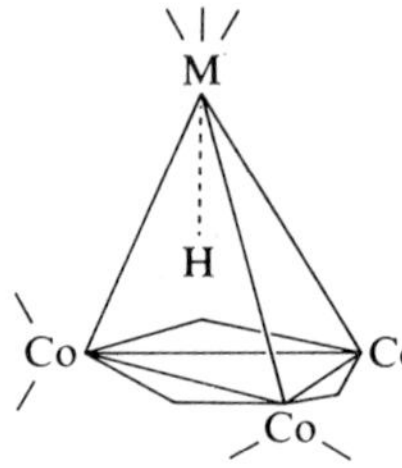

(16)

Chatt *et al.*[403, 404] reported the first unambiguous isolation of crystalline paramagnetic transition-metal hydrides which are stable in air. The ν(M—H) stretching vibrations occur as strong bands in the range 1725–2250 cm.$^{-1}$ which is within the range found for the closely similar diamagnetic hydride-complexes. The compound $[OsHCl_2(Bu_2{}^nPhP)_3]$ is a red crystalline solid with ν(Os—H) at 2064 cm.$^{-1}$. On boiling this solid in carbon tetrachloride for 4 hours, a solid isomeric hydride was obtained with ν(Os—H) 1915 cm.$^{-1}$.

The assignment of ν(M—N) stretching and δ(M—NO) rocking vibrations in nitrosyl complexes of general formula $[M(NO)X_5]^{n-}$ is still controversial. Sabatine[405] has measured the i.r. spectra in polarised light of single crystals of $Na_2[Fe(NO)(CN)_5],2H_2O$ and he assigns ν(Fe—N) to 650 cm.$^{-1}$ and Fe—NO rock to 662 cm.$^{-1}$. Hydroxo-bridged complexes of Fe^{III} with 2,2′-bipyridyl and 1,10-phenanthroline are said to show ν(Fe—N) vibrations in the range 260–294 cm.$^{-1}$.[381] The solid-state spectra of $[Ru^{III}(NH_3)_5X]X_2$ (X = Cl, Br, I) show ν(Ru—N) bands between 424–485 cm.$^{-1}$.[406] In contrast $(Ru^{II}\ pic_3]Cl_2$ (pic = β- or γ- picoline) has weak absorptions in the region 500–540 cm.$^{-1}$ which are tentatively assigned to ν(Ru—N) vibrations.[407]

The ν(Fe—O—Fe) vibration has been assigned to bands in the region 820–840 cm.$^{-1}$ for the compounds $Fe_2O\ phen_4Cl_4,6H_2O$; $Fe_2O\ phen_4Cl_2(ClO_4)_2,6H_2O$; $Fe_2O\ phen_4(SO_4)_2,8\frac{1}{2}H_2O$; $Fe_2O\ phen_4(ClO_4)_4,8H_2O$; and $Fe_2O\ bipy_4Cl_2(ClO_4)_2,7H_2O$.[408]

Metal–halogen frequencies have been assigned in numerous compounds. The i.r. spectra of di- and tri-2-pyridylamine complexes with Fe^{II} yield the

403 J. Chatt, G. J. Leigh, D. M. P. Mingos, and R. J. Paske, *Chem. and Ind.*, 1967, 1324.
404 J. Chatt, G. J. Leigh, and R. J. Paske, *Chem. Comm.*, 1967, 671.
405 A. Sabatini, *Inorg. Chem.*, 1967, **6**, 1756.
406 A. D. Allen and C. V. Senoff, *Canad. J. Chem.*, 1967, **45**, 1337.
407 F. G. Nasouri, M. W. Blackmore, and R. J. Magee, *Austral. J. Chem.*, 1967, **20**, 1291.
408 A. V. Khedekar, J. Lewis, F. E. Mabbs, and H. Weigold, *J. Chem. Soc.* (*A*), 1967, 1561.

following values:[409, 410]

	ν(Fe—X) (cm.$^{-1}$)		ν(Fe—X) (cm.$^{-1}$)
Fe dpa Cl_2	342, 313	Fe dpa_2Cl_2	253
Fe dpa Br_2	260, 244	Fe dpa_2Br_2	~ 210
Fe dpa I_2	228, 217	Fe tpa_2FeCl_4	289
		Fe tpa_2FeBr_4	<222

dpa = Di-2-pyridylamine; tpa = tri-2-pyridylamine.

The three compounds on the left-hand side have distorted tetrahedral structures and the value of 289 cm.$^{-1}$ for ν(Fe—Cl) in Fe tpa_2FeCl_4 correlates well with the reported value for $FeCl_4^{2-}$ (282 cm.$^{-1}$).

Several Fe^{III} complexes have also been examined. Some 1,10-phenanthroline and bipyridyl complexes have bands at *ca.* 250 cm.$^{-1}$ which can be assigned to ν(Fe—Cl).[408] $(pyH)_3Fe_2X_9$ and (quinolinium)FeX_4 (X = Cl, Br) salts contain the anion FeX_4^- and the ν_3 fundamental occurs at 368 and 369 cm.$^{-1}$ respectively for the chloride salts and at 281 and 291sh, and at 280 and 291sh cm.$^{-1}$ for the bromides.[411] $Cs_3Fe_2Cl_9$ exists in two forms (α and β); the α-form contains μ-trichloro-$Fe_2Cl_9^{3-}$ and the β-form is essentially tetrahedral $FeCl_4^-$.

α-$RuCl_3$ shows an intense ν(Ru—Cl) stretching band at 315 cm.$^{-1}$ and deformation modes at 169 and 188 cm.$^{-1}$. β-$RuCl_3$ has two strong bands in the ν(Ru—Cl) region (286 and 376) and two deformations (170 and 188) and the results can be compared with the corresponding f_{1u} vibrations which occur at 346 and 188 cm.$^{-1}$ in K_2RuCl_6.[198] The authors also find a single Ru—Cl stretching vibration at 328 for $K_4[Ru_2OCl_{10}]$ and at 320 cm.$^{-1}$ for $Cs_2[RuO_2Cl_4]$.

The Ru—Cl stretching vibrations for the *cis*- and *trans*-isomers (17) and (18) of trichlorotriaquoruthenium(III) fall in the range 291–315 cm.$^{-1}$ and the i.r. spectra are consistent with the previous structures proposed for these isomers.[412]

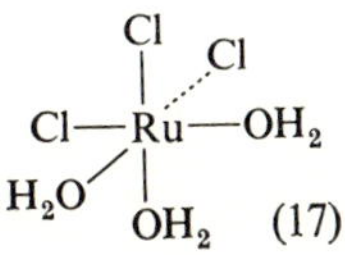
(17)

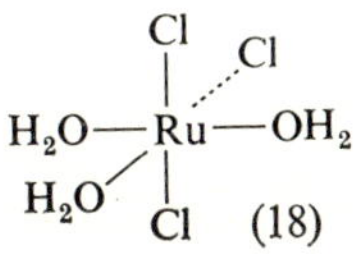
(18)

For $[Ru^{III}(NH_3)_5Cl]Cl_2$, ν(Ru—Cl) occurs at 301 cm.$^{-1}$.[406] Other Ru^{II} and Ru^{III} complexes with nitrogen ligands have been investigated and the

[409] W. R. McWhinnie, R. C. Poller, and M. Thevarasa, *J. Chem. Soc.* (*A*), 1967, 1671.
[410] C. D. Burbridge and D. M. L. Goodgame, *J. Chem. Soc.* (*A*), 1967, 694.
[411] R. J. H. Clark and F. B. Taylor, *J. Chem. Soc.* (*A*), 1967, 693.
[412] J. R. Durig, W. A. McAllister, and E. E. Mercer, *J. Inorg. Nuclear Chem.*, 1967, **29**, 1441.

metal–chlorine stretching modes assigned as:[413]

	ν(Ru—Cl) (cm.$^{-1}$)		ν(Ru—Cl) (cm.$^{-1}$)
$RuCl_2$,4py	295, 335	$RuCl_3$,3py	301, 346
$RuCl_2$,4-β-pic	280, 312	$RuCl_3$,3-β-pic	*ca.* 290, 336

Values for the corresponding Os^{III} complexes are:[413]

	ν(Os—Cl) (cm.$^{-1}$)
$OsCl_3$ 3-py	289, 313
$OsCl_3$,3-γ-pic	287, 312

F. Cobalt Rhodium, and Iridium.—Metal–halogen stretching frequencies have been studied for all three elements in this group. The i.r. spectrum of tris(triphenylphosphine)cobalt dihydride, $(Ph_3P_3)_3CoH_2$, dissolved in benzene, exhibits three bands in the region 1700–2000 cm.$^{-1}$ (1755, 1900, and 1935 cm.$^{-1}$). The absorption at 1755 cm.$^{-1}$ is tentatively assigned to ν(Co—H) but the origin of the other two is still under investigation.[414] The KBr disc spectrum of the pentacyanohydridocobalt(III) ion, $[Co(CN)_5H]$, prepared from H_2O and D_2O showed strong bands at 1840 and 1340 (Co—H/D stretch), and at 774 and 610 cm.$^{-1}$ (Co—H/D bend).

Some Rh—H stretching vibrations are as follows, the symbol dmpe representing the bidentate phosphine ligand $Me_2P(CH_2)_2PMe_2$:

	ν(Rh—H) (cm.$^{-1}$)	*Ref.*
$(Ph_3P)_3RhH$	2020	415
cis-$[RhH_2\ dmpe_2]Cl$	1870, 1900	416
trans-$[RhHCl\ dmpe_2]Cl$	2050	416
trans-$[RhHBr\ dmpe_2]Cl$	2030	416
$(Ph_3E)_2RhXH(SiR_3)$ (X = Cl or Br; E = P, As or Sb)	2040–2080	417
$RhCl_2H(Ph_3P)_2,\frac{1}{2}CH_2Cl_2$	2105	418
$RhCl_2H(CO)(Ph_3P)_2$	2122	418

Hydridopentacyanoiridate(II) complexes have been synthesised and ν(Ir—H) occurs at 2043 cm.$^{-1}$. The authors[419] also assign δ(Ir—H) at 810 cm.$^{-1}$. All the data are consistent with a C_{4v} symmetry for the molecule. Chlorotris(triphenylphosphine)iridium(I) provides an example of hydrogen transfer to a metal from a co-ordinated ligand.[420] When $Ir(Ph_3P)_3Cl$ is

413 J. Lewis, F. E. Mabbs, and R. A. Walton, *J. Chem. Soc.* (*A*), 1967, 1366.
414 A. Misono, Y. Uchida, T. Saito, and K. M. Song, *Chem. Comm.*, 1967, 419.
415 W. Keim, *J. Organometallic Chem.*, 1967, **8**, P25.
416 J. Chatt and S. A. Bulter, *Chem. Comm.*, 1967, 501.
417 R. N. Haszeldine, R. V. Parish, and D. J. Parry, *J. Organometallic Chem.*, 1967, **9**, P13.
418 M. C. Baird, J. T. Mague, J. A. Osborn, and G. Wilkinson, *J. Chem. Soc.* (*A*), 1967, 1347.
419 K. Krogmann and W. Binder, *Angew. Chem. Internat. Edn.*, 1967, **6**, 881.
420 M. A. Bennett and D. L. Milner, *Chem. Comm.*, 1967, 581.

heated in cyclohexane, benzene, C_6D_6, acetone, or $(CD_3)_2Cl$, $Ir(Ph_3P_3H_xCl$ is obtained [ν(Ir—H) = 2190 cm.$^{-1}$ in benzene solution]. If $Ir[(C_6D_5)_3P]_3Cl$ is heated in benzene, the resulting complex shows ν(Ir—D) at 1600 and 1540 cm.$^{-1}$ but there are no ν(Ir—H) absorptions. An intense band in the i.r. spectrum of the hydride at 728 cm.$^{-1}$, which is absent from the spectra of triphenylphosphine and its complexes, can be assigned to the C—H out-of-plane deformation mode of an *ortho*-disubstituted benzene, and the authors therefore suggest the metal–carbon σ-bonded structure (19) (or possible isomers).

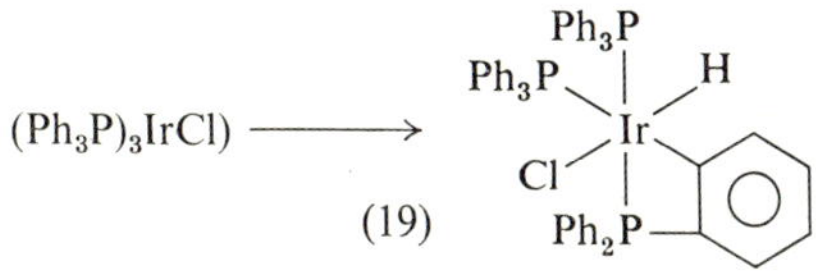

(19)

Metal–carbon vibrations have been assigned for a series of methyl derivatives of Ir^{III} containing dimethylphenylphosphine or dimethylphenylarsine ligands:[421] ν(Ir—C) stretching vibrations occur in the region 488–543 cm.$^{-1}$. For the dimethyl complexes two bands are observed; these are below 518 cm.$^{-1}$ if the iridinium–carbon bands are *trans* to P or As atoms, but lie in the region 521–543 cm.$^{-1}$ when halogen atoms are in the *trans* position.

The cobalt–nitrogen stretching frequency in a series of cobalt carbonyl–phosphorus trifluroride complexes $Co(NO)(CO)_x(PF_3)_{3-x}$ has been assigned at 595 cm.$^{-1}$ and it is suggested that a reasonable assignment for the Co—N—O bending mode would be 440 cm.$^{-1}$.[422] The detailed i.r. spectra of some nitroammine Co^{III} complexes $[Co(NO_2)_n(NH_3)_{6-n}]^{(3-n)+}$ (n = 0–6) have been measured down to 80 cm.$^{-1}$.[423] The normal modes of vibration in the low-frequency region have been discussed for the series on the basis of a normal-co-ordinate analysis. Specific assignment for Co—N stretching and bending modes are in Table 36.

Table 36 *Assignments for* Co—N *stretching and bending modes; frequencies in* cm.$^{-1}$

Compound	ν(Co—N)	δ(Co—N)	*Ref.*
$Co(NH_3)_6Cl_3$	(498)	339, 328, 291	376
$Co(NH_3)_6CuCl_5$	463, 483	333	376
$[Co(NH_3)_5NO]Cl_2$	441, 477	289	424
$[Co(NH_3)_5NO](NO_3)_2$	446, 471	285	424
$[Co(NH_3)_5NO]SO_4$	440, 485	289	424
$[Co_2(NH_3)_{10}N_2O_2](NO_3)_4$	459, 489	310	424

421 B. L. Shaw and A. C. Smithies, *J. Chem. Soc.* (*A*), 1967, 1047.
422 R. J. Clark, *Inorg. Chem.*, 1967, **6**, 299.
423 I. Nakagawa and T. Shimanouchi, *Spectrochim. Acta*, 1967, **23**, *A*, 2099.
424 P. Gans, *J. Chem. Soc.* (*A*), 1967, 943.

Table 36 (*cont.*)

Compound	ν(Co—N)	δ(Co—N)	*Ref.*
$[Co_2(NH_3)_{10}N_2O_2H](NO_3)_5$	454		424
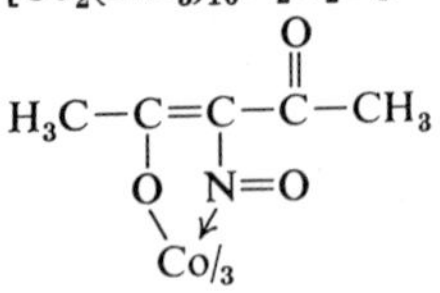	587,593		425
py_2CoCl_2	190, 254	142	426
py_2CoBr_2	179, 244	134	426
$[Et_4N[py\ CoCl_3]$	162, 186, 226	128	426
$[Et_4N[py\ CoBr_3]$	174, 150	96	426
$[Co\ dcppa\ Cl_2]^a$	247	(~ 153)	427

[a] dcppa = Dichlorodi-(2-pyridyl)phenylamine.

Watt *et al.*[428] have observed ν(Rh—N) for a series of *N*-methyl substituted ethylenediamine complexes of Rh^{III} in the region 450–600 cm.$^{-1}$. The relative thermodynamic stabilities of the complexes are indicated by a correlation found between the nephelauxetic series of ligands and ν(Rh—N).

Cobalt–sulphur modes have been assigned for some essentially tetrahedrally co-ordinated cobalt thiourea complexes.[130]

	ν(Co—tu) (cm.$^{-1}$)		ν(Co—tu) (cm.$^{-1}$)
$Co\ tu_2(OAc)_2$	240	$Co\ tu_3OSeO_3$	265, 284
$Co\ tu_3OSO_3$	260, 279	$Co\ tu_3(S_2O_3)$	(241, 255, 275)

The bands shown in brackets probably arise from strongly coupled metal–thiourea and metal–anion modes.

Cobalt–phosphorus stretching frequencies in some triphenylphosphine-cobalt(II) complexes have been assigned:[426]

	ν(Co—P) (cm.$^{-1}$)
$(Ph_3P)_2CoCl_2$	151, 187
$(Ph_3P)_2CoBr_2$	142, 187
$[Et_4N][Ph_3PCoCl_3]$	167

Clark and Williams[429] suggest that $CoCl_2,bipy_{1\cdot 33}$,β-$CoCl_2$,bipy and $(CoCl_4)(bipy\ H_2)$ all contain tetrahedral $CoCl_4^{2-}$ ions since they have strong bands at *ca.* 290 cm.$^{-1}$ and this corresponds to the $\nu_3(f_2)$ vibrations of that anion. The bipyridyl complexes are formulated as $[bipy_2CoCl_2Co\ bipy_2]^{2+}CoCl_4^{2-}$. ν(Co—Cl) for $[Co\ dppa\ Cl_2]$ in $CHCl_3$ solution occurs at 320 and 347 cm.$^{-1}$ and the corresponding bromide bands are at 250 and 269 cm.$^{-1}$ [427] [dppa = di-(2-pyridyl)phenylamine].

425 N. J. Patel and B. C. Haldar, *J. Inorg. Nuclear Chem.*, 1967, **29**, 1037.
426 J. Bradbury, K. P. Forest, R. H. Nuttall, and D. W. A. Sharp, *Spectrochim. Acta*, 1967, **23**, *A*, 2701.
427 G. C. Kulasingam and W. R. McWhinnie, *Spectrochim. Acta*, 1967, **23**, *A*, 1601.
428 G. W. Watt and P. W. Alexander, *J. Amer. Chem. Soc.*, 1967, **89**, 1814.
429 R. J. H. Clark and C. S. Williams, *Spectrochim. Acta*, 1967, **23**, *A*, 1055.

The far-i.r. spectra of some L_2MX_2 (M = Co or Zn) and LMX_3 (M = Co, Ni, or Zn) complexes of pseudo-tetrahedral symmetry have been examined for L = py or Ph_3P and X = Cl or Br.[426] The results are best described in terms of either C_{2v} or C_{3v} symmetries, and the metal–halogen stretching and deformation modes for all the complexes can be assigned as:

	(cm.$^{-1}$)		(cm.$^{-1}$)
ν(M—Cl)	~ 320	δM—Cl	~ 100
ν(M—Br)	~ 250	δM—Br	~ 80

Rhodium–chlorine frequencies have been investigated more extensively.[418, 430–433] For twenty-seven complexes of Rh^{III} with dimethylphenylphosphine and -arsine, ν(Rh—Cl) falls into two ranges: 293–345 cm.$^{-1}$ when two chlorines are mutually *trans*, and 264–278 cm.$^{-1}$ when chlorine is *trans* to phosphorus or arsenic.[431] Square planar and octahedral complexes of Rh and Ir with π-bonding ligands have also been examined:[433] for the square planar complexes ν(M—Cl) = 250–310 cm.$^{-1}$ whereas for octahedral complexes ν(M—Cl) = 250–350 cm.$^{-1}$ and the frequencies depend mainly on the ligand *trans* to the chlorine atom. The ligands can be ordered according to their effect on ν(M—Cl-*trans*). The sequence of decreasing values of ν(M—Cl) being:

$$Cl > Br > I \sim CO > CH_3 \sim R_3P \sim R_3As > H.$$

Wilkinson and co-workers[418] find that ν(Rh—Br)/ν(Rh—Cl) = 0·66 for a series of Rh^I complexes and assign ν(Rh—I) = 95 cm.$^{-1}$ in $Rh_2I_2Me_4$. Rhodium(III)–chlorine stretching vibrations in complexes with 1,4-thioxan (which is sulphur bonded) and ligands of the type $RSCH_2CH_2SR$ (R = Me or Ph) have been investigated.[432] $[RhCl_3,dpte]_n$ (dpte = 1,2-di(phenylthio)-ethane) is a 1 : 1 complex and has two broad bands at *ca.* 280 and *ca.* 330 cm.$^{-1}$ which can be assigned to terminal ν(Rh—Cl) and bridging ν(Rh—Cl) which would occur in a dimer such as (20).

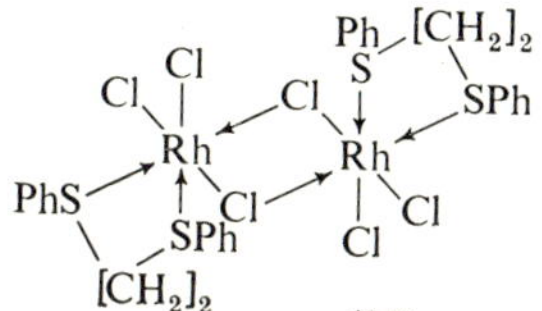

(20)

Some typical ν(Ir—Cl) frequencies are assigned as:[430]

	ν(Ir—Cl) (cm.$^{-1}$)
$[IrCl_3L]_2$	277, 312
$[IrCl_3L,p\text{-}MeC_6H_4\cdot NH_2]$	276, 314
$[IrCl_3L,py]$	275, 312

[430] D. C. Goodall, *J. Chem. Soc.* (*A*), 1967, 203.
[431] P. R. Brookes and B. L. Shaw, *J. Chem. Soc.* (*A*), 1967, 1079.
[432] R. A. Walton, *J. Chem. Soc.* (*A*), 1967, 1852.
[433] M. A. Bennett, R. J. H. Clark, and D. L. Milner, *Inorg. Chem.*, 1967, **6**, 1647.

The complexes (21) are dimeric and non-conducting in dimethylsulphoxide. The ligand (L = 1,2-dithiocyanatoethane) is present in its gauche form and the bonding is via the S atom. The decrease in ν(Ir—Cl) from 254–310 cm.$^{-1}$ in going from $IrCl_2Me(PhPMe_2)_3$ to $IrClMe_2(PhPMe_2)_3$ suggests that the methyl group has a strong *trans* bond-weakening effect as in Pt^{II} complexes.[421]

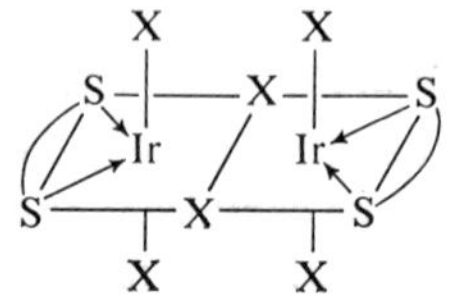

(21)

G. Nickel, Palladium, and Platinum.—Metal–hydrogen frequencies for palladium and platinum complexes are shown in Table 37. Morelli *et al.*[434] have investigated the co-ordination of small molecules to bistriphenylphosphineplatinum(0). The i.r. spectra of $(Ph_3P)_2Pt{\cdot}SH_2$ and

Table 37 *Some* Pd—H *and* Pt—H *stretching frequencies* (cm.$^{-1}$)

Compound	ν(M—H)	*Ref.*
$(Et_3P)_2PdHCl$[a]	2035	435
$(Et_3P)_2PdHBr$[b]	2029	435
$(Et_3P)_2PdHI$	2004	435
$PtH(SiPh_3)(PEt_3)_2$	2056	342
$[\{PtH(SnCl_3)(PPh_3)_2\}_2(1,5\text{-cod})]$	2100, 2200	436
	ν(Pt—H)	
$PtH(CN)(PPh_3)_2$	2075	437
$PtH(SnCl_3)(PPh_3)_2$	2080, 2120	487
$PtH(NO_2)(PPh_3)_2$	2180	437
$PtH(Cl)(PPh_3)_2$	2220	437
$PtH(SCN)(PPh_3)_2$	2250	437
$PtH(OCN)(PPh_3)_2$	2260	437
$PtH(Br)(PPh_3)_2$	2280	437

[a] δ(Pd—H)721.
[b] δ(Pd—H)712.

$(Ph_3P)_2Pt{\cdot}SeH_2$ reveal ν(Pt—H) modes at 2116 and 2140 cm.$^{-1}$ respectively and the authors suggest the presence of two types of product, (22) and (23) based mainly on n.m.r. evidence.

The low-frequency fundamentals of $Ni(CO)_4$, which involve Ni—CO motions, have been revised;[438] for gaseous $Ni(CO)_4$, $\nu_4(E) = 62$ and

[434] (The late) D. Morelli, A. Segre, R. Ugo, G. La Monica, S. Cenini, F. Conti, and F. Bonati, *Chem. Comm.*, 1967, 524.
[435] E. H. Brooks and F. Glockling, *J. Chem. Soc.* (*A*), 1967, 1030.
[436] H. A. Tayim and J. C. Bailar, jun., *J. Amer. Chem. Soc.*, 1967, **89**, 3420.
[437] J. C. Bailar, jun., and H. Itatani, *J. Amer. Chem. Soc.*, 1967, **89**, 1592.
[438] L. H. Jones and R. S. McDowell, *J. Chem. Phys.*, 1967, **46**, 1536.

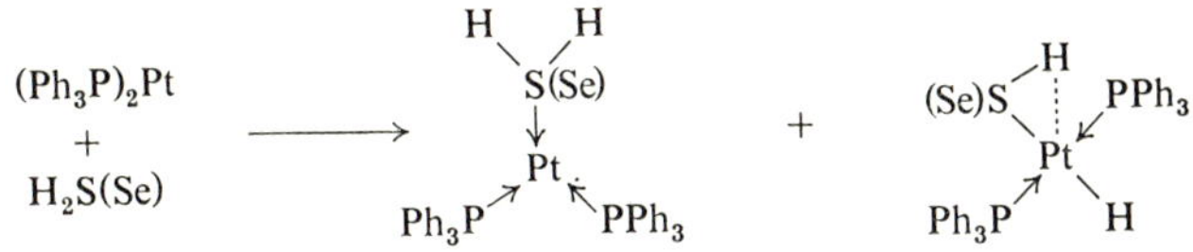

(22) (23)

$\nu_8(F_2)$ = 80 cm.$^{-1}$ whereas for $Ni(CO)_4$ in carbon tetrachloride solution the modes occur at 78 and 91 cm.$^{-1}$ respectively.

Far-i.r. spectra have been measured and normal-co-ordinate analyses have been carried out for $KPtCl_3,C_2H_4$ [439] (Zeise's salt) and for sodium dichlorobis(γ-acetylacetonato)platinum(II).[440] Metal–ligand stretching force-constants were calculated in both cases. Denning and Venanzi have looked at the i.r. spectra of some platinum complexes with unsaturated amines;[441] for $[all\cdot NH_3]^+PtCl_3^-$, ν(Pt—olefin) = 417 cm.$^{-1}$ and for $[all\cdot NH_3]^+PtBr_3^-$, ν(Pt—olefin) = 420 cm.$^{-1}$ (all = $CH_2{:}CH\cdot CH_2\cdot$). Thus the Pt–olefin bond is more stable in the bromo- than in the chloro-complex, a conclusion also reached from an analysis of enthalpy data.

Compounds containing M—N *Bonds* (M = Ni, Pd, *and* Pt). For $Ni(NH_2CH_2CO_2)_2,2H_2O$ and $Ni(NH_2(CH_2)_2CO_2)_2,2H_2O$ ν(Ni—N) has been assigned to 435 and 366 cm.$^{-1}$ respectively.[442] The corresponding vibration for the Ni–alanine chelate (24) has been assigned to a band at 328 cm.$^{-1}$ on the basis of a normal-co-ordinate analysis.[443] Six other M^{II}–alanine complexes (24) were also examined (M = Co, Cu, Zn, Cd, Pd, and Pt) and the values for the M—N bond-stretching force-constants vary in the order $Pt^{II} > Pd^{II} > Cu^{II} > Zn^{II} > Cd^{II} > Ni^{II} \approx Co^{II}$: the positions of ν(M—N) are 422, 419, 335, 328, 305, 282, and 322 cm.$^{-1}$. The order for the bond-stretching force-constants is also observed when comparing the relative frequency values among the different complexes, for the ν(N—H) of the co-ordinated $—NH_2$ groups. Similar relationships hold for M—O force-constants and for the carboxylate stretching frequencies.

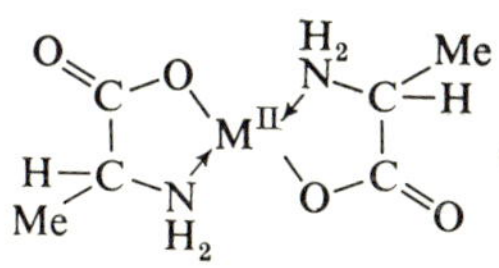

(24)

Metallo-porphyrin complexes, containing a series of divalent metals (Co, Ni, Cu, Zn, Pd, Ag, Cd, and Mg), show strong absorptions at *ca.* 350 cm.$^{-1}$ which are assigned to ν(M—N) coupled to porphyrin skeletal deformation modes.[444]

439 J. Pradilla-Sorzano and J. P. Fackler, jun., *J. Mol. Spectroscopy*, 1967, **22**, 80.
440 G. T. Behnke and K. Nakamoto, *Inorg. Chem.*, 1967, **6**, 440.
441 R. G. Denning and L. M. Venanzi, *J. Chem. Soc.* (*A*), 1967, 336.
442 G. W. Watt and J. F. Knifton, *Inorg. Chem.*, 1967, **6**, 1010.
443 J. F. Jackovitz, J. A. Durkin, and J. L. Walter, *Spectrochim. Acta*, 1967, **23**, *A*, 67.
444 L. J. Boucher and J. J. Katz, *J. Amer. Chem. Soc.*, 1967, **89**, 1340.

For a series of *meta*-hydrazine complexes (Ni, U, Th, and Ce), ν(M—N) occurs at *ca.* 550 cm.$^{-1}$ and the actual positions reflect the bond strengths. The order Ni > UO_2 > Th ≈ Ce is confirmed by thermochemical data.[445]

Pd complexes of the type *cis*-$Pd(NH_3)_2X_2$ (X = halogen) are unstable at room temperature and isomerise to the corresponding *trans*-isomer. For *cis*-$Pd(NH_3)_2Cl_2$, ν(Pd—N) occurs at 476 and 495 cm.$^{-1}$ and for *cis*-$Pd(NH_3)_2Br_2$, ν(Pd—N) at 460 and 480 cm.$^{-1}$. The compounds *trans*-$Pd(NH_3)_2Cl_2$ and *trans*-$Pd(NH_3)_2Br_2$ exhibit ν(Pd—N) at 490 and 496 cm.$^{-1}$ respectively.[153a, 446, 446a] Hendra[447] has calculated approximate force-constants for both the Pd and Pt systems and, as expected, the values for Pt—N are greater than those for Pd—N. Adams and Chandler give the following Pt—N stretching and bending frequencies:[448]

	$\nu(Pt—N_2)$ (cm.$^{-1}$)	$\delta(Pt—N_2)$ (cm.$^{-1}$)
trans-$PtCl_4(NH_3)_2$	517, 554	253
trans-$PtBr_4(NH_3)_2$	506, 531	228
trans-$PtI_4(NH_3)_2$	507, 516	249
cis-$PtCl_4(NH_3)_2$	518, 558	240
cis-$PtBr_4(NH_3)_2$	485, 492	219
cis-$PtI_4(NH_3)_2$	429, 440	220
cis-$PtCl_4\,py_2$	267, 280	
cis-$PtBr_4\,py_2$	263, 267, 282	

It is notable that the *trans*-series show two ν(Pt—N) bands but these are lowered by 38 and 10 cm.$^{-1}$ only on replacing chloride by iodide. The far-i.r. spectra of several square planar *cis*- and *trans*-Pd^{II} and Pt^{II} complexes of the type ML_2X_2 (X = Cl, Br, and I; L = py or bipy) have been recorded.[449] The authors discuss possible assignments for ν(M—N) and finally conclude that the 250–300 cm.$^{-1}$ region is the most reasonable. Methylamine complexes of Pt^{II} show Pt—N stretching vibrations in the region 490–526 cm.$^{-1}$ and $\delta(Pt—N_2)$ occurs at *ca.* 305 cm.$^{-1}$.[450, 451] Similarly a weak band at *ca.* 550 cm.$^{-1}$ has been assigned to ν(Pt—N) in K[$PtCl_2$,glycine].[452] The Raman spectrum of [$PtMe_3(NH_3)_3$]Cl (25) shows ν(Pt—N) at 390 cm.$^{-1}$ and this is moved to 364 cm.$^{-1}$ in the deuterio-compound [$PtMe_3(ND_3)_3$]Cl.[453]

445 V. T. Athavale and C. S. Padmanabha Iyer, *J. Inorg. Nuclear Chem.*, 1967, **29**, 1003.
446 J. R. Durig and B. R. Mitchell, *Appl. Spectroscopy*, 1967, **21**, 221.
446a J. S. Coe and A. A. Malik, *Inorg. Nuclear Chem. Letters*, 1967, **3**, 99.
447 P. J. Hendra, *Spectrochim. Acta*, 1967, **23**, *A*, 1275.
448 D. M. Adams and P. J. Chandler, *J. Chem. Soc.* (*A*), 1967, 1009.
449 J. R. Durig, B. R. Mitchell, D. W. Sink, J. N. Willis, jun., and A. S. Wilson, *Spectrochim. Acta*, 1967, **23**, *A*, 1121.
450 Yu. Ya. Khamtorov, I. K. Dymina, and T. N. Leonova, *Zhur. Neorg. Khim.*, 1967, **12**, 830; see also *Chem. Abs.*, 1967, **67**, 37,881*x*.
451 G. W. Watt, B. B. Hutchinson, and D. S. Klett, *J. Amer. Chem. Soc.*, 1967, **89**, 2007.
452 J. A. Kieft and K. Nakamoto, *J. Inorg. Nuclear Chem.*, 1967, **29**, 2561.
453 D. E. Clegg and J. R. Hall, *Spectrochim. Acta*, 1967, **23**, *A*, 263.

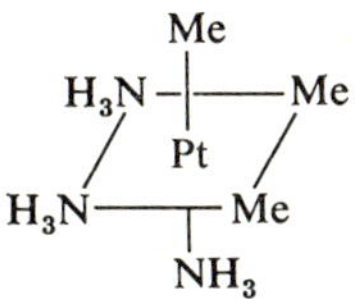

(25)

Compounds containing M—P *Bonds* (M = Ni, Pd, *and* Pt). ν(Ni—P) varies in the following manner for a series of tricarbonyl (organometallophosphine)nickel(0) complexes of a general formula $(CO)_3NiP(EMe_3)_3$:[454]

E	Si	Ge	Sn	Pb
ν(Ni—P) (cm.$^{-1}$)	478	454	448	446

The compounds are all colourless, air stable crystals, but they decompose at fairly low temperatures.

Pd—P stretching vibrations have been assigned at 382 and 431 cm.$^{-1}$ in $(Me_3P)_2Pd_2Cl_4$ and $(Et_3P)_2Pd_2Cl_4$.[343] Pt—P stretching and bending vibrations are shown in Table 38.

Table 38 *Some* $Pt—P_2$ *stretching and bending vibrations* (cm.$^{-1}$)

Compound	ν(Pt—P_2)	δ(Pt—P_2)	*Ref.*
trans-$(PEt_3)_2PtCl_4$	410	179	448
trans-$(PEt_3)_2PtBr_4$	411	182	448
trans-$(PEt_3)_2PtI_4$	412	181	448
cis-$(PEt_3)_2PtCl_4$	428, 444	183	448
cis-$(PEt_3)_2PtBr_4$	425, 444	182	448
$(Et_3P)_2Pt(GePh_3)GeMe_3$	415		341
$(Et_3P)_2Pt(GePh_3)SiMe_3$	402		341
$(Et_3P)_2(Cl)SiMe_3$	417		341
$(Et_3P)_2Pt(Cl)GeMe_3$	412		341
$(Et_3P)_2Pt(Br)GeMe_3$	398		341
$(Et_3P)_2Pt(I)GeMe_3$	407		341
$(Et_3P)_2Pt(Ph)GeMe_3$	406, 420		341
$(Et_3P)_2(CN)GeMe_3$	413, 433, 441		341

Data for the corresponding ν(Pt—As_2) and δ(Pt—As_2) modes are:[448]

	ν(Pt—As_2) (cm.$^{-1}$)	δ(Pt—As_2) (cm.$^{-1}$)
trans-$(Et_3As)_2PtCl_4$	328	137
trans-$(Et_3As)_2PtBr_4$	325	142
trans-$(Et_3As)_2PtI_4$	321, 328	144
cis-$(Et_3As)_2PtCl_4$	285, 346	

Adams and Chandler[448] assign ν(Pt—S_2) at 311, 321, and 319 cm.$^{-1}$ and δ(Pt—S_2) at 122, 122, and 128 cm.$^{-1}$ for the complexes *trans*-$(Me_2S)_2PtX_4$ where X = Cl, Br, and I. For *cis*-$(Me_2S)_2PtCl_4$, ν(Pt—S) occurs at

[454] H. Schumann and O. Stelzer, *Angew. Chem. Internat. Edn.*, 1967, **6**, 701.

298–350 cm.$^{-1}$. The vibrational spectra of some *trans*-$Pt^{III}X_2Y_2$ complexes (X = Cl or Br; Y = Me_2S, Me_2Se, and Me_2Te) have been assigned.[455]

Symmetric Pt—L modes: ν(Pt—S)345, ν(Pt—Se)175, ν(Pt—Te)165 cm.$^{-1}$.
Antisymmetric Pt—L modes: ν(Pt—S)312, ν(Pt—Se)230, ν(Pt—Te)230 cm.$^{-1}$.

These frequencies remain relatively unchanged if the central atom is altered to Pd.

Compounds containing M–*Halogen Bonds* (M = Ni, Pd, *and* Pt). Nickel–chlorine and nickel–bromine stretching and bending modes in py–NiX_3 and Ph_3P,NiX_3 complexes (X = Cl or Br) have been assigned.[426]

[PdCl all L] complexes (all = allylic group; L = tertiary phosphine, Ph_3As, Ph_3Sb, or CO) show ν(Pd—Cl) at 274–288 cm.$^{-1}$. This suggests a rather weak Pd—Cl bond because of a large σ-bond weakening effect of the allylic group [456] [cf. ν(Pd—Cl (*trans* to Cl) usually occurs 350–360 cm.$^{-1}$].

The Pd–halogen vibrations for the *cis*- and *trans*-isomers of square planar complexes PdL_2X_2 have been examined thoroughly and some values are given in Table 39. The i.r.-active modes for the molecules belong to the following character species:

		ν(M—X)	ν(M—L)	*Bending modes*
trans-ML_2X_2	D_{2h}	B_{3u}	B_{2u}	$2B_{1u}+B_{2u}+B_{3u}$
cis-ML_2X_2	C_{2v}	A_1+B_1	A_1+B_1	$2A_1+B_1+B_2$

Table 39 *Vibrational frequencies* (cm.$^{-1}$) *of some* cis *and* trans *square planar complexes of palladium*

trans-*Complexes*	ν(Pd—X)	*Ref.*
$Pd(NH_3)_2Cl_2$	333	153*a*, 447
$Pd(ND_3)_2Cl_2$	329	153*a*
Pd py_2Cl_2	353	449
$Pd(C_4H_8OS)_2Cl_2$	350	432
$Pd[(C_6F_5)_3P]_2Cl_2$	365	457
Pd py_2Br_2	256	449
$Pd[(C_6F_5)_3P]_2Br_2$	275	457
$Pd(NH_3)_2I_2$	191	153*a*
Pd py_2I_2	176	449
cis-*Complexes*		
$Pd(NH_3)_2Cl_2$	306, 327	153*a*
Pd py_2Cl_2	333, 342	449
$Pd[Pr^iSe(CH_2)_2SePr^i]Cl_2$ [a]	298, 319	458
$Pd(NH_3)_2Br_2$	258	153*a*
$Pd[Pr^iSe(CH_2)_2SePr^i]Br_2$ [a]	176, 188, 254	458

[a] These complexes crystallise as monomers from acetone and dimers from $CHCl_3$ (26). The dimeric complexes give ν(M—X) bands in the same positions as for the monomers.

455 J. R. Allkins and P. J. Hendra, *J. Chem. Soc.* (*A*), 1967, 1325.
456 J. Powell and B. L. Shaw, *J. Chem. Soc.* (*A*), 1967, 1839.
457 R. D. W. Kemmitt, D. I. Nichols, and R. D. Peacock, *Chem. Comm.*, 1967, 599.
458 N. N. Greenwood and G. Hunter, *J. Chem. Soc.* (*A*), 1967, 1520.

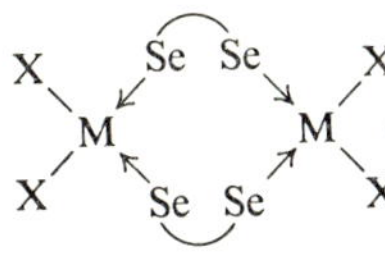 (26)

The stereochemistry of some organophosphorus derivatives of palladium and platinum dichlorides and di-iodides have been examined[459] and ν(M—Cl) for *trans*-complexes of both metals occurs in the range 329–362 cm.$^{-1}$ while ν(M—Cl) for *cis*-complexes occurs in the range 289–334 cm.$^{-1}$. ν(M—I) is more difficult to assign but the authors suggest 151 and 158 cm.$^{-1}$ for two *trans*-complexes and 150, 157 cm.$^{-1}$ for one *cis*-complex. The results in Table 39 confirm the two distinct regions for the Pd—Cl complexes but the distinction disappears for the bromide and iodide complexes. Metal–halogen bending modes for the diammine complexes occur at 135 and 160 cm.$^{-1}$ for δ(Pd—Cl), 100 and 120 cm.$^{-1}$ for δ(Pd—Br), and 109 cm.$^{-1}$ for δ(Pd—I).

Tetrakis(n-butylthiomethyl)methane and tetrakis(phenylthiomethyl)-methane form complexes with Pd and Pt halides of general formula M_2X_4L (27). The ligand acts as a doubly bidentate chelate.[460] ν(M—Cl) for both Pd and Pt complexes occurs in the range 300–330 cm.$^{-1}$ and indicates the presence of mutually *cis*-chlorine atoms.

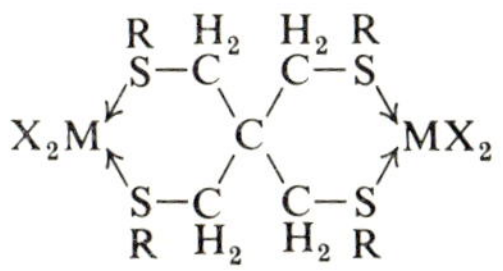

(27) R = Bu^n or Ph

The palladium chloride complex with the di-t-butyl *N*-oxide radical is dimeric $[ClPdON(CMe_3)_2]_2$ and contains Pd—Cl—Pd bridging bonds which give a strong i.r. absorption at 254 cm.$^{-1}$.[461]

The i.r.-active skeletal stretching frequencies of some bridged Pd and Pt chloro-complexes, $[M_2Cl_4L_2]$, have been examined[343] where L = PMe_3, PEt_3, PPr_3^n, PPh_3, $AsMe_3$, $AsEt_3$, SMe_2, or C_2H_4, [(28)–(31)]. The authors give the following ranges (cm.$^{-1}$) for the metal–halogen vibrations, and conclude that there is no justification for assuming that all M—Cl bridging bonds are equivalent.

	ν(M—Cl)term.	ν(M—Cl)bridging (*trans* to Cl)	ν(M—Cl)bridging (*trans* to L)
Pd	339–360	294–308	241–283
Pt	347–368	317–331	257–301

[459] J. M. Jenkins and J. G. Verkade, *Inorg. Chem.*, 1967, **6**, 2250.
[460] D. C. Goodall, *J. Chem. Soc.* (*A*), 1967, 1387.
[461] W. Beck and K. Schmidtner, *Chem. Ber.*, 1967, **100**, 3363.

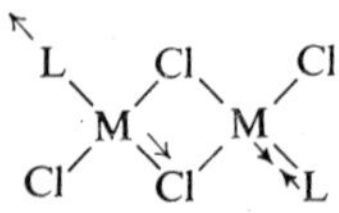

ν(M-L)

(28)

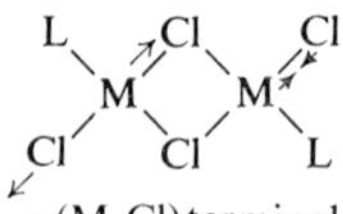

ν(M-Cl)terminal

(29)

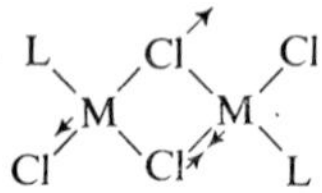

ν(M-L)bridging (*trans* to Cl)

(30)

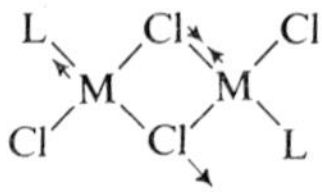

ν(M-L)bridging (*trans* to L)

(31)

Some silyl derivatives of Pt^{II} have been examined and the platinum–halogen stretch is very sensitive to the nature of the substituent on the Si atom.[462] For $[PtCl(SeMePh_2)(PMe_2Ph)_2]$, ν(Pt—Cl) = 242 cm.$^{-1}$ and for $[PtCl(SeCl_3)(PMe_2Ph)_2]$, ν(Pt—Cl) = 274 cm.$^{-1}$. This effect is being studied for any light it may throw on the *trans*-effect. The vibrational spectra of *cis*- and *trans*-square planar complexes of platinum have received considerable attention throughout the year and some results are given in Table 40.

Table 40 *Vibrational frequencies* (cm.$^{-1}$) *of some* cis *and* trans *square planar complexes of platinum*

trans-*Complexes*	ν(Pt—X)	*Ref.*
$Pt[(C_6F_5)_3P]_2Cl_2$	351	457
$Pt(NH_3)_2Cl_2$	330	447
$Pt(NH_3)_2Br_2$	260	447
Pt py_2Br_2	252	449
$Pt(C_4H_8OS)Br_2$	278	432
$Pt[(C_6F_5)_3P]_2Br_2$	255	457
Pt py_2I_2	183	449
cis-*Complexes*		
Pt py_2Cl_2	329, 343	449
$Pt(C_4H_8OS)_2Cl_2$	315, 327	432
$Pt[Pr^iSe(CH_2)_2SePr^i]Cl_2$	304, 324	458
Pt cod Cl_2	318, 338	463
Pt cot Cl_2	325, 345	463
Pt py_2Br_2	235, 252	449
$Pt[Pr^iSe(CH_2)_2SePr^i]Br_2$	204, 234	458
Pt cod Br_2	218	463
Pt cot Br_2	223	463
Pt py_2I_2	167, 178	449
Pt cod I_2	175	463
Pt cot I_2	176	463

Platinum–halogen vibrations for a series of *cis*- and *trans*-complexes of general formula PtX_4L_2 (X = Cl, Br, or I; L = NH_3, py, SMe_2, PEt_3, or

[462] J. Chatt, C. Eaborn, S. Ibekwe, and P. N. Kapoor, *Chem. Comm.*, 1967, 869.
[463] P. Fritz and D. Sellmann, *Z. Naturforsch.*, 1967, 22, *B*, 20.

$AsEt_3$) have also been extensively investigated.[448] For the *trans*-complexes the halogen-sensitive modes occur near those of the related $[PtX_4]^{2-}$ species whilst the ν(Pt—X) pattern for *cis*-complexes covers a wide range which includes that of ν(Pt—X) in bridging situations in Pt^{II} compounds.

H. Copper, Silver, and Gold.—Little vibrational data have been published on these elements during the year except for modes involving metal–halogen vibrations. It has, however, been noted that $CuCl_2$, 2L and $CuBr_2$,2L complexes (L = pyridine *N*-oxide or substituted-pyridine *N*-oxides) exhibit relatively intense absorptions in the range 350–450 cm.$^{-1}$ and these are assigned to Cu—O stretching vibrations.[464]

Several workers have reported copper–halogen vibrations in a wide variety of structures.[116, 432, 464–468] Adams and Lock[116] investigated the far-i.r. spectra of some $[Cu_2X_6]^{2-}$, $CsCuX_3$, $M_2{}^ICuCl_4,2H_2O$, and CuX_2L_2 complexes (X = Cl or Br; L = pyO). Their results indicate that ν(Cu—Cl) lies in the range 222–328 cm.$^{-1}$. If this range is extended (222–368 cm.$^{-1}$) then it includes all the published values for ν(Cu—Cl). Bridging ν(Cu—X—Cu) modes are also assigned but the region is severely overlapped by ν(Cu—X) terminal vibrations and the distinction is of no diagnostic value. Copper–bromine stretching vibrations occur in the range 168–278 cm.$^{-1}$.

Adams, in an earlier paper, suggested that only the higher frequency *X*-sensitive band observed in the spectra of Cu py_2X_2 (X = Cl, Br) complexes arises from a copper–halogen stretching vibration and that the other has a different, but obscure, origin. Campbell and co-workers[466] report two such bands in twelve CuL_2X_2 compounds and conclude that it is normal for such halogen-bridged systems to exhibit two ν(Cu—X) absorptions. When the groups L are very bulky, however, the halogen bridging is inhibited and only one ν(Cu—X) is observed.

Whyman *et al.*[464, 468] have examined the low-frequency i.r. spectra of Cu^{II} complexes with substituted-pyridine *N*-oxides and quinoline *N*-oxides. Both form 1 : 1 and 1 : 2 complexes, CuX_2,L and CuX_2,2L (X = Cl, Br). The values of ν(M—Cl) are very sensitive to the ligand used, even when the structures are thought to be the same. The 1 : 1 complexes with substituted-pyridine *N*-oxides probably possess binuclear oxygen-bridged structures and ν(Cu—Cl)term. lie in the range 305–342 cm.$^{-1}$. The 1 : 2 compounds with substituted-pyridine *N*-oxides can be divided into two types: (*a*) green complexes where ν(Cu—X) is at higher wavenumbers than in the corresponding 1 : 1 complex; and (*b*) yellow complexes where ν(Cu—X) is at a lower position than in the corresponding 1 : 1 complex.

464 R. Whyman and W. E. Hatfield, *Inorg. Chem.*, 1967, **6**, 1859.
465 R. W. Adams, C. G. Barraclough, R. L. Martin, and G. Winter, *Austral. J. Chem.*, 1967, **20**, 2351.
466 M. J. Campbell, M. Goldstein, and R. Grzeskowiak, *Chem. Comm.*, 1967, 778.
467 C. W. Bird and E. M. Briggs, *J. Chem. Soc.* (*A*), 1967, 1004.
468 R. Whyman, D. B. Copley, and W. E. Hatfield, *J. Amer. Chem. Soc.*, 1967, **89**, 3135.

For quinoline *N*-oxide complexes values of ν(Cu—Cl) > 320 cm.$^{-1}$ indicate the presence of a terminal Cu—Cl bond and hence an oxygen-bridged structure; when ν(Cu—Cl) < 320 cm.$^{-1}$ then bridging Cu—Cl—Cu bonds are present.

The i.r. spectra of a number of crystalline compounds containing anionic silver halide complexes have been examined in the 30–400 cm.$^{-1}$ region at room temperature and at *ca.* 143°K. These compounds were of the types RAg_2X_3, $RAgX_2$, and R_2AgI_3 where R is a univalent cation.[469] The metal–halogen frequencies (cm.$^{-1}$) of $[Ag_4X_6]_n^{2n-}$ anionic chains are:[469]

<table>
<tr><th></th><th>ν′</th><th>ν″</th><th>ν‴</th><th>δ′</th><th colspan="2">δ″</th></tr>
<tr><td>$Et_4NAg_2Cl_3$</td><td>184</td><td>155</td><td>134</td><td>89</td><td>75(?)</td><td>82</td></tr>
<tr><td>$Et_4NAg_2Br_3$</td><td colspan="2">139</td><td>109</td><td>74</td><td colspan="2">64</td></tr>
<tr><td>$Et_4NAg_2I_3$</td><td colspan="2">108</td><td>86</td><td></td><td colspan="2">53</td></tr>
</table>

The metal–halogen frequencies (cm.$^{-1}$) of $[Ag_2X_4]_n^{2n-}$ anionic chains are:[469]

	ν'	ν''	δ
Me_4NAgCl_2	170	144	109
Me_4NAgBr_2	124	104	87

Some pertinent lattice modes have also been assigned.

I. Zinc, Cadmium, and Mercury.—The i.r. spectrum of $(Hg(B_{10}H_{12})_2)^{2-}$ in the range 400–4000 cm.$^{-1}$ is identical to that reported previously for $[Zn(B_{10}H_{12})_2]^{2-}$ and $[Cd(B_{10}H_{12})_2]^{2-}$. In the range 200–400 cm.$^{-1}$ each anion exhibits a single definitive absorption which may be assigned to ν_{asym} (MB_4) of the tetrahedral MB_4 unit:[470] $[Zn(B_{10}H_{12})_2]^{2-}$ 278; $[Cd(B_{10}H_{12})_2]^{2-}$ 235; and $[Hg(B_{10}H_{12})_2]^{2-}$ 225 cm.$^{-1}$.

Clarke and Woodward[471] have examined the vibrational spectra of the tris(methylmercuric)sulphonium and tris(methylmercuric)oxonium ions. While it is concluded that the $[(MeHg)_3S]^+$ ion has a pyramidal (C_{3v}) skeletal structure, the results obtained for the $[(MeHg)_3O]^+$ ion are best explained on the basis of D_{3h} skeletal symmetry. It is suggested that the latter ion has a planar or very nearly planar structure:

	ν (cm.$^{-1}$)	*Assignment*
$[(MeHg)_3S]^+$	544	ν(Hg—C) sym. str.
	540	ν(Hg—C) asym. str.
	281	$\nu(Hg_3S)$ sym. str.
	332	$\nu(Hg_3S)$ asym. str.
$[(MeHg)_3O]^+$	563	ν(Hg—C) sym. str.
	599	ν(Hg—C) asym. str.
	132	$\nu(Hg_3O)$ sym. str.
	549	$\nu(Hg_3O)$ asym. str.

[469] G. L. Bottger and A. L. Geddes, *Spectrochim. Acta*, 1967, **23**, *A*, 1551.
[470] N. N. Greenwood and N. F. Travers, *Chem. Comm.*, 1967, 216.
[471] J. H. R. Clarke and L. A. Woodward, *Spectrochim. Acta*, 1967, **23**, *A*, 2077.

Metal–nitrogen modes have been assigned for zinc halide complexes with pyridine, 2,2′-bipyridyl, and 2,2′,2″-terpyridine,[472] for tetrahedral anionic complexes of zinc with NCO^-, NCS^-, and $NCSe^-$,[473] and for the Raman spectra of mixed complexes of glycine with zinc and cadmium in aqueous solution[474], as indicated in Table 41. For the latter complexes, the corresponding Cd—N modes are in the range 400–415 cm.$^{-1}$,[474] and the Hg—N mode for the monoglycine complexes is at 464 cm.$^{-1}$.[475]

Table 41 *Some* Zn—N *vibrational frequencies* (cm.$^{-1}$) [472, 473, 474]

Compound	ν(Zn—N)	δ(Zn—N)
Zn py_2Cl_2	218	154
Zn py_2Br_2	219	153
Zn py_2I_2	222	147, 159
Zn bipy Cl_2	241	192
Zn bipy Br_2	250	190
Zn bipy I_2	250	194
Zn terpy Cl_2	244	167
Zn terpy Br_2	243	167
Zn terpy I_2	245	167
	ν_{sym}	ν_{asym}
$Zn(NCO)_4^{2-}$	~ 330	326
$Zn(NCS)_4^{2-}$	255	285
$Zn(NCSe)_4^{2-}$		~ 235
Zn–monoglycine complexes		~ 445
Zn–bisglycine complexes		~ 430
Zn–trisglycine complexes		~ 425

Some Zn—P vibrations are:[426] $(Ph_3P)_2ZnCl_2$, 140, 166; $(Ph_3P)_2ZnBr_2$, 128, 138, 156; $[Et_4N][Ph_3PZnCl_3]$, 146; and $[Et_4N][Ph_3PZnBr_3]$, 141 cm.$^{-1}$.

For a series of anions, $M(NO_2)_4^{2-}$ (M = Mn^{II}, Co^{II}, Cu^{II}, and Zn^{II}), ν(M—O) have been assigned to a strong band at 300 ± 50 cm.$^{-1}$.[476] For the corresponding Cd and Hg anions, bands at 230–40 cm.$^{-1}$ are assigned to modes with appreciable M—O stretching character.

Flint and Goodgame report the following assignments for thiourea complexes:[130]

Compound	ν(Zn—tu) (cm.$^{-1}$)	*Compound*	ν(Cd—tu) (cm.$^{-1}$)
Zn $tu_2(OAc)_2$	248	Cd $tu_2(OAc)_2$	(221, 228, 238)
Zn $tu_3(OSO_3)$	249	Cd $tu_3(OSO_3)$	(200, 218, 228)
Zn $tu_3(OSeO_3)$	245, 251, 270	Cd $tu_3(S_2O_3)$	(215, 236)
Zn $tu_3(S_2O_3)$	(235, 254)	Cd $tu_4(ClO_4)_2$	209, 237
Zn $tu_4(ClO_4)_2$	238, 250	Cd $tu_4(NO_3)_2$	237
Zn $tu_4(NO_3)_2$	238, 258		

472 C. Postmus, J. R. Ferraro, and W. Wozniak, *Inorg. Chem.*, 1967, **6**, 2030.
473 D. Forster and W. de W. Horrocks, jun., *Inorg. Chem.*, 1967, **6**, 339.
474 K. Krishnan and R. A. Plane, *Inorg. Chem.*, 1967, **6**, 55.
475 T. V. Long, jun., and C. M. Yoshida, *Inorg. Chem.*, 1967, **6**, 1754.
476 D. M. L. Goodgame and M. A. Hitchman, *J. Chem. Soc.* (*A*), 1967, 612.

The bands shown in parentheses probably arise from strongly coupled metal–thiourea and metal–anion modes.

The tetrahedral diphenylcyclopropenone (dpcp) complexes of ZnX_2 show metal–halogen stretching vibrations for $dpcp_2ZnCl_2$ at 303 and 327 cm.$^{-1}$ and for $dpcp_2ZnBr_2$ at 247 cm.$^{-1}$.[467] Some other Zn—X vibrational frequencies are shown in Table 42.[472].

Table 42 *Some* Zn–*halogen vibrational frequencies* (cm.$^{-1}$)

Compound	ν(Zn—Cl)	ν(Zn—Br)	ν(Zn—I)	δ(Zn—Cl)	δ(Zn—Br)	δ(Zn—I)
py_2ZnCl_2	293, 326			200		
py_2ZnBr_2		254, 260			182	
py_2ZnI_2			210			167
bipy $ZnCl_2$	323					
bipy $ZnBr_2$		261				
bipy ZnI_2			217			
terpy $ZnCl_2$	278, 287					
terpy $ZnBr_2$		213, 222				
terpy ZnI_2			187			

The presence of an ν(Hg—Cl) vibration in some novel organo-mercury complexes provides evidence for a covalent structure (32) [477] rather than an ionic one (33): [447]

	ν(Hg—Cl) (cm.$^{-1}$)
C_6F_5HgCl,bipy	274
C_6F_5HgCl,dimethylphen	*ca.* 245
C_6F_5HgCl,phen	263

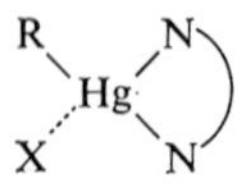

(32)

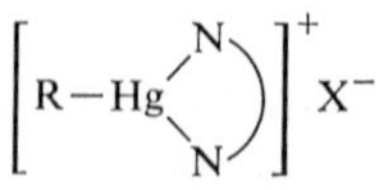

(33)

$HgCl_2(C_4H_8OS)_2$ shows ν(Hg—Cl) at 235 and 253 cm.$^{-1}$. These values are considerably lower than for related complexes of $HgCl_2$ and the usual criterion of decreasing values of ν(M—X) with increasing co-ordination number is of limited value in such systems, cf. $HgCl_4^{2-}$ 228 cm.$^{-1}$).[432] The pyridine *N*-oxide complex $HgCl_4,(pyO)_2$ must have a bridge-bonded structure since ν(HgCl) at 278, 313, and 340 cm.$^{-1}$ can only be explained in terms of bridging-chlorine modes.[478]

[477] A. J. Canty, G. B. Deacon, and P. W. Felder, *Inorg. Nuclear Chem. Letters*, 1967, **3**, 263.
[478] G. Schmauss and H. Specker, *Naturwiss.*, 1967, **54**, 248.

J. Lanthanides and Actinides.—The i.r. spectra of some tris(acetylacetonato)-lanthanide(III) complexes have been examined and force constant calculations have been carried out. ν(M—O) are at 420–432 and 304–322 cm.$^{-1}$, and the stretching force-constants increase with increasing atomic number. As a result of the lanthanide contraction, the force-constants increase with a decrease in the M—O bond-length.

The metal–oxygen vibrations in some protoactinium oxide halides have been assigned as:[479]

$PaOBr_3$	515	476	364	303 cm.$^{-1}$
$PaOI_3$	480		339	276 cm.$^{-1}$
PaO_2Br	575		386	286 cm.$^{-1}$
PaO_2I	555	469	386	281 cm.$^{-1}$

Sandhu and Sandu[480] have assigned ν(M—O) for some complexes of U^{VI} with oxides of ditertiary phosphines; ν(U=O) occurs in the range 912–930 cm.$^{-1}$ for UO_2^{2+} compounds but falls by 20–30 cm.$^{-1}$ on complex formation. Ohwada[481] has measured the i.r. spectrum of the uranyl ion attached to ion-exchange resins in order to determine accurately the U—O bond-length; the force-constant *k'* was determined from ν_{asym}(U—O) and then the U—O bond-length was calculated using Badger's rule.

In the americium complex, $Cs_2(Am^{VI}O_2)Cl_4$, ν(Am=O) occurs at 902 cm.$^{-1}$, and in $Cs_8(Am^{V}O_2)_3Cl_{11}$ it occurs at 800 cm.$^{-1}$.[482]

PART III: Vibrational Spectra of Some Co-ordinated Ligands

The sequence of presentation follows as closely as possible the ordering of donor-atoms according to the vertical Groups of the Periodic Table though in some cases the actual donor atom is not known with certainty. Some ligands bond via one atom in some complexes and via a second atom in other complexes and it has proved convenient to treat both sets of complexes in the same subsection, *e.g.* nitro- and nitrito-complexes, cyanato- and isocyanato-complexes, etc.

1 Olefins, Acetylenes, and Cyclic Polyenes

Information about the vibrational spectra of the co-ordinated ligands in the following complexes has been reported during the year:

$AgClO_4,C_{10}H_{18}$ and $2AgClO_4,3C_6H_{10}$;[483]

complexes of $PdCl_2$ with vinylcyclopropanes;[484]

$[LPt(Ph_3P)_2]$ where L is CH_2:CHCN, CH_2:CHCOMe, *trans*-MeO_2C·CH:CH·CO_2Me, and CO·CH:CH·CO·O (cyclic, CO bonded to O);[485]

479 D. Brown, J. F. Easey, and P. J. Jones, *J. Chem. Soc.* (*A*), 1967, 1698.
480 S. S. Sandhu and S. S. Sandhu, *Chem. and Ind.*, 1967, 1405.
481 K. Ohwada, *J. Inorg. Nuclear Chem.*, 1967, **29**, 1802.
482 K. W. Bagnall, J. B. Laidler, and M. A. A. Stewart, *Chem. Comm.*, 1967, **1**, 241.
483 G. Bressan, R. Broggi, M. P. Lachi, and A. L. Segre, *J. Organometallic Chem.*, 1967, **9**, 355.
484 A. D. Ketley and J. A. Braatz, *J. Organometallic Chem.*, 1967, **9**, P5.
485 S. Cenini, R. Ugo, F. Bonati, and G. La Monica, *Inorg. Nuclear Chem. Letters*, 1967 **3**, 191.

[$LCoX_2$], [$LNiX_2$], and [$LCuX_2$] where L is 1,2-di-(2-pyridyl)ethylene or 1,2-di(4-pyridyl)ethylene;[486]

trans-dichloro-*cis*-2-butene(4-Z-pyridine)platinum(II) complexes;[486a]

dimethallylzinc and dicrotylzinc;[487]

π-allylic complexes of Pd;[488]

acetylene complexes of Ir^I and Rh^I;[489] and

mono- and di-phenylacetylene complexes with I_2 and $SbCl_3$.[490]

Several papers[491–496a] are concerned with cyclopentadienyl complexes of the transition elements but to understand many of the structural inferences made the reader is referred to an extensive review by Fritz (*Adv. Organometallic Chem.*, 1964, **1**, 240) and also to a paper by Rosenblum (*Chem. & Ind.*, 1958, 953). The latter has shown, for example, that ferrocene derivatives possessing an unsubstituted ring exhibit i.r. absorptions near 1000–1100 cm.$^{-1}$ whereas those derivatives in which both rings are either singly or multiply substituted do not possess such an absorption. Sullivan and Little have shown that this useful rule for substituted ferrocenes appears to have its counterpart in some mono- and di-alkyltitanocene dichlorides.[493]

2 Carbonyls

The carbonyl stretching frequencies of the following vanadium complexes have been given:[497]

$$(\pi\text{-}C_5H_5)V(CO)_4,\ (\pi\text{-}C_5H_5)V(CO)_3Bu_3P,\ (\pi\text{-}C_5H_5)V(CO)_3PhEt_2P,$$
$$(\pi\text{-}C_5H_5)V(CO)_3Ph_3P,\ (\pi\text{-}C_5H_5)V(CO)_2(Bu_3P)_2,\ (\pi\text{-}C_5H_5)V(CO)_2(PhEt_2P)_2,$$
$$(\pi\text{-}C_5H_5)V_2(CO)_5,\ \text{and}\ (\pi\text{-}C_5H_5)V_2(CO)_4Ph_3P.$$

Ang and West[498] concluded that carbonyl complexes of chromium, molybdenum, and tungsten, which also contain a monodentate phosphine ligand [$LM(CO)_5$], show: (*a*) an intense absorption band at 1930–1960 cm.$^{-1}$; (*b*) a sharp band of medium intensity 2070–2085 cm.$^{-1}$; and (*c*) a band on the high side of the 1930–1960 cm.$^{-1}$. When the phosphine group behaves as a bidentate ligand [$L^*M(CO)_4$] the following features were observed:

486 (Mrs.) M. Brierley and W. J. Geary, *J. Chem. Soc.* (*A*), 1967, 963.
486a P. Schmidt and M. Orchin, *Inorg. Chem.*, 1967, **6**, 1260.
487 K.-H. Thiele, G. Engelhardt, J. Köhler, and M. Arnstedt, *J. Organometallic Chem.*, 1967, **9**, 385.
488 F. A. Cotton, J. W. Faller, and A. Musco, *Inorg. Chem.*, 1967, **6**, 179.
489 J. P. Collman and J. W. Kang, *J. Amer. Chem. Soc.*, 1967, **89**, 844.
490 J. Gerbier and V. Lorenzelli, *Compt. rend.*, 1967, **264**, *B*, 690.
491 G. Marr, *J. Organometallic Chem.*, 1967, **9**, 147.
492 R. E. Moore, B. W. Rockett, and D. G. Brown, *J. Organometallic Chem.*, 1967, **9**, 141.
493 M. F. Sullivan and W. F. Little, *J. Organometallic Chem.*, 1967, **8**, 277.
494 H. A. Martin and F. Jellinek, *J. Organometallic Chem.*, 1967, **8**, 115.
495 M. Van den Akker and F. Jellinek, *Rec. Trav. chim.*, 1967, **89**, 897.
496 I. J. Hyams, R. T. Bailey, and E. R. Lippincott, *Spectrochim. Acta*, 1967, 23, *A*, 273.
496a R. Hüttel, U. Raffay, and H. Reinheimer, *Angew. Chem. Internat. Edn.*, 1967, **6**, 862.
497 E. O. Fischer and R. J. J. Schneider, *Angew. Chem. Internat. Edn.*, 1967, **6**, 569.
498 H. G. Ang and B. O. West, *Austral. J. Chem.*, 1967, **20**, 1133.

(*a*) a sharp band at 2010–2040 cm.$^{-1}$; (*b*) a strong peak *ca.* 1900 cm.$^{-1}$; (*c*) a shoulder 1920–1928 cm.$^{-1}$; (*d*) an inflection 1890 cm.$^{-1}$.

King[499] has calculated approximate force constants from the $\nu(C{\equiv}O)$ values using the equations developed by Cotton and Kraihanzel. A comparison of the calculated force constants indicate that the π-acceptor strengths of HNC- and Me_3MNC-ligands (M = Si or Sn) are approximately equal to the π-acceptor strength of MeCN.

For *cis*-$L_2Mo(CO)_4$ derivatives,[500] the group lability decreases with changes in L in the order: $PCl_3 \gg py \gg C_8H_{12} \sim AsPh_3 \geqslant PPh_3 \geqslant SbPh_3 \gg CO$. A comparison between the order of lability and $\nu(C{\equiv}O)$ is presented. The order of reactivity for analogous carbonyl derivatives involving the same ligands with different metals depends strongly on the nature of the central atom, the order being Mo > Cr > W. For the hexacarbonyl compounds this is related to the order of the strengths of the M—C bonds as deduced from i.r. studies.

Calculation based on a recent analysis of the force constants for CO vibrations in $M(CO)_5X$ complexes (M = Mo or Re; X = halogen) confirm that the observed intensity ratio for the A_1 modes cannot be accounted for solely through vibrational coupling. Braterman *et al.*[501], using improved force-constant values, have calculated from the intensity data both an angle between oscillators, θ, and the effective dipole-moment derivative for each of the chemically different CO groups. They also calculate from the intensities of the all [^{12}C] carbonyl derivatives the expected intensities of [^{13}C] substituted carbonyl molecules, from which the concentration of various specifically labelled species can be deduced.

Braterman and Thompson[502] also determine the structures of metal carbonyl derivatives in solution by making use of combination spectra to assess the symmetry and to furnish evidence for metal–metal bonding. Theory predicts that when two conditions are satisfied, binuclear metal carbonyls will have a 'high-frequency inactive mode ν_1' in the CO stretching region, such that the following relationships hold:

$$\nu_{max}(c) = \nu_1 + \nu_2$$
$$\nu_{max}(f) = \nu_2$$
$$\nu_1 > \nu_2.$$

The conditions are that (*a*) the point group of the molecule must be of suitably high symmetry, and (*b*) there must exist a direct interaction between CO groups on different metals, such as could arise from a metal–metal bond.

The effects of solvents on the i.r. spectra of $Cr(CO)_6$, $Mo(CO)_6$, and $W(CO)_6$ have been investigated over the range 70–2100 cm.$^{-1}$ for twelve solvents.[503]

499 R. B. King, *Inorg. Chem.*, 1967, **6**, 25.
500 M. Graziani, F. Zingales, and U. Belluco, *Inorg. Chem.*, 1967, **6**, 1582.
501 P. S. Braterman, R. Bau, and H. D. Kaesz, *Inorg. Chem.*, 1967, **6**, 2097.
502 P. S. Braterman and D. T. Thompson, *J. Organometallic Chem.*, 1967, **10**, P11.
503 R. J. H. Clark and B. Crociani, *Inorg. Chim. Acta*, 1967, **1**, 12.

Complete vibrational assignments have been made for $Re_2(CO)_{10}$ and $Re(CO)_5I$ based on i.r. and Raman data.[504] The data are consistent with D_{4d} symmetry for $Re_2(CO)_{10}$ and C_{4v} for $Re(CO)_5I$ as shown in (34) and (35). The ν(Re—Re) vibration is assigned to a Raman line near 105 cm.$^{-1}$ and ν(Re—I) to a Raman line near 165 cm.$^{-1}$.

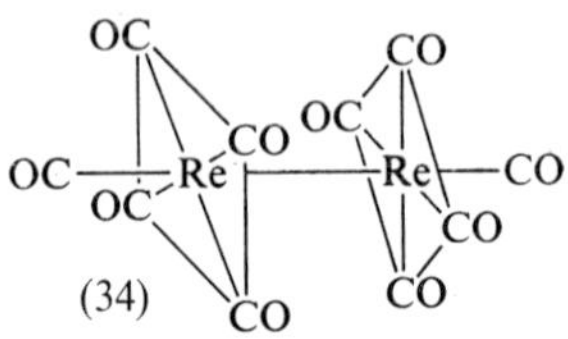

(34)

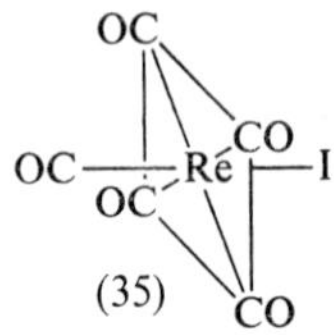

(35)

Some phosphorus halide ligands in metal carbonyl derivatives have been classified according to the π-acceptor strength of the P atom in the ligands as derived from the changes in position of ν(C≡O).[505] The order is independent of the metal atom (M = V, Cr, Mn, Fe, Co, Ni, Mo, or W) and can be classified as: $PR_3 < PPh_3 < PClPh_2 < PCl_2Ph < PCl_3 < PF_3$.

The carbonyl stretching frequencies are assigned for some pentafluorophenyltin derivatives of manganese pentacarbonyl.[506] A linear relation between ν(C≡O) and the electronegativity of the substituents on the tin atom suggests a Pauling electronegativity of *ca.* 2·4 for the C_6F_5 group.

The intensities of carbonyl stretching modes have been examined in detail.[507, 508] In numerous instances intensity variations, for a particular band within a series of compounds are more evident than are frequency shifts for the same band. The intensity of the ν_3(CO) stretch of $Mn_2(CO)_{10}$ and $Re_2(CO)_{10}$ is very sensitive to structural changes and Wing and Crocker[507] estimate the distortion of the radial carbonyl groups from planarity, from their intensity data.

Several workers have studied isotopically substituted metal carbonyls (^{13}CO[509–512] and C^{18}O[513]) and detailed assignments are proposed, *e.g.* for $LMn(CO)_5$ complexes.[512]

	Radial ^{12}CO *groups*			*Axial* ^{12}CO *groups*	*Acetyl* ^{12}CO
L	A_1	E	A'	A_1	
Me	2110	2012	(1976)	1991 (1949)	
MeCO	2115	2011	(1970)	2003 (1963)	1664 (1625)

Numbers in parentheses denote ^{13}CO frequencies (cm.$^{-1}$).

[504] I. J. Hyams, D. Jones, and E. R. Lippincott, *J. Chem. Soc.* (*A*), 1967, 1987.
[505] W. Strohmeier and F.-J. Müller, *Chem. Ber.*, 1967, **100**, 2812.
[506] J. A. J. Thompson and W. A. G. Graham, *Inorg. Chem.*, 1967, **6**, 1875.
[507] R. M. Wing and D. C. Crocker, *Inorg. Chem.*, 1967, **6**, 289.
[508] E. W. Abel and I. S. Butler, *Trans. Faraday Soc.*, 1967, **63**, 45.
[509] P. S. Braterman, R. W. Harrill, and H. D. Kaesz, *J. Amer. Chem. Soc.*, 1967, **89**, 2851.
[510] H. D. Kaesz, R. Bau, D. Hendrickson, and J. M. Smith, *J. Amer. Chem. Soc.*, 1967, **89**, 2844.
[511] F. A. Cotton, A. Musco, and G. Yagupsky, *Inorg. Chem.*, 1967, **6**, 1357.
[512] K. Noack and F. Calderazzo, *J. Organometallic Chem.*, 1967, **10**, 101.
[513] B. F. G. Johnson, J. Lewis, J. R. Miller, B. H. Robinson, P. W. Robinson, and A. Wojcicki, *Chem. Comm.*, 1967, 379.

The structure of the anion $[Fe(CO)_4I]^-$ is probably trigonal bipyramidal with the iodine atom in an apex position.[514] Octacarbonyldi-iododi-iron, $[Fe_2(CO)_8I_2]$, has also been investigated[515] and the staggered (D_{4d}) structure (36) has been proposed.

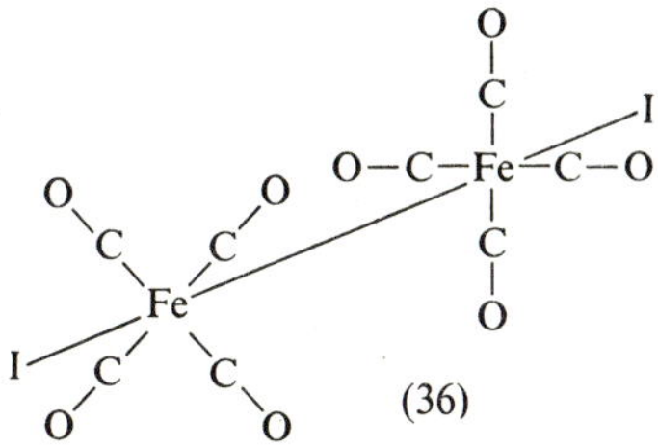

(36)

Two groups of workers,[516-518] have examined the dimeric complexes $[(\pi\text{-}C_5H_5)Fe(CO)_2]_2$ and $[(\pi\text{-}C_5H_5)Ru(CO)_2]_2$ in solution. The structure of solid $(\pi\text{-}C_5H_5)_2Fe_2(CO)_4$ is known from *X*-ray work to be (37). In solution, however, the i.r. spectrum is anomalous but may be explained in terms of a tautomerisation to structure (39) via structure (38). The structure which is not carbonyl bridged is only present in minute concentrations for the iron complex but in solution $[(\pi\text{-}C_5H_5)_2Ru_2(CO)_4]$ exists as a mixture of tautomers (38) and (39) and it exhibits six ν(CO) bands in the i.r. region.[518] The i.r. spectrum is given in Figure 2. Enthalpy and entropy differences between the bridged and non-bridged forms have also been calculated:[517]

$[(\pi\text{-}C_5H_5)Fe(CO)_2]_2$ in nonane solution, $\Delta H = 4$ kcal mole^{-1}, $\Delta S = 3$ cal deg.$^{-1}$ mole^{-1}. At 30°, 0·6% of (38) is present in the equilibrium.

$[(\pi\text{-}C_5H_5)Ru(CO)_2]_2$ in CS_2 solution, $\Delta H = 1{\cdot}56$ kcal. mole^{-1} $\Delta S = 5{\cdot}5$ cal deg.$^{-1}$ mole^{-1}. At 30°, 45% of (39) and 55% of (38) are present in the equilibrium.

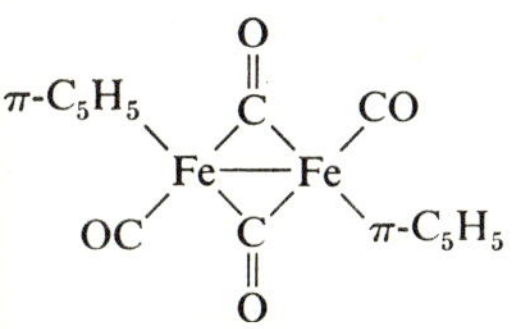

(37)

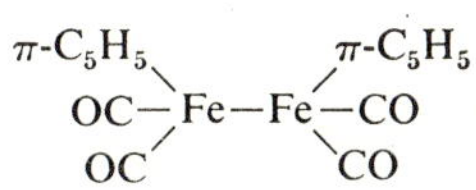

(38)

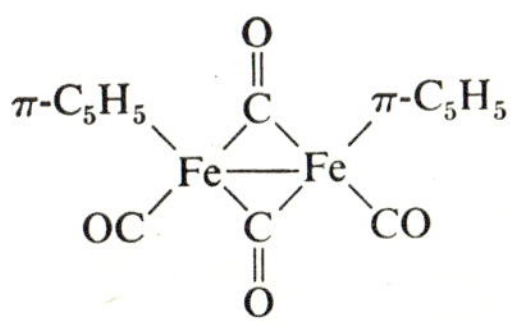

(39)

$(MeCl_2Si)Fe(CO)_2(\pi\text{-}C_5H_5)$ also gives an anomalous i.r. spectrum in the carbonyl region.[519] The authors suggest the presence of two conformers (40) and (41). A similar effect is observed for $MeCl_2SnFe(CO)_2(\pi\text{-}C_5H_5)$.

514 E. W. Abel, I. S. Butler, and C. R. Jenkins, *J. Organometallic Chem.*, 1967, **8**, 382.
515 F. A. Cotton and B. F. G. Johnson, *Inorg. Chem.*, 1967, **6**, 2113.
516 R. D. Fischer, A. Vogler, and K. Noack, *J. Organometallic Chem.*, 1967, **7**, 135.
517 K. Noack, *J. Organometallic Chem.*, 1967, **7**, 151.
518 F. A. Cotton and G. Yagupsky, *Inorg. Chem.*, 1967, **6**, 15.
519 W. Jetz and W. A. G. Graham, *J. Amer. Chem. Soc.*, 1967, **89**, 2773.

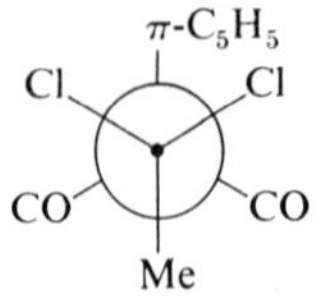

(40)

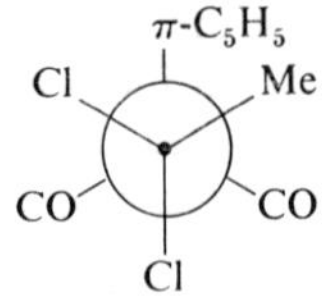

(41)

The tendency of molecules to be stable without CO bridges increases as the group is descended; thus $Co_4(CO)_{12}$ and $Rh_4(CO)_{12}$ contain CO bridges whereas $Ir_4(CO)_{12}$ does not. For this reason the complex $(\pi\text{-}C_5H_5)_2Os_2(CO)_4$ is expected to have a non-bridged structure and this is now confirmed both in solution and in the solid state.[516]

A symmetrical linear structure is proposed[520] for $Re_2Fe(CO)_{14}$ which would be analogous to $Mn_2Fe(CO)_4$, *i.e.* $(CO)_5Re\cdot Fe(CO)_4\cdot Re(CO)_5$ (D_{4h}).

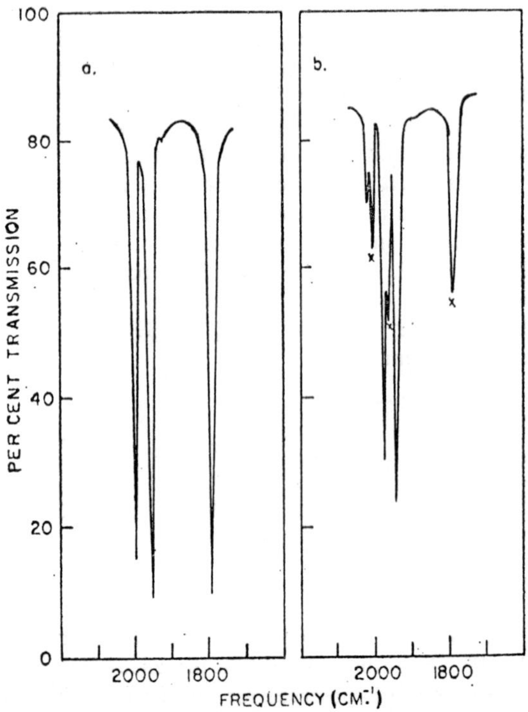

Figure 2 *The spectra of* (a) $(\pi\text{-}C_5H_5)_2Fe_2(CO)_4$ *and* (b) $(\pi\text{-}C_5H_5)_2Ru_2(CO)_4$ (*in* CS_2 *solution at* 25°) *in the* CO *stretching region*
(Reproduced by permission from *Inorg. Chem.*, 1967, **6**, 15)

[520] G. O. Evans, J. P. Hargaden, and R. K. Sheline, *Chem. Comm.*, 1967, 186.

Finally, structure (42) has been confirmed for tricarbonylcyclo-octatetraene-iron in solution, although two other groups of workers have recently disagreed with this. N.m.r. data also favours this structure.[520a]

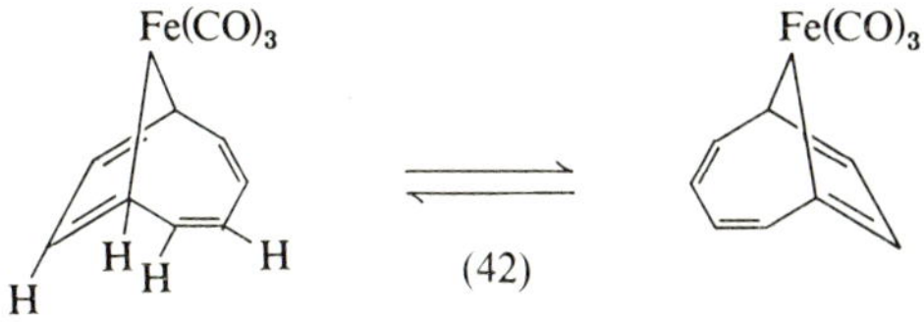

(42)

The 3,3,3-trifluoropropyne complex of cobalt carbonyl[521] shows three terminal carbonyl stretching bands at 2028, 2040, and 2056 cm.$^{-1}$ and the proposed structure is (43). The C—C link lies above and normal to the Co—Co axis.

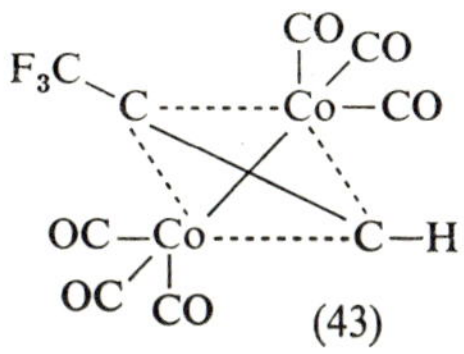

(43)

The first reported examples (44) and (45) of a metal–metal bond supported by bridging tin and carbonyl groups is given by Patmore and Graham;[522, 523] (45) is the first example with $x = 1$, cf. $Os_3(CO)_{12}$, $x = 3$; $Fe_3(CO)_{12}$, $x = 2$.

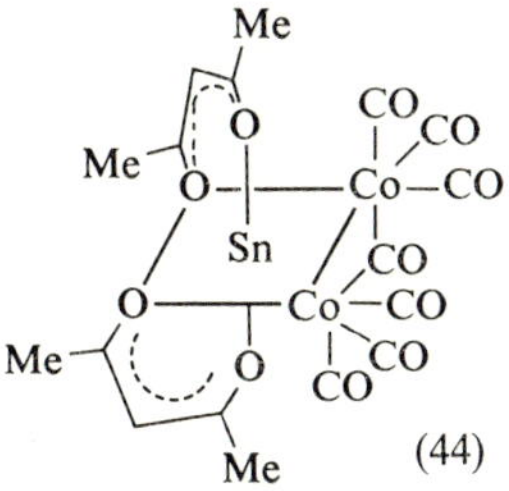

(44)

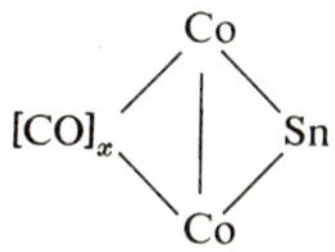

(45)

3 Nitrogen Donors

Considerable interest still centres on complexes of molecular nitrogen and these will be considered first, together with azido-complexes. Sections follow on amine and polyamine ligands, Schiff bases, nitrites, cyanides, and nitrosyls. Ligands in which more than one element can act as donor atom,

[520a] F. A. L. Anet, H. D. Kaesz, A. Maasbol, and S. Winstein, *J. Amer. Chem. Soc.*, 1967, **89**, 2489.
[521] R. S. Dickson and D. B. W. Yawney, *Austral. J. Chem.*, 1967, **20**, 77.
[522] D. J. Patmore and W. A. G. Graham, *Chem. Comm.*, 1967, **1**, 7.
[523] D. J. Patmore and W. A. G. Graham, *Inorg. Chem.*, 1967, **6**, 1879.

e.g. NO_2^-, NCO^-, NCS^-, and $NCSe^-$, will then be discussed. Amide ligands have been studied by several authors[524a–e, 525a–d] who infer the site of donation by examining the shifts induced in ν(C=O) and ν(N—H) as the ligand is complexed. These papers will not be considered further; the spectra of many other nitrogen-containing ligands have also been published during the year.[526a–n, 527a–o]

Molecular Nitrogen and Azido-Complexes.—Data on the N≡N stretching vibration are summarized in Table 43. The shoulder which is present in the

Table 43 *Molecular nitrogen complexes; frequencies in* $cm.^{-1}$

Compound	ν(N≡N)	*Ref.*
$[Ru(NH_3)_5\cdot N_2]Cl_2$	2105	528
$[Ru(NH_3)_5\cdot N_2]Br_2$	2114	528
$[Ru(NH_3)_5\cdot N_2]I_2$	2129	528
$[Ru(NH_3)_5\cdot N_2](BF_4)_2$	2144	528
$[Ru(NH_3)_5\cdot N_2](PF_6)_2$	2167	528

524 (*a*) S. K. Madan, *Inorg. Chem.*, 1967, **6**, 421; (*b*) C. S. Kraihanzel and D. N. Stehly, *Inorg. Chem.*, 1967, **6**, 277; (*c*) S. C. Jain and R. Rivest, *Canad. J. Chem.*, 1967, **45**, 139; (*d*) D. Schwartz and R. Heyer, *J. Inorg. Nuclear Chem.*, 1967, **29**, 1384; (*e*) E. Uhlig and V. Neugebauer, *Z. anorg. Chem.*, 1967, **351**, 286.

525 (*a*) R. Nast and P. Dilly, *Angew. Chem. Internat. Edn.*, 1967, **6**, 357; (*b*) C. Ringel and H.-A. Lehmann, *Z. anorg. Chem.*, 1967, **353**, 158; (*c*) J. E. Schwarberg, D. R. Gere, R. E. Sievers, and K. J. Eisentraut, *Inorg. Chem.*, 1967, **6**, 1933; (*d*) G. Kaufmann, J. F. Leroy, and R. Rohmer, *Bull. Soc. chim. France*, 1967, 900.

526 (*a*) R. W. Hay and L. J. Porter, *Austral. J. Chem.*, 1967, **20**, 675; (*b*) M. K. Kim and A. E. Martell, *J. Amer. Chem. Soc.*, 1967, **89**, 5138; (*c*) O. Dahl and P. H. Nielsen, *Acta Chem. Scand.*, 1967, **21**, 1153; (*d*) W. Beck, H. S. Smedal, and H. Köhler *Z. anorg. Chem.*, 1967, **354**, 69; (*e*) D. St. C. Black, *Austral. J. Chem.*, 1967, **20**, 2275; (*f*) J. H. Weber, *Inorg. Chem.*, 1967, **6**, 258; (*g*) N. F. Curtis and D. A. House, *J. Chem. Soc.* (*A*), 1967, 537; (*h*) Miss P. R. Shukla, *J. Inorg. Nuclear Chem.*, 1967, **29**, 1800; (*i*) M. B. Koromantzou and G. A. Pneumaticakis, *Chem. and Ind.*, 1967, 1089; (*j*) A. Corsini, J. Abraham, and M. Thompson, *Chem. Comm.*, 1967, 1101; (*k*) A. D. Ahmed and P. K. Mandal, *J. Inorg. Nuclear Chem.*, 1967, **29**, 2759; (*l*) C. A. McAuliffe, *J. Chem. Soc.* (*A*), 1967, 641; (*m*) G. W. A. Fowles and B. J. Russ, *J. Chem. Soc.* (*A*), 1967, 517; (*n*) L. E. Sutton and M. E. Kenney, *Inorg. Chem.*, 1967, **6**, 1869.

527 (*a*) L. F. Lindoy, S. E. Livingstone, and T. N. Lockyer, *Austral. J. Chem.*, 1967, **20**, 471; (*b*) E. G. Jäger, *Z. anorg. Chem.*, 1967, **349**, 139; (*c*) R. C. Paul and S. L. Chadha, *Spectrochim. Acta*, 1967, **23**, *A*, 1249; (*d*) R. C. Aggarwal and R. C. Makhija, *Z. anorg. Chem.*, 1967, **349**, 101; (*e*) J. A. Scruggs and M. A. Robinson, *Inorg. Chem.*, 1967, **6**, 1007; (*f*) N. K. Dutt and S. Upadhyaya, *J. Inorg. Nuclear Chem.*, 1967, **29**, 1368; (*g*) G. E. Batley and D. P. Graddon, *Austral. J. Chem.*, 1967, **20**, 877; (*h*) G. E. Batley and D. P. Graddon, *Austral. J. Chem.*, 1967, **20**, 1749; (*i*) J. D. Curry, M. A. Robinson, and D. H. Busch, *Inorg. Chem.*, 1967, **6**, 1570; (*j*) G. J. Sutton, *Austral. J. Chem.*, 1967, **20**, 1859; (*k*) C. H. Misra, S. S. Parmar, and R. C. Arora, *Inorg. Nuclear Chem. Letters*, 1967, **3**, 603; (*l*) R. M. Issa, M. F. Iskander, and M. F. El-Shazly, *Z. anorg. Chem.*, 1967, **354**, 98; (*m*) R. M. Issa, M. F. El-Shazly, and M. F. Iskander, *Z. anorg. Chem.*, 1967, **354**, 90; (*n*) K.-H. Thiele and H. Rau, *Z. anorg. Chem.*, 1967, **353**, 127; (*o*) M. B. Ćelap, S. R. Niketić, T. J. Janjić, and V. N. Nikolić, *Inorg. Chem.*, 1967, **6**, 2063.

528 A. D. Allen, F. Bottomley, R. O. Harris, V. P. Reinsalu, and C. V. Senoff, *J. Amer. Chem. Soc.*, 1967, **89**, 5595.

Table 43 (*cont.*)

Compound	$\nu(N\equiv N)$	*Ref.*
$[Os(NH_3)_5 \cdot N_2]Cl_2$	2010, 2011sh	529
$[Os(NH_3)_5 \cdot N_2]Br_2$	2028, 2035sh	529
$[Os(NH_3)_5 \cdot N_2]I_2$	2033, 2043sh	529
$[Os(NH_3)_5 \cdot N_2](ClO_4)_2$	2051, 2062sh	529
$[Os(NH_3)_5 \cdot N_2](BF_4)_2$	2055, 2064sh	529
$[Os(NH_3)_5 \cdot N_2](BPh_4)_2$	2061	529
$(Ph_3P)_3Co \cdot N_2$	2088	414, 530, 531
$(Ph_3P)_3CoH \cdot (N_2)$	2080–2084	532
$(EtPh_2P)_3CoH \cdot (N_2)$	2080–2084	532

solid phase spectra of the osmium compounds is probably due to a crystal effect. Co-ordination to osmium has lowered the N_2 stretching frequency by about 300 cm.$^{-1}$ from that observed for the free N_2 molecule. This is the largest shift yet observed in complexes containing molecular nitrogen and reflects the strength of the osmium–nitrogen bond.

Numerous papers have been published on the vibrational spectra of the azide ion and compounds in which an azido-group is bonded to a metal atom.[533–549] Beck *et al.*[544] have examined an extensive range of complexes with both main-group and transition elements. The azido-group exhibits strong characteristic absorptions in the ranges 2000–2200 and 1200–1300 cm.$^{-1}$ which may be assigned to $\nu_{asym}(N_3)$ and $\nu_{sym}(N_3)$ vibrations. Examples of some particular assignments are in Table 44.

529 A. D. Allen and J. R. Stevens, *Chem. Comm.*, 1967, 1147.
530 A. Yamamoto, S. Kitazume, L. S. Pu, and S. Ikeda, *Chem. Comm.*, 1967, **2**, 79.
531 A. Yamamoto, L. S. Pu, S. Kitazume, and S. Ikeda, *J. Amer. Chem. Soc.*, 1967, **85**, 3071.
532 A. Sacco and M. Rossi, *Chem. Comm.*, 1967, 316.
533 T. Theophanides and G. C. Turrell, *Spectrochim. Acta*, 1967, **23**, *A*, 1927.
534 R. J. Shozda, *Inorg. Chem.*, 1967, **6**, 1919.
535 L. K. Dyall and J. E. Kemp, *Austral. J. Chem.*, 1967, **20**, 1395.
536 H. Müller and K. Dehnicke, *J. Organometallic Chem.*, 1967, **10**, P1.
537 J. Müller and K. Dehnicke, *J. Organometallic Chem.*, 1967, **7**, P1.
538 N. Wiberg, W.-Ch. Joo, and H. Henke, *Inorg. Nuclear Chem. Letters*, 1967, **3**, 267.
539 N. Wiberg and K. H. Schmid, *Chem. Ber.*, 1967, **100**, 741.
540 N. Wiberg and K. H. Schmid, *Chem. Ber.*, 1967, **100**, 748.
541 U. Müller and K. Dehnicke, *Z. anorg. Chem.*, 1967, **350**, 113.
542 H. W. Roesky, F. N. Tebbe, and E. L. Muetterties, *J. Amer. Chem. Soc.*, 1967, **89**, 1272.
543 H. W. Roesky, *Chem. Ber.*, 1967, **100**, 2138.
544 W. Beck, W. P. Fehlhammer, P. Pöllmann, E. Schuierer, and K. Feldl, *Chem. Ber.*, 1967, **100**, 2335.
545 A. Schmidt, *Chem. Ber.*, 1967, **100**, 3725.
546 P. I. Paetzold, P. P. Habereder, and R. Müllbauer, *J. Organometallic Chem.*, 1967, **7**, 45.
547 H. Gorth and M. C. Henry, *J. Organometallic Chem.*, 1967, **9**, 117.
548 J. P. Collman, M. Kubota, J. Y. Sun, and F. Vastine, *J. Amer. Chem. Soc.*, 1967, **89**, 169.
549 H.-H. Schmidtke and D. Garthoff, *J. Amer. Chem. Soc.*, 1967, **89**, 1317.

Table 44 *Vibrational assignments* (cm.$^{-1}$) *for some azido-complexes*

Compound	$\nu_{asym}(N_3)$	$\nu_{sym}(N_3)$	$\delta(N_3)$	$\gamma(N_3)$	*Ref.*
$MeZnN_3$	2106	1230	636	558	536
$EtZnN_3$	2105 2118	1280	699	621	536
$PhZnN_3$	2108	1276	639	590	536
$PhZnN_3$,2py	2065	1295	640		536
$[Et_2GaN_3]_3$	2108	1241	715	660	537

It is notable that $[Et_2GaN_3]_3$, which has been assigned the structure (46), has such similar vibrational characteristics to those complexes in which the azide group is a simple monodentate ligand rather than being part of a heterocyclic ring.

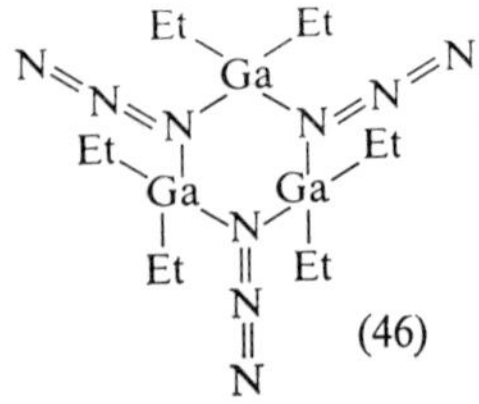

(46)

Amines and Related Ligands.—Primary aromatic amine complexes have been reported[550–553] and Haigh *et al.* show that for thirty-seven complexes of zinc and mercury halides the symmetric and antisymmetric N—H stretching frequencies are lowered by 100–150 cm.$^{-1}$ on co-ordination and they follow a relation of the type previously reported only for free amines:

$$\nu_{sym} = 345{\cdot}5 + 0{\cdot}876\ \nu_{asym}$$

Relations between ν(N—H) and Hammett σ-functions are also discussed.

Baldwin (*J. Chem. Soc.*, 1960, 4369) concluded that the most significant spectral difference between *cis*- and *trans*-$[Co\ en_2Cl_2]X$ complexes was the 850–900 cm.$^{-1}$ region (CH_2 rocking vibrations). For complexes with simple and small anions, the *cis*-complexes all exhibit two bands in this region, while the *trans*-complexes show only one. Several authors have confirmed and used this conclusion.[554–556] The criterion also applies to $[RhCl_2\ en_2]NO_3$.[557] Among other ethylenediamine complexes reported,[349, 558, 559] Clark *et al.* conclude that some complexes of Ti^{III}, V^{III}, and Cr^{III}, $[M\ en_3Cl_3]$, contain tris-bidentate cations and all the chlorine

550 J. M. Haigh, Miss M. A. van Dam, and D. A. Thornton, *Inorg. Nuclear Chem. Letters*, 1967, **3**, 7.
551 J. M. Haigh, M. A. van Dam, and D. A. Thornton, *Z. anorg. Chem.*, 1967, **355**, 94.
552 C. H. Misra, S. S. Parmar, and S. N. Shukla, *Canad. J. Chem.*, 1967, **45**, 2459.
553 C. H. Misra, S. S. Parmar, and S. N. Shukla, *J. Inorg. Nuclear Chem.*, 1967, **29**, 2589.
554 S. K. Madan and J. Peone, jun., *Inorg. Chem.*, 1967, **6**, 463.
555 S. C. Chan, C. Y. Cheng, and F. Leh, *J. Chem. Soc.* (*A*), 1967, 1586.
556 S. C. Chan and F. Leh, *J. Chem. Soc.* (*A*), 1967, 1730.
557 J. A. Broomhead and L. A. P. Kane-Maguire, *J. Chem. Soc.* (*A*), 1967, 546.
558 K. Brodersen and Th. Kahlert, *Z. anorg. Chem.*, 1967, **355**, 323.
559 P.-C. Kong and T. Theophenides, *Canad. J. Chem.*, 1967, **45**, 3192.

atoms are ionic. The corresponding propylenediamine complexes are interpreted similarly and they suggest that both ligands adopt a gauche configuration in all the complexes.

Some diethylenetriamine (dien) complexes of Group IIA salts have been prepared and examined in the i.r. region.[560] They are shown to be nine-co-ordinate, *e.g.* [Ca dien$_3$](ClO_4)$_2$

For complexes of Co^{III} and Ru^{III} with triethylenetetramine, three geometric isomers are expected: *cis*-α, *cis*-β, and *trans*. The isomers can be distinguished by examining the 990–1100 cm.$^{-1}$ region.[557, 561–564] Complexes with the *cis*-α-configuration of the chelate ring show two strong absorption bands in the region, the β-isomers are more complex and have at least four intense bands, and the *trans*-isomers exhibit a single absorption. Complexes of tetraethylenepentamine with transition metals have also been studied.[565]

Several workers have examined complexes formed by metals with Schiff bases [566–570] and ν(C=N) occurs at *ca.* 1600 cm.$^{-1}$.

Marks *et al.*[571] have examined twenty Ni^{II} complexes containing uni- and bi-dentate *o*-phenylenediamine and 4-methyl-*o*-phenylenediamine. A strong band between 1277 and 1282 cm.$^{-1}$ appears to be diagnostic of the presence of unco-ordinated amine groups in *o*-phenylenediamine complexes. The range 1230–1260 cm.$^{-1}$ is characteristic of co-ordinated amine groups and both ranges are associated with ν(C—N) which occurs at 1276 cm.$^{-1}$ in the free ligand.

It is well known that free and co-ordinated pyridine can be distinguished by shifts of bands as follows: 1578 $\rightarrow$ *ca.* 1600; 601 $\rightarrow$ *ca.* 625; and 403 $\rightarrow$ *ca.* 420 cm.$^{-1}$. On this basis it has been concluded that Ni py$_6$(NCO)$_2$ should be formulated as [Ni py$_4$(NCO)$_2$],2py where the two extra pyridine molecules are not co-ordinated to the metal.[572]

The far i.r. spectra of 8-aminoquinoline complexes with lanthanide halides have been well characterized [573] and a useful 'stick-diagram' of the 300–380 cm.$^{-1}$ region has been given.

Definitive spectra have been recorded for trimethylamine and its deuteriomethyl derivatives.[574] Among several papers concerned with P

560 P. S. Gentile, J. Carlotto, and T. A. Shankoff, *J. Inorg. Nuclear Chem.*, 1967, **29**, 1427.
561 D. A. Buckingham, J. P. Collman, D. A. R. Happer, and L. G. Marzilli, *J. Amer. Chem. Soc.*, 1967, **89**, 1082.
562 L. G. Marzilli and D. A. Buckingham, *Inorg. Chem.*, 1967, **6**, 1042.
563 A. M. Sargeson and G. H. Searle, *Inorg. Chem.*, 1967, **6**, 787.
564 S. Yoshikawa, T. Sekihara, and M. Goto, *Inorg. Chem.*, 1967, **6**, 169.
565 D. A. House and C. S. Garner, *Inorg. Chem.*, 1967, **6**, 272.
566 J. E. Kovacic, *Spectrochim. Acta*, 1967, **23**, *A*, 183.
567 M. J. O'Connor and B. O. West, *Austral. J. Chem.*, 1967, **20**, 2077.
568 N. Sadasivan, J. A. Kernohan, and J. F. Endicott, *Inorg. Chem.*, 1967, **6**, 770.
569 E. M. Dudek and G. Dudek, *Inorg. Nuclear Chem. Letters*, 1967, **3**, 241.
570 G. E. Batley and D. P. Graddon, *Austral. J. Chem.*, 1967, **20**, 884.
571 D. R. Marks, D. J. Phillips, and J. P. Redfern, *J. Chem. Soc.* (*A*), 1967, 1464.
572 A. H. Norbury, E. A. Ryder. and R. F. Williams, *J. Chem. Soc.* (*A*), 1967, 1439.
573 L. J. Basile, D. L. Kovacic, and J. R. Ferraro, *Inorg. Chem.*, 1967, **6**, 406.
574 J. N. Gayles, *Spectrochim. Acta*, 1967, 23, *A*, 1521.

and As ligands,[575-580] the deformation modes of Me_3P and Me_3As have been reassigned and force constants have been calculated.

	$\nu_1(A_1)$	$\nu_2(A_1)$	$\nu_3(E)$	$\nu_4(E)$
Me_3P	653	305	708	263 (cm.$^{-1}$)
Me_3As	567	236	582	221 (cm.$^{-1}$)

A very strong band near 1040 cm.$^{-1}$ is characteristic of the P—O—CH_3 group but the precise assignment has been somewhat confused in the literature; Chatt and Heaton now assign it to ν(C—OP) rather than ν(P—OC) on the basis of their spectra for some platinum complexes involving phosphorus ligands.[576]

Nitriles and Cyanides.—A definitive spectrum for acetonitrile, MeCN, has been published[581] and the entropy discrepancy at 298·16°K was shown to result partly from the use of liquid-state frequencies. $(MeCN)_2ReCl_4$ and $(PhCN)_2ReCl_4$ are the first simple nitrile complexes of Re^{IV}; $\nu(C\equiv N)$ is at 2292 cm.$^{-1}$ for the first complex and at 2251 and 2260 cm.$^{-1}$ for the second.[582]

Jain *et al.*[583] have examined the co-ordination between Group IV halides and dialkylaminoacetonitriles. For the 1 : 1 complex of Et_2NCH_2CN with $SnCl_4$, $TiCl_4$, $TiBr_4$, and $ZrCl_4$, $\nu(C\equiv N)$ is *lowered* by 40–80 cm.$^{-1}$ from the position in the free nitrile ligand (2220 cm.$^{-1}$). This indicates co-ordination through the triple bond. This may be compared with $2BCl_3,Et_2NCH_2CN$ [$\nu(C\equiv N)$ 2280 cm.$^{-1}$] where the ligand co-ordinates via the nitrogen atom. $Me_2N(CH_2)_2CN,TiCl_4$ exhibits two cyanide stretching frequencies at 2205 and 2280 cm.$^{-1}$ [free ligand $\nu(C\equiv N)$ 2240 cm.$^{-1}$]. This suggests a competition between two types of bonding $C\equiv N{:}\rightarrow$ and $\underset{\downarrow}{C\equiv N}{:}$, and the intensities of the bands show that co-ordination through the triple bond predominates.

Complexes involving other nitrile ligands have been studied: acetonitrile, propionitrile, crotononitrile,[584] cinnamonitrile, acrylonitrile,[585] and (2-cyanoethyl)tin halides.[586, 587]

575 G. Bouquet and M. Bigorgne, *Spectrochim. Acta*, 1967, **23**, *A*, 1231.
576 J. Chatt and B. T. Heaton, *Spectrochim. Acta*, 1967, 23, *A*, 2220.
577 M. A. Bennett, J. Chatt, G. J. Erskine, J. Lewis, R. F. Long, and R. S. Nyholm, *J. Chem. Soc.* (*A*), 1967, 501.
578 A. N. Nesmeyanov, Y. A. Chapovsky, and Y. A. Ustynyuk, *J. Organometallic Chem.*, 1967, **9**, 345.
579 A. J. Carty, *Canad. J. Chem.*, 1967, **45**, 345.
580 G. A. Barclay, I. K. Gregor, M. J. Lambert, and S. B. Wild, *Austral. J. Chem.*, 1967, **20**, 1571.
581 G. A. Crowder and B. R. Cook, *J. Phys. Chem.*, 1967, **71**, 914.
582 G. Rouschias and G. Wilkinson, *Chem. Comm.*, 1967, 442.
583 S. C. Jain and R. Rivest, *Inorg. Chem.*, 1967, **6**, 467.
584 C. Krüger, *J. Organometallic Chem.*, 1967, **9**, 125.
585 J. F. Guttenberger and W. Strohmeier, *Chem. Ber.*, 1967, **100**, 2807.
586 G. H. Reifenberg and W. J. Considine, *J. Organometallic Chem.*, 1967, **10**, 285.
587 G. H. Reifenberg and W. J. Considine, *J. Organometallic Chem.*, 1967, **9**, 505.

The nitrile stretching frequencies for some ruthenium chloride complexes of acrylonitrile, $CH_2{:}CH{\cdot}CN$, occur at *ca.* 2230 cm.$^{-1}$. The band for the free ligand occurs at 2214 cm.$^{-1}$ and this increase on co-ordination indicates bonding through the nitrogen atom.[588]

Kubota *et al.*[589] have examined ten dicyanide complexes of copper(I) perchlorate. The spectra suggest that the ligand serves as a bridge between two adjacent copper atoms. Bonding through the olefinic π-system as well as through the nitrogen atom is indicated for perchloratobis(acrylonitrile)-copper. In this compound ν(C≡N) (2264 and 2276 cm.$^{-1}$) is shifted to higher wavenumbers [free ligand, ν(C≡N) 2230 cm.$^{-1}$] while ν(C=C) decreases from 1610 to 1512 cm.$^{-1}$.

Purcell[590] has utilized the results of normal-co-ordinate analysis and MO calculations to explain the increase which occurs in ν(C≡N) when either the carbon or nitrogen atoms of the cyanide group act as a Lewis base.

Dows *et al.* (*J. Inorg. Nuclear. Chem.*, 1961, **21**, 33, and refs. therein) have concluded that the ratio of co-ordination number to oxidation number for transition-metal cyanide complexes determines the frequency of ν(C≡N) within a narrow range. For all transition-metal cyanide complexes in which the co-ordination number is twice the oxidation number of the metal, an i.r. absorption occurs at 2135 ± 15 cm.$^{-1}$ and as the ratio increases further, the value for ν(C≡N) decreases. In addition, they found that bridging cyanide groups absorb some 20–40 cm.$^{-1}$ higher than terminal cyanide groups. Using these criteria Banks *et al.*[591] have suggested the presence of bridging cyanide groups in the complexes $K_3Cr(CN)_4$, $K_2Fe(CN)_4$, and $K_5V(CN)_5$. Rupp *et al.*[592] have prepared eleven new addition compounds by the addition of Group IVB halides to dicyanobis(1,10-phenanthroline)iron(II). The 1 : 1 adducts are probably cyclic structures involving cyanide bridging groups, *e.g.* $[Fe\,phen_2(CN)_2(SiF_4)]_2$. Data on some other nitrile and cyanide complexes which have appeared during the year are shown in Table 45.

Table 45 *Vibrational data* (cm.$^{-1}$) *for some nitrile and Cyanide Complexes*

Compound	ν(C≡N)	*Ref.*
Mo_6Cl_{12},2MeCN	2278	593
Mo_6Cl_{12},2EtCN	2286	593
Mo_6Cl_{12},2Pr^{n}CN	2284	593
BF_3,MeCN	2359	594
BF_3,PhCN	2336	594
BCl_3,PhCN	2304	594
$AlMe_3$,PhCN	2272	594

[588] A. Misono, Y. Uchida, M. Hidai, and H. Kanai, *Chem. Comm.*, 1967, 357.
[589] M. Kubota and D. L. Johnston, *J. Inorg. Nuclear Chem.*, 1967, **29**, 769.
[590] K. F. Purcell, *J. Amer. Chem. Soc.*, 1967, **89**, 247.
[591] D. F. Banks and J. Kleinberg, *Inorg. Chem.*, 1967, **6**, 1849.
[592] J. J. Rupp and D. F. Shriver, *Inorg. Chem.*, 1967, **6**, 755.
[593] W. M. Carmichael and D. A. Edwards, *J. Inorg. Nuclear Chem.*, 1967, **29**, 1535.
[594] D. Hall, P. K. Ummat, and K. Wade, *J. Chem. Soc.* (*A*), 1967, 1612.

Table 45 (*cont.*)

Compound	ν(C≡N)	*Ref.*
$PhCNMe^+BCl_4^-$	2380	594
$PhCNEt^+BCl_4^-$	2365	594
$PhCNPh^+BCl_4^-$	2318	594
$MeCNMe^+SbCl_6^-$	2416	594
$(CN)_3CPCl_3(OEt)$	2190	595
$Ag^+B(CN)_4^-$	2195	596
$K_4[MoO_2(CN)_4],6H_2O$	2060	386
$K_4[WO_2(CN)_4],6H_2O$	2060	386
$K_3[MoO(OH)(CN)_4],2H_2O$	2095	386
$K_3[MoO(OD)(CN)_4],2D_2O$	2090	386
$K_3[WO(OH)(CN)_4]$	2075	386
$Na_2[Fe(CN)_5NO],2H_2O$	2173(A_1), 2160(A_1) 2156(B_1), 2143(E)	597
$(Ph_4P)_2[Fe(CN)_4(C{\equiv}CPh)NO]$	~ 2150	598
$K(Ph_2Tl)[Fe(CN)_2(C{\equiv}CPh)_3NO]\cdot NH_3$		
$(Ph_4P)_2[Fe(CN)_2(C{\equiv}CPh)_3NO]$		
$Co(PhNC)_5(ClO_4)_2,1{\cdot}5H_2O$	2195, 2220	599
$Co(PhNC)_5(ClO_4)_2$	2180, 2220	599

Latka [600] has reported some cyano–isocyano complexes of tungsten [(ν(C≡N) *ca.* 2137 and ν(N≡C) *ca.* 2210 cm.$^{-1}$) and Schilt and Leman [601] have examined substituent effects on proton affinities of cyanide ligands in some mixed ligand Fe^{II} complexes. The substituent groups have very little effect on ν(C≡N).

Finally Firth *et al.*[602] have studied metal–ligand interactions in Co^{III} complexes derived from vitamin B_{12}. In particular, they examine the effect of changing the axial ligand on the i.r. stretching frequency of cyanide co-ordinated in the *trans*-position. Their results show that, as the axial ligand becomes more polarisable (H_2O, N≡C, ... , $CH_3{\cdot}CH_2{\cdot}$) so the *trans*-metal–ligand bond-length increases, the *trans*-ligand itself becomes more ionic and the values of the formation constants approach those expected for ion-pairs with non-transition metals.

Nitrosyls.—Dibromodinitrosyl- and di-iododinitrosyl-compounds of molybdenum and dibromodinitrosyl compounds of tungsten readily react with ligands (Ph_3P, Ph_3As, or py) to give monomeric octahedral complexes $M(NO)_2Br_2L_2$. All the complexes give two strong bands in the region 1600–1800 cm.$^{-1}$ which have been assigned to $N{-}O^+$ stretching frequencies and the NO groups are in *cis*-positions.[603] $[(\pi\text{-}C_5H_5)Mo(NO)I_2]_2$

595 A. D. F. Toy and H. J. Emeleus, *J. Inorg. Nuclear Chem.*, 1967, **29**, 269.
596 E. Bessler and J. Goubeau, *Z. anorg. Chem.*, 1967, **352**, 67.
597 L. Tosi, *Compt. rend.*, 1967, **264**, *B*, 1313.
598 R. Nast, K. W. Krüger, and G. Beck, *Z. anorg. Chem.*, 1967, **350**, 177.
599 J. M. Pratt and P. R. Silverman, *Chem. Comm.*, 1967, 117.
600 H. Latka, *Z. anorg. Chem.*, 1967, **353**, 243.
601 A. A. Schilt and T. W. Leman, *J. Amer. Chem. Soc.*, 1967, **89**, 2012.
602 R. A. Firth, H. A. O. Hill, J. M. Pratt, R. G. Thorp, and R. J. P. Williams, *Chem. Comm.*, 1967, 400.
603 B. F. G. Johnson, *J. Chem. Soc.* (*A*), 1967, 475.

exhibits ν(N—O) at 1670 cm.$^{-1}$.[604] The following molybdenum complexes have also been prepared and ν(N—O) values reported: (π-C_5H_5)Mo(NO)I_2,L (L = Ph_3P, 1660; $(PhO)_3P$, 1679; C_5H_5N, 1661; $C_{10}H_8N_2$, 1674 cm.$^{-1}$).

The i.r. spectrum of $[(\pi\text{-}C_5H_5)Mn(NO)_2]_n$ suggests that the complex contains a bridging nitrosyl group. The strong band occurs at 1525 cm.$^{-1}$, cf. ν(N—O)term. *ca.* 1760 cm.$^{-1}$.[604] Elder *et al.*[605] have shown that tris(cyclopentadienyl)trimanganese tetranitrosyl, $(\pi\text{-}C_5H_5)_3Mn_3(NO)_4$, is a metal cluster compound (47) with double and triple-bridging nitrosyl groups. The i.r. spectrum (KBr disc) shows N—O stretching vibrations at 1313, 1475, and 1520 cm.$^{-1}$ and the former has been assigned to the triple-bridging NO group.

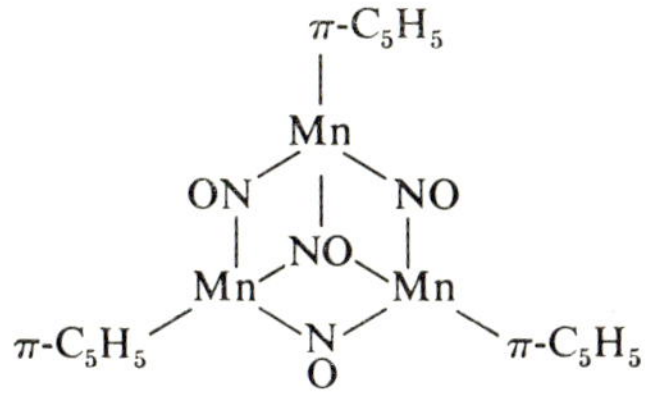

(47)

The cations of the following compounds have been shown, by a variety of methods, to be monomeric with the nitrosyl group *trans* to the halide [values of ν(N—O) cm.$^{-1}$) in parentheses]: [606]

trans-[FeClNO das_2]ClO_4 (1620); *trans*-[FeBrNO das_2]Br (1625); *trans*-[FeINO das_2]I (1640); and [Fe(NO) das_2]$(ClO_4)_2$ (1760).

Some nitrosyl dithiolene complexes of iron have also been assigned.[607]

Mercer *et al.*[608] have investigated the two isomers of the penta-ammine-nitrosylcobalt ion. On the basis of all the evidence they conclude that the black isomer has a monomeric structure $[Co(^{14}NO)(NH_3)_5]Cl_2$ [ν(N—O) 1610 cm.$^{-1}$] and that the red isomer is dimeric and contains a hyponitrite moiety, $[Co(^{14}NO)(NH_3)_5]_2(NO_3)_4$ [ν_{asym}(N—O) 1046; ν_{sym}(N—O) 932 cm.$^{-1}$].

4 Nitro- and Nitrito-complexes

I.r. measurements on solid sodium nitrite show that the three fundamental vibrational modes of the NO_2^- group are at 1325 (ν_{sym}), 1270 (ν_{asym}), and 829 (δ_{sym}) cm.$^{-1}$. On co-ordination of a single nitro-group to a transition metal via the nitrogen, ν_{asym} rises to *ca.* 1400 cm.$^{-1}$ while ν_{sym} and δ_{sym} are little changed: thus $\nu_{sym} > \nu_{asym}$ for the free ion and $\nu_{asym} > \nu_{sym}$ for the

[604] R. B. King, *Inorg. Chem.*, 1967, **6**, 30.
[605] R. C. Elder, F. A. Cotton, and R. A. Schunn, *J. Amer. Chem. Soc.*, 1967, **89**, 3645.
[606] W. Silverthorn and R. D. Feltham, *Inorg. Chem.*, 1967, **6**, 1662.
[607] J. A. McCleverty, N. M. Atherton, J. Locke, E. J. Wharton, and C. J. Winscom, *J. Amer. Chem. Soc.*, 1967, **89**, 6082.
[608] E. E. Mercer, W. A. McAllister, and J. R. Durig, *Inorg. Chem.*, 1967, **6**, 1816.

co-ordinated group. In addition three new modes appear as a result of such co-ordination: an out-of-plane wag ρ_w, an in-plane rock ρ_r, and a metal–ligand stretch ν(M—N); there will also be a torsional mode but this should be active only in the Raman spectrum and will be of very low frequency. Examples of this type of co-ordination are shown in Table 46.

Table 46 *References to vibrational data on nitro-compounds*

Compound	*Ref.*
$NaNO_2$ and KNO_2 in solid solutions	609
$H_2C(NO_2)_2$	610
$K[HC(NO_2)_2]$	610
$Na_4Fe(CN)_5NO_2$	611
$[Co(NO_2)_n(NH_3)_{6-n}]^{(3-n)+}$	612
cis-$[Co(NH_3)_4(HNO_2)(NO_2)](NO_3)_2$	613
[Co bipy$_2$(NO$_2$)$_2$]	614
cis-Na[Co acac$_2$(NO$_2$)$_2$]	615
trans-[Co acac$_2$(NO$_2$)(H$_2$O)]	615
$CoL_2(NO_2)_2$ (L = subst. pyridine N-oxide)	616
$NiL_2(NO_2)_2$ (L = subst. pyridine *N*-oxide)	617
$M(NO_2)_6^{n-}$ (M = Fe^{2+}, Co^{2+}, Ni^{2+}, Cu^{2+}, Co^{3+}, or Rh^{3+})	618
$K[Rh^V(NO_2)_6]$	619

Cleare and Griffith[620] have examined the i.r. spectra of normal, ^{15}N-substituted, and, in some cases, deuteriated complexes of Cr, Os, Rh, Ir, and Pt containing unidentate nitro-groups. They assign the fundamentals to the following approximate regions: ν_{asym} 1345–1412; ν_{sym} 1300–1346; δ_{sym} 821–844; ρ_w 500–660; ν(M—N) 325–374; and ρ_r 287–304 cm.$^{-1}$.

For nitrile complexes, where the nitro-group is co-ordinated through a single oxygen atom, six modes are expected. These are approximately two nitrogen–oxygen stretches [ν(N=O) and ν(N—O)], an ONO deformation, δ_{sym}, a metal–ligand stretch, ν(M—O), a metal–oxygen bend, δ(MONO), and an out-of-plane deformation ρ_w. The following complexes have been

609 W. A. Morgan, E. Silberman, and H. W. Morgan, *Spectrochim. Acta*, 1967, **23**, *A*, 2855.
610 K. Singh, *Spectrochim. Acta*, 1967, **23**, *A*, 1089.
611 D. X. West, *J. Inorg. Nuclear Chem.*, 1967, **29**, 1163.
612 I. Nakagawa and T. Shimanouchi, *Spectrochim. Acta*, 1967, **23**, *A*, 2099.
613 R. Ugo and R. D. Gillard, *J. Chem. Soc.* (*A*), 1967, 2078.
614 A. A. Vlček, *Inorg. Chem.*, 1967, **7**, 1425.
615 B. P. Cotsoradis and R. D. Archer, *Inorg. Chem.*, 1967, **6**, 800.
616 L. El-Sayed and R. O. Ragsdale, *Inorg. Chem.*, 1967, **6**, 1644.
617 L. El-Sayed and R. O. Ragsdale, *Inorg. Chem.*, 1967, **6**, 1640.
618 K. G. Caulton and R. F. Fenske, *Inorg. Chem.*, 1967, **6**, 562.
619 M. Le Postollec and J.-P. Mathieu, *Comp. rend.*, 1967, **265**, *B*, 138.
620 M. J. Cleare and W. P. Griffith, *J. Chem. Soc.* (*A*), 1967, 1144.

examined: *cis*- and *trans*-[Cr $en_2(ONO)_2]ClO_4$,[621] $[Cr(ONO)_3(NH_3)_3]$,[622] $[Cr(ONO)_3(H_2O)_3],1{\cdot}5KNO_2$,[622] and $[Ni(NO_2)_4(ONO)_2]^{4-}$.[623]

Cleare and Griffith,[620] have assigned the fundamentals to the approximate regions: ν(N=O) 1412–1461; ν(N—O) 1046–1065, δ_{sym} 825–840, and ν(M—ONO) 339–355 cm.$^{-1}$.

The nitrite ion can also behave as a chelating ligand and the following are examples for which vibrational data have been published during the year:[476] $[Mn(NO_2)_4]^{2-}$, $[Co(NO_2)_4]^{2-}$, $[Cu(NO_2)_4]^{2-}$, $[Zn(NO_2)_4]^{2-}$, $[Cd(NO_2)_4]^{2-}$, and $[Hg(NO_2)_4]^{2-}$. It is suggested that the nitrite ions are preserved as chelating groups in the manganese, cadmium, and mercury compounds and as unsymmetrically bound chelate groups in $[Co(NO_2)_4]^{2-}$.[476] Data have also appeared for $[NiL_2(NO_2)]X$[624] and [Ni $en_2(ONO)]X$;[625] the symmetrical NO_2 bending mode for this latter complex shifts from *ca.* 830 to 855–860 cm.$^{-1}$ and this is consistent with the presence of a bridging nitrito-group.

Goodgame *et al.*[476] suggests that the nitrite ions are present as chelating groups in the manganese, cadmium, and mercury compounds, and as unsymmetrically bound chelate groups in $Co(NO_2)_4{}^{2-}$. In the copper and zinc complexes they are bonded through oxygen.

5 Cyanato-, Thiocyanato-, and Selenocyanato-complexes and Their Respective Iso-complexes

Metal cyanato-complexes contain the grouping M—O—C≡N whereas isocyanato-complexes contain the grouping M—N=C=O. Several workers have examined cyanato- and isocyanato-complexes and some examples are:

Compound	*Ref.*	*Compound*	*Ref.*
MeHgONC	626	$(o\text{-}C_6H_4Me)_3SnNCO$	629
BuHgONC	626	$(p\text{-}C_6H_4Me)_3SnNCO$	629
PhHgONC	626	$(PhCH_2)_3SnNCO$	629
HNBNCO	627	$Zn(NCO)_4{}^{2-}$	473
MeHBNCO	627	$(Ph_3P)_2Pd(NCO)_2$	630
$(Me_2O)_2BNCO$	627	$(Ph_3P)_2Pt(NCO)_2$	630
FCONCO	628		
F_2PONCO	628		

[621] W. W. Fee, C. S. Garner, and J. N. MacB. Harrowfield, *Inorg. Chem.*, 1967, **6**, 87.
[622] A. Garnier, *Compt. rend.*, 1967, **265**, *B*, 198.
[623] D. M. L. Goodgame and M. A. Hitchman, *Inorg. Chem.*, 1967, **6**, 813.
[624] E. Uhlig, H.-J. Bergmann, and U. Schneider, *Z. anorg. Chem.*, 1967, **354**, 130.
[625] B. J. Hathaway and R. C. Slade, *J. Chem. Soc.* (*A*), 1967, 952.
[626] M. V. Kashutina and O. Yu. Okhlobystin, *J. Organometallic Chem.*, 1967, **9**, 5.
[627] M. F. Lappert and H. Pyszora, *J. Chem. Soc.* (*A*), 1967, 854.
[628] O. Glemser, A. Biermann, and M. Fild, *Chem. Ber.*, 1967, **100**, 1082.
[629] T. N. Srivastava and S. N. Bhattacharya, *J. Inorg. Nuclear Chem.*, 1967, **29**, 1873.
[630] W. Beck and W. P. Fehlhammer, *Angew. Chem. Internat. Edn.*, 1967, **6**, 169.

Lappert and Pyszora[627] have examined the i.r. spectra of thirty-seven iso- and isothio-cyanatoboranes, having one, two, or three isocyanate or isothiocyanate groups attached directly to boron. From the frequency, intensity, and shape of the absorption band associated with $\nu_{asym}(\cdot NCX)$, the iso structure $>B{-}N{=}C{=}X$ is proposed for all compounds.

$$\nu_{asym}(-N{=}C{=}O) = 2273 \pm 30, \quad \nu_{asym}(-N{=}C{=}S) = 2089 \pm 31 \text{ cm.}^{-1}.$$

The NCS group has three fundamental frequencies whether it is N- or S-bonded: ν_1(C—N str.), ν_2(N—C—S bend), and ν_3(C—S str.). The following criteria may be applied to differentiate between the two possibilities: for an N-bonded group, the fundamentals lie in the ranges: ν_1 2040–2080; ν_2 465–480; and ν_3 780–860 cm.$^{-1}$, whereas for an S-bonded group the ranges are: ν_1 2080–2120; ν_2 410–470; and ν_3 690–720 cm.$^{-1}$. Examples of thiocyanato- and isothiocyanato-complexes which have been studied during the year are given in Tables 47 and 48.

A comparison of solid and liquid state spectra of several alkyl isothiocyanates has resolved some of the interpretative difficulties in assigning ν_{sym}(NCS) and ν(Cα—N).[631] A complete vibrational assignment is given for methyl, ethyl, isopropyl, and t-butyl isothiocyanates and an extensive discussion is given for the observed rotational isomerism about the Cα—N bond.

Table 47 *Sulphur-bonded thiocyanato-complexes*

Compound	*Ref.*	*Compound*	*Ref.*
KSCN	632	$Cs_2[Re(SCN)_6]$	635, 636
$[Et_2AlSCN]_3$	633	$Tl_2[Re(SCN)_6]$	635
$[Et_2GaSCN]_3$	633	$Ag_2[Re(SCN)_6]$	635
$[Et_2InSCN]_3$	633	$[Pt(NH_3)_4(SCN)Cl]^{2+}$	637
$[Co\ dmg_2SCN]^{-a}$	634	$Hg(SCN)_4^{2-}$	638
$[PdL(SCN)_2]^b$	312	$[PhHgSCN]_2$	639

[a] dmg = Dimethylglyoxime.

[b] L = 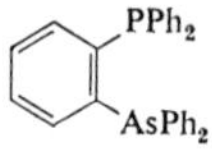

and the complex contains S- and N-bonded SCN groups.

[631] R. N. Kniseley, R. P. Hirochmann, and V. A. Fassel, *Spectrochim. Acta*, 1967, **23**, *A*, 109.
[632] R. Savoie and M. Pézolet, *Canad. J. Chem.*, 1967, **45**, 1677.
[633] K. Dehnicke, *Angew. Chem. Internat. Edn.*, 1967, **6**, 947.
[634] K. Burger and B. Pinter, *J. Inorg. Nuclear Chem.*, 1967, **29**, 1717.
[635] R. A. Bailey and S. L. Kozak, *Inorg. Chem.*, 1967, **6**, 419.
[636] R. A. Bailey and S. L. Kozak, *Inorg. Chem.*, 1967, **6**, 2155.
[637] W. R. Mason, jun., E. R. Berger, and R. C. Johnson, *Inorg. Chem.*, 1967, **6**, 248.
[638] K. A. Taylor, T. V. Long, jun., and R. A. Plane, *J. Chem. Phys.*, 1967, **47**, 138.
[639] K. Dehnicke, *J. Organometallic Chem.*, 1967, **9**, 11.

Table 48 *Nitrogen–bonded isothiocyanato-complexes*

Compound	*Ref.*
$P(NCS)_3$	640
$PO(NCS)_3$	640
$OPF_2(NCS)$	641, 309
$OPF(NCS)_2$	641, 309
$SPF_2(NCS)$	642, 309
$SPF(NCS)_2$	642, 309
$PF_2(NCS)$	309
$PF(NCS)_2$	309
$P_3N_3(NCS)_6$	643
$Me_3Sn(NCS)$	644
$Me_3Sn(NCS)$,py	644
$(Me_4N)^+[Me_3Sn(NCS)_2]^-$	644
$(Me_4N)_2{}^+[Me_2Sn(NCS)_4]^{2-}$	644
$VCl(NCS)_2L_3{}^a$	645
$[VCl_2(NCS)L_3],2L^a$	645
$[V(NCS)_3L_3],2L^a$	645
$M_3{}^I[V(NCS)_6],2L^a$	645
$[Cr(NCS)_4L_2]^{-b}$	378
M paphy X_2 (M = Mn, Co, Ni, Cu, Zn, Cd, or Hg)	646
$Fe(NCS)_2$ bipy	647
$Fe(NCS)_2$ $phen_2$	647, 648, 649
$[Fe(NCS)\ phen_2]ClO_4$	647
Co(diamine)$(NCS)_2$	650
$Co(Me_4daes)(NCS)_2$	654
$Co(Me_4daeo)(NCS)_2$	653
$Co(Me_5dien)(NCS)_2$	653
[Co tren NCS]NCS	660
Ni(diamine)$(NCS)_2$	650
Ni $bipy_2(NCS)_2$	651
[Ni $bipy_4(NCS)_2](SCN)_2$	651
Ni tpt $(NCS)_2$	652
Ni tpt $(NCS)_2$,L (L = H_2O, py, or 4Me—py)	652
[Ni $tpt_2](NCS)_2$,1·5H_2O	652
$Ni(Me_4daeo)(NCS)_2$	653

[a] L = N- or O-donors of MeCN, heterocyclic amines, or tetrahydrofuran (THF).
[b] L = NH_3, py, ½(2,2′-bipy), PMe_3, PEt_3, $EtPh_2P$, etc.

640 K. Oba, F. Watari, and K. Aida, *Spectrochim. Acta*, 1967, **23**, *A*, 1515.
641 H. W. Roesky, *Angew. Chem. Internat. Edn.*, 1967, **6**, 90.
642 H. W. Roesky, *Angew. Chem. Internat. Edn.*, 1967, **6**, 90.
643 R. Stahlberg and E. Steger, *Spectrochim. Acta*, 1967, **23**, *A*, 2185.
644 M. Wada and R. Okawara, *J. Organometallic Chem.*, 1967, **8**, 261.
645 H. Böhland and P. Malitzke, *Z. anorg. Chem.*, 1967, **350**, 70.
646 F. Lions, I. G. Dance, and J. Lewis, *J. Chem. Soc.* (*A*), 1967, 565.
647 R. Driver and W. R. Walker, *Austral. J. Chem.*, 1967, **20**, 1375.
648 E. König and K. Madeja, *Spectrochim. Acta*, 1967, **23**, *A*, 45.
649 E. König and K. Madeja, *Inorg. Chem.*, 1967, **6**, 48.
650 L. Sacconi, I. Bertini, and F. Mani, *Inorg. Chem.*, 1967, **6**, 262.
651 C. M. Harris and E. D. McKenzie, *J. Inorg. Nuclear Chem.*, 1967, **29**, 1047.
652 R. S. Vagg, R. N. Warrener, and E. C. Watton, *Austral. J. Chem.*, 1967, **20**, 1841.
653 M. Ciampolini and N. Nardi, *Inorg. Chem.*, 1967, **6**, 445.
654 M. Ciampolini and J. Gebsomini, *Inorg. Chem.*, 1967, **6**, 1821.

Table 48 (*cont.*)

Compound	*Ref.*
$Ni(Me_5dien)(NCS)_2$	653
$Ni(Me_4daes)(NCS)_2$	654
Ni(4-vinyl py)2 or 4$(NCS)_2$	655
$Ni[Ph_2(o\text{-}Ph_2AsC_6H_4)P](NCS)_2$	656
$Ni[Ph_2(o\text{-}Ph_2AsC_6H_4)P]_2NCS$	656
$Zn(NCS)_4{}^{2-}$	473, 638
$KNb(NCS)_6$	657
$RhL_2(CO)(NCS)$	658
$RhL_3(NCS)^c$	658
$Rh_2L_4(NCS)_2{}^c$	658
$(R_4N)[Rh(CO)_2(NCS)_2]$	658
$KTa(NCS)_6$	657
$(Ph_4As)_2Re(NCS)_6$	659
$(Ph_4As)_2Re_2(NCS)_8$	659
$(Ph_4As)_3[Re_2(NCS)_{10}(CO)_2]$	659

[c] L = a phosphine, arsine, stibine, or a phosphite.

tpt = 2,4,6-tri-(2-pyridyl)-1,3,5-triazine; Me_4daeo = bis-(2-methylaminoethyl) oxide; Me_5dien = $MeN[CH_2{\cdot}CH_2{\cdot}NMe_2]_2$; and Me_4daes = bis-(2-methylaminoethyl) sulphide.

The thiocyanato-group is able to form a bridge between two metal atoms and a characteristic band is then observed in the range 2150–2182 cm.$^{-1}$. Three papers [661–663] describe complexes of this type, *e.g.* $Ni(NCS)_2$,2tu.

Cotton and McCleverty [664] have investigated the chemistry of the so-called 'dithiocyanate ion', $S_2C_2N_2{}^{2-}$. It behaves as a bidentate chelating ligand but the ion is identical with the *N*-cyanodithiocarbimate ion $S_2C{=}N{-}C{\equiv}N^{2-}$.

Goodall [430] has examined some Co^{II}, Rh^{III}, and Ir^{III} complexes of 1,2-dithiocyanatoethane. For (CoX_2L) (X = Cl, Br, or I) the ligand is probably in a slightly distorted *trans*-form and the bonding is through nitrogen atoms. ν(CN) is altered by 50–60 cm.$^{-1}$ on complex formation while ν(C—S) hardly alters. In contrast, the Rh^{III} and Ir^{III} complexes, (MX_3,L), probably contain the ligand in the gauche form and the bonding is via sulphur atoms.

Selenocyanato- and isoselenocyanato-complexes have been examined and typical examples are:

[655] R. N. Patel and D. V. R. Ras, *Z. anorg. Chem.*, 1967, **351**, 68.
[656] T. D. Dubois and D. W. Meek, *Inorg. Chem.*, 1967, **6**, 1395.
[657] T. M. Brown and G. F. Knox, *J. Amer. Chem. Soc.*, 1967, **89**, 5296.
[658] M. A. Jennings and A. Wojcicki, *Inorg. Chem.*, 1967, **6**, 1854.
[659] F. A. Cotton, W. R. Robinson, R. A. Walton, and R. Whyman, *Inorg. Chem.*, 1967. **6**, 929.
[660] M. Ciampolini and P. Pasletti, *Inorg. Chem.*, 1967, **6**, 1261.
[661] J. M. Rowe and D. A. White, *J. Chem. Soc.* (*A*), 1967, 1451.
[662] C. Puglisi and R. Levitus, *J. Inorg. Nuclear Chem.*, 1967, **29**, 1069.
[663] S. M. Nelson and J. Rodgers, *Inorg. Chem.*, 1967, **6**, 1390.
[664] F. A. Cotton and J. A. McCleverty, *Inorg. Chem.*, 1967, **6**, 229.

Compound	*Ref.*
Co(DMG)SeCN$^-$	634
PdL(SeCN)$_2$	312[a]
[Pd($C_{12}H_{29}N_3$)SeCN]BPh_4	665
($Bu^n_4N)_3$[M(NCSe)$_6$] (M = Ti or V)	666
($Bu^n_4N)_3$[VO(NCSe)$_4$]	666
[Zn(NCSe)$_4$]$^{2-}$	473
R_3SiNCSe (R = H, Me, Et, and Ph)	667

L = 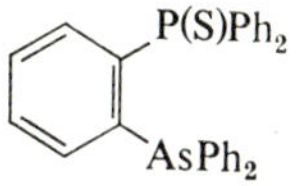

or 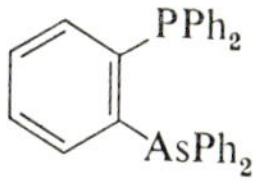

6 Oxygen Donors

Nitrito-complexes have already been mentioned in section 4, and cyanato-complexes in section 5 of Part III of this chapter.

Complexes which exhibit bands characteristic of an M—O_2 group are shown in Table 49.

Table 49 *Molecular oxygen complexes*[a]*; frequencies in* cm.$^{-1}$

Compound	ν	*Ref.*
Nb_2O_7,$4H_2O$	870	668
[Cr en (OH_2)(O_2)$_2$],H_2O	885	669, 670
[Cr en (NH_3)(O_2)$_2$],H_2O	860, 875, 885	669, 670
[Cr pn (OH_2)(O_2)$_2$],$2H_2O$	880	669, 670
[Cr ibn (OH_2)(O_2)$_2$],H_2O	865	669, 670
[Cr dien (O_2)$_2$],H_2O	870, 885	669, 670
Cs_2[W(O_2)OCl_4]	899, 948	671
($Ph_3P)_2PtO_2$	830	672, 673
($Ph_3P)_2PdO_2$	*ca.* 875	673

[a] In addition there are two reports [674, 675] of other complexes which may contain coordinated oxygen.
ibn = Isobutylenediamine.

Acac complexes have been the subject of many papers because a confusion has arisen over the assignment of bands in the 1500–1600 cm.$^{-1}$ region. Nakamoto *et al.* examined Cu acac$_2$ in 1959 and made the following assignments on the basis of Urey–Bradley force-field calculations: ν(C=C) *ca.* 1580; ν(C=O) *ca.* 1520; ν(M—O) coupled to ring deformation *ca.* 680; ν(M—O) coupled to C—Me bend *ca.* 660; and ν(M—O) *ca.* 450 cm.$^{-1}$

665 J. L. Burmeister and H. J. Gysling, *Chem. Comm.*, 1967, 543.
666 J. L. Burmeister and L. E. Williams, *J. Inorg. Nuclear Chem.*, 1967, **29**, 839.
667 J. S. Thayer, *J. Organometallic Chem.*, 1967, **9**, P30.
668 D. N. Sathyanarayana and C. C. Patel, *Z. anorg. Chem.*, 1967, **353**, 103.
669 E. A. V. Ebsworth, C. S. Garner, D. A. House, and R. G. Hughes, *Inorg. Nuclear Chem. Letters*, 1967, **3**, 61.
670 D. A. House, R. G. Hughes, and C. S. Garner, *Inorg. Chem.*, 1967, **6**, 1077.
671 J. E. Guerchais and M. T. Youinou, *Compt. rend.*, 1967, **264**, *C*, 1389.
672 C. J. Nyman, C. E. Wymore, and G. Wilkinson, *Chem. Comm.*, 1967, 407.
673 G. Wilke, H. Schott, and P. Heimbach, *Angew. Chem. Internat. Edn.*, 1967, **6**, 92.
674 G. L. Johnson and W. D. Beveridge, *Inorg. Nuclear Chem. Letters*, 1967, **3**, 323.
675 H. P. Fritz and W. Gretner, *Inorg. Nuclear Chem. Letters*, 1967, **3**, 141.

(*J. Chem. Phys.*, 1959, **32**, 588). Pinchas *et al.*[676] have examined the spectra of ^{18}O-labelled acetylacetone complexes of Cr^{III} and Mn^{III} and make the assignments: ν(C=C) 1520; ν(C=O) *ca.* 1570; $\delta_{asym}CH_3$ *ca.* 1425, out-of-plane deformation *ca.* 590; and ν(M—O) (almost pure) *ca.* 590 cm.$^{-1}$. They conclude that Urey–Bradley force-field calculations as usually applied are in many cases very inadequate and misleading. Mikami *et al.*[677] have measured the i.r. spectra (60–4000 cm.$^{-1}$) of M $acac_2$ (M = Cu, Pd, Pt, Mn, and Cd) and M $acac_3$ (M = Fe, Cr, Co, Rh, and Mn) complexes and they have carried out a normal co-ordinate analysis using a modified Urey–Bradley force-field. Their results suggest that the 1577 cm.$^{-1}$ band of Cu $acac_2$ is 72% C—C and 14% C—O stretching vibration in which the two C—O and the two C—C bonds, belonging to one ligand molecule, stretch antisymmetrically. The 1529 cm.$^{-1}$ band is 46% C—O and 22% C—C stretching vibration in which the two C—O and two C—C bonds stretch symmetrically.

Behnke and Nakamoto[678] have examined the i.r. spectra of K[Pt^{II} acac Cl_2] and three of its deuteriated analogues; the results suggest a revision of several band-assignments published previously when the force-field ignored interactions between the ligands. The reassignment gives ν(C=O) 1563, ν(C=C) 1538 cm.$^{-1}$. For europium acetylacetonate (Eu $acac_3$,$2H_2O$), the band at 1600 cm.$^{-1}$ is essentially the ν(C=C) stretching mode (B_2) while 1515 cm.$^{-1}$ is essentially the ν(C=O) stretching mode (A_1).[679] Other rare-earth complexes may be assigned similarly.[680, 681]

Hancock *et al.*[682, 683] have examined forty-eight acetylacetone complexes with metal ions and they recognise three different types of spectra: type 1 have ν(C=C) > ν(C=O) and are shown by ionic complexes, *e.g.* with alkali metals; type 2 have ν(C=O) > ν(C=C) and are given by covalent complexes, *e.g.* with Al, Ga, or In; type 3 show a considerable scatter in the positions of ν(C=C) and ν(C=O) as the transition element is varied, and they exemplify covalent complexes involving π-bonding, *e.g.* transition-metal derivatives. They also report a correlation between ν(C=O), ν(C=C) and the ligand-field parameter $10Dq$ for the type 3 complexes and on this basis the authors prefer to assign ν(C=O) *ca.* 1600 and ν(C=C) *ca.* 1500 cm.$^{-1}$. The authors treat some trivalent transition-metal complexes with dibenzoyl methanates in an analogous manner.[684]

676 S. Pinchas, B. L. Silver, and I. Laulicht, *J. Chem. Phys.*, 1967, **46**, 1506.
677 M. Mikami, I. Nakagawa, and T. Shimanouchi, *Spectrochim. Acta*, 1967, **23**, *A*, 1037.
678 G. T. Behnke and K. Nakemoto, *Inorg. Chem*,, 1967, **6**, 433.
679 C. Y. Liang, E. J. Schimitschek, D. H. Stephens, and J. A. Trias, *J. Chem. Phys.*, 1967, **46**, 1588.
680 A. I. Byrke, N. N. Magdesieve, L. I. Martynenko, and V. I. Spitsyn, *Zhur. neorg. Khim.*, 1967, **12**, 666; see also *Chem. Abs.*, 1967, **67**, 48,650*x*.
681 G. A. Domrachev and V. P. Ippolitoua, *Zhur. neorg. Khim.*, 1967, **12**, 459; see also *Chem. Abs.*, 1967, **66**, 120,347*c*.
682 R. D. Hancock, H. W. Sacks, R. Thornton, and D. A. Thornton, *Inorg. Nuclear Chem. Letters*, 1967, **3**, 51.
683 R. D. Hancock and D. A. Thornton, *Inorg. Nuclear Chem. Letters*, 1967, **3**, 419.
684 R. D. Hancock and D. A. Thornton, *Inorg. Nuclear Chem. Letters*, 1967, **3**, 423.

Among several other papers concerned with acetylacetone complexes,[685–693] Gibson *et al.* give examples of platinum–carbon-bonded acetylacetone complexes. Finally, some more complex carbonyl ligands such as thenoyltrifluoroacetone,[694] α-hydroxyarylcarbonyl,[695] and 1,1,1,2,2,3,3-heptafluoro-7,7-dimethyl-4,6-octanedione [696] have been examined.

Many papers have been concerned with the co-ordination of acetates, halogenoacetates, benzoates, etc.[697–712]. The carboxylate ion may co-ordinate to a metal either with a symmetrical bridged structure or as an unidentate ligand and the two possibilities can be differentiated using i.r. spectroscopy. For a series of salts having structure (48), the antisymmetric

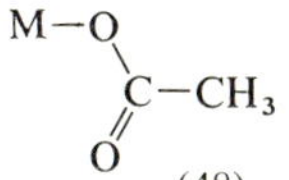

(48)

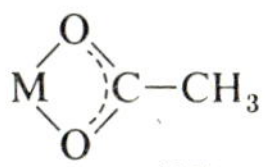

(49)

COO stretching frequency (ω_2) will increase and the symmetric O=C—O stretching frequency (ω_1) will decrease, as the M—O bond order becomes stronger. For the symmetrical bridged structure (49), both the O=C—O stretching bands are shifted in the same direction upon changing the metal. Furthermore, for symmetrical bridged structures, the separation values of ω_1 and ω_2 are comparable to those of the free ion, *e.g.* sodium acetate

685 D. A. Buckingham, R. C. Gorges, and J. T. Henry, *Austral. J. Chem.*, 1967, **20**, 281.
686 R. W. Hay and B. P. Caughley, *Austral. J. Chem.*, 1967, **20**, 1829.
687 G. Doyle and R. S. Tobias, *Inorg. Chem.*, 1967, **6**, 1111.
688 P. R. Singh and R. Sahai, *Austral. J. Chem.*, 1967, **20**, 639.
689 M. Idelson, I. R. Karady, B. H. Mark, D. O. Rickter, and V. H. Hooper, *Inorg. Chem.*, 1967, **6**, 450.
690 D. Gibson, J. Lewis, and C. Oldham, *J. Chem. Soc. (A)*, 1967, 72.
691 S. C. Chattoraj and R. E. Sievers, *Inorg. Chem.*, 1967, **6**, 408.
692 M. Kilner, F. A. Hartman, and A. Wojcicki, *Inorg. Chem.*, 1967, **6**, 406.
693 L. J. Boucher, *Inorg. Chem.*, 1967, **6**, 2162.
694 K. Ohwada, *J. Inorg. Nuclear Chem.*, 1967, **29**, 833.
695 D. P. Graddon and G. M. Mockler, *Austral. J. Chem.*, 1967, **20**, 21.
696 C. S. Springer, jun., D. W. Meek, and R. E. Sievers, *Inorg. Chem.*, 1967, **6**, 1105.
697 T. A. Stephenson and G. Wilkinson, *J. Inorg. Nuclear Chem.*, 1967, **29**, 2122.
698 A. V. R. Warrier and P. S. Narayanan, *Spectrochim. Acta*, 1967, **23**, *A*, 1061.
699 A. B. P. Lever and D. Ogden, *J. Chem. Soc. (A)*, 1967, 2041.
700 R. S. P. Coutts and P. C. Wailes, *Austral. J. Chem.*, 1967, **20**, 1579.
701 G. B. Deacon, *Austral. J. Chem.*, 1967, **20**, 459.
702 P. B. Simons and W. A. G. Graham, *J. Organometallic Chem.*, 1967, **10**, 457.
703 G. Winkhaus and P. Ziegler, *Z. anorg. Chem.*, 1967, **350**, 51.
704 K. Kuroda and P. S. Gentile, *J. Inorg. Nuclear Chem.*, 1967, **29**, 1963.
705 N. J. Rose, C. A. Root, and D. H. Busch, *Inorg. Chem.*, 1967, **6**, 1431.
706 G. H. Reifenberg and W. J. Considine, *J. Organometallic Chem.*, 1967, **9**, 495.
707 T. N. Srivastava and S. K. Landon, *Z. anorg. Chem.*, 1967, **353**, 87.
708 L. Beyer, E. Hoyer, and G. Kühn, *Z. anorg. Chem.*, 1967, **350**, 27.
709 G. B. Deacon and P. W. Felder, *Austral. J. Chem.*, 1967, **20**, 1587.
710 H. Kurosawa and R. Okawara, *J. Organometallic Chem.*, 1967, **10**, 211.
711 J. Gopalakrishnan and C. C. Patel, *Inorg. Chem.*, 1967, **6**, 2111.
712 H. A. Goodwin and R. N. Sylva, *Austral. J. Chem.*, 1967, **20**, 217.

164 cm.$^{-1}$, while the separation for unidentate carboxylate complexes is usually much larger. The acetato-complexes of Pd^{II} are useful examples:[697]

Compound	*Bridging*		*Unidentate*	
	ω_2	ω_1	ω_2	ω_1
$[Pd(OCOMe)_2PPh_3]_2$	1580	1411	1629	1314
$[Pd(OCOMe)_2AsPh_3]_2$	1582	1410	1631	1314
$[Pd(OCOCF_3)_2Me_2CO]_2$	1541	1429	1650 or 1626	1362
$[Pd(OCOC_2F_5)_2Me_2CO]_2$	1543	1414	1653 or 1629	1364

Goldsmith and Ross[713] have studied the i.r. spectra of some rare-earth carbonates and Narvor *et al.*[714] have examined hydrogeno- and deuterio-carbonates. Pedersen[715] and Gruen and Plane[716] have interpreted the spectra of some metal oxalate complexes.

The vibrational spectrum of the nitrate ion (D_{3h} symmetry) consists of four bands:

N—O stretching modes, ν_1, A_1', *ca.* 1050; ν_3, E', *ca.* 1380;
N—O bending modes, ν_2, A_2'', *ca.* 830; and ν_4, E', *ca.* 720 cm.$^{-1}$.

Several workers have examined aqueous solutions of nitrate salts: $Ca(NO_3)_2$,[717] $In(NO_3)_3$,[718] $Cr(NO_3)_2$, $Ni(NO_3)_3$, $Co(NO_3)_2$, and NH_4NO_3.[719] For $Ca(NO_3)_2$ the removal of the degeneracy of the E' modes, even in dilute solutions, and the activity of the A' mode in the i.r. region suggest that the symmetry of the nitrate ion has been lowered by solvation with water. The perturbation is enhanced by ionic interaction with hydrated calcium ions, and the symmetry of NO_3^- is probably lowered to C_{2v}.

The vibrational spectra of some molten nitrate salts have been interpreted[720–721] and the changing appearance of the spectra for $Ca(NO_3)_2$ dissolved in molten KNO_3 and molten $NaNO_3$ is a function of the calcium content of the system. This indicates clearly an asymmetric perturbation of the NO_3^- ion by the Ca^{2+} ions in these mixtures. The perturbation is also consistent with the existence, in both types of systems, of contact ion-pair complexes.

Among several other papers,[722–726] Bonn *et al.* examined the spectra of NO_3^- and NO_2^- ions as impurities in single crystals of KCl and showed

713 J. A. Goldsmith and S. D. Ross, *Spectrochim. Acta*, 1967, **23**, *A*, 1909.
714 A. Le Narvor, P. Sanmagne, and A. Novak, *J. Chim. phys.*, 1967, **64**, 1643.
715 B. F. Pedersen, *Acta Chem. Scand.*, 1967, **21**, 801.
716 E. C. Gruen and R. A. Plane, *Inorg. Chem.*, 1967, **6**, 1123.
717 D. E. Irish and G. E. Walrafen, *J. Chem. Phys.*, 1967, **46**, 378.
718 R. E. Hester and W. E. L. Grossman, *Spectrochim. Acta*, 1967, **23**, *A*, 1945.
719 W. Krasser and H. W. Nürnberg, *Naturwiss.*, 1967, **54**, 134.
720 R. E. Hester and K. Krishnan, *J. Chem. Phys.*, 1967, **46**, 3405.
721 R. E. Hester and K. Krishnan, *J. Chem. Phys.*, 1967, **47**, 1747.
722 R. E. Hester and R. A. Plane, *Spectrochim. Acta*, 1967, **23**, *A*, 2289.
723 G. J. Janz, M. J. Tait, and J. Meier, *J. Phys. Chem.*, 1967, **71**, 963.
724 R. Kato and J. Rolfe, *J. Chem. Phys.*, 1967, **47**, 1901.
725 R. Bonn, R. Metselaar, and J. van der Elsken, *J. Chem. Phys.*, 1967, **46**, 1988.
726 J. D. Donaldson and B. J. Senior, *J. Chem. Soc. (A)*. 1967, 1821.

that the fine structure present must be attributed to rotational motions of these ions.

The nitrate ion is known to be capable of co-ordination as either a unidentate (50), bidentate (51), or bridging group (52). Under C_{2v} symmetry

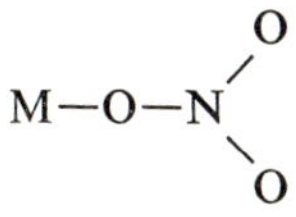
(50)

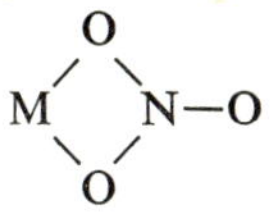
(51)

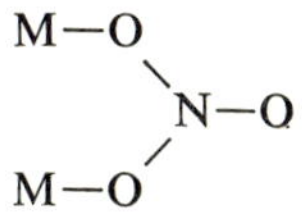
(52)

the ionic E' modes split into A_1 and B_1 components, so that the spectrum now ideally is one of six bands, all being active in both i.r. and Raman. Thus it is not possible to distinguish the above three possibilities unless the actual frequency positions are different. The normal-co-ordinate calculations of Hester and co-workers (*Inorg. Chem.*, 1966, **5**, pp. 980, 1308) suggest that the sequence of polarised and depolarised Raman lines should not, in fact, be identical (see Table 50).

Table 50 *Pattern of Raman polarisation for ionic, unidentate, and bidentate nitrate groups*

	Increasing Frequency			
Ionic (D_{3h})	E' (d)	A_1' (p)	A_2'' (d)	E' (d)
Unidentate (C_{2v})	B_1 (d), A_1 (p)	A_1 (p)	[a]	[b] [b]
Bidentate (C_{2v})	A_1 (p), B_1 (d)	A_1 (p)	[a]	A_1 (p), B_1 (d)

[a] Out-of-plane modes have not been calculated.
[b] Relative positions of A_1 and B_1 components uncertain.
d = Depolarised; p = polarised.

The arguments regarding bidentate bonding apply equally well to bidentate bridging nitrato-groups. There is rather more certainty attached to the stretching than to the bending modes. Thus for unidentate complexes the highest N—O stretching mode is depolarised while the converse is true for a bidentate nitrate group.

Complexes reported to contain bidentate nitrate groups are: $Sn(NO_3)_4$,[727] $Ti(NO_3)_4$,[727, 728] $VO(NO_3)_3$,[727] $M(NO_3)_3$,3(tri-n-butylphosphate) (M = rare-earth element),[729] $UO_2(NO_3)_2,2H_2O$,[730] $L_2UO_2(NO_3)_2$ (L = py, Ph_3PO, etc.).[730]

Other data reported include complexes of Ph_3PO with lanthanide and yttrium nitrates,[731] nickel nitrate with a tridentate *NNN*-Schiff base,[732] and

[727] C. C. Addison, D. W. Amos, D. Sutton, and W. H. H. Hoyle, *J. Chem. Soc.* (*A*), 1967, 808.
[728] C. D. Garner, D. Sutton, and S. C. Wallwork, *J. Chem. Soc.* (*A*), 1967, 1949.
[729] J. R. Ferraro, C. Cristallini, and I. Fox, *J. Inorg. Nuclear Chem.*, 1967, **29**, 139.
[730] J. I. Bullock, *J. Inorg. Nuclear Chem.*, 1967, **29**, 2257.
[731] D. R. Cousins and F. A. Hart, *J. Inorg. Nuclear Chem.*, 1967, **29**, 1745.
[732] L. Sacconi, I. Bertini, and R. Morassi, *Inorg. Chem.*, 1967, **6**, 1548.

the cerium(IV) ion–nitrate ion–water system.[733] Reference [728] gives a very extensive table of ν(N—O) frequencies for sixty compounds and further examples of nitrate complexes are given in references.[734–739]

The uncomplexed perchlorate ion has a regular tetrahedral structure and exhibits two characteristic i.r. absorption bands at 1050–1150 (ν_3) and 630 (ν_4) cm.$^{-1}$. The non-degenerative symmetrical stretch ν_1, which is Raman-active, is usually observed in the i.r. spectrum as a weak band at *ca.* 930 cm.$^{-1}$. If the perchlorate ion becomes co-ordinated to a metal ion by a single oxygen atom, then the symmetry is lowered to C_{3v}. As a result, the broad, degenerate, ν_3 band splits into two well defined bands with maxima between 1000 and 1200 cm.$^{-1}$. Likewise, the chlorine-co-ordinated oxygen frequency ν_2 becomes i.r. active and appears as a relatively intense band at about 925–950 cm.$^{-1}$. The degenerate bending mode ν_4 also splits into two bands. If a complex contains bidentate co-ordinated groups then the symmetry of the latter is C_{2v}, and the two i.r.-active bands ν_3 and ν_4 in the free ion are each split into three bands. Some examples of compounds containing ionic perchlorate groups are: Cu pmt$_2$ClO$_4$ (pmt = pentamethylenetetrazole),[740, 741] Cu pmt$_4$(ClO$_4$)$_2$,[740, 741] Cu pmt$_6$(ClO$_4$)$_2$,[740, 741] [Ni(4-vinylpyridine)$_4$](ClO$_4$)$_2$,[655] M(ClO$_4$)$_3$,8DMSO (M = La, Ce, Pr, or Nd; DMSO = dimethyl sulphoxide),[742, 743] M(ClO$_4$)$_3$,7DMSO (M = Sm, Gd, Y).[742, 743]

Rodley and Smith [744] report two different forms of *trans*-Co(*o*-phenylenebisdimethylarsine)$_2$(ClO$_4$)$_2$, monoclinic and orthorhombic. *X*-ray analysis of the orthorhombic form has shown that it has a *trans*-configuration with weak association of perchlorate ions in the axial positions. The i.r. spectrum of this form has a single broad ν(Cl—O) absorption which indicates that the perchlorate ion is not significantly distorted from T_d symmetry. By contrast, the monoclinic form reveals a marked splitting of the perchlorate ν(Cl—O) band. The two strong absorptions at 1034 and 1116 cm.$^{-1}$ can be assigned as the *E* and *A* modes of a unidentate perchlorate ion with C_{3v} symmetry.

The infrared spectrum of bis-(2-benzoylpyridine)copper(II) perchlorate suggests the perchlorate groups are weakly co-ordinated to the copper.[745] Farago *et al.*[746] suggest that blue [Ni en$_2$(ClO$_4$)$_2$] contains bidentate

[733] J. T. Miller and D. E. Irish, *Canad. J. Chem.*, 1967, **45**, 148.
[734] C. D. Cook and G. S. Janhal, *J. Amer. Chem. Soc.*, 1967, **89**, 3066.
[735] F. A. Hart, J. E. Newbery, and D. Shaw, *Chem. Comm.*, 1967, **1**, 45.
[736] C. M. Harris, H. R. H. Patil, and E. Sinn, *Inorg. Chem.*, 1967, **6**, 1102.
[737] C. C. Addison, R. Davis, and N. Logan, *J. Chem. Soc.* (*A*), 1967, 1449.
[738] S. K. Ramalingam and S. Soundararajan, *J. Inorg. Nuclear Chem.*, 1967, **29**, 1763.
[739] (Mrs.) J. H. Hickford and J. E. Fergusson, *J. Chem. Soc.* (*A*), 1967, 113.
[740] F. M. D'Itri and A. I. Popov, *Inorg. Chem.*, 1967, **6**, 597.
[741] F. M. D'Itri and A. I. Popov, *Inorg. Chem.*, 1967, **6**, 1591.
[742] V. N. Krishnamurthy and S. Soundararajan, *J. Inorg. Nuclear Chem.*, 1967, **29**, 517.
[743] V. N. Krishnamurthy and S. Soundararajan, *Z. anorg. Chem.*, 1967, **349**, 220.
[744] G. A. Rodley and P. W. Smith, *J. Chem. Soc.* (*A*), 1967, 1580.
[745] R. R. Osborne and W. R. McWhinnie, *J. Chem. Soc.* (*A*), 1967, 2075.
[746] M. E. Farago, J. M. James, and V. C. G. Trew, *J. Chem. Soc.* (*A*), 1967, 820.

co-ordinated perchlorate. Similar conclusions are drawn from the i.r. spectra of some solid adducts of $UO_2(ClO_4)_2$.[747]

ν(N—O) occurs at 1265 cm.$^{-1}$ for pyridine *N*-oxide and it is lowered by 40–50 cm.$^{-1}$ when the ligand is co-ordinated through the oxygen atom;[748–750] δ(N—O) (840 cm.$^{-1}$) is relatively unaffected. For complexes with *p*-bromo-*NN*-dimethylaniline *N*-oxide,[751] ν(N—O) is shifted very little. This may be attributed to the single bond character of the N—O bond, cf. pyridine *N*-oxide. Co-ordination relieves the electron density around the oxygen atom but it has little effect on the N—O bond of an aliphatic amine *N*-oxide.

The addition compounds of aluminium and gallium trihalides with ethers have been examined.[752] The B_1 ring-stretching mode in tetrahydrofuran which involves ν(C—O), moves to lower wavenumbers on co-ordination (C_4H_8O = 1070, $C_4H_8O,AlCl_3$ = 985, $C_4H_8O,GaCl_3$ = 990 cm.$^{-1}$). From a study of the B_1 mode of diethyl ether (1135 cm.$^{-1}$), the order of acceptor strengths towards Et_2O as a reference donor is

$$AlBr_3 > AlCl_3 > GaCl_3 \simeq GaBr_3.$$

The order for tetrahydrofuran is $AlCl_3 > GaCl_3$. The authors stress, however, that no particular emphasis is placed upon this correlation because major assumptions are always made in attempting to correlate i.r. data with bond strengths.

References [753–762] discuss a variety of other oxygen-donor ligands such as anthrone, tropone, etc.

The electronic structure of sulphoxides may be represented by a resonance hybrid of the structures [(53)–(55)]. If co-ordination occurs through

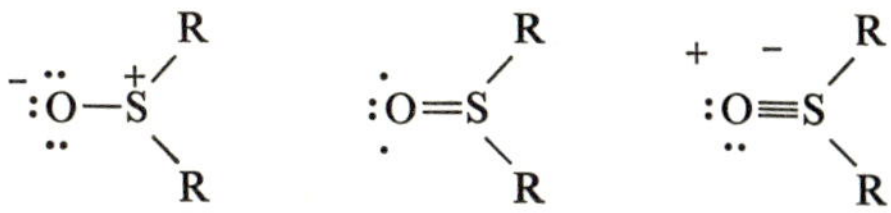

(53) (54) (55)

[747] V. M. Vdovenko, L. G. Mashinov, and D. N. Sugloblov, *Radiokhimiya*, 1967, **9**, 37; see also *Chem. Abs.*, 1967, **67**, 68,920*p*.
[748] V. N. Krishnamurthy and S. Soundararajan, *Canad. J. Chem.*, 1967, **45**, 189.
[749] R. G. Garvey and R. O. Ragsdale, *J. Inorg. Nuclear Chem.*, 1967, **29**, 1527.
[750] G. Schmauss and H. Specker, *Naturwiss.*, 1967, **54**, 442.
[751] C. J. Popp and R. O. Ragsdale, *Inorg. Chem.*, 1967, **6**, 2123.
[752] R. L. Richards and A. Thompson, *J. Chem. Soc.* (*A*), 1967, 1244.
[753] P. W. N. M. van Leeuwen, *Rec. Trav. chim.*, 1967, **86**, 247.
[754] L. L. Quill and G. L. Clink, *Inorg. Chem.*, 1967, **7**, 1433.
[755] H. D. Lutz, *Z. anorg. Chem.*, 1967, **353**, 207.
[756] E. Uhlemann and F. Dietze, *Z. anorg. Chem.*, 1967, **353**, 26.
[757] R. C. Paul, R. Parkash, and S. S. Sandhu, *J. Inorg. Nuclear Chem.*, 1967, **29**, 1915.
[758] F. Larèze, *Compt. rend.*, 1967, **265**, *C*, 307.
[759] P. S. Gentile, P. Jargiello, T. A. Shankoff, and J. Carlotto, *J. Inorg. Nuclear Chem.*, 1967, **29**, 1685.
[760] W. Dembínski and A. Deptula, *J. Inorg. Nuclear Chem.*, 1967, **29**. 799.
[761] A. W. McLellan and G. A. Melson, *J. Chem. Soc.* (*A*), 1967, 137.
[762] R. C. Paul and S. L. Chadha, *Spectrochim. Acta*, 1967, **23**, *A*, 1243.

oxygen, contributions of structures (54) and (55) will decrease and the result will be a decrease of the S—O stretching frequency. If co-ordination occurs through sulphur, the contribution of structure (53) will decrease and the result may be an increase in ν(S—O) [ν(S—O) for free dimethyl sulphoxide occurs at 1055–1100 cm.$^{-1}$]. Six authors [742, 763–767] have used this criterion for a wide range of complexes of Mn^{II}, Fe^{II}, Co^{II}, Ni^{II}, Zn^{II} and rare-earth chlorides, etc., and they deduce co-ordination via the oxygen atom. In contrast, ν(S—O) for $PtCl_2,(Me_2SO)_2$ consists of two bands at 1134 and 1157 cm.$^{-1}$. This is consistent with sulphur donation and the presence of two bands suggests a *cis*-structure.[768]

Dithian monosulphoxide (56) forms a 1 : 1 adduct with Ph_3SnCl and 2 : 1 adducts with Ph_2SnCl_2 and SnI_4. *cis*- and *trans*-dithian disulphoxide [(57) and (58)] form 1 : 1 complexes with all three.[769] I.r. spectra of the ν(S—O) region indicate that whereas the *trans*-disulphoxide is a bidentate ligand, the *cis*-form is a unidentate ligand.

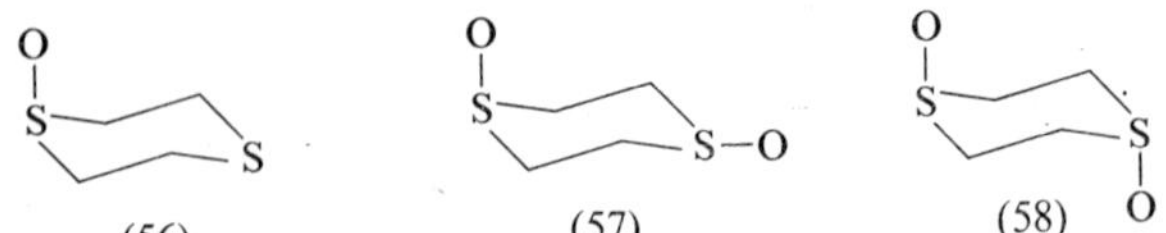

(56) (57) (58)

Van Leeuwen and Groeneveld [770] have prepared the tetramethylene sulphoxide and pentamethylene sulphoxide complexes of Pt^{II} and Cu^{II} and they find that ν(S=O) can be qualitatively related to the electronegativity of metal ions.

Dimethyl selenoxide and some of its complexes have been examined [771, 772] and ν(Se=O) occurs at 800vs (KBr). For a series of transition-metal complexes ($HgCl_2$, $CdCl_2$, $PdCl_4$, $CuCl_2$, $NiCl_2$, $NiBr_2$, $CoCl_2$, and $FeCl_3$) the band at 800 cm.$^{-1}$ is lowered and in all cases except the $HgCl_2$ complex it appears as a doublet. A similar doubling has been found for Ph_2Se=O complexes and it is due to the coupling of the Se=O vibrations through the metal atom.

Complexes of alkyl sulphinamides $RS(O)NR_2$ are also found to complex via the oxygen atom.[773]

763 W. F. Currier and J. H. Weber, *Inorg. Chem.*, 1967, **6**, 1539.
764 V. G. Kumar Das and W. Kitching, *J. Organometallic Chem.*, 1967, **10**, 59.
765 S. K. Ramalingam and S. Soundararajan, *Z. anorg. Chem.*, 1967, **353**, 216.
766 P. W. N. M. Van Leeuwen, *Rec. Trav. chim.*, 1967, **86**, 201.
767 T. Tanaka, *Inorg. Chim. Acta*, 1967, **1**, 217.
768 D. A. Langs, C. R. Hare, and R. G. Little, *Chem. Comm.*, 1967, 1080.
769 R. C. Poller and D. L. B. Toley, *J. Chem. Soc. (A)*, 1967, 2035.
770 P. W. N. M. van Leeuwen and W. L. Groeneveld, *Rec. Trav. chim.*, 1967, **86**, 721.
771 R. Paetzold, U. Lindner, G. Bochmann, and P. Reich, *Z. anorg. Chem.*, 1967, **352**, 295.
772 K. A. Jensen and V. Krishnan, *Acta Chem. Scand.*, 1967, **21**, 1988.
773 K. M. Nykerk, D. P. Eyman, and R. L. Smith, *Inorg. Chem.*, 1967, **6**, 2262.

Levison and Robinson[774] have isolated $Pt(SO_2)(Ph_3P)_3$. This is a red brown solid and it exhibits bands due to co-ordinated sulphur dioxide at 1195 and 1045 cm.$^{-1}$.

A normal-co-ordinate analysis[775] has been carried out for dimethyl sulphone (Me_2SO_2) and ν(S—O) and OSO scissoring motions are found to be pure modes [ν_{sym}(S—O) = 1167, ν_{asym}(S—O) = 1352, and OSO scissoring = 493 cm.$^{-1}$].

7 Sulphur and Selenium Donors

Thiocyanato- and selenocyanato-complexes have already been discussed in section 5 of Part III.

Carbon disulphide can act either as a simple donor ligand, through sulphur, or it can form a three-membered ring with the metal atom bound both to sulphur and carbon (π-CS_2 complexes). It can also bridge between two different metal atoms and these different possibilities [(59)–(61)] have been examined in three important papers by Wilkinson and co-workers.[776–778]

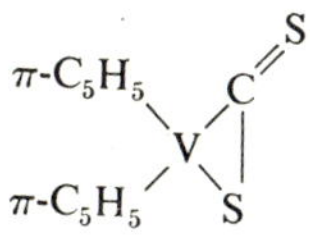

(59)

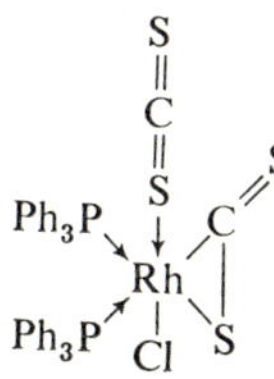

(61)

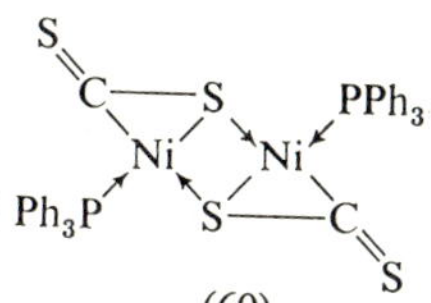

(60)

In addition to bands due to triphenylphosphine, the rhodium complex has two strong bands at 1510 and 1028 cm.$^{-1}$. The latter is attributable to a π-bonded CS_2 molecule and the other band is assigned to a carbon disulphide molecule co-ordinated through a sulphur atom to the sixth position in the rhodium(III) complex.

Adams and Cornell[779] have examined the far-i.r. spectra of thirty-six thiourea and ethylenethiourea complexes, and the metal–sulphur stretching frequencies lie in the range 205–298 cm.$^{-1}$ for the transition-metal complexes and vary with co-ordination number and electron configuration. Complexes with thiourea have been well characterised by other

774 J. J. Levison and S. D. Robinson, *Chem. Comm.*, 1967, 198.
775 J. H. Carter, J. M. Freeman, and T. Henshall, *J. Mol. Spectroscopy*, 1967, **22**, 18.
776 M. Baird, G. Hartwell, R. Mason, A. I. M. Rae, and G. Wilkinson, *Chem. Comm.*, 1967, **2**, 92.
777 M. C. Baird and G. Wilkinson, *J. Chem. Soc.* (*A*), 1967, 865.
778 M. C. Baird, G. Hartwell, jun., and G. Wilkinson, *J. Chem. Soc.* (*A*), 1967, 2037.
779 D. M. Adams and J. B. Cornell, *J. Chem. Soc.* (*A*), 1967, 884.

workers,[780–787a] and Cotton *et al.* find that the following bands are characteristic for sulphur co-ordination in some [ReX_3 tu_3] complexes: (*a*) The N—C—N stretching vibration (B_1) which occurs at 1470 cm.$^{-1}$ in the free ligand is increased to *ca.* 1505 cm.$^{-1}$; (*b*) the band at 730 cm.$^{-1}$ (assigned to C=S+N—C—N bending vibrations) is decreased to *ca.* 680 cm.$^{-1}$; (*c*) the intensity of the band at *ca.* 1090 cm.$^{-1}$ [a mode involving-ν(N—C—N)+$\rho(NH_2)$+ν(C=S)] is drastically reduced on co-ordination; (*d*) the band at *ca.* 1410 cm.$^{-1}$ does not split on co-ordination and so the thiourea is not behaving as a bridging ligand.

Cotton and co-workers[781] have also examined the corresponding complexes with 2,5-dithiahexane [ReX_3 dth]$_n$ (X = Cl or Br). In the gaseous or liquid states, 2,5-dithiahexane exists in both *trans* and gauche forms. The ligand adopts a gauche configuration in these complexes because the band most sensitive to the presence or absence of the *trans*-isomer is a strong absorption at 1205 cm.$^{-1}$ (CH_2 wag). This band is absent in all the complexes. Also bands at 840 and *ca.* 1030 cm.$^{-1}$, which are characteristic of the gauche isomer, increase in intensity on complex formation. Finally, ν(C—S) at 739 and 688 cm.$^{-1}$ in the *trans*-isomer disappear but the analogous band at 650 cm.$^{-1}$, due to ν(C—S) of the gauche form, is still observed.

Among several other papers on sulphur-bonded ligands[430, 788–800] Richards *et al.* find a correlation of thermodynamic data with i.r. data for some addition compounds of aluminium and gallium trihalides with organic sulphides. For Me_2S complexes, two bands which occur at 741 and 691 cm.$^{-1}$ in the free ligand [ν(C—S),B_1 and A_1] move to lower frequencies in the order $GaBr_3 > AlBr_3 > GaCl_3 > AlCl_3$.

780 A. Merijanian and H. M. Neumann, *Inorg. Chem.*, 1967, **6**, 165.
781 F. A. Cotton, C. Oldham, and R. A. Walton, *Inorg. Chem.*, 1967, **6**, 214.
782 L. V. Borisova and A. V. Karyakin, *Zhur. strukt. Khim.*, 1967, **8**, 359; see also *Chem. Abs.*, 1967, **67**, 58,943*j*.
783 K. C. Dash and D. V. R. Ras, *Z. anorg. Chem.*, 1967, **350**, 207.
784 R. K. Gosavi and C. N. R. Rao, *J. Inorg. Nuclear Chem.*, 1967, **29**, 1937.
785 R. A. Bailey and T. R. Peterson, *Canad. J. Chem.*, 1967, **45**, 1135.
786 R. K. Gosavi, U. Agarwala, and C. N. R. Rao, *J. Amer. Chem. Soc.*, 1967, **89**, 235.
787 A. D. Ahmed and P. K. Mandal, *J. Inorg. Nuclear Chem.*, 1967, **29**, 2347.
787a K. C. Dash, M. Ali, R. N. Patel, and D. V. Raman, *J. Indian Chem. Soc.*, 1967, **44**, 246.
788 R. L. Richards and A. Thompson, *J. Chem. Soc.* (*A*), 1967, 1248.
789 M. G. King and G. P. McQuillan, *J. Chem. Soc.* (*A*), 1967, 898.
790 D. M. Wiles, B. A. Gingras, and T. Suprunchuk, *Canad. J. Chem.*, 1967, **45**, 469.
791 M. J. Campbell and R. Grzeskowiak, *J. Chem. Soc.* (*A*), 1967, 396.
792 F. Bonati and R. Ugo, *J. Organometallic Chem.*, 1967, **10**, 257.
793 H. O. Desseyn and M. A. Herman, *Spectrochim. Acta*, 1967, 23, *A*, 2457.
794 S. H. H. Chaston and S. E. Livingstone, *Austral. J. Chem.*, 1967, **20**, 1065.
795 U. Agarwala, V. A. Narayan, and S. K. Dikshit, *Canad. J. Chem.*, 1967, **45**, 1057.
796 H. Groeger and A. Schmidpeter, *Chem. Ber.*, 1967, **100**, 3216.
797 S. K. Madan and M. Sulich, *J. Inorg. Nuclear Chem.*, 1967, **29**, 2765.
798 G. B. Deacon, *Austral. J. Chem.*, 1967, **20**, 1367.
799 Y. Matsumura, M. Shindo, and R. Okawara, *Inorg. Nuclear Chem. Letters*, 1967, **3**, 219.
800 F. Bonati, S. Cenini, and R. Ugo, *J. Organometallic Chem.*, 1967, **9**, 395.

King and McQuillan[789] have examined the complexing behaviour of Ph_3PS and Ph_3PSe. For Ph_3PS complexes, ν(P—S) falls by 40–50 cm.$^{-1}$ on co-ordination, *i.e.* 637 → *ca.* 590 cm.$^{-1}$, and for Ph_3PSe complexes, ν(P—Se) falls by *ca.* 20 cm.$^{-1}$ on co-ordination (562 → 543). Both these results may be compared with Ph_3PO complexes where ν(P—O) is lowered by *ca.* 38–70 cm.$^{-1}$.

Wiles *et al.*[790] have examined the spectra of some aldehyde and ketone thiosemicarbazones and the corresponding selenosemicarbazones. A band in the former at 805–830 cm.$^{-1}$ is replaced in the latter at 775–800 cm.$^{-1}$ [ν(C=Se)]. Formation of 1 : 1 CuI complexes of thiosemicarbazone results in the removal of bands in two regions (1075–1110 and 805–830 cm.$^{-1}$) and the authors conclude that there is a contribution from ν(C=S) in both regions but the lower region represents the more nearly pure ν(C=S).

PART IV Appendices

Part IV comprises two long Appendices containing references to further compounds for which partial vibrational data are available. These compounds have not been referred to in the main text but the tabulations are presented here to facilitate the location of further vibrational spectroscopic data. The first Appendix refers to transition-metal carbonyl complexes and the second contains references to data on all other compounds arranged according to the Periodic Table.

Appendix 1 Additional References to Metal Carbonyl Complexes

Chromium Carbonyl Complexes

		Ref.
$Cr(CO)_5X$	X = I, CN, NCS, CNH, $CNSnMe_3$	801, 499
$[Cr(CO)_5X]^-$	X = Cl, Br, I, CN, NCS	508, 801, 499
$[Cr(CO)_5L]^-$	L = $SnCl_3$, $GeCl_3$, C(O)Me, $NCC(CN)_2$	802, 803, 804
$[Cr(CO)_5InBr_3]^{2-}$		805
$Cr(CO)_5(PR)_5$	R = Me, Et, Ph	498
$Cr(CO)_5(R_nPPh_{3-n})$	$n = 0,1,2,3$	806
$Cr(CO)_5Ph_2P[CH_2]_2NEt_2$		807
$Cr(CO)_5Ph_2PH$		305
$Cr(CO)_5BuPEt_2$		808
$Cr(CO)_5Se(Ph)CH(OMe)Me$		809
$Cr(CO)_5NCCNCr(CO)_5$		810
$Cr(CO)_4(PR_3)_2$	R = Me, CH_2Cl, NMe_2	416
$Cr(CO)_4[P(CH_2O)_3CR]_2$	R = Me, Et, $[CH_2]_4Me$	416
$Cr(CO)_4C_4H_7PEt_2$		808
$Cr(CO)_4$ ffos		811
$Cr(CO)_4C_2H_4(PPh_2)_2$		811

801 E. Lindner and H. Behrens, *Spectrochim. Acta*, 1967, **23**, *A*, 3025.
802 J. K. Ruff, *Inorg. Chem.*, 1967, **6**, 1502.
803 E. O. Fischer and A. Maasböl, *Chem. Ber.*, 1967, **100**, 2445.
804 W. Beck, R. E. Nitzschmann, and H. S. Smedal, *J. Organometallic Chem.*, 1967, **8**, 547.
805 J. K. Ruff, *Inorg. Chem.*, 1967, **6**, 2080.
806 S. O. Grim, D. A. Wheatland, and W. McFarlane, *J. Amer. Chem. Soc.*, 1967, **89**, 5573.
807 G. R. Dobson, R. C. Taylor, and T. D. Walsh, *Inorg. Chem.*, 1967, **6**, 1929.
808 K. Issleib and M. Haftendorn, *Z. anorg. Chem.*, 1967, **351**, 9.
809 E. O. Fischer and V. Kiener, *Angew. Chem. Internat. Edn.*, 1967, **6**, 961.
810 J. F. Guttenberger, *Angew. Chem. Internat. Edn.*, 1967, **6**, 1081.
811 W. R. Cullen, P. S. Dhaliwal, and C. J. Stewart, *Inorg. Chem.*, 1967, **6**, 2256.

Chromium Carbonyl Complexes *(cont.)*

		Ref.
$Cr(CO)_4$*o*-$C_6H_4(PEt_2)_2$		811
$Cr(CO)_4Ph_2P(CH_2)_2NEt_2$		807
$Cr(CO)_4$(X-*o*-phen)	X = Cl, NO_2, Ph	812
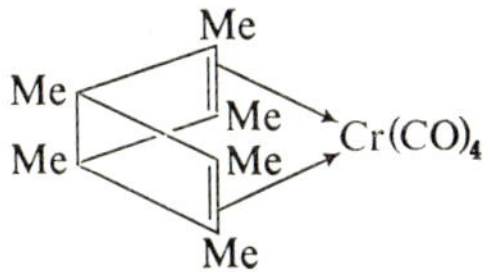		813
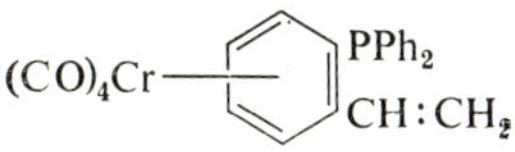		814
$(CO)_4Cr(PMe_2)_2Cr(CO)_4$		502
$Cr(CO)_3[B_3N_3Me_6]$		815
$Cr(CO)_3[P(OCH_2)_3CMe]$(X-*o*-phen)	X = Cl, NO_2, Ph	812
$Cr(CO)_3MeCNC_4H_7PEt_3$		808
$Cr(CO)_3C_7H_8$		816
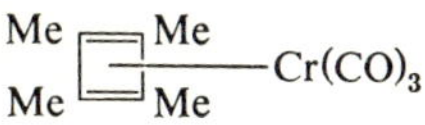		817
$Cr(CO)_3(C_6H_{11}NH_2)_3$		818
$Cr(CO)_3$ cot		819
RPh$Cr(CO)_3$(1,3,5-trinitrobenzene)	R = Me, MeO, Me_2N	820
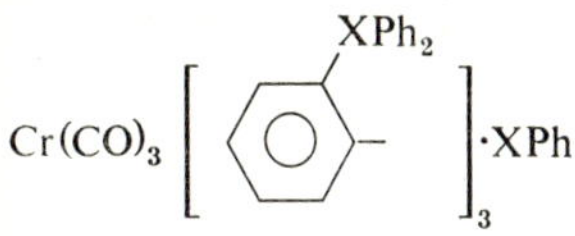	X = P or As	821
$[Cr(CO)_2(NH_3)_4]I$		822
$[Cr(CO)_2(NH_3)_4][Cr(CO)_3(NH_3)_2I]$		822

[812] R. J. Angelici and J. R. Graham, *Inorg. Chem.*, 1967, **6**, 988.
[813] E. O. Fischer, C. G. Kreiter, and W. Barngruber, *Angew. Chem. Internat. Edn.*, 1967, **6**, 634.
[814] M. A. Bennett, R. S. Nyholm, and J. D. Saxby, *J. Organometallic Chem.*, 1967, **10**, 301.
[815] R. Prinz and H. Werner, *Angew. Chem. Internat. Edn.*, 1967, **6**, 91.
[816] A. Pidcock and B. W. Taylor, *J. Chem. Soc.* (*A*), 1967, 877.
[817] R. G. Amiet, P. C. Reeves, and R. Pettit, *Chem. Com.*, 1967, 1208.
[818] H. Werner and R. Prinz, *Chem. Ber.*, 1967, **100**, 265.
[819] R. B. King, *J. Organometallic Chem.*, 1967, **8**, 139.
[820] G. Huttner and E. O. Fischer, *J. Organometallic Chem.*, 1967, **8**, 299.
[821] I. V. Howell and L. M. Venanzi, *J. Chem. Soc.* (*A*), 1967, 1007.
[822] H. Behrens and D. Herrmann, *Z. anorg. Chem.*, 1967, **351**, 225.

Chromium Carbonyl Complexes (*cont.*)

		Ref.
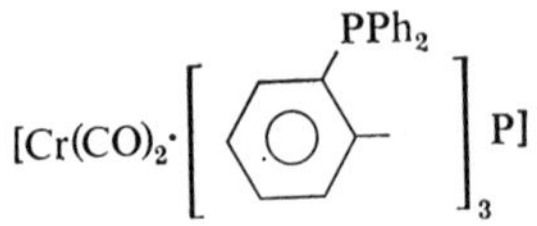		821
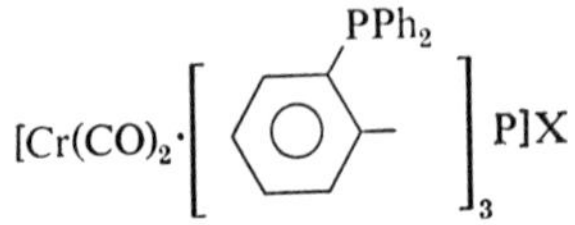	X = BPh_4, $SbCl_6$	821
$Cr_2(CO)_{10}MI_2$	M = Ge, Sn	805
$Cr_2(CO)_{10}SO_2{}^{2-}$		805
$Cr_2(CO)_{10}X$	X = I, SCN	801
$[Cr_2(CO)_{10}X]$	X = CN, I, SCN	801
$[(CO)_5M{-}Cr(CO)_5]^-$	M = Mn, Re	823

Molybdenum Carbonyl Complexes

$Mo(CO)_5L$	L = $CNSiMe_3$, $CNSnMe_3$, CNH $Ph_2P(CH_2)_2NEt_2$, PPh_3, NEt_3 Se(Ph)CH(OMe)Me, PPh_2H	499, 498 807, 809 305
$Mo(CO)_5X^-$	X = CN, $SnCl_3$, $GeCl_3$	499, 802
$Mo(CO)_5(alkyl_nPPh_{3-n})$	$n = 0,1,2,3$	806
$Mo(CO)_5(PX)_5$	X = Me, Et, Ph	498
$(NEt_4)Mo(CO)_5NCC(CN)_2$		804
$Mo(CO)_4[PR_3]_2$	R = Me, CH_2Cl_3, NMe_2, Cl, n-C_4H_9, OPh, Ph	500, 824 459
cis-$Mo(CO)_4(PCl_2Ph)_2$		824
$Mo(CO)_4[P(CH_2O)_3CR]_2$	R = Me, Et, $(CH_2)_4Me$	459
cis-$Mo(CO)_4L_2$	L = py, 3-pic, 4-pic, 3,4-lut, 3-Clpy	824
cis-$Mo(CO)_4C_8H_{12}$		824
cis-$Mo(CO)_4(MPh_3)_2$	M = P, As, Sb	500
cis-$Mo(CO)_4$bipy		500
$Mo(CO)_4(PR)_5$	R = Me, Et, Ph	498
$Mo(CO)_4(AsPh)_6$		498
$Mo(CO)_4$(X-*o*-phen)	X = Cl, NO_2Ph	812
$Mo(CO)_4(GeCl_3)_2{}^{2-}$		802

[823] U. Anders and W. A. G. Graham, *J. Amer. Chem. Soc.*, 1967, **89**, 539.
[824] F. Zingales, F. Canziana, and F. Basolo, *J. Organometallic Chem.*, 1967, **7**, 461.

Molybdenum Carbonyl Complexes (*cont.*)

		Ref.
$Mo(CO)_4(GeCl_3)PPh_3$		802
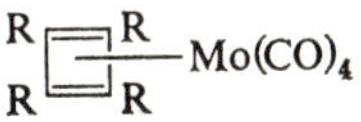	R = H or Me	817
$Mo(CO)_4L$	L = diphos, ffos, $C_2H_4(PPh_2)_2$, *o*-$C_6H_4(PEt_2)_2$, ffars, *o*-$C_6H_4(AsMe_2)_2$, *o*-$C_6F_4(AsMe_2)_2$	811, 825
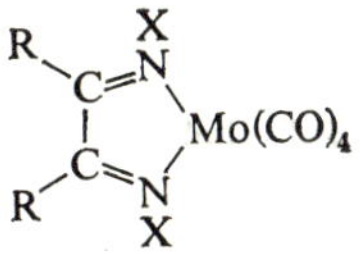	R = Me or Ph, X = NH_2, NMe_2, Me, Ph, OH	826
cis-$Mo(CO)_4(PPh_2H)_2$	(asym and sym)	305
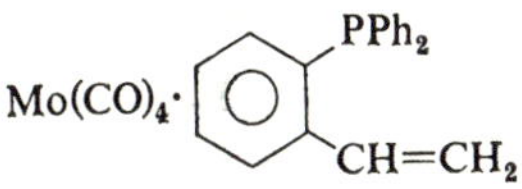		814
$Mo(CO)_3Cl_2$,2py		827
$Mo(CO)_3X_2L$	L = bipy, phen X = Cl, Br	827
(π-C_5H_5)$Mo(CO)_3X$,	X = BCl_2, BPh_2, CH_2Cl, CH_2Br, CH_2OMe, $SiCl_3$, COC_6F_5, $SiMe_3$	828, 829, 519 830, 831
$MoCO_3L_3$	L = PF_3, $MePF(NMe_2)$, $C_5H_{10}NPF_2$, $PhPF_2$, PPh_2H	832, 305
$Mo(CO)_3(PR)_5$	R = Me, Et, Ph	498
(π-C_5H_5)$Mo(CO)_3Br(CH_2)_n$	n = 3 or 4	833
$Mo(CO)_3Ph(NH_2)_n$	n = 1 or 3	818
$Mo(CO)_3(C_6H_{11}NH_2)_3$		818
$(Ph_3PH)[Mo(CO)_3(PPh_3)I_3]$		834
Et_3NCl_2B-$Mo(CO)_3$(π-C_5H_5)		828
$Mo(CO)_3C_7H_8$		816
$Mo(CO)_3Me(Me_5C_5)$		835

[825] R. J. Haines, R. S. Nyholm, and M. H. B. Stiddard, *J. Chem. Soc.* (*A*), 1967, 94.
[826] H. Bock and H. tom Dieck, *Chem. Ber.*, 1967, **100**, 228.
[827] R. Cotton and I. B. Tomkins, *Austral. J. Chem.*, 1967, **20**, 13.
[828] G. Schmid and H. Nöth, *J. Organometallic Chem.*, 1967, **7**, 129.
[829] M. L. H. Green, M. Ishaq, and R. N. Whiteley, *J. Chem. Soc.* (*A*), 1967, 1508.
[830] D. J. Cardin, S. A. Keppie, B. M. Kingston, and M. F. Lappert, *Chem. Comm.*, 1967, 1035.
[831] P. M. Treichel and R. L. Shubkin, *Inorg. Chem.*, 1967, **6**, 1328.
[832] G. S. Reddy and R. Schmutzler, *Inorg. Chem.*, 1967, **6**, 823.
[833] R. B. King and M. B. Bisnette, *J. Organometallic Chem.*, 1967, **7**, 311.
[834] J. Lewis and R. Whyman, *J. Chem. Soc.* (*A*), 1967, 77.
[835] R. B. King and M. B. Bisnette, *J. Organometallic Chem.*, 1967, **8**, 287.

Molybdenum Carbonyl Complexes (*cont.*)

		Ref.
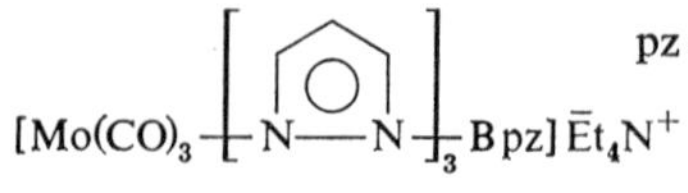	pz = 1-pyrazolyl	836
$(\pi\text{-}C_5H_5)Mo(CO)_2PPh_3R$	R = Me, $COCH_3$, COC_6F_5	837, 831
$[(\pi\text{-}C_5H_5)Mo(CO)_2COCH_3]_2(\mu\text{-diphos})$		837
$Mo(CO)_2[Me_5C_5]$		835
$[Mo(CO)_2L_2I]I$	L = bipy, phen	838
$(\pi\text{-}C_5H_5)Mo(CO)_2PR_3I$	R = OPh, OMe, OEt, Ph, Et, Bu^n	839
$(\pi\text{-}C_5H_5)Mo(CO)_2(MPh_3)Cl$	M = P, As, Sb	825
$[(\pi\text{-}C_5H_5)Mo(CO)_2Cl]_2$diphos		825
$(\pi\text{-}C_5H_5)Mo(CO)_2(AsMePh_2)I$		825
$[(\pi\text{-}C_5H_5)Mo(CO)_2\text{diphos}]ClO_4$		825
$[(\pi\text{-}C_5H_5)Mo(CO)_2\text{diars}]I$		825
$Mo(CO)_2$ bipy(π-allyl)X	X = Cl, Br, I, NCS, SC_6F_5	840
$Mo(CO)_2$ phen(π-allyl)X	X = Cl, Br, I, NCS	840
$Mo(CO)_2$ phen py(π-allyl)BF_4		840
$(NEt_4)[Mo(CO)_2X(\text{maleimide})_3]$	X = Cl, Br	841
$(NEt_4)[Mo(CO)_2Br(\text{maleic anhydride})_3]$		841
$(\pi\text{-}C_5H_5)Mo(CO)_2(MPh_3)X$	M = P, As, Sb; X = Cl, Br, I	842
$(\pi\text{-}C_5H_5)Mo(CO)(MPh_3)X$	M = P, As, Sb; X = Cl, Br, I	825, 842
$(\pi\text{-}C_5H_5)Mo(CO)(PEt_3)_2X$	X = Cl, I	825
$(\pi\text{-}C_5H_5)Mo(CO)LX$	L = diphos, diars, X = Cl, I	825, 842
$(\pi\text{-}C_5H_5)Mo_2(CO)_5PPh_3$		837
$(CO)_5MoNC{-}CNMo(CO)_5$		810
$Mo_2(CO)_6(PMe_2)L_2$	L = PPh_3, PEt_3, $P(C_6H_{11})_3$, $Ph_2P(CH_2)_nPPh_2$ where $n = 1, 2$ or 4	843
$Mo_2(CO)_6\,en_3$		818
$[(CO)_5M{-}Mo(CO)_5]$	M = Mn, Re	823

[836] S. Trofimenko, *J. Amer. Chem. Soc.*, 1967, **89**, 3904.
[837] K. W. Barnett and P. M. Treichel, *Inorg. Chem.*, 1967, **6**, 294.
[838] H. Behrens and J. Rosenfelder, *Z. anorg. Chem.*, 1967, **352**, 61.
[839] A. R. Manning, *J. Chem. Soc.* (*A*), 1967, 1984.
[840] C. G. Hall and M. H. B. Stiddard, *J. Organometallic Chem.*, 1967, **9**, 519.
[841] H. D. Murdoch, R. Henzi, and F. Calderazzo, *J. Organometallic Chem.*, 1967, **7**, 441.
[842] P. M. Treichel, K. W. Barnett, and R. L. Shubkin, *J. Organometallic Chem.*, 1967, **7**, 449.
[843] R. H. B. Mais, P. G. Owston, and D. T. Thompson, *J. Chem. Soc.* (*A*), 1967, 1735.

Tungsten Carbonyl Complexes

		Ref.
$W(CO)_5L$	L = py, Me_3N, Me_2NH, $PhNH_2$, 4-$BrC_6H_4NH_2$, 4-$MeOC_6H_4NH_2$, 4-MeC_5H_4N, $CNSnMe_3$, CNH	844
$[W(CO)_5X]^-$	X = Cl, Br, I, CN	508, 499
$[W(CO)_5L]^-$	L = $SnCl_3$, $GeCl_3$, $NCC(CN)_2$	802, 804
$W(CO)_5(PR)_5$	R = Me, Et, Ph	498
$W(CO)_5(R_nPPh_{3-n})$	$n = 0, 1, 2, 3$; R = alkyl or H	806, 844, 305
$W(CO)_5InBr_3{}^{2-}$		805
$W(CO)_5Se(Ph)CH(OMe)Me$		809
cis-$W(CO)_4$ py_2		500
cis-$W(CO)_4$ bipy		500
$W(CO)_4(PR_3)_2$	R = Me, Ph, NMe_2, CH_2Cl	500, 459
$W(CO)_4[P(CH_2O)_3CR]_2$	R = Me, Et, $(CH_2)_4Me$	459
$W(CO)_4$(X-*o*-phen)	X = Cl, NO_2, Ph	812
$W(CO)_4(PR)_5$	R = Me, Et, Ph	498
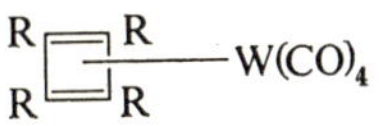	R = H or Me	817
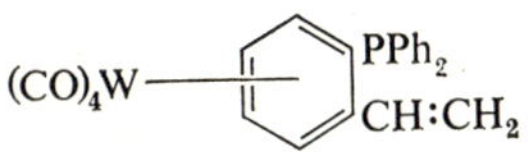		814
$(\pi-C_5H_5)W(CO)_3L$	L = BPh_2, BCl_2, BCl_2NEt_3, $(CH_2)_3Br$, $(CH_2)_4Br$, $SiMe_3$, CH_2OMe, CH_2Cl, CH_2Br	833, 828 830, 829
$W(CO)_3(PR)_5$	R = Me, Et, Ph	498
$W(CO)_3Cl_2(MPh_3)_2$	M = P, As, Sb	845
$[Ph_3PH][W(CO)_3PPh_3X_3]$	X = I, Br	834
$W(CO)_3C_7H_8$		816
$(NEt_4)[W(CO)_2XL_3]$	X = Cl, Br, L = maleimide, maleic anhydride	841
$W(CO)_2Cl_2(PPh_3)_2$		845
$W(CO)_2Br_2(MPh_3)_2$	M = P, As, Sb	845
$(\pi-C_5H_5)W(CO)_2PPh_3X$	X = Br, Me	837, 831
$[W(CO)_2$ $phen_2I]I$		838

[844] R. J. Angelici and Sr. M. D. Malone, *Inorg. Chem.*, 1967, **6**, 1731.
[845] M. W. Anker, R. Colton, and I. B. Tomkins, *Austral. J. Chem.*, 1967, **20**, 9.

Tungsten Carbonyl Complexes (*cont.*)

		Ref.
$(\pi\text{-}C_5H_5)W(CO)_2MPh_3COC_6F_5$	M = P, As	831
$W(CO)_2$ bipy(π-allyl)X	X = Cl, Br, I, NCS, SC_6F_5	840
$W(CO)_2$ phen(π-allyl)X	X = Cl, Br, I, NCS	840
$[W(CO)_2L\ py(\pi\text{-allyl})]BPh_4$	L = bipy, phen	840
$(\pi\text{-}C_5H_5)W(CO)_2(MPh_3)Cl$	M = P, As, Sb	842
$[(\pi\text{-}C_5H_5)W(CO)_2L]^+PF_6^-$	L = bipy, phen, diphos	842
$[(\pi\text{-}C_5H_5)W(CO)_2Cl]_2\text{-}\mu\text{-diphos}$		842
$W_2(CO)_{10}MI_2^{2-}$	M = Ge, Sn	805
$W_2(CO)_{10}SO_2^{2-}$		805
$(CO)_5WNC\cdot CNW(CO)_5$		810
$[(CO)_5M{-}W(CO)_5]^-$	M = Mn, Re	823

Manganese Carbonyl Complexes

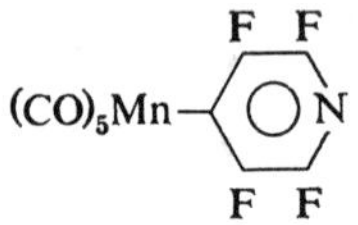

846

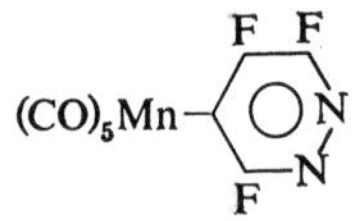

846

$Mn(CO)_5X$	X = H, D, Cl, Br, I, CH_3, CF_3, SiH_3, $SiCl_3$	509, 847, 513, 510, 508, 401, 511, 396, 512, 848, 519
$Mn(CO)_5X$	X = CH_2F, CHF_2, Me_3CCO, $Me_2CHCOPhCH_2CO$	849
$[Mn(CO)_5]X$	X = Cl, $NO_3CF_3CO_2$, SO_4H	850
$[Mn(CO)_5]_3SnR$	R = Cl, I, Me, Bu^n, Ph, $CH_2{:}CH$	851, 852
$[Mn(CO)_5]_2SnR_2$	R = Cl, I, Me, Bu^n, Ph, $CH_2{:}CH\cdot$	851

[846] J. Cooke, M. Green, and F. G. A. Stone, *Inorg. Nuclear Chem. Letters*, 1967, **3**, 47.
[847] W. G. McDugle, jun., A. F. Schreiner, and T. L. Brown, *J. Amer. Chem. Soc.*, 1967, **89**, 3114.
[848] B. J. Aylett and J. M. Campbell, *Chem. Comm.*, 1967, 159.
[849] K. Noack, U. Schaerer, and F. Calderazzo, *J. Organometallic Chem.*, 1967, **8**, 517.
[850] M. L. H. Green, A. G. Massey, J. T. Moelwyn-Hughes, and (the late) P. L. I. Nagy, *J. Organometallic Chem.*, 1967, **8**, 511.
[851] J. A. J. Thompson and W. A. G. Graham, *Inorg. Chem.*, 1967, **6**, 1365.
[852] J. H. Tsai, J. J. Flynn, and F. P. Boer, *Chem. Comm.*, 1967, 702.

Manganese Carbonyl Complexes (*cont.*)

		Ref.
$Mn(CO)_5SiH_32L$	$L = NMe_3$, py	848
$[(\pi\text{-}CH_3CH{=}CH_2)Mn(CO)_5]^+ClO_4^-$		850
$Mn(CO)_5SnPh_n(C_6F_5)_{3-n}$	$n = 0, 1, 2$	506
$Mn(CO)_5X$	$X = CFCFCF_2CF_2H$, $CF_2HCCFCF_3$, $CF_3CHCFCF_2$, $COCFCFCF_2H$	853
$Mn(CO)_5SO_2R$	$R = C_3H_5, C_4H_7, C_5H_9$	854
$Mn(CO)_4X_2$	X = Cl, Br, I, IBr, BrCl	855, 508
$[Mn(CO)_4SeR]_2$	R = Me, Et, Ph	856
$Me_2[Mn(CO)_4PPh_3]_2$		833
$[Mn(CO)_4PPh_2]_2$		857
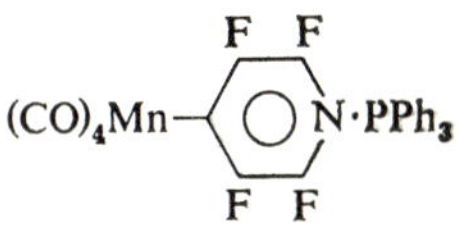		846
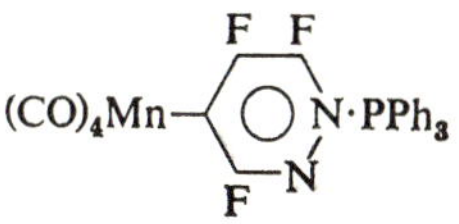		846
$Mn(CO)_3X(PR_3)_2$	X = H, Br, I; R = Ph or OPh	858
$Mn(CO)_3L$	L = thiophene and methyl-substituted thiophenes	859
$Mn(CO)_3L$ hfac	$L = py, PPh_3, AsPh_3, Bu^n$, $P(C_6H_{11})_3$, $4\text{-}CH_3C_5H_4N$	860
$[Mn(CO)_3(NH_3)_3]BPh_4$		861
$Mn(CO)_3X\,py_2$	X = Cl, Br, I	855
$Mn(CO)_3ClL_2$	$L = \gamma$-pic, 3-chloropyridine	855
$Mn(CO)_3NOPR_3$	$R = Ph, OPh, Bu^n$	862

[853] B. W. Tattershall, A. J. Rest, M. Green, and F. G. A. Stone, *Angew. Chem. Internat. Edn.*, 1967, **6**, 878.
[854] F. A. Hartiman, P. J. Pollick, R. L. Downs, and A. Wojcicki, *J. Amer. Chem. Soc.*, 1967, **89**, 2493.
[855] F. Zingales and U. Sartorelli, *Inorg. Chem.*, 1967, **6**, 1243.
[856] E. W. Abel, B. C. Crosse, and G. V. Hutson, *J. Chem. Soc.* (*A*), 1967, 2014.
[857] E. W. Abel and I. H. Sabherwal, *J. Organometallic Chem.*, 1967, **10**, 491.
[858] R. Ugo and F. Bonati, *J. Organometallic Chem.*, 1967, **8**, 189.
[859] H. Singer, *J. Organometallic Chem.*, 1967, **9**, 135.
[860] F. A. Hartman, M. Kilner, and A. Wojcicki, *Inorg. Chem.*, 1967, **6**, 34.
[861] H. Behrens, E. Ruzter, and H. Wakamatsu, *Z. anorg. Chem.*, 1967, **349**, 241.
[862] H. Wawersik and F. Basolo, *J. Amer. Chem. Soc.*, 1967, **89**, 4626.

Manganese Carbonyl Complexes (*cont.*)

		Ref.
$Mn(CO)_3NOAsPh_3$		862
$[Mn(CO)_3SeR]_n$	R = Me, Et, Ph; n = 3 or 4	856
$[Mn(CO)_3PPh_2]_3$		857
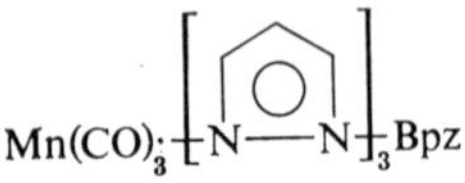		836
$(\pi\text{-}C_5H_5)Mn(CO)_3$		946
$K_2[Mn(CO)_3(CN)_3]$		863
$K_3[Mn(CO)_2(CN)_4]$		863
$(\pi\text{-}C_5H_5)Mn(CO)_2MPh_3$	M = P, As, Sb, Bi	864
$Mn(CO)_2NO(PR_3)_2$	R = Ph, OPh, Bu^n	862
$Mn(CO)_2\ hfac(PR_3)_2$	R = Ph, Bu^n, OMe, OPh, OBu^n	860
$Mn(CO)_2\ hfac(PPh_2Me)_2$		860
$Mn(CO)_2\ hfac(PPhMe_2)_2$		860
$(\pi\text{-}C_5H_5)Mn(CO)_2L$	L = propylene, pentene-1, cyclopentene, cycloheptene, *cis*-cyclo-octene, norbornylene, endic anhydride	865
$(\pi\text{-}C_5H_5)MnCO(MPh_3)_2$	M = P, As, Sb	864
Mn_2CO_{10}		396
$Mn_2(CO)_8X_2$	X = Cl, Br, I, PMe_2	508, 502

Rhenium Carbonyl Complexes

$Re(CO)_5X$	X = H, D, Cl, Br, I, $SiCl_3$	396, 504, 507, 508, 509, 510, 519
$[Re(CO)_5]_2SnPh_2$		851
$[Re(CO)_5]_3SnPh$		851
$[Re(CO)_5Cl]_2SbCl_5$		866

[863] H. Behrens, E. Ruzter, and E. Lindner, *Z. anorg. Chem.*, 1967, **349**, 251.
[864] C. Barbeau, *Canad. J. Chem.*, 1967, **45**, 162.
[865] R. J. Angelici and W. Loewen, *Inorg. Chem.*, 1967, **6**, 682.
[866] R. B. King, *J. Inorg. Nuclear Chem.*, 1967, **29**, 2119.

Rhenium Carbonyl Complexes (*cont.*)

		Ref.
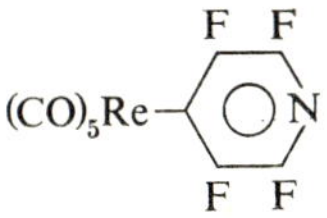		846
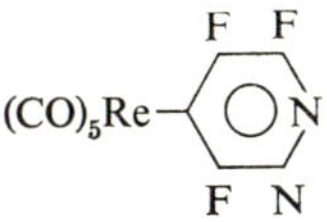		846
$[Re(CO)_4PPh_3]_2$		867
$Re(CO)_4PPh_3I$		867
$Re(CO)_4BrPPh_2Et$		868
$Re(CO)_4BrP(Bu^n)_3$		868
$[Re(CO)_4SeR]_2$	R = Me, Et, Ph	856
$[Re(CO)_4PPh_2]_2$		857
$Re(CO)_4ClL$	L = py, PPh_3, γ-pic, α-chloropyridine	869
$Re(CO)_4Br$ py		869
$Re(CO)_3(PPh_3)_2$		867
trans-$Re(CO)_3Ph_2P(CH_2)_2PPh_2I$		867
$Re(CO)_3BrL_2$	L = PPh_3, γ-pic	868
$[(\pi\text{-}C_5H_5)Re(CO)_3Cl]SbCl_6$		866
$[Re(CO)_3SeR]_n$	R = Me, Et, Ph; n = 3, 4	856
$[Re(CO)_3PPh_2]_3$		857
$Re(CO)_3Me_5C_5$		835
cis-$Re(CO)_2(Ph_2P(CH_2)_2PPh_2)_2$		867
$Re(CO)(Ph_2P(CH_2)_2PPh_2)_2$		867
$Re_2(CO)_{10}$		504, 396
$[Re(CO)_3(PPh_3)_2]_2$		867
cis- and *trans*-$[Re(CO)_3Ph_2P(CH_2)_2PPh_2]_2$		867

[867] M. Freni, D. Giusto, and P. Romiti, *J. Inorg. Nuclear Chem.*, 1967, **29**, 761.
[868] F. Zingales, U. Sartorelli, and A. Trovati, *Inorg. Chem.*, 1967, **6**, 1246.
[869] F. Zingales, U. Sartorelli, F. Canziani, and M. Raveglia, *Inorg. Chem.*, 1967, **6**, 155.

Iron Carbonyl Complexes

		Ref.
$Fe(CO)_5$		870
$[Fe(CO)_4X]^-$	X = I, $SnCl_3$, $GeCl_3$	514, 802
$Fe(CO)_4L$	L = PF_3, PCl_3, PPh_2H, MeCOCHCHCl, C_4F_6	871, 305, 872, 873
$Fe(CO)_4R_2$	R = CF_3CCH, $SiCl_3$, Et_2SnCl, Bu_2, SnCl	521, 519, 874
$[Fe(CO)_4SiCl_2]_2$		519
$[Fe(CO)_4SnR_2]_2$	R = Me, Et, Pr^n, Bu^n, Ph	875, 876, 874
$Sn[Fe(CO)_4]_4$		874
$[Fe(CO)_4]_4Sn_3R_4$	R = Me, Et, Bu	874
$Fe(CO)_3L$	L = PhCH:CHCH:NPh, MeCH:CHCH:NC_4H_9, cot, *exo*-(1-C_6F_5)(π-C_5H_5)	877, 878, 831
$Fe(CO)_3PhCH$:$CHCHNPhPPh_3$		877

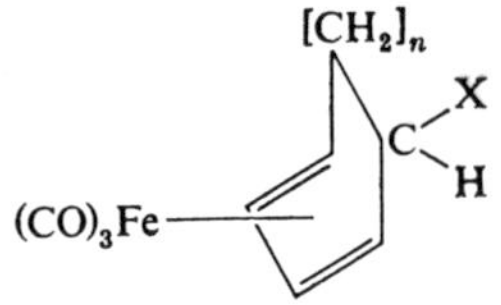

X = OMe; $n = 1$
X = OEt; $n = 2$ — 879
X = $CH(CO_2Et)_n$; $n = 2$ or 1

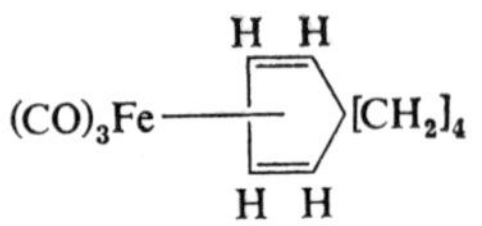

880

[870] J. Brunvoll, *Acta Chem. Scand.*, 1967, **21**, 1390.
[871] J. B. Pd. Tripathi and M. Bigorgne, *J. Organometallic Chem.*, 1967, **9**, 307.
[872] A. N. Nesmeyanov, K. Ahmed, L. V. Rybin, M. I. Rybinskaya, and Yu. A. Ustynyuk, *J. Organometallic Chem.*, 1967, **10**, 121.
[873] R. L. Hunt, D. M. Roundhill, and G. Wilkinson, *J. Chem. Soc.* (*A*), 1967, 982.
[874] J. D. Cotton, S. A. R. Knox, I. Paul, and F. G. A. Stone, *J. Chem. Soc.* (*A*), 1967, 264.
[875] R. M. Sweet, C. J. Fritchie, jun., and R. A. Schunn, *Inorg. Chem.*, 1967, **6**, 749.
[876] S. D. Ibekwe and M. J. Newlands, *J. Chem. Soc.* (*A*), 1967, 1783.
[877] S. Otsuka, T. Yoshida, and A. Nakamura, *Inorg. Chem.*, 1967, **6**, 21.
[878] F. A. L. Anet, H. D. Kaesz, A. Maasbol, and S. Winstein, *J. Amer. Chem. Soc.*, 1967, **89**, 2489.
[879] M. A. Hashini, J. D. Manro, P. L. Pausan, and (the late) J. M. Williamson. *J. Chem. Soc.* (*A*), 1967, 240.
[880] G. F. Grant and P. L. Pauson, *J. Organometallic Chem.*, 1967, **9**, 553.

Iron Carbonyl Complexes (*cont.*)

		Ref.
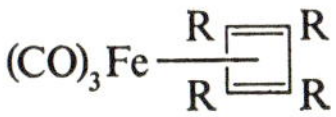	R = H or Me	881, 817
$Fe(CO)_3(PPh_2H)_2$		305
$Fe(CO)_3(PF_3)_2$		871
$Fe(CO)_3NiC_8Me_8$		881
$[Fe(CO)_3]_2(PR)_4$	R = Me, Et, Ph	498
$[Fe(CO)_3SnBu^n{}_2PPh_3]_2$		876
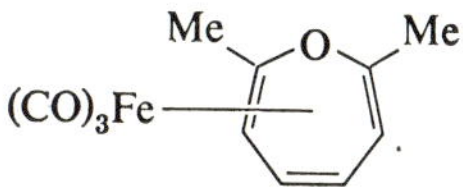		882
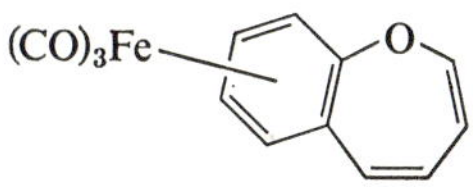		882
$(\pi\text{-}C_5H_5)Fe(CO)_2L$	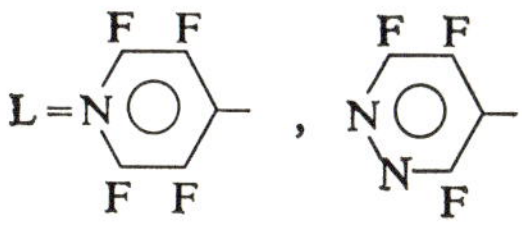	846, 804, 519
$\{[Fe(CO)_2(\pi\text{-}C_5H_5)_2]C_3H_5\}PF_6$		833
$Fe(CO)_2[PREt_2]Br_2$		883
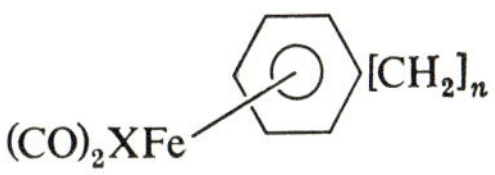	X = I or CN; *n* = 1 or 2	879
$(\pi\text{-}C_5H_5)Fe(CO)_2L$	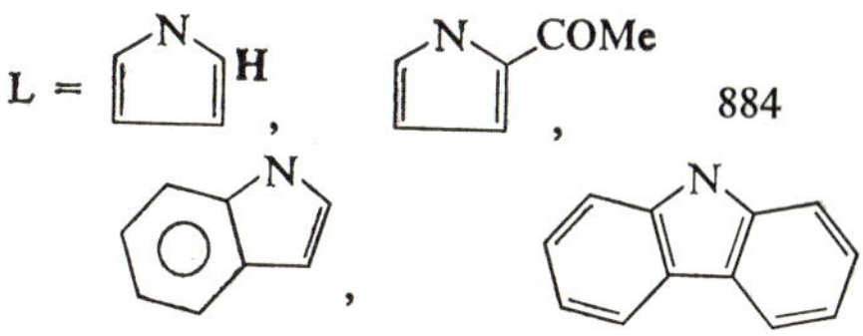	884
$(\pi\text{-}C_5H_5)Fe(CO)_2PPh_3[exo\text{-}(1\text{-}C_6X_5)]$	X = H, F	831
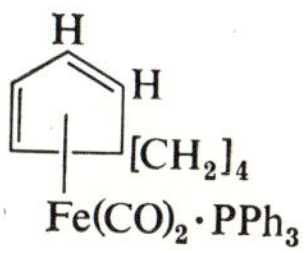		880

[881] R. Bruce, K. Moseley, and P. M. Maitlis, *Canad. J. Chem.*, 1967, **45**, 2011.
[882] E. O. Fischer, C. G. Kreiter, H. Rühle, and K. E. Schwarzhans, *Chem. Ber.*, 1967, **100**, 1905.
[883] W. Hieber, V. Frey, and P. John, *Chem. Ber.*, 1967, **100**, 1961.
[884] P. L. Pauson and A. R. Qazi, *J. Organometallic Chem.*, 1967, **7**, 321.

Iron Carbonyl Complexes (*cont.*)

		Ref.
$Fe(CO)_2(PF_3)_3$		871
$(\pi\text{-}C_5H_5)Fe(CO)_2CH_2R$	R = Cl, Br, OMe, OEt, SEt, OCOMe	829
$(\pi\text{-}C_5H_5)Fe(CO)_2NCMe]^+PF_6^-$		829
$(\pi\text{-}C_5H_5)Fe(CO)P(OPh)_3C_6X_5$	X = H, F	578
$(\pi\text{-}C_5H_5)_2Fe_2(CO)_4$		516, 517, 518
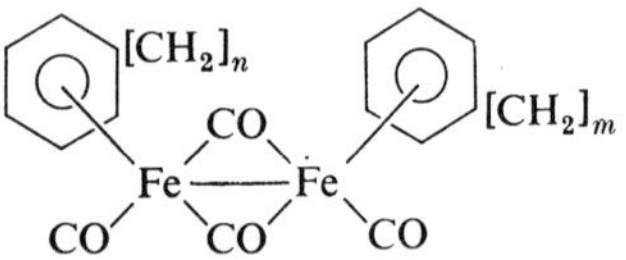	$n = m = 1$ $n = 1, m = 0$ $n = 2, m = 0$	879
$[Fe(CO)_2Me_5C_5]_2$		835
$Ph_2C{:}C{:}NRFe_2(CO)_5PPh_3$	R = Me, Ph	885
$Ph_2C{:}C{:}NRFe_2(CO)_6$	R = Me, Ph	885
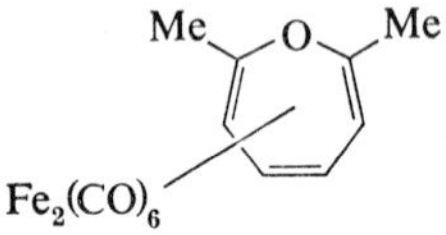		882
$Fe_2(CO)_7(RN)_2$	R = Et, Ph	886
$Fe_2(CO)_8I_2$		515
$Fe_4(CO)_{16}Sn_3Me_4$		875
$[ReFe_2(CO)_{12}]^-NEt_4^+$		520
$Re_2Fe(CO)_{14}$		520

Ruthenium Carbonyl Complexes

$Ru(CO)_5$		887
$Ru(CO)_4X_2$	X = Br, I	888
$Ru(CO)_4(Me_3Sn)_2$		889
$Ru(CO)_3C_6H_8$		888
$Ru(CO)_3C_6H_7(OMe)$		888
$Ru(CO)_3C_6H_6(OMe)_2$		888

[885] S. Otsuka, A. Nakamura, and T. Yoshida, *J. Organometallic Chem.*, 1967, **7**, 339.
[886] W. T. Flannigan, G. R. Knox, and P. L. Pauson, *Chem. and Ind.*, 1967, 1094.
[887] F. Calderazzo and F. L'Eplattenier, *Inorg. Chem.*, 1967, **6**, 1220.
[888] B. F. G. Johnson, R. D. Johnston, P. L. Josty, J. Lewis, and I. G. Williams, *Nature*, 1967, **213**, 901.
[889] J. D. Cotton, S. A. R. Knox, and F. G. A. Stone, *Chem. Comm.*, 1967, 965.

Ruthenium Carbonyl Complexes (*cont.*)

		Ref.
$Ru(CO)_3$ cot		890
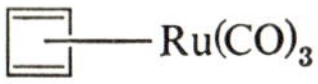		817
$Ru(CO)_2X_2$	X = Cl, Br, I	891
$Ru(CO)_2Cl_22L$	L = PPh_3, $AsPh_3$, py, C_7H_9N, $PhNH_2$	892, 891, 893
$Ru(CO)_2Cl_2L$	L = dipy, diphos	892
$Cs_2Ru(CO)_2X_4$	X = Cl, Br, I	891, 894
$Ru(CO)_2\,py_2(SnX_3)_2$		893
$[Ru(CO)_2(Et_2S)_3SnCl_3]X$		893
$Cs_2Ru(CO)(H_2O)Cl_4$		894
$[Ru(CO)_3SR]_2$	R = Et, Bu, Ph	888
$[Ru(CO)_3PR_3]_2$	R = Ph, OPh, C_4H_9	895
$[Ru(CO)_3]X$		896
$[Ru(CO)_3Cl_2]_2$		892
$[(\pi\text{-}C_5H_5)Ru(CO)_2]_2$		516, 517, 518
$[Ru(CO)_2X_2]_n$	X = Cl, Br, I, SMe, SEt, SPh	888
$[Ru(CO)_4SiMe_3]_2$		889
$Ru_2(CO)_6X_4$	X = Cl, Br, I	888
$Ru_3(CO)_{12}X_6$	X = Cl, Br	888
$Ru_2(CO)_6Sn_4Me_{10}$		889
$Ru_3(CO)_9(PPh_3)_3$		888
$Ru_6C(CO)_{17}$		897

Osmium Carbonyl Complexes

		Ref.
$Os(CO)_5$		887
$Os(CO)_4X_2$	X = H, Br, I	898, 899, 900, 901

890 M. Green and D. C. Wood, *Chem. Comm.*, 1967, 1062.
891 R. Colton and R. H. Farthing, *Austral. J. Chem.*, 1967, **20**, 1283.
892 M. I. Bruce and F. G. A. Stone, *J. Chem. Soc.* (*A*), 1967, 1238.
893 J. V. Kingston, J. W. S. Jamieson, and G. Wilkinson, *J. Inorg. Nuclear Chem.*, 1967, **29**, 133.
894 M. J. Cleare and W. P. Griffith, *Chem. & Ind.*, 1967, 1705.
895 F. Piacenti, M. Bianchi, E. Benedetti, and G. Sbrana, *J. Inorg. Nuclear Chem.*, 1967, **29**, 1389.
896 F. Piacenti, M. Bianchi, and E. Benedetti, *Chem. Comm.*, 1967, 775.
897 B. F. G. Johnson, R. D. Johnston, and J. Lewis, *Chem. Comm.*, 1967, 1057.
898 F. L'Eplattenier and F. Calderazzo, *Inorg. Chem.*, 1967, **6**, 2092.
899 L. A. W. Hales and R. J. Irving, *J. Chem. Soc.* (*A*), 1967, 1389.
900 L. A. W. Hales and R. J. Irving, *J. Chem. Soc.* (*A*), 1967, 1932.
901 L. A. W. Hales and R. J. Irving, *Spectrochim. Acta*, 1967, **23**, *A*, 2981.

Osmium Carbonyl Complexes *(cont.)*

		Ref.
$Os(CO)_3X_2$	X = Cl, Br, I	900
$Os(CO)_3PPh_3X_2$	X = H, Cl, Br, I	902, 898
$Os(CO)_3PPh_3ClHgCl$		902
$Cs[Os(CO)_3X_3]$	X = Cl, Br	894
$Os(CO)_2I_2$		900
$Os(CO)_2(PPh_3)_2Cl_2$		900, 901
$Cs_2[Os(CO)_2X_4]$	X = Cl, Br	894
$Cs_3[Os(CO)Cl_5]$		894
$[Os(CO)_2$ diars$]_3$		902
$Os_3(CO)_{12}$		902
$[Os(CO)_3PPh_3]_3$		902

Cobalt Carbonyl Complexes

$Co(CO)_4X$	X = Cl, Br, I	903
$Co(CO)_4SiX_3$	X = H, F, Cl, Me	887, 904
$Co(CO)_4SiH_32L$	L = NMe_3, py	887
$Co(CO)_4SiH_2Me$		904
$Co(CO)_4GeX_3$	X = Cl, Br, I, Ph	905
$Co(CO)_4GePh_nX_{3-n}$	X = Cl, I; $n = 1, 2$	905
$Co(CO)_4GeMe_nX_{3-n}$	X = Cl, I; $n = 1, 2$	905
$Co(CO)_4SnX_3$	X = Cl, I, Me, Ph	905
$Co(CO)_4SnCl(C_5H_7O_2)_2$		523
$Co(CO)_4PbPh_3$		905
$[Co(CO)_4]_2GeX_2$	X = Cl, I, Me	905
$[Co(CO)_4]_2SnX_2$	X = Cl, Br, I, Me, Ph, $CH_2{:}CH$, $(C_5H_7O_2)$	905, 523
$[Co(CO)_4]_2GeXR$	X = Cl, I; R = Me, Ph	905
$[Co(CO)_4]_2SnClR$	R = Bu^n, $CH_2{:}CH$	905
$M[Co(CO)_4]_2$	M = Zn, Cd	906
$[Co(CO)_4]_nSnCl_{4-n}$	$n = 4, 3, 2, 1$	907

902 C. W. Bradford and R. S. Nyholm, *Chem. Comm.*, 1967, 384.
903 M. Pankowski and M. Bigorgne, *Compt. rend.*, 1967, **264**, *C*, 1382.
904 A. P. Hagen and A. G. MacDiarmid, *Inorg. Chem.*, 1967, **6**, 686.
905 D. J. Patmore and W. A. G. Graham, *Inorg. Chem.*, 1967, **6**, 981.
906 J. M. Burlitch, *J. Organometallic Chem.*, 1967, **9**, P9.
907 M. Bigorgne and A. Quintin, *Compt. rend.*, 1967, **264**, *C*, 2055.

Cobalt Carbonyl Complexes (*cont.*)

		Ref.
$Co(CO)_3L$	$L = NO, PEt_3, PPh_3$	903, 422
$Co(CO)_2NOPF_3$		422
$Co(CO)_2(PR_3)_2$	R = Et, Ph	903
$(\pi\text{-}C_5H_5)Co(CO)_2$		908
$(\pi\text{-}C_5H_5)Co(CO)_2HgX_2$	X = Cl, Br, I	908
$(\pi\text{-}C_5H_5)Co(CO)_22HgX_2$	X = Cl, Br	908
$Co(CO)_2NOL$	$L = Et_2NC_2H_4PPh_2$, $Et_2NC_2H_4PPhC_2H_4NEt_2$	909
$[Co(CO)_2NO]_2Ph_2PC_2H_4CN$		909
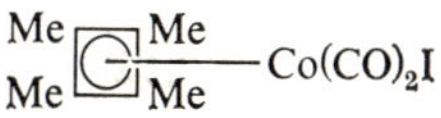		910
$Co(CO)NOL_2$	$L = PF_3, Et_2NC_2H_4PPh_2$, $Ph_2PC_2H_4CN$	909, 422
$Co_2(CO)_6C_4F_6$		873
$Co_2(CO)_6C_4F_6L_2$	$L = PPh_3PBu_3{}^n, AsPh_3$	873
$Co_2(CO)_6C_2R_2$	$R = CH_2Cl, CH_2NEt_2, COOH$	911
$Co_2(CO)_6C_2MeX$	X = H, Ph	911
$Co_2(CO)_6CF_3C{:}CH$		521
$Co_2(CO)_6C_6F_5C{:}CH$		912
$Co_2(CO)_6(PEt_3)_2$		913
$Co_2(CO)_6C_{14}X_{10}$	X = H, F	914
$Co_2(CO)_6[C_{14}F_8(OMe)_2]$		914
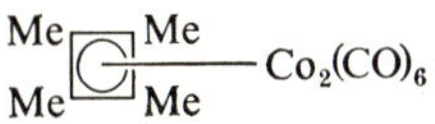		910
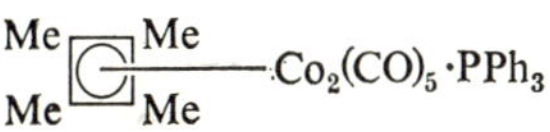		910
$Co_2(CO)_7L$	$L = PEt_3, C_2F_4, Sn(C_5H_7O_2)_2$	913, 915, 523, 522

[908] D. J. Cook, J. L. Dawes, and R. D. W. Kemmitt, *J. Chem. Soc.* (*A*), 1967, 1547.
[909] E. P. Ross and G. R. Dobson, *Inorg. Chem.*, 1967, **6**, 1256.
[910] R. Bruce and P. M. Maitlis, *Canad. J. Chem.*, 1967, **45**, 2017.
[911] G. Cetini, O. Gambino, R. Rossetti, and E. Sappa, *J. Organometallic Chem.*, 1967, **8**, 149.
[912] R. S. Dickson and D. B. W. Yawney, *Inorg. Nuclear Chem. Letters*, 1967, **3**, 209.
[913] G. Capron-Cotigny and R. Poilblanc, *Bull. Soc. chim. France*, 1967, 1440.
[914] J. M. Birchall, F. L. Bowden, R. N. Haszeldine, and A. B. P. Lever, *J. Chem. Soc.* (*A*), 1967, 747.
[915] B. L. Booth, R. N. Haszeldine, P. R. Mitchell, and (in part) J. J. Cox, *Chem. Comm.*, 1967, 529.

Cobalt Carbonyl Complexes (*cont.*)

		Ref.
$Co_3(CO)_4(SPh)_5$		916
$Co_3(CO_9)SR$	R = Ph, C_6H_4Me	916
$Co_3(CO)_9C{:}CH_2X$	X = H, CF_3, C_6F_5	912
$Co_6(CO)_{16}$		917
$[Co_6(CO)_{15}]^{2-}$		918
$Co_6(CO)_{10}S(SPh)_5$		916

Rhodium Carbonyl Complexes

$[Rh(CO)_3PPh_2]_2$		919
$Rh_6(CO)_{16}$		917, 920

Iridium Carbonyl Complexes

$Ir(CO)Cl(C_8H_{14})_3$		921
$Ir(CO)ClR_2$	R = C_6H_{10}, C_8H_{14}, PEt_2, Ph, PPh_3	921, 922
$Ir(CO)Cl_2(allyl)C_8H_{14}$		921
$Ir(CO)Cl_2(C_3H_5)L_2$	L = C_8H_{14}, PEt_2Ph	921, 922
$[Ir(CO)_2Cl_2Me]_2$		921
$Ir(CO)_2Cl_2$ py		921
$[Ir(CO)Cl_2COMePMe_2Ph]_2$		921
$Ir(CO)Cl_2COR(PR_2Ph)_2$	R = Me, Et	921, 922
$[Ir(CO)_2Cl_2Et]_2$		921
$Ir(CO)_2Cl_2Et$ py		921
$[Ir(CO)_2HCl_2]_2$		921
$Ir(CO)HCl_2L_2$	L = py, PMe_2Ph	921
$Ir(CO)BrL_2$	L = C_8H_{14}, PEt_2Ph, PEt_3, PPh_3, PBu^nPh_2, $AsEt_2Ph$	921, 922
$Ir(CO)Br_2MeCO(PEt_3)_2$		922
$Ir(CO)Br_2C_3H_5(PEt_2Ph)_2$		922
$Ir(CO)I(PEt_2Ph)_2$		922

916 E. Klumpp, G. Bor, and L. Marko, *Chem. Ber.*, 1967, **100**, 1451.
917 P. Chini, *Chem. Comm.*, 1967, 440.
918 P. Chini, *Chem. Comm.*, 1967, 1, 29.
919 W. Hieber and R. Kummer, *Chem. Ber.*, 1967, **100**, 148.
920 S. H. H. Chaston and F. G. A. Stone, *Chem. Comm.*, 1967, 964.
921 B. L. Shaw and E. Singleton, *J. Chem. Soc.* (*A*), 1967, 1683.
922 J. Chatt, N. P. Johnson, and B. L. Shaw, *J. Chem. Soc.* (*A*), 1967, 604.

Iridium Carbonyl Complexes (*cont.*)

		Ref.
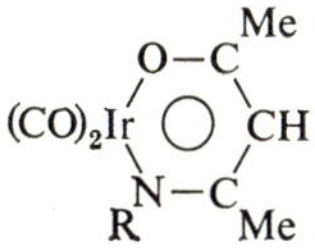	R = Ph, α-naphthyl, *p*-tolyl	923
$[Ir(CO)_{11}H]^-$		924
$[Ir(CO)_3PPh_2]_2$		919
$Ir_4(CO)_{12}$		920
$[Ir_4(CO)_{10}]_2^-$		924
$Ir_4(CO)_{10}(PPh_3)_2$		924
$Ir_4(CO)_9(PPh_3)_3$		924

Nickel Carbonyl Complexes

$Ni(CO)L_3$	$L = PF_3, PMe_3, P(OMe)_3, AsMe_3$	925
$(\pi\text{-}C_5H_5)Ni(CO)SiCl_3$		519
$Ni(CO)_2L_2$	$L = PF_3, PMe_3, P(OMe)_3, AsMe_3, MePFNMe_2, PF(NMe_2)_2, PF_2(Me_2N), PF_2(Et_2N)$, PF_2(piperidine)	832, 925
$Ni(CO)_3(PPh)_4$		498
$Ni(CO)_3L$	$L = PF_3, PMe_3, P(OMe)_3, AsMe_3$	925
$Ni(CO)_3P(MMe_3)_3$	M = Si, Ge, Sn, Pb	454
$Ni(CO)_4$		438, 925, 926, 927, 928

Appendix 2 Additional References to Vibrational Data on Inorganic Compounds

Beryllium

		Ref.
$Be(OH)_2$		929
$R^1BeNR_2{}^2$	R^1 = Me, Et, Pr^i; R^2 = Me, Et, Pr^i, Ph	930
Be py_2NR_2Me	R = Me, Ph	930
Be py_2NPh_2Et		930

[923] F. Bonati and R. Ugo, *J. Organometallic Chem.*, 1967, **7**, 167.
[924] L. Malatesta and G. Caglio, *Chem. Comm.*, 1967, 420.
[925] M. Bigorgne and G. Bouquet, *Compt. rend.*, 1967, **264**, *C*, 1485.
[926] L. H. Jones, *J. Chem. Phys.*, 1967, **47**, 1196.
[927] G. Bor, *J. Organometallic Chem.*, 1967, **10**, 343.
[928] H. Haas and R. K. Sheline, *J. Chem. Phys.*, 1967, **47**, 2996.
[929] T. P. Prasad and M. N. Sastri, *J. Inorg. Nuclear Chem.*, 1967, **29**, 246.
[930] G. E. Coates and A. H. Fishwick, *J. Chem. Soc.* (*A*), 1967, 1199.

Beryllium (*cont.*)

		Ref.
[Be pyNMe$_2$Me]$_2$		930
R$_2$BeNPh$_2$ bipy	R = Me, Et	930
Bu$^t{}_2$BeL	L = Me$_2$NH, *NNN*-trimethylethylenediamine	930

Boron

BHS		931
BH$_3$,PH$_2$Me		932
BH$_3$,PHF$_2$		933
M$^+$[(BH$_3$)$_2$PH$_2$]	M = Na, K, NH$_4$, Me$_2$NH$_2$	934
M$^+$[(BD$_3$)$_2$PH$_2$]	M = K, Me$_2$NH$_2$	934
BX(NMe$_2$)$_2$	X = H, Cl, Me, CN	935
BMe$_2$(ON:CMe$_2$)		936
BMe$_2$(HN:CPh$_2$)		936
BMe$_2$(N:CPh$_2$)		936
BMe$_2$OH		937
BMe(OH)$_2$		937
BF$_3$,NMe$_3$		938, 939, 940
BF$_3$,RNH$_2$	R = Me, Et, Bu	938, 939, 940
BF$_3$,NMe$_2$H		938, 939, 940
BF$_3$,SMe$_2$		941
BF$_2$CF$_3$		942
BX$_2$NEt$_2$	X = F, Cl, Br	943
BF$_2$Bu^n		942
BF,Bu$_2{}^n$		942

931 R. W. Kirk and P. L. Timms, *Chem. Comm.*, 1967, **1**, 18.
932 L. J. Malone and R. W. Parry, *Inorg. Chem.*, 1967, **6**, 176.
933 R. W. Rudolph and R. W. Parry, *J. Amer. Chem. Soc.*, 1967, **89**, 1621.
934 J. W. Gilje, K. W. Morse, and R. W. Parry, *Inorg. Chem.*, 1967, **6**, 1761.
935 J. Goubeau, E. Bessler, and D. Wolff, *Z. anorg. Chem.*, 1967, **352**, 285.
936 I. Pattison and K. Wade, *J. Chem. Soc.* (*A*), 1967, 1098.
937 J. E. de Moor, G. P. Van der Kelen, and Z. Eeckhaut, *J. Organometallic Chem.*, 1967, **9**, 31.
938 J. Delaule, E. Taillandier, and M. Taillandier, *Compt. rend.*, 1967, **265**, *B*, 921.
939 W. R. Hertler, F. Klanberg, and E. L. Muetterties, *Inorg. Chem.*, 1967, **6**, 1696.
940 A. D. H. Clague and A. Danti, *Spectrochim. Acta*, 1967, **23**, *A*, 2359.
941 G. M. Begun and A. A. Palko, *J. Chem. Phys.*, 1967, **47**, 967.
942 T. D. Parsons, J. M. Self, and L. H. Schaad, *J. Amer. Chem. Soc.*, 1967, **89**, 3446.
943 N. N. Greenwood and J. Walker, *J. Chem. Soc.* (*A*), 1967, 959.

Boron (*cont.*)

		Ref.
B_3F_5		944
B_4F_6,L	L = CO, PF_3	944
BCl_2OMe		937
$BCl(OMe)_2$		937
$BCl_3,3(NH_2C_6H_4CO_2H)$		255
$BCl_2NMe_2,2py$		254
$BCl_2(H)CCHBCl_2$		945
$H_3B_3N_3H_3$		946, 947
$H_3B_3N_3D_3$		946
$(Me_2N)_3B_3N_3R_3$	R = H, Me, Ph	948
$[(CD_3)_2N]_3B_3N_3R_3$	R = H, Me, Ph	948
$Me_2(CF:CF_2)B_3N_3Me_3$		949
$(CF:CF_2)_3B_3N_3Me_3$		949
$B_3N_3H_6O$		950
$B_6N_6H_{10}O$		950
$MO,3B_2O_3,nH_2O$	M = Mg, Ca, Sr; n = 4–7·5	951
$B_3O_3ClF_2$		952
$B_3O_3Cl_2F$		952
$2\text{-}MeC_3B_3H_6$		953
$2,3\text{-}Me_2C_3B_3H_5$		953
$2,4\text{-}Me_2C_3B_3H_5$		953
Alkyl-substituted pentaboranes		954
$H_2C_2B_5Me_5$		955
$B_9H_{11}S$		939
$B_9H_{11}SL$	L = MeCN, Et_3N, Me_2NCHO, Ph_2PH	939

[944] P. L. Timms, *J. Amer. Chem. Soc.*, 1967, **89**, 1629.
[945] R. W. Rudolph, *J. Amer. Chem. Soc.*, 1967, **89**, 4216.
[946] E. Silberman, *Spectrochim. Acta*, 1967, **23**, *A*, 2021.
[947] K. Niedenzu, W. Sawodny, H. Watanabe, J. W. Dawson, T. Totani, and W. Weber, *Inorg. Chem.*, 1967, **6**, 1453.
[948] A. Meller and E. Schaschel, *Monatsh. Chem.*, 1967, **98**, 1358.
[949] A. J. Klanica, J. P. Faust, and C. S. King, *Inorg. Chem.*, 1967, **6**, 840.
[950] G. H. Lee, jun., and R. F. Porter, *Inorg. Chem.*, 1967, **6**, 648.
[951] H.-A. Lehmann and G. Kessler, *Z. anorg. Chem.*, 1967, **354**, 30.
[952] B. Latimer and J. P. Devlin, *Spectrochim. Acta*, 1967, **23**, *A*, 81.
[953] R. N. Grimes and C. L. Bramlett, *J. Amer. Chem. Soc.*, 1967, **89**, 2557.
[954] T. Onak, G. B. Dunks, I. W. Searcy, and J. Spielman, *Inorg. Chem.*, 1967, **6**, 1465.
[955] H. V. Seklemian and R. E. Williams, *Inorg. Nuclear Chem. Letters*, 1967, **3**, 289.

Boron (*cont.*)

		Ref.
$B_9H_{11}NHMeCN$		939
$Cs^+B_9H_{12}S^-$		939
$B_9H_{12}NNMe_3$		939
$(Me_4N)B_9H_{12}NH$		939
1,10-$B_{10}H_8(CO)_2$		956
[1,10-$B_{10}H_8(CO)CN]Me_4N^+$		956
$B_{10}H_8(CO)(1\text{-}Me_2S)$		956
2,4- and 2,7,(8)-$B_{10}H_8(CO)NMe_3$		956
$Cs_2(1,10\text{-}B_{10}H_8Me_2)$		956
$B_{10}H_{10}CH^-$		957
$B_{10}H_{10}CNMe_3$		957
$B_{10}H_{10}CNMe_2C_3H_7$		957
$X_2B_{10}H_{10}CH^-$	X = Cl, Br	957
$(Me_4N)B_{10}H_{10}SRe(CO)_3$		939
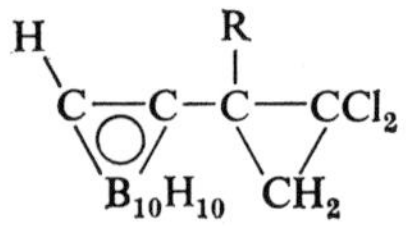	R = H, Me	958
$BrB_{10}H_{11}CH^-$		957
$(Me_4N)B_{10}H_{11}S$		939
6-$XB_{10}H_{11}CNMe_3$	X = Cl, Br	957
6-$BrB_{10}H_{11}CNMe_2Pr$		957
$B_{10}H_{12}S$		939
$B_{10}H_{12}CH^-$		957
$Ph_2B_{11}H_9S$		939
1,7-$B_{12}H_{10}(CO)_2$		956
1,12-$B_{12}H_{10}(CO)_2$		956
$Me_4N^+[B_{12}H_{11}(CO)]$		956
$C_2B_{14}D_4H_8$		959

[956] W. H. Knoth, J. C. Sauer, J. H. Balthis, H. C. Miller, and E. L. Muetterties, *J. Amer. Chem. Soc.*, 1967, **89**, 4842.
[957] D. E. Hyatt, F. R. Scholer, L. J. Todd, and J. L. Warner, *Inorg. Chem.*, 1967, **6**, 2229.
[958] D. Seyferth and B. Prokai, *J. Organometallic Chem.*, 1967, **8**, 366.
[959] R. E. Williams and F. J. Gerhart, *J. Organometallic Chem.*, 1967, **10**, 168.

Boron (*cont.*)

		Ref.
$CsB_{24}H_{23},3H_2O$		960
$B_{10}Cl_8L_2$	$L = CH_2NH_3, CH_2NMe_2H, CH_2PPh_3$	961, 955
$(Me_4N)_2{}^+1,10\text{-}B_{10}Cl_8L_2$	$L = CH_2OH, CH_2CO_2H, CH_2OCOMe, CH_2Br, CH_2I, (CH_2)_2SOMe$	961
$Cs_2B_{10}Cl_8(CH_2CN)_2$		961
$(Me_4N)_2[B_{10}Br_8(COPh)_2]$		956
$(Me_4N)_2[B_{12}Br_{10}(CO_2H)_2]$		956
$(Me_4N)_2[B_{12}Br_{11}(CO_2H)]$		956
$Cs_2B_{10}I_8(CO_2H)_2$		956
$Cs\text{-}1,12\text{-}B_{12}I_{10}(CO_2H)_2$		956

Aluminium, Gallium, Indium

$AlCl_3,COCl_2$		962
AlX_3,THF	X = Cl, Br	963
$AlX_3,N(SiMe):PR_3$	X = Cl, Br, I, Me, Et; R = Me, Et, Ph	964, 965
$(AlR_2N:PMe_3)_2$	R = Br, Me	964
$[Al(CX_2CX_3)_3]_2$	X = H, D	966
$Ph_2C:NH,AlR_3$	R = Me, Et, Ph	967
$Ph_3MN(SiMe_3):PPh_3$	M = Al, Ga	968
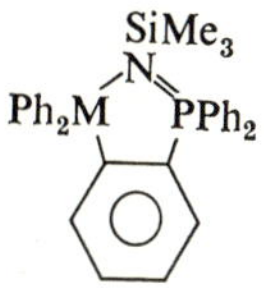	M = Al, Ga	968
Al_4C_3		969
Al_4O_4C		969

960 R. J. Wiersema and R. I. Middaugh, *J. Amer. Chem. Soc.*, 1967, **89**, 5078.
961 W. H. Knoth, *J. Amer. Chem. Soc.*, 1967, **89**, 4850.
962 K. O. Christe, *Inorg. Chem.*, 1967, **6**, 1706.
963 J.-L. Calvé, J. Derouault, M.-T. Forel, and J. Lascombe, *Compt. rend.*, 1967, **264**, *B*, 611.
964 H. Schmidbaur, W. Wolfsberger, and H. Kröner, *Chem. Ber.*, 1967, **100**, 1023.
965 H. Schmidbaur and W. Wolfsberger, *Chem. Ber.*, 1967, **100**, 1000.
966 K. H. Reichert and H. Sinn, *J. Organometallic Chem.*, 1967, **7**, 189.
967 K. Wade and B. K. Wyatt, *J. Chem. Soc.* (*A*), 1967, 1339.
968 H. Schmidbaur and W. Wolfsberger, *Chem. Ber.*, 1967, **100**, 1016.
969 R. Juza, U. Landt, and H. Hansen, *Z. anorg. Chem.*, 1967, **352**, 77.

Aluminium, Gallium, Indium (*cont.*)

		Ref.
$CaOAl_2O_3$		969
$7Al_2O_3,12CaO$		969
ClAl pc		970
Cl_2Al pc,$2H_2O$		970
(HO)Al pc,H_2O		970
$(HSO_4)AlCl$ pc		970
pc Al-O-Al pc		970
$[Al(C_{10}H_8ON)_2OH]_2$		971
GaF_3,NH_3		972
$(NH_4)(GaF_4)$		972
$(NH_4)(GaF_4,NH_3)$		972
$GaCl_3,N(SiMe_3):PMe_3$		964
$Ph_2C:NHGaR_3$	R = Me, Et, Ph	973
$(Ph_2C:NGaR_2)_2$	R = Me, Et, Ph	973
R^1CN,GaR^2_3	R^1 = Me, Bu^t, Ph; R^2 = Me, Et	974
$Cl_3CCN,GaMe_3$		974
$InX_3,N(SiMe_3):PMe_3$	X = Cl, Br, Me	964, 965

Silicon

$MeCH_2SiH_3$		975
$EtSiD_3$		975
$M(SiH_3)_3$	M = As, Sb	976
$Me_2(SiH_3)N$		977
$Me(SiH_3)_2N$		977
$CH_3OSiH_2CH_3$		978
$Me_2NC(O)OSiX_3$	X = H, D	979
$Me_2NC(S)OSiX_3$	X = H, D	979
$Me_2NC(S)SSiX_3$	X = H, D	979

970 M. Starke and C. Hamann, *Z. anorg. Chem.*, 1967, **354**, 1.
971 P. R. Scherer and Q. Fernando, *Chem. Comm.*, 1967, 1107.
972 G. Mermant, C. Belinski, and F. Lalau-Keraly, *Compt. rend.*, 1967, **265**, *C*, 238.
973 J. R. Jennings, I. Pattison, K. Wade, and B. K. Wyatt, *J. Chem. Soc.* (*A*), 1967, 1608.
974 J. R. Jennings and K. Wade, *J. Chem. Soc.* (*A*), 1967, 1222.
975 K. M. Mackay and R. Watt, *Spectrochim. Acta*, 1967, **23**, *A*, 2761.
976 G. Davidson, L. A. Woodward, E. A. V. Ebsworth, and G. M. Sheldrick, *Spectrochim. Acta*, 1967, **23**, *A*, 2609.
977 T. D. Goldfarb and B. N. Khare, *J. Chem. Phys.*, 1967, **46**, 3384.
978 J. T. Wang and C. H. Van Dyke, *Inorg. Chem.*, 1967, **6**, 1741.
979 E. A. V. Ebsworth, G. Rocktäschel, and J. C. Thompson, *J. Chem. Soc.* (*A*), 1967, 362.

Silicon (*cont.*)

		Ref.
R_3SiF	R = Me, Et, Pr, Bu, C_5H_{11}	980, 981
Me_2EtSiF		981
$MeEt_2SiF$		981
$Cl_nSi(NMe_2)_{4-n}$	$n = 3, 2, 1$	982
$ClMeSi(NEt_2)_2$		982
Ph_2SiCl_2		983
		983
$(Cl_2SiNH)_3$		984
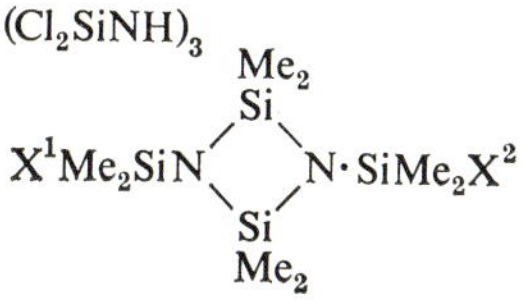	$X^1 = X^2$ = H, Cl, Br, I, Me, OMe, NMe_2, NHMe, C_4H_9; X^1 = Cl, X^2 = $NHSiMe_2Cl$	985
$Me_3Si(NEt_2)$		982
$Me_2Si(NEt_2)_2$		982
$Me_3SiNCSe$		986
$(Me_3Si)_3SiC_6Cl_5$		987
$[(Me_3Si)_3Si]_2C_6Cl_4$		987
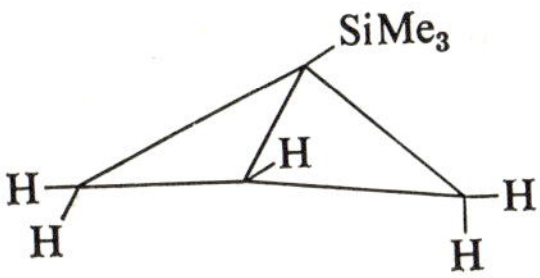		988
$Me_3SiN{:}PR_3$	R = Me, Et, Ph	965, 989
$Me_3Si(C\vdots CH)$		990
$Me_2Si(C\vdots CH)_2$		990
$R_3^1SiC\vdots CMR_3^2$	$R^1 = R^2$ = Me, Ph; M = Si, Sn, Pb	990, 991
$Me_nPh_{3-n}SiCH{:}CHSiPh_3$	$n = 3, 2, 1$	990

[980] K. Licht, *Z. Chem.*, 1967, **7**, 321; see also *Chem. Abs.*, 1967, **67**, 95,398*f*.
[981] K. Licht, *Z. Chem.*, 1967, **7**, 242; see also *Chem. Abs.*, 1967, **67**, 69,207*y*.
[982] H. Bürger and W. Sawodny, *Spectrochim. Acta*, 1967, **23**, *A*, 2827.
[983] K. A. Andrianov and V. M. Kotov, *J. Organometallic Chem.*, 1967, **7**, 211.
[984] H. Bürger, M. Schulze, and U. Wannagat, *Inorg. Nuclear Chem. Letters*, 1967, **3**, 43.
[985] H. Bürger, E. Bogusch, and P. Geymayer, *Z. anorg. Chem.*, 1967, **349**, 124.
[986] H. Bürger and U. Goetze, *J. Organometallic Chem.*, 1967, **10**, 380.
[987] H. Gilman and K. Shuna, *J. Organometallic Chem.*, 1967, **8**, 369.
[988] H. M. Cohen, *J. Organometallic Chem.*, 1967, **9**, 375.
[989] H. Schmidbaur and W. Tronich, *Chem. Ber.*, 1967, **100**, 1032.
[990] C. S. Kraihanzel and M. L. Losee, *J. Organometallic Chem.*, 1967, **10**, 427.
[991] W. Findeiss, W. Davidsohn, and M. C. Henry, *J. Organometallic Chem.*, 1967, **9**, 435,

Silicon (*cont.*)

Compound		*Ref.*
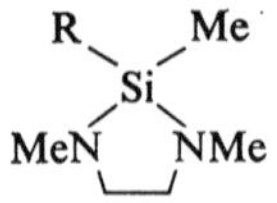	R = Me, Ph	992
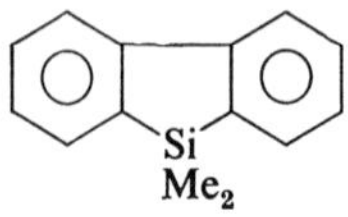		993
$R_3P\!:\!CH(SiMe_3)$	R = Me, Ph	989
$Me_3P\!:\!C(SiMe_3)(MMe_3)$	M = Si, Ge, Sn	989
$Me_3SiNH(SiMe_3)$		989
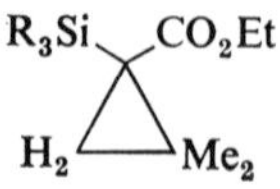	R = Me, Ph	994
$Me_3SiOMMe_3$	M = C, Si	995
Et_3SiCCl_2H		996
$(C_6Cl_5)_2SiPh_2$		997
$M(C\vdots CH)_4$	M = Si, Ge, Sn	998

Germanium, Tin, Lead

GeH_3OMe		999
GeH_3CH_2OMe		999
GeX_3CH_2Me	X = H, D	975
$Ph_3P\!:\!C(GeMe_3)_2$		989
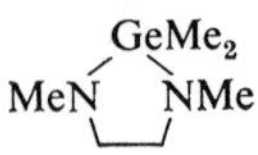		992

[992] E. W. Randall, C. H. Yoder, and J. J. Zuckerman, *Inorg. Chem.*, 1967, **6**, 744.
[993] R. W. Coutant and A. Levy, *J. Organometallic Chem.*, 1967, **10**, 175.
[994] U. Schöllkopf and N. Rieber, *Angew. Chem. Internat. Edn.*, 1967, **6**, 884.
[995] I. F. Kovalev, L. Ozolins, M. G. Voronkov, and L. Zagata, *Opt. Spektrosk, Akad. Nauk. S.S.S.R. Otd. Fiz.-Mat. Nauk, Sb, Statei*, 1967, **3**, 301; see also *Chem. Abs.*, 1967, **67**, 68,944*z*.
[996] D. Seyferth, J. M. Burlitch, H. Dertousos, and H. D. Simmons, jun., *J. Organometallic Chem.*, 1967, **7**, 405.
[997] H. Gilman and S.-Y. Sim, *J. Organometallic Chem.*, 1967, **7**, 249.
[998] R. E. Sacher, D. H. Lemmon, and F. A. Miller, *Spectrochim. Acta*, 1967, **23**, *A*, 1169.
[999] G. A. Gibbon, J. T. Wang, and C. H. van Dyke, *Inorg. Chem.*, 1967, **6**, 1989.

Germanium, Tin, Lead (*cont.*)

		Ref.
$[MeGeS_{1\cdot5}]_4$		1000
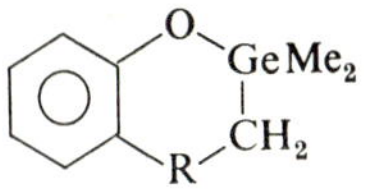	R = O, NH	1001
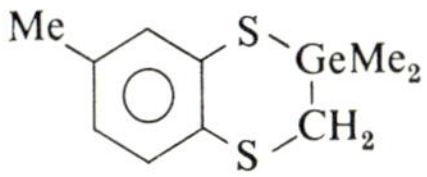		1001
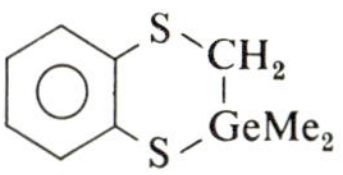		1001
$(R_3Ge)_3P$	R = Me, Ph	1002
$Bu_3{}^nGeNPhCO_2R$	R = Me, Et	1003
Ph_3GeCCl_2H		996
$(C_{12}F_8)_2Ge$		1004
$(C_{12}F_8)GePh_2$		1004
Ge phthalocyanines		1005
Ge hemiporphyrazines		1005
$HN(R_2P{:}O)_2SnX_4$	X = Cl, Br	1006
Ph_3SnCl_2		1007
Ph_3SnCl		1008
$SnCl_2\ acac_2$		1009
$R^1R^2SnX_2$ } $RSnX_3$ }	$R^1 = (\pi\text{-}C_5H_5)Fe(CO)_2$, $(\pi\text{-}C_5H_5)Mo(CO)_3$, $(\pi\text{-}C_5H_5)W(CO)_3$, $Co(CO)_4$, $Co(CO)_3(PBu_3{}^n)$; R^2 = R or Cl; X = halogen	1010
$Ph_3P{:}C(SnMe_3)_2$		989
$(R_3Sn)_3P$	R = Me, Ph	1002

[1000] K. Moedritzer, *Inorg. Chem.*, 1967, **6**, 1248.
[1001] M. Wieber and C. D. Frohning, *J. Organometallic Chem.*, 1967, **8**, 459.
[1002] G. Engelhardt, P. Reich, and H. Schumann, *Z. Naturforsch.*, 1967, **22**, *B*, 352.
[1003] Y. Ishii, K. Itoh, A. Nakamura, and S. Sakai, *Chem. Comm.*, 1967, 224.
[1004] S. C. Cohen and A. G. Massey, *J. Organometallic Chem.*, 1967, **10**, 471.
[1005] J. N. Esposito, L. E. Sutton, and M. E. Kenney, *Inorg. Chem.*, 1967, **6**, 1116.
[1006] A. Schmidpeter and K. Stoll, *Angew. Chem. Internat. Edn.*, 1967, **6**, 252.
[1007] T. S. Srivastava, *J. Organometallic Chem.*, 1967, **10**, 375.
[1008] T. S. Srivastava, *J. Organometallic Chem.*, 1967, **10**, 373.
[1009] J. W. Faller and A. Davison, *Inorg. Chem.*, 1967, **6**, 182.
[1010] F. Bonati, S. Cenini, and R. Ugo, *J. Chem. Soc.* (*A*), 1967, 932.

Germanium, Tin, Lead (*cont.*)

		Ref.
$Me_3SnCClXSnMe_3$	X = Cl, Br	1011
Ph_3SnPPh_2		1002
$(Ph_3Sn)_2PPh$		1002
$Ph_3SnN{=}PR_3$	R = Bu^n, Ph, C_8H_{17}	1012
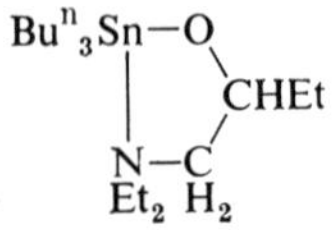		1013
$(C_{12}F_8)_2Sn$		1004
$PbEt_4$		1014
$Et_2Pb[CH_2]_n$	$n = 4, 5$	1014
		1014
		1014
Ph_4Pb		1014
$Ph_2Pb[CH_2]_4$		1014

Phosphorus

CX_3PY_2	X = H, D; Y = H, D	1015
XPF_2H	X = O, S	1016
R_4PI	R = Me, Et	1017
$CX_3OP(S)Cl_2$	X = H, D	1018
CH_2ClPXF_2	X = O, S	1019
$CH_2XCH_2POCl_2$	X = H, Cl	1019
$CF_3P(PMe_2)_2$		1020

1011 D. Seyferth and F. M. Armbrecht, jun., *J. Amer. Chem. Soc.*, 1967, **89**, 2790.
1012 W. L. Lehn, *Inorg. Chem.*, 1967, **6**, 1061.
1013 A. Tzschach and E. Reiss, *J. Organometallic Chem.*, 1967, **8**, 225.
1014 E. C. Juenge and S. Gray, *J. Organometallic Chem.*, 1967, **10**, 465.
1015 J. A. Lannon and E. R. Nixon, *Spectrochim. Acta*, 1967, **23**, *A*, 2713.
1016 T. L. Charlton and R. G. Cavell, *Inorg. Chem.*, 1967, **6**, 2204.
1017 J. A. Creighton, G. B. Deacon, and J. H. S. Green, *Austral. J. Chem.*, 1967, **20**, 583.
1018 R. A. Nyquist, *Spectrochim. Acta*, 1967, **23**, *A*, 1499.
1019 E. Steger and M. Kuntze, *Spectrochim. Acta*, 1967, **23**, *A*, 2189.
1020 A. H. Cowley, *J. Amer. Chem. Soc.*, 1967, **89**, 5990.

Phosphorus (*cont.*)

		Ref.
$C[CH_2P(X)Ph_2]_4$	X = O, S, Se, Te	1021
PF_2NCS		1022
$PF(NCS)_2$		1022
XPF_2NCS	X = O, S	1022
$XPF(NCS)_2$	X = O, S	1022
$M^+(SPOClX)^-$	X = Cl, F; M = Ph_4P, Ph_4As	1023
$R_2PC(CF_3){:}C(CF_3)H$	R = CF_3, Et, Ph	1024
$[(MeO)_3POMe]SbCl_6^-$		1025
Me_2NPOF_2		1026
$(Me_2N)_2POF$		1026
Me_2PF		1027
H_2NPOF_2		1028
$Me_3SiOPOF_2$		1028
$Bu_3SnOPOF_2$		1028
$Si(OPOF_2)_4$		1028
$K(PO_3)_x,K_2O$		1029
$Na_3P_3O_9,6H_2O$		1030
$[P_4O_{12}]^{4-}$		1030
$M_2^I(PHO_2S)^{2-}$	M^I = K, NH_4	1031
$M(H_2PO_4)$	M^I = K, NH_4	1032, 1033
$M(O_2PCl_2)_3$	M = Al, Ga, In, Fe	1034
$CaXPO_4$	X = H, D	1035
$CaXPO_42X_2O$	X = H, D	1035

[1021] J. Ellermann and D. Schirmacher, *Chem. Ber.*, 1967, **100**, 2220.
[1022] H. W. Roesky, *Chem. Ber.*, 1967, **100**, 2142.
[1023] H. W. Roesky, *Chem. Ber.*, 1967, **100**, 1447.
[1024] W. R. Cullen and D. S. Dawson, *Canad. J. Chem.*, 1967, **45**, 2887.
[1025] J. S. Cohen, *J. Amer. Chem. Soc.*, 1967, **89**, 2543.
[1026] R. G. Cavell, *Canad. J. Chem.*, 1967, **45**, 1309.
[1027] F. Seel, K. Rudolph, and W. Gombler, *Angew. Chem. Internat. Edn.*, 1967, **6**, 708.
[1028] H. W. Roesky, *Chem. Ber.*, 1967, **100**, 2147.
[1029] U. Stahlberg and E. Steger, *Z. anorg. Chem.*, 1967, **349**, 50.
[1030] W. P. Griffith, *J. Chem. Soc.* (*A*), 1967, 905.
[1031] J. D. Murray, G. Nickless, and (the late) F. H. Pollard, *J. Chem. Soc.* (*A*), 1967, 1726.
[1032] H. Ratajczak, *Compt. rend.*, 1967, **265**, *B*, 1023.
[1033] H. Ratajczak, *Compt. rend.*, 1967, **265**, *B*, 807.
[1034] H. Müller and K. Dehnicke, *Z. anorg. Chem.*, 1967, **350**, 231.
[1035] I. Petrov, B. Soptrajanov, N. Fuson, and J. R. Lawson, *Spectrochim. Acta*, 1967, **23**, *A*, 2637.

Phosphorus (*cont.*)

		Ref.
$Ca_8H_2(PO_4)_6,5H_2O$		1036
$C_6F_5P(OR)_2$	R = Me, Et, Ph	1037
$(C_6F_5)_2P(OR)$	R = Me, Et, Ph	1037
$M(P_2O_7)$	M = Si, Sn, Pb, Ti, Zr, Hf, U	1038
$M(P_2O_7)_2$	M = Zn, Be	1039
$(C_6F_5P)_4$		1037

Phosphorus–Nitrogen Compounds

$P_3N_3F_nBr_{6-n}$	n = 0–6	1040
$(PNX_2)_3$	X = F, Cl, Br	1041, 1042
$(F_3PNMe)_2$		1043, 1044, 1045
F_2OPNH_2		1046, 1028
F_4PNMe_2		1046
$F_2OPN{:}PX_3$	X = Cl, Ph	1046
$Ph_2PF{:}NCN$		1047
$PhPF_2{:}NCN$		1047
$F_2SPNHNHPh$		1048
$P_3N_3X_6$	X = Cl, Br	1049
$(PNX_2)_4$	X = Cl, Br	1041, 1049
$P_3N_3X_{6-n}(NMe_2)_n$	n = 1, 2, 3; X = Cl, Br	1050
$P_3N_3Cl_5Br$		1051
$P_3N_3Cl_nBr_{6-n}$	n = 1–5	1052

[1036] E. E. Berry and C. B. Baddiel, *Spectrochim. Acta*, 1967, **23**, *A*, 1781.
[1037] M. Fild, I. Hollenberg, and O. Glemser, *Naturwiss.*, 1967, **54**, 89.
[1038] R. Hubin and P. Tarte, *Spectrochim. Acta*, 1967, **23**, *A*, 1815.
[1039] H. Brintzinger and R. A. Plane, *Inorg. Chem.*, 1967, **6**, 623.
[1040] E. Steger and D. Klemm, *J. Inorg. Nuclear Chem.*, 1967, **29**, 1812.
[1041] T. R. Manley and D. A. Williams, *Spectrochim. Acta*, 1967, **23**, *A*, 149.
[1042] T. R. Manley and D. A. Williams, *Spectrochim. Acta*, 1967, **23**, *A*, 1221.
[1043] G. C. Demitras and A. G. MacDiarmid, *Inorg. Chem.*, 1967, **6**, 1903.
[1044] A. J. Downs, *Chem. Comm.*, 1967, 628.
[1045] M. P. Yagupsky, *Inorg. Chem.*, 1967, **23**, *A*, 1770.
[1046] S. Kongpricha and W. C. Preusse, *Inorg. Chem.*, 1967, **6**, 1915.
[1047] O. Glemser, E. Niecke, and J. Stenzel, *Angew. Chem. Internat. Edn.*, 1967, **6**, 709.
[1048] H.-G. Horn and O. Glemser, *Chem. Ber.*, 1967, **100**, 2258.
[1049] U. Stahlberg and E. Steger, *Spectrochim. Acta*, 1967, **23**, *A*, 627.
[1050] R. Stahlberg and E. Steger, *Spectrochim. Acta*, 1967, **23**, *A*, 2005.
[1051] U. Stahlberg, R. Stahlberg, and E. Steger, *Spectrochim. Acta*, 1967, **23**, *A*, 2691.
[1052] R. Stahlberg, and E. Steger, *Spectrochim. Acta*, 1967, **23**, *A*, 2057.

Phosphorus–Nitrogen Compounds (*cont.*)

		Ref.
$P_3N_3Cl_6$–$P_3N_3Br_6$ mixed crystal		1049
$P_3N_3Cl_4(NHMe)_2$		1053
$P_3N_3Cl_2(NMe_2)_4$		1054
$(Cl_3P:NCH_2Cl)_2$		1055
$P_3N_3Br_n(NMe_2)_{6-n}$	$n = 5, 4, 3$	1054
$P_6N_6Br_{12}$		1056
$R_3P:NMMe_3$	M = Si, Ge, Sn; R = Me, Et, Ph	1057
$(Me_nPPh_{3-n}NH_2)Cl$	$n = 3, 2, 1$	1058
$(Me_2NPPh_2NH_2)Cl$		1058
$[(Me_2N)_2PRNH_2]Cl$	R = Me, Ph	1058
$[(Me_2N)_3PNH_2]X$	X = Cl, PF_6, BPh_4	1059
$[(Me_2N)_3PMe]Cl$		1059
$[(Me_2N)_2PMe_2]Cl$		1059
$SP(NH\cdot NH_2)_3$		1048
$(RO)_2P(S)NDX$	X = H, D, Me	1060
$Ph_2PSNHCSNHR$	R = H, Me	1061
$Ph_2PSNHCSNMe_2$		1061
$Ph_2P\cdot NH\cdot PPh_2$		1062
$[Ph_2P\cdot PPH_2:N\cdot PPh_2\cdot PPh_2]Cl$		1062
$[Ph_2P\cdot PPh_2\cdot NH\cdot PPh_2]Cl$		1062
$[Ph_2P\cdot PPh_2:N\cdot PPh_2]$		1062
$Ph_2(S)P\cdot PPh_2:N\cdot P(S)Ph_2$		1062
$Ph_2P(SMe):N\cdot CS\cdot NMe_2$		1061
$Ph_2PS\cdot N:CSMe\cdot NMe_2$		1061
$K[Ph_2PS\cdot N:CS\cdot NMe_2]MeOH$		1061

[1053] W. Lehr, *Z. anorg. Chem.*, 1967, **352**, 27.
[1054] R. Stahlberg and E. Steger, *J. Inorg. Nuclear Chem.*, 1967, **29**, 961.
[1055] T. Moeller and A. H. Westlake, *J. Inorg. Nuclear Chem.*, 1967, **29**, 957.
[1056] G. E. Coxon and D. B. Sowerby, *J. Chem. Soc.* (*A*), 1967, 1566.
[1057] H. Schmidbaur and G. Jonas, *Chem. Ber.*, 1967, **100**, 1120.
[1058] S. R. Jain, W. S. Brey, jun., and H. H. Sisler, *Inorg. Chem.*, 1967, **6**, 515.
[1059] S. R. Jain, L. K. Krannich, R. E. Highsmith, and H. L. Sisler, *Inorg. Chem.*, 1967, **6**, 1058.
[1060] R. A. Nyquist, E. H. Blair, and D. W. Osborne, *Spectrochim. Acta*, 1967, **23**, *A*, 2505.
[1061] A. Schmidpeter and H. Groeger, *Chem. Ber.*, 1967, **100**, 3052.
[1062] H. Nöth and L. Meinel, *Z. anorg. Chem.*, 1967, **349**, 225.

Phosphorus–Nitrogen Compounds (*cont.*)

		Ref.
$Ph_2P(S)NH\cdot P(O)Ph_2$		1062a
$P[NMeNMe_2]_3$		1063
$PhP[NMeNMe_2]_2$		1063
$[PhP(NH_2)(NMeNMe_2)_2]Cl$		1063
$[Ph_2PNH_2)_2NNMe_2]Cl_2$		1063
$[(Ph_2PNH_2)_2NNMe(PPh_2)]Cl$		1063
$R^1P[C(S)\cdot NHR^2]_2$ } $R_2{}^1P[C(S)\cdot NHR^2]$	$R^1 = C_6H_{11}$, Ph; R^2 = Me, Et, $CH_2{:}CH\cdot CH_2$ Ph, p-Br$\cdot C_6H_4$, β-$C_{10}H_7$	1064
$K_3(PO_2NH)_3$		1065

Arsenic, Antimony

$Ph_3As(OH)X$	X = Cl, Br, OH	1066
$[Ph_3AsOH^+]X^-$	X = Br_3, ICl_2, ClO_4	1066
$[Ph_3AsOH^+]_2X^-$	X = Br_3Br, $HgCl_4$, $HgBr_4$	1066
$SbCl_3L$	L = benzene, toluene, *o*-, *m*-, or *p*-xylene, mesitylene	1067
$SbEt_5$		1068

Sulphur

$MeSC{\vdots}CX$	X = H, D	1069
Me_2SO_2		1070
$MeSO_2X$	X = F, Cl	1070
$[M^{II}(SC_6F_5)_4]^{2-}$	M^{II} = Co, Pd, Pt, Zn, Cd, Hg	1071
$[M^I(SC_6F_5)_4]^-$	M^I = Cu, Ag, Au	1071
$(\pi$-$C_5H_5)MS_2C_2(CF_3)_2$	M = Co, Rh, Ir	1072
$[Et_4N(\pi$-$C_5H_5)WS_2C_2(CF_3)_2]_2$		1072

1062a A. Schmidpeter and H. Groeger, *Chem. Ber.*, 1967, **100**, 3979.
1063 J. M. Kanamueller and H. H. Sisler, *Inorg. Chem.*, 1967, **6**, 1765.
1064 K. Issleib and G. Harzfeld, *Z. anorg. Chem.*, 1967, **351**, 18.
1065 K. Lunkwitz and E. Steger, *Spectrochim. Acta*, 1967, **23**, *A*, 2593.
1066 G. S. Harris and F. Inglis, *J. Chem. Soc.* (*A*), 1967, 497.
1067 J. Gerbier, *Compt. rend.*, 1967, **264**, *B*, 444.
1068 Y. Takashi, *J. Organometallic Chem.*, 1967, **8**, 225.
1069 A. G. Moritz, *Spectrochim. Acta*, 1967, **23**, *A*, 167.
1070 M. Spoliti, S. M. Chackalackal, and F. E. Stafford, *J. Amer. Chem. Soc.*, 1967, **89**, 1092.
1071 W. Beck, K. H. Stetter, S. Tadros, and K. W. Schwarzhans, *Chem. Ber.*, 1967, **100**, 3944.
1072 R. B. King and M. B. Bisnette, *Inorg. Chem.*, 1967, **6**, 469.

Sulphur (*cont.*)

		Ref.
$(Et_4N)_2[S_2C_2(CF_3)_2]_2MnNO$		1072
$[MeSFe(CO)(\pi\text{-}C_5H_5)]_2(SbF_6)$		1072
$[(\pi\text{-}C_5H_5)W(CO)S_2C_2(CF_3)_2]_2$		1072
$(\pi\text{-}C_5H_5)W[S_2C_2(CF_3)_2]_2$		1072
$[(\pi\text{-}C_5H_5)Mo(NO)S_2C_2(CF_3)_2]_2$		1072
$(\pi\text{-}C_5H_5)Mn(NO)S_2C_2(CF_3)_2$		1072
$S_4O_{11}Cl_2$		1073
PSF_2OSO_2F		1074
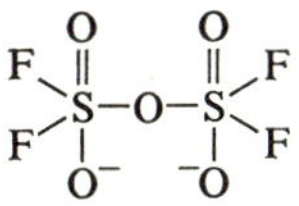 $(NO^+)_2$		1075
S:CF·NCS		1076
$RNSF_2$	R = Cl, CF_3, C_2F_5, CNN, C_6F_5, OCF, SF_5, CF_3S	1077
$(CF_3SNCO)_2$		1078
FSO_2N:S:O		1079
$FSO_2N:SCl_2$		1079
$OPF_2N:SF_2$		1080
$F_2SO_2N:SOF_2$		1080
$FCONSF_2$		1081
XSO_2NSF_2	X = F, Cl	1081, 1082
FCOSX	X = SCN, NCO	1083
$(XCOSNH)_2CO$	X = F, Cl	1083
ClCOSNCO		1083
$H(NSF_2O)$		1084

[1073] K. Stopperka and V. Grove, *Z. anorg. Chem.*, 1967, **353**, 72.
[1074] M. Lustig, *Angew. Chem. Internat. Edn.*, 1967, **6**, 959.
[1075] O. Glemser and U. Biermann, *Chem. Ber.*, 1967, **100**, 1184.
[1076] A. Haas and W. Klug, *Angew. Chem. Internat. Edn.*, 1967, **6**, 940.
[1077] A. Mueller, O. Glemser, and B. Krebs, *Z. Naturforsch.*, 1967, **22**, *B*, 550.
[1078] A. J. Downs and A. Haas, *Spectrochim. Acta.*, 1967, **23**, *A*, 1023.
[1079] H. W. Roesky, *Angew. Chem. Internat. Edn.*, 1967, **6**, 711.
[1080] O. Glemser, H. W. Roesky, and P. R. Heinze, *Angew. Chem. Internat. Edn.*, 1967, **6**, 710.
[1081] U. Biermann and O. Glemser, *Chem. Ber.*, 1967, **100**, 3795.
[1082] O. Glemser, H. W. Roesky, and P. R. Heinze, *Angew. Chem. Internat. Edn.*, 1967, **6**, 179.
[1083] A. Haas and H. Reinke, *Angew. Chem. Internat. Edn.*, 1967, **6**, 705.
[1084] H. W. Roesky, O. Glemser, A. Hoff, and W. Koch, *Inorg. Nuclear Chem. Letters*, 1967, **3**, 39.

Sulphur (*cont.*)

		Ref.
$M^+(NSF_2O)^-$	$M^+ = Ph_4P, Ph_4As$	1084
$CF_3SN{:}SF_2$		1085
$CF_3SN{:}S{:}NSCF_3$		1085
$SF_2{:}N{\cdot}CN$		1086, 1087
$SF_2NCF_2N{:}SF_2$		1087
$O{:}SF_2{:}NCN$		1088
$CF_3N{:}SF_2{:}O$		1088
$XSO_2{\cdot}N{:}CCl_2$	X = F, Cl	1089
$OSNSO_2Cl$		1090
$HN(SO_2NH_2)_2$		1091

Selenium, Tellurium

$(CD_3)_2Se$		1092
H_2SeO_3(aq.)		1093
$FSeO_2OR$	R = Me, Et	1094
$(SeS_2N_2Cl_2)_n$		1095
$Ph_2Se(OR)_2$	R = Me, Et	1096
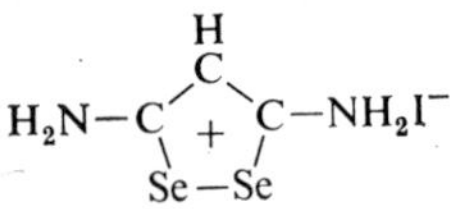		1097
$TeF_{6-n}(OH)_n$	n = 1–5	1098
$pyH[Te(OH)_3F_2]$,2py,7HF		1098
$pyH[Te(OH)F_4]$		1098
$Te[Te(OH)_3F_2]HF$		1098
$[TeCl_32L]^+Cl^-$	L = py, pyO, Ph_3PO	1099
$[TeCl_3L]^+Cl^-$	L = tmen, bipy, terpy	1099

1085 A. Haas and P. Schott, *Angew. Chem. Internat. Edn.*, 1967, **6**, 370.
1086 W. Sundermeyer, *Angew. Chem. Internat. Edn.*, 1967, **6**, 90.
1087 O. Glemser and U. Biermann, *Inorg. Nuclear Chem. Letters*, 1967, **3**, 223.
1088 M. Lustig and J. K. Ruff, *Inorg. Nuclear Chem. Letters*, 1967, **3**, 531.
1089 H. W. Roesky and U. Biermann, *Angew. Chem. Internat. Edn.*, 1967, **6**, 882.
1090 J. K. Ruff, *Inorg. Chem.*, 1967, **6**, 2108.
1091 W. Schneider and H.-A. Lehmann, *Z. anorg. Chem.*, 1967, **354**, 235.
1092 J. R. Allkins and P. J. Hendra, *Spectrochim Acta*, 1967, **23**, *A*, 1671.
1093 G. E. Walraffen, *J. Chem. Phys.*, 1967, **46**, 1870.
1094 R. Paetzold, R. Kurze, and G. Engelhardt, *Z. anorg. Chem.*, 1967, **353**, 62.
1095 A. J. Banister and J. S. Padley, *J. Chem. Soc.* (*A*), 1967, 1437.
1096 R. Paetzold and U. Lindner, *Z. anorg. Chem.*, 1967, **350**, 295.
1097 K. A. Jensen and U. Henriksen, *Acta Chem. Scand.*, 1967, **21**, 1991.
1098 L. Kolditz and I. Fitz, *Z. anorg. Chem.*, 1967, **349**, 175.
1099 D. A. Couch, P. S. Elmes, J. E. Fergusson, M. L. Greenfield, and C. J. Wilkins, *J. Chem. Soc.* (*A*), 1967, 1813.

Fluorine, Chlorine, Iodine

		Ref.
CF_3OF		1100
$CF_2(OF)_2$		1101, 1102, 1103, 1104
$CF_3XC(OF)_2$	X = F, CF_3	1105
FC(O)OF		1106
CF_3OOOR	R = CF_3, C_2F_5	1107, 1108
FCO(X)	X = CN, NCS	1083, 1109, 1110
NF_2CXO	X = F, Cl	1111
$(NF_2)_2CO$		1111
FCl_2CSCl		1112
$(FCl_2CS)_2$		1112
$(FCl_2CSNCO)_3$		1112
$(FCl_2CSNH)_2CO$		1112
$FCl_2CSNHCO_2CH_2Ph$		1112
FCl_2CSX	X = Cl, NCO, SCN	1112
$(F_2ClCS)_2$		1112
$(F_2ClCSNCO)_n$	$n = 2, 3$	1112
$F_2ClCSNHCO_2CH_2X$	X = H, Ph	1112
$FO(CF_2)_nOF$	$n = 4, 5$	1113
$CF_2(NF_2)CF_2OF$		1113
$CF_3CF(NF_2)CF_2OF$		1113
$RC{:}CH(CF_2)_nCF_2$ (ring: C bonded to terminal CF_2)	R = F, Cl, Et_2N, Me_2As; n = 1 or 2	1114
$(CF_3)_2C(OPF_2)X$	X = H, Br, I	1115

[1100] P. M. Wilt and E. A. Jones, *J. Inorg. Nuclear Chem.*, 1967, **29**, 2108.
[1101] R. L. Cauble and G. H. Cady, *J. Amer. Chem. Soc.*, 1967, **89**, 1962.
[1102] R. W. Mitchell and J. A. Merritt, *J. Mol. Spectroscopy*, 1967, **24**, 128.
[1103] P. G. Thompson, *J. Amer. Chem. Soc.*, 1967, **89**, 1811.
[1104] F. A. Hahorst and J. M. Shreeve, *J. Amer. Chem. Soc.*, 1967, **89**, 1809.
[1105] P. G. Thompson and J. H. Prager, *J. Amer. Chem. Soc.*, 1967, **89**, 2263.
[1106] R. L. Cauble and G. H. Cady, *J. Amer. Chem. Soc.*, 1967, **89**, 5161.
[1107] P. G. Thompson, *J. Amer. Chem. Soc.*, 1967, **89**, 4316.
[1108] L. R. Anderson and W. B. Fox, *J. Amer. Chem. Soc.*, 1967, **89**, 4313.
[1109] W. Verbeek and W. Sundermeyer, *Angew. Chem. Internat. Edn.*, 1967, **6**, 871.
[1110] R. L. Cauble and G. H. Cady, *Inorg. Chem.*, 1967, **6**, 2117.
[1111] G. W. Fraser and J. M. Shreeve, *Inorg. Chem.*, 1967, **6**, 1711.
[1112] A. Haas and D.-Y. Oh, *Chem. Ber.*, 1967, **100**, 480.
[1113] M. Lustig, A. R. Pitochelli, and J. K. Ruff, *J. Amer. Chem. Soc.*, 1967, **89**, 2841.
[1114] W. R. Cullen and P. S. Dhaliwal, *Canad. J. Chem.*, 1967, **45**, 719.
[1115] M. Lustig and W. E. Hill, *Inorg. Chem.*, 1967, **6**, 1448.

Fluorine, Chlorine, Iodine (*cont.*)

		Ref.
4-$BrC_{12}F_9$		1116
2-H-2′-Br-$C_{12}F_8$		1116
C_6F_5IO		1117
$C_6F_5I(OCOR)_2$	R = CF_3, C_3F_7, C_6F_5	1117
$I(OCOR)_3$	R = CF_3, C_3F_7, C_6F_5	1117

Titanium

$(\pi\text{-}C_5H_5)_2Ti(C{\vdots}CPh)_2$		1118
$(\pi\text{-}C_5H_5)_2TiC_{10}H_8N_2$		1119
$[TiO(OCOCF_3)_2]_n$		1120
Na_xTiO_2	$0{\cdot}20 < x < 0{\cdot}25$	1121
$Ti(OMe)_2[OPMe_2O]_2$		1122
$Ti(O\cdot Pr^i)_2[OPPh_2O]_2$		1122
$TiO[OPMePhO]_2$		1122
$TiO[OPPh_2O]_n$	$n = 2, 4$	1122
$M_2[TiOoCl_4,H_2O]$	M = alkali metal; o = ‘active oxygen’	1123

Chromium, Molybdenum, Tungsten

$Cr(C_6H_6)_2$		1124
$[Cr(C_6H_6)_2]I$		1125
$[Cr(PhMe)_2]I$		1125
$M^{III}(C_5H_6NO_4)_3$	M^{III} = Cr, Co, Fe, Mn, Al	1126
$M^{III}(C_{15}H_{10}NO_4)_3$	M^{III} = Cr, Al	1126
$K[(CrO_3)(IO_3)]$		1127
$Li[MX(C_6H_4O_2)_3]$,2Diglyme	M = Mo, W; X = Cl, Br	1128

1116 S. C. Cohen, D. E. Fenton, D. Shaw, and A. G. Massey, *J. Organometallic Chem.*, 1967, **8**, 1.
1117 M. Schmeisser, K. Dahmen, and P. Sartori, *Chem. Ber.*, 1967, **100**, 1633.
1118 H. Köpf and M. Schmidt, *J. Organometallic Chem.*, 1967, **10**, 383.
1119 E. O. Fischer and R. Amtmann, *J. Organometallic Chem.*, 1967, P15.
1120 P. Sartori and M. Weidenbruch, *Chem. Ber.*, 1967, **100**, 2049.
1121 A. F. Reid and M. J. Sienko, *Inorg. Chem.*, 1967, **6**, 321.
1122 G. H. Dahl and B. P. Block, *Inorg. Chem.*, 1967, **6**, 1439.
1123 E. Wendling, *Bull. Soc. chim. France*, 1967, 16.
1124 H. P. Fritz and E. O. Fischer, *J. Organometallic Chem.*, 1967, **7**, 121.
1125 H. Saito and Y. Kakiuti, *Spectrochim. Acta*, 1967, **23**, *A*, 3013.
1126 P. R. Singh and R. Sahai, *Austral. J. Chem.*, 1967, **20**, 649.
1127 T. Dupuis, *Mikrochim. Acta*, 1967, 461; see also *Chem. Abs.*, 1967, **67**, 48,674*h*.
1128 F. Calderazzo and R. Henzi, *J. Organometallic Chem.*, 1967, **10**, 483.

Chromium, Molybdenum, Tungsten (*cont.*)

		Ref.
$(NH_4)_2MoSe_4$		1129
$M(MoO_2oCl_4)$	M = alkali metal; o = 'active oxygen'	1123
$W(C_9H_6NO)_4$		1130
$W_2Cl_6L_4$	L = py, isopropylpyridine	1131

Cobalt, Rhodium

$[Bis(6,6'\text{-diphenylfulvene})M]^+PF_6^-$	M = Co, Rh, Ir	1132
$Co(imidazole)_2$		1133
$Co(imidazole)_6X_2$	X = Cl, Br, I, NO_3	1133
$Co(thiazole)_2Cl_2$		1133
$[(\pi\text{-}C_5H_5)CoC_3F_7X_2]^-$	X = I, CN	1134
$[(\pi\text{-}C_5H_5)CoC_3F_7IF]^-$		1134
$[(\pi\text{-}C_5H_5)CoC_3F_7SCN]_3$		1134
$[(\pi\text{-}C_5H_5)CoC_3F_7CN]_x$	$x = 4$–5	1134
$[(\pi\text{-}C_5H_5)CoC_3F_7I]_2$		1134
$[C_{10}H_{16}RhCl]_2$		1135
acac $Rh(R)_2$	R = CH_2:CHCl, C_3H_6	1136
$[O\cdot Np\cdot O\cdot Rh(H_2O)_5]^{4+}$		
$M(RH)_2Cl_3$	M = Ru, Rh, Ir; RH = 2 mercaptobenzothiazole	1137

Nickel, Palladium, Platinum

$Ni(NH_3)_2Ni(CN)_4L$	L = C_6H_6, $PhNH_2$	1138
$M(NH_3)_2Ni(CN)_4L$	L = C_6H_6, $PhNH_2$; M = Cu, Zn, Cd	1138
$Ni(S_2CNH_2)_2L_2$	L = py, γ-pic	1139
$(Et_3P)_2M(CF{:}CF_2)_2$	M = Ni, Pd, Pt	1140
$(Et_3P)_2Ni(CF{:}CF_2)Br$		1140

[1129] A. Müller, B. Krebs, and E. Diemann, *Angew. Chem. Internat. Edn.*, 1967, **6**, 257.
[1130] R. D. Archer and W. D. Bonds, jun., *J. Amer. Chem. Soc.*, 1967, **89**, 2236.
[1131] R. Saillant, J. L. Hayden, and R. A. D. Wentworth, *Inorg. Chem.*, 1967, **6**, 1497.
[1132] E. O. Fischer and B.-J. Weimann, *J. Organometallic Chem.*, 1967, **8**, 535.
[1133] W. J. Eilbeck, F. Holmes, and A. E. Underhill, *J. Chem. Soc.* (*A*), 1967, 757.
[1134] P. M. Treichel and G. P. Werber, *J. Organometallic Chem.*, 1967, **7**, 157.
[1135] J. C. Trebellas, J. R. Olechowski, H. B. Jonassen, and D. W. Moore, *J. Organometallic Chem.*, 1967, **9**, 153.
[1136] R. Cramer, *J. Amer. Chem. Soc.*, 1967, **89**, 4621.
[1137] R. F. Wilson and P. Merchant, jun., *J. Inorg. Nuclear Chem.*, 1967, **29**, 1993.
[1138] T. Miyoshi, T. Iwamoto, and Y. Sasaki, *Inorg. Chim. Acta*, 1967, **1**, 20.
[1139] D. Coucouvanis and J. P. Fackler, jun., *Inorg. Chem.*, 1967, **6**, 2047.
[1140] A. J. Rest, D. T. Rosevear, and F. G. A. Stone, *J. Chem. Soc.* (*A*), 1967, 66.

Nickel, Palladium, Platinum *(cont.)*

		Ref.
Ni(imidazole)$_2$		1133
Ni(imidazole)$_6$X$_2$	X = Cl, Br, I	1133
Ni(thiazole)$_4$Cl$_2$		1133
[C$_7$H$_8$(OMe)PdCl]$_2$		1141
[PdC$_6$H$_{13}$N$_5$O][Pd(SCN)$_4$]		1142
M(RH)$_2$Cl$_2$	M = Pd, Pt; RH = 2-mercaptobenzothiazole	1137
Pt(RH)$_2$Br$_2$		1137
Pt(RH)$_2$Cl$_4$		1137
[{PtH(PPh$_3$)$_2$(SnCl$_3$)}$_2$L]	L = cod, C$_7$H$_8$	1143
PtH(SnCl$_3$) cot(PPh$_3$)$_2$		1143
[{PtCl(PPh$_3$)}PPh$_3$(CO)]Cl$_2$		1143
[Pt(C$_5$H$_{10}$)$_2$PPh$_3$][Sn$_2$Cl$_6$]		1143
[PtH(C$_6$H$_8$(PPh$_3$)$_2$]SnCl$_3$(CH$_2$Cl$_2$)		1143
(Ph$_3$P)$_2$Pt(SO$_2$C$_6$H$_4$Me)Cl		1144
(Ph$_3$P)$_2$Pt(R)Br	R = Me, OMe, C⋮CPh, CH:CHPh	1144
(Ph$_3$P)$_2$Pt(C⋮CI)I		1144
Pt$_2$(C$_5$H$_5$)$_4$		1145
Bis-(1,5-cod)Pt(O)		1146

Zinc, Cadmium

β-Zn(OH)X	X = F, Cl	1147
ε-Zn(OH)$_2$		1147
(Bu^{t}O)$_2$Zn		1148
(Bu^{t}ZnOBu^t)$_2$		1148
Zn$_5$(OH)$_8$Cl$_2$		1147
Zn$_5$(OH)$_8$Cl$_2$,H$_2$O		1147
Cd(OH)X	X = F, Cl, OH	1149
Cd(OD)X	X = F, Cl, OD	1149
Cd(C$_6$F$_5$)$_2$		1150

[1141] M. Green and R. I. Hancock, *J. Chem. Soc.* (*A*), 1967, 2054.
[1142] P. Spacu and C. Gheorghiu, *Z. anorg. Chem.*, 1967, **351**, 201.
[1143] H. A. Tayim and J. C. Bailar, jun., *J. Amer. Chem. Soc.*, 1967, **89**, 4330.
[1144] C. D. Cook and G. S. Janhal, *Canad. J. Chem.*, 1967, **45**, 301.
[1145] E. O. Fischer and H. Schuster-Woldan, *Chem. Ber.*, 1967, **100**, 705.
[1146] J. Müller and P. Goser, *Angew. Chem. Internat. Edn.*, 1967, **6**, 364.
[1147] O. K. Srivastava and E. A. Secco, *Canad. J. Chem.*, 1967, **45**, 585.
[1148] G. E. Coates and P. D. Roberts, *J. Chem. Soc.* (*A*), 1967, 1233.
[1149] O. K. Srivastava and E. A. Secco, *Canad. J. Chem.*, 1967, **45**, 3198.
[1150] M. Schmeisser and M. Weidenbruch, *Chem. Ber.*, 1967, **100**, 2306.

6
Electronic Spectra

This chapter is divided into four main sections dealing with the electronic spectra of:

The main-group elements; the transition elements; the lanthanide and actinide elements; and optically active co-ordination compounds.
The account is not fully comprehensive, and in particular, data contained in papers dealing with reaction kinetics have been excluded when the spectra were used primarily to follow changes of concentration in the reaction system.

The task of assessing the most significant advances is not easy. Whilst there do not appear to have been any particularly sensational advances during the year, a number of significant developments have been noted. The great potential of magnetic dichroism, which has been apparent for some time, is slowly being exploited. It is now clear that magnetic circular dichroism measurements provide a very direct and sensitive method for investigating both the angular momentum of d^n states, and also the assignment of spin-forbidden transitions in chromium(III) species. Magnetic circular dichroism also provides information on the intensity-gaining mechanism of certain *d–d* transitions.[1]

It has also become increasingly clear during the last few years that the magnitude of an extinction coefficient is not an entirely reliable criterion for differentiating between charge-transfer and *d–d* transitions. Thus, it has been shown that several, very intense bands are almost certainly predominantly due to ligand-field transitions.[2,3]

1 The Main-group Elements

Little work of interest to inorganic chemists was published during the year on compounds of Groups I and II. However, the predictions of the Franck–Condon principle have been checked [4] by comparing the theoretical intensities with those observed experimentally for the transitions from the ground-state, ${}^1\Sigma_g^+$, to the $1s2p$, $3p$, and $4p\,{}^1\Pi_u$ excited-states of the hydrogen molecule. The observed intensities were found to differ by less than 10% from the predicted values.

[1] A. J. McCaffery, P. J. Stephens, and P. N. Schatz, *Inorg. Chem.*, 1967, **6**, 1614.
[2] M. J. Norgett, J. H. M. Thornley, and L. M. Venanzi, *J. Chem. Soc.* (*A*), 1967, 540.
[3] G. Dyer and D. W. Meek, *J. Amer. Chem. Soc.*, 1967, **89**, 3982.
[4] S. Rothenberg and E. R. Davidson, *J. Mol. Spectroscopy*, 1967, **22**, 1.

Group III.—*Boron.* Experimentally, very little is known about the BN molecule. The nature of the ground-state is uncertain and it has been investigated[5] by calculating some low-lying valence levels using the LCAO MO SCF method. Variational calculations were carried out for three different configurations, and term values were obtained from the minima of the computed potential curves together with estimates of pair-correlation energy differences between the various configurations. The calculations indicated a $^3\Pi$ ground-state with low-lying $^3\Sigma^+$, $^1\Sigma^+$, and $^1\Pi$ states. The isoelectronic molecules BeO and C_2 are known to have closed shell $^1\Sigma^+$ ground-states.

The π electronic structures of a number of boron–oxygen containing compounds have been calculated[6,7] by the Pople SCF method. A satisfactory account of the electronic spectra was given by this method.

Electronic spectral data have also been reported for the following boron-containing compounds: $(Me_4N)_2[(B_{10}H_{12})_2Hg]$,[8] carbonyl derivatives of $B_{10}H_{10}^{2-}$ and $B_{12}H_{12}^{2-}$,[9] 1,10-$B_{10}Cl_8(CH_2X)_2^{2-}$ (X = Br, I, or CN), 1,10-$B_{10}Cl_8(CH_2Y)_2$ (Y = NH_3 or NMe_2H),[10] 2-ethynyl-4,4,6-trimethyl-1,3,2-dioxaborinane, and dimethyl acetylene boronate.[11]

Thallium. Thallium atoms, Tl^0, and halogen molecule ions, X_2^-, have been produced in KCl–TlCl, KBr–TlBr, and KI–TlI crystals at 77°K by exposure to γ-rays. Two bands with very low oscillator strengths in the i.r. absorption spectrum of Tl^0 are thought to arise from the forbidden $^2P_{\frac{3}{2}} \leftarrow {}^2P_{\frac{1}{2}}$ transition of the free atom;[12] the two bands, rather than one, are presumed to arise from a splitting of the $^2P_{\frac{3}{2}}$ excited state caused by a dynamical Jahn–Teller effect. The spectrum in KCl–TlCl was studied at three different temperatures and it was noted that both the splitting and the oscillator strengths of these two transitions increase with increasing temperature. Knox[13] has developed a semi-classical theory to account for these features and in particular to account for the temperature dependence. It is proposed that this temperature-dependent splitting may occur rather generally and independently of any explicit dynamical Jahn–Teller effect in forbidden transitions at impurity centres.

The u.v. spectra of methyl- and ethyl-thallium dioxinates have been recorded[14] in dry benzene and from these results it is concluded that the thallium is five-co-ordinate in solution.

5 G. Verhaegen, W. G. Richards, and C. M. Moser, *J. Chem. Phys.*, 1967, **46**, 160.
6 D. R. Armstrong and P. G. Perkins, *J. Chem. Soc.* (*A*), 1967, 123.
7 D. R. Armstrong and P. G. Perkins, *J. Chem. Soc.* (*A*), 1967, 790.
8 N. N. Greenwood and N. F. Travers, *Chem. Comm.*, 1967, 216.
9 W. H. Knoth, J. C. Sauer, J. H. Balthis, H. C. Miller, and E. L. Muetterties, *J. Amer. Chem. Soc.*, 1967, **89**, 4842.
10 W. H. Knoth, *J. Amer. Chem. Soc.*, 1967, **89**, 4850.
11 W. G. Woods and P. L. Strong, *J. Organometallic Chem.*, 1967, **7**, 371.
12 C. J. Delbecq, A. K. Ghosh, and P. H. Yuster, *Phys. Rev.*, 1967, **154**, 797.
13 R. S. Knox, *Phys. Rev.*, 1967, **154**, 799.
14 H. Kurosawa and R. Okawara, *J. Organometallic Chem.*, 1967, **10**, 211.

Group IV.—The extent of *d*-orbital participation by the Group IV elements has been discussed [15] on the basis of an analysis of the change in intensity of the symmetry-forbidden $B_{2u} \leftarrow A_{1g}$ transition in the u.v. spectra of two series of compounds:

(M = C, Si, Ge, or Sn)

The extent of overlap between the *d* orbitals of the Group IV metals and the π orbitals of the benzene ring in *para*-substituted anisoles appears to decrease in the series Si > Ge > Sn. Nagy *et al.*[16] studied the same question for the compounds trimethylphenyl–M and trimethylbenzyl–M (M = C, Si, Ge, or Sn). U.v. spectra and dipole moments were measured and again it was concluded that the d_π–p_π bonds were strongest for silicon; the spectra are discussed in some detail. The u.v. spectra of the compounds 4-$Me_3MC_6H_4Ph$ and 4,4′-$Me_3MC_6H_4 \cdot C_6H_4MMe_3$ (M = C, Si, Ge, and Sn) have also been discussed.[17] The spectra are characterised by a broad band at approximately 40,000 cm.$^{-1}$ (*p* band) and a strong band at about 50,000 cm.$^{-1}$ (β band).

Carbon. The absorption spectra of CO_3^{2-}, CS_3^{2-}, and CSe_3^{2-} in aqueous solution have been measured [18] and discussed in some detail. Simple HMO calculations were carried out for the three ions. Several organo-tin(IV) *NN*-disubstituted-dithiocarbamates have been isolated, and characterised [19] by three absorption bands. Two of the bands are assigned to $\pi \rightarrow \pi^*$ transitions of the dithiocarbamate moiety (about 38,500 and 35,700 cm.$^{-1}$), whilst the third weak band (*ca.* 29,500–30,000 cm.$^{-1}$) is assigned to the $n \rightarrow \pi^*$ transition. The u.v. spectra of silyl *NN*-dimethylcarbamate, silyl *NN*-dimethyldithiocarbamate, and *O*-silyl *NN*-dimethylmonothiocarbamate have also been recorded.[20]

The electronic structure of inorganic fulminates has been investigated by Iqbal and Yoffe.[21] The fulminate ion, CNO^-, is linear with $C_{\infty v}$ symmetry, and is isoelectronic with the azides, cyanates, and cyanamides. The absorption spectra of thin films of sodium and potassium fulminate have peaks at *ca.* 51,300 and 50,400 cm.$^{-1}$, at $-176°$, which have been attributed to $n = 1$ exciton transitions. Bands have also been resolved at about 43,500 cm.$^{-1}$, which may be attributed to a $\pi \rightarrow \pi^*$ transition, within the fulminate ion by analogy with alkali azides.

[15] W. K. Muster and G. B. Savitsky, *J. Phys. Chem.*, 1967, **71**, 431.
[16] J. Nagy, J. Réffy, A. Kuszmann-Borbély, and K. Pálossy-Becker, *J. Organometallic Chem.*, 1967, **7**, 393.
[17] M. D. Curtis, R. K. Lee, and A. L. Allred, *J. Amer. Chem. Soc.*, 1967, **89**, 5150.
[18] A. Müller, H. Seidel, and W. Rittner, *Spectrochim. Acta*, 1967, **23**, *A*, 1619.
[19] F. Bonati and R. Ugo, *J. Organometallic Chem.*, 1967, **10**, 257.
[20] E. A. V. Ebsworth, G. Rocktäschel, and J. C. Thompson, *J. Chem. Soc.* (*A*), 1967, 362.
[21] Z. Iqbal and A. D. Yoffe, *Proc. Roy. Soc.*, 1967, *A*, **302**, 35.

Silicon. The electronic spectrum of gaseous SiF_2 has been studied.[22] The principal feature of the spectrum is the presence of a long progression in the bending frequency of the upper electronic state, which suggests that the bond angle undergoes a considerable change during the transition.

Organo-silicon derivatives continue to attract considerable interest, and the electronic spectra have been recorded in many cases to provide 'finger-prints' of these compounds. The electronic spectra of twenty-eight silyl ketimines and several germanium and tin analogues have been studied in some detail[23] to estimate the d_{π}–p_{π} interaction between the empty d orbitals of silicon and the π system of the C:N bond. The spectral data indicate a bent structure for the C:N·Si linkage and hence very little multiple bonding in the Si—N bond. The u.v. spectra of the compounds $Me_nSi(OPh)_{4-n}$ (n = 0–3) have also been investigated[24] in some detail. A small bathochromic shift is found for the phenoxy–silicon bond as compared with the phenoxy-group in phenyl ethers, and the position of the maximum is independent of the number of phenoxy-groups. The $\pi \rightarrow \pi^*$ transition of the phenoxy-groups in the phenyl ethers and of the phenoxysilicon groups in the phenoxysilanes have been calculated by the LCAO–MO one-electron method, and the results are in accordance with the observed maxima.

The following types of compound have also been characterised by means of their electronic spectra: $(C_6F_5)_nSiPh_{4-n}$ (n = 1–4)[25], $MeO(SiMe_2)_nOMe$,[26] pentachlorophenyl derivatives of Si,[27] tetrazadisilacyclohexanes,[28] silyl-hydrazines,[29] and several adducts of halogenosilanes.[30]

Germanium. The vapour phase u.v. spectrum of trigermylphosphine, $P(GeH_3)_3$, has been recorded[31] at frequencies below 50,000 cm.$^{-1}$; a single maximum was observed at 46,300 cm.$^{-1}$ with $\varepsilon \sim 10^4$, together with increasing absorption near 50,000 cm.$^{-1}$. The absorption maxima in the near u.v. region for fourteen Ge—Sn derivatives have been tabulated.[32] The strong bonds in this region are primarily due to the metal–metal system.

The reactions between Ge^{IV} and a series of 2,5-dihydroxy *p*-benzo-quinones, (1),[33] and between germanic acid and 8-mercaptoquinoline[34] have been studied spectrophotometrically.

22 V. M. Khanna, G. Besenbruch, and J. L. Margrave, *J. Chem. Phys.*, 1967, **46**, 2310.
23 L.-H. Chan and E. G. Rochow, *J. Organometallic Chem.*, 1967, **9**, 231.
24 J. Nagy and P. Hencsei, *J. Organometallic Chem.*, 1967, **9**, 57.
25 F. M. G. Fearon and H. Gilman, *J. Organometallic Chem.*, 1967, **10**, 409.
26 W. H. Atwell and D. R. Wayenberg, *J. Organometallic Chem.*, 1967, **7**, 71.
27 H. Gilman and S.-Y. Sim, *J. Organometallic Chem.*, 1967, **7**, 249.
28 S. S. Dua and M. V. George, *J. Organometallic Chem.*, 1967, **10**, 219.
29 S. S. Dua and M. V. George, *J. Organometallic Chem.*, 1967, **9**, 413.
30 H. J. Campbell-Ferguson and E. A. V. Ebsworth, *J. Chem. Soc.* (*A*), 1967, 705.
31 S. Cradock, E. A. V. Ebsworth, G. Davidson, and L. A. Woodward, *J. Chem. Soc.* (*A*), 1967, 1229.
32 H. M. J. C. Creemers and J. G. Noltes, *J. Organometallic Chem.*, 1967, **7**, 237.
33 A. Beauchamp and R. Benoit, *Bull. Soc. chim. France*, 1967, 672.
34 P.-J. Clerc, *Bull. Soc. chim. France*, 1967, 394.

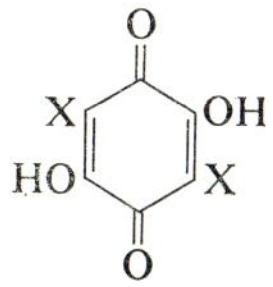

(1)

Tin. The structures of several six-co-ordinate tin complexes of the type SnX_2L_2 have been studied[35] [X = Cl, Br, or I; L = acetylacetonate or dibenzoylmethanate]. The electronic spectra indicate that the acetylacetonate and dibenzoylmenthanate groups are not strongly perturbed by the central tin atom; this has been interpreted as indicating that the tin–oxygen bonds are primarily ionic in character.

Group V.—*Nitrogen.* Potassium amide in liquid ammonia exhibits an absorption band at about 29,900 cm.$^{-1}$ and this band has been assigned[36] to the amide ion and to ion aggregates containing this species. The position and intensity of this band are temperature dependent which suggests that it arises from a charge-transfer to solvent transition.

A study[37] of the u.v. spectra of yellow etheral solutions of bromamine, NH_2Br, has shown that there is an equilibrium mixture of dibromamine, free ammonia, and bromamine in the solution.

$$2NH_2Br \rightleftharpoons NH_3 + NHBr_2$$
$$2MeNHBr \rightleftharpoons MeNH_2 + MeNBr_2$$

Similar conclusions were also reached for *N*-bromomethylamine. Data from the spectra of twelve *N*-halogenoamines, in a variety of solvents, are collected from the literature and presented in tabular form.

The electronic absorption spectra of aqueous solutions of Na, K, Ba, Ag, and Pb azides, perchlorates, and nitrates have been measured. The observed absorption was attributed to excitations from the highest-occupied to the lowest-unoccupied molecular orbital. The spectra of single crystals of these salts were also measured, and it was shown that there is a direct correlation between the absorption spectra of the solids and their respective sensitivities. Those azides, perchlorates, and nitrates which, in the solid state show absorption bands in the long wavelength side of the anionic excitation band in solution, are the most unstable members of their respective series.[38]

Antimony. The diffuse reflectance spectra are reported for seven new antimonates:[39] $CaSb_2O_4$, $Ca_2Sb_2O_5$, $Ca_3Sb_2O_6$, $SrSb_2O_4$, $Sr_2Sb_2O_5$, $CdSb_2O_4$, $Cd_3Sb_2O_6$, and $Mn_3Sb_2O_6$. No comment on the spectra is given beyond the

[35] W. H. Nelson, *Inorg. Chem.*, 1967, **6**, 1509.
[36] R. E. Cuthrell and J. J. Lagowski, *J. Phys. Chem.*, 1967, **71**, 1298.
[37] J. Jander and C. Lafrenz, *Z. anorg. Chem.*, 1967, **349**, 57.
[38] J. N. Maycock, V. R. Pai Verneker, and C. S. Gorzynski, *Spectrochim. Acta*, 1967, **23**, *A*, 2849.
[39] W. L. Wanmaker, A. H. Hoekstra, and J. G. Verrier, *Rec. Trav. chim.*, 1967, **86**, 537.

fact that they are quite different from those of the starting materials used in their preparation (Sb_2O_3 and MO or $MnCO_3$). The absorption spectra [40] of $SbCl_5$,8-hydroxyquinoline in carbon tetrachloride, chloroform, and acetone have been recorded.

Bismuth. Some fascinating chemistry of bismuth in its lower oxidation states has been studied using spectrophotometric techniques. Bi^+ was prepared [41] by the hydrogen reduction of dilute solutions of $BiCl_3$ in a molten $AlCl_3$–NaCl eutectic at 310°, and its absorption spectrum was later [42] rationalised in terms of intraconfigurational $6p^2 \rightarrow 6p^2$ transitions, split by a ligand field with a symmetry lower than cubic. Since the ligands play an important role in determining the details of the spectrum, it was of interest to observe the effects of changing from a chloride to a bromide environment. Consequently [43] the preparation of Bi^+ was repeated using a molten $AlBr_3$–NaBr eutectic at 250° and the observed spectrum is shown in Figure 1, where it is compared with the strikingly similar spectrum obtained from the corresponding chloride eutectic.

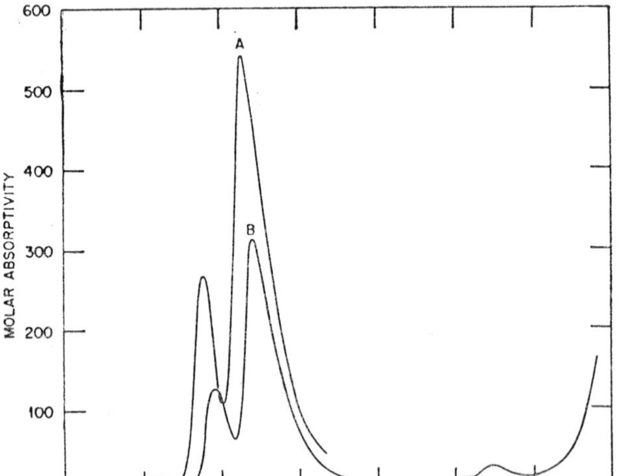

Figure 1 *Spectra of* Bi^+ *in halogenoaluminate media.* (A) *Solvent was molten* $AlBr_3$–NaBr *eutectic at* 250°; (B) *Solvent was molten* $AlCl_3$–NaCl *eutectic at* 310°
(Reproduced by permission from *Inorg. Chem.*, 1967, **6**, 1603)

The ion $[Bi_5]^{3+}$ was identified spectrophotometrically [325] when bismuth metal reacts with dilute solutions of $BiCl_3$ in liquid eutectic mixtures of

[40] S. S. M. A. Khorasani and A. Rauf, *Z. anorg. Chem.*, 1967, **350**, 326.
[41] N. J. Bjerrum, C. R. Boston, and G. P. Smith, *Inorg. Chem.*, 1967, **6**, 1162.
[42] H. L. Davis, N. J. Bjerrum, and G. P. Smith, *Inorg. Chem.*, 1967, **6**, 1172.
[43] N. J. Bjerrum, H. L. Davis, and G. P. Smith, *Inorg. Chem.*, 1967, **6**, 1603.

$AlCl_3$–NaCl or $ZnCl_2$–KCl. This species could not be obtained free from Bi^+, but its absorption spectrum was obtained by subtracting the spectrum due to Bi^+. When a sufficient excess of bismuth metal is added to form a separate phase, the absorption spectrum [44, 45] of the reaction mixture is quite different from that of either Bi^+ or $[Bi_5]^{3+}$. This third product contains eight bismuth atoms with an average oxidation state of $+\frac{1}{4}$, and is represented by the formula $[Bi_8]^{2+}$, but no information was obtained as to the number of ligands attached to this entity. The $AlCl_4^-$ anion is known to stabilise the lower oxidation states of bismuth, and the solid phases $Bi_5(AlCl_4)_3$ and $Bi_4(AlCl_4)$ have been isolated. The reflectance spectra of these materials have been recorded [46] and discussed in relation to the molten-state spectra of $[Bi_5]^{3+}$ and $[Bi_8]^{2+}$ mentioned above.

A spectrophotometric investigation of bismuth iodide complexes in aqueous solution is reported.[47] It was established that with decreasing iodide concentration the complexes formed are $[BiI_7]^{4-}$, $[BiI_6]^{3-}$, $[BiI_4]^-$, and a solvated BiI_3 species. The successive formation constants of these complexes were established, and individual absorption spectra presented.

Group VI.—*Sulphur.* Of the Group VI elements, sulphur is by far the most soluble in liquid ammonia; green solutions are obtained which become red on cooling to $-84{\cdot}6°$. A spectrophotometric study [48] has been carried out in an attempt to establish the identity of the species present. Sulphur–ammonia solutions have a spectrum different from those of sulphur in several other solvents, and the temperature dependence of the spectrum suggests that at least two species are present. In an attempt to identify these, the spectrum of tetrasulphur tetranitride, S_4N_4, was also studied in liquid ammonia. This spectrum changes with time and was interpreted in terms of a reaction of the type:

$$S_4N_4 + 2NH_3 \rightleftharpoons 2S_2N_2,NH_3$$

It was concluded that the species H_2S, S_2N_2,NH_3, S_4N_4, or S_8 are not present in appreciable concentrations in sulphur–ammonia solutions.

Thiourea and substituted thioureas exhibit two characteristic charge-transfer bands in their u.v. spectra, and it is known that co-ordination via the sulphur atom lowers the wavelength and intensity of both bands; *N*-co-ordination, on the other hand, would be expected to have little effect. This criteria for *N*-co-ordination has been confirmed by Greenwood and Robinson [49] who showed that the BF_3 adducts formed with a number

[44] N. J. Bjerrum and G. P. Smith, *Inorg. Nuclear Chem. Letters*, 1967, **3**, 165.
[45] N. J. Bjerrum and G. P. Smith, *Inorg. Chem.*, 1967, **6**, 1968.
[46] J. D. Corbett, *Inorg. Nuclear Chem. Letters*, 1967, **3**, 173.
[47] A. J. Eve and D. N. Hume, *Inorg. Chem.*, 1967, **6**, 331.
[48] J. T. Nelson and J. J. Lagowski, *Inorg. Chem.*, 1967, **6**, 1292.
[49] N. N. Greenwood and B. H. Robinson, *J. Chem. Soc.* (*A*), 1967, 511.

of thiourea derivatives were *N*-bonded. The u.v. spectra of isothiocyanatoboranes[50] were found to have absorption maxima at $40,000 \pm 800$ cm.$^{-1}$ whereas the isocyanatoborane chromophore does not absorb in the range 28,500–45,500 cm.$^{-1}$. It is suggested that these absorptions probably arise from $n \rightarrow \pi^*$ transitions of the NCS chromophore, and involve sulphur lone-pairs.

Tellurium. From the u.v. spectra of solutions of $TeCl_6^{2-}$ and $TeCl_4$ in propionitrile, it is concluded[51] that a new species is formed at a mole ratio of 1 : 1 and is apparently the $TeCl_5^-$ ion. Unfortunately, a crystalline derivative could not be isolated, and the i.r. and Raman studies on the solutions were difficult to interpret.

Group VII.—*Chlorine.* The magnetically induced circular birefringence and dichroism of several bands in the electronic spectrum of ICl have been studied[52] in order to test the reliability of predictions based upon the classical theory. The frequency dependence agrees with the predictions based upon the semi-classical dispersion formulas provided that each Zeeman component of a single zero-field line is treated separately. The line-centre is displaced by the Zeeman shift and the intensity is determined by first-order perturbation by the magnetic dipole-moment operator which mixes neighbouring *J* states. Thus, although the qualitative aspects of the semi-classical treatment of the Faraday effect may be extended into regions of the spectrum in the immediate vicinity of sharp-line molecular spectra, the details of the spectrum depend in a very sensitive way upon the molecular parameters and the rotational structure of the band.

Several organic salts containing the new pentahalide anions $I_2Cl_3^-$ and $I_2Cl_2Br^-$ have been reported.[53] However, attempts to obtain the electronic spectra of these species in 1,2-dichloroethane suggest that the pentahalide ions dissociate completely to the trihalide ion, ICl_2^-, and the corresponding interhalogen compound.

Bromine. The absorption spectrum of gaseous bromine has been reinvestigated[54] in order to obtain molar absorptivities suitable for quantitative use. Molar absorptivities have been determined from 33,300–13,300 cm.$^{-1}$ at various temperatures from 25–440°. A marked temperature and concentration dependence between 33,300 and 50,000 cm.$^{-1}$ has been studied in some detail and evidence for Br_4 is discussed.

Iodine. Spectroscopic evidence has been advanced for the formation of I_3^- in AgI at high pressures[55] and the kinetics of the reaction in a number

[50] M. F. Lappert and H. Pyszora, *J. Chem. Soc.* (*A*), 1967, 854.
[51] I. R. Beattie and H. Chudzynska, *J. Chem. Soc.* (*A*), 1967, 984.
[52] A. F. Stalder and W. H. Eberhardt, *J. Chem. Phys.*, 1967, **47**, 1445.
[53] Y. Yagi and A. I. Popov, *J. Inorg. Nuclear Chem.*, 1967, **29**, 2223.
[54] A. A. Passchier, J. D. Christian, and N. W. Gregory, *J. Phys. Chem.*, 1967, **71**, 937.
[55] M. J. Moore and D. W. Skelly, *J. Chem. Phys.*, 1967, **46**, 3676.

of solvents:

$$I_3^- + HCO_2^- \longrightarrow 3I^- + CO_2 + H^+$$

have been studied [56] spectrophotometrically.

Spectrophotometric studies of the 1 : 1 iodine complexes of cyclic selenides (selenacyclopentane and selenacyclohexane) have also been reported.[57] Equilibrium constants were calculated from the absorbance data and the small dissociation constants show these complexes, L,I_2, to be among the most stable compounds of this type.

2 The Transition Elements

General Introduction.—The contributions to the Symposium on Theoretical and Spectroscopic Aspects of the Chemistry of Co-ordination Compounds, held at Venice during September 1966, have now been published,[58] and several excellent reviews are included.

Although the ligand-field theory for systems of cubic symmetry is now fairly well understood, few attempts have been made to extend this theory to include symmetries lower than cubic. However, a detailed account has now appeared for d^3 and d^7 electronic configurations in quadrate, trigonal, and cylindrical fields.[59] The symmetry-adapted strong-field wavefunctions were derived and the energy matrices constructed as functions of Dq and $D\mu$, $D\nu$, the cubic and axial ligand-field parameters, and A, B, and C, the Racah electron-correlation parameters. The spin–orbit coupling perturbation was not included in these calculations. The quadrate energy-matrices for the d^3 configuration were solved [60] and the appropriate energy-diagrams were constructed to interpret the spectra of various mono- and *trans*-di-substituted complexes of chromium(III). It is concluded that at the present time there is no unique way of assigning all the observed bands in the systems considered. In all the complexes discussed, the first spin-allowed cubic band splits into two components which are given the definitive quadrate assignments, and the corresponding Dt values are evaluated. As the second spin-allowed cubic band does not split in most of these systems, alternative quadrate assignments were found possible for different choices of values of B and κ. Independently Macfarlane [61] has investigated d^3 systems, and considered the possibility of obtaining good approximate analytical expressions for the zero-field splittings of cubic terms belonging to the t_2^3 configuration. Spin–orbit, and trigonal- and tetragonal-field perturbations were considered, and a perturbation expansion based on a strong cubic-field zeroth-order Hamiltonian was used. Only in the case of $^4A_2$ and a^2E terms does the

[56] F. W. Hiller and J. H. Krueger, *Inorg. Chem.*, 1967, **6**, 528.
[57] J. D. McCullough and A. Brunner, *Inorg. Chem.*, 1967, **6**, 1251.
[58] *Co-ordination Chem. Rev.*, 1967, **2**, No. 1.
[59] J. R. Perumareddi, *J. Phys. Chem.*, 1967, **71**, 3144.
[60] J. R. Perumareddi, *J. Phys. Chem.*, 1967, **71**, 3155.
[61] R. M. Macfarlane, *J. Chem. Phys.*, 1967, **47**, 2066.

expansion converge sufficiently rapidly for a single-order perturbation theory to provide a useful approximation for octahedral Cr^{III} or tetrahedral Co^{II} complexes.

In an elegant piece of work, Fenske[62] has considered the relative intensities of Laporte forbidden transitions in octahedral complexes, in which the charge-transfer transitions are from the ligand to the metal. Within the framework of a one-electron approximation, it was shown from theoretical considerations that the ligand-field transitions in such complexes vibronically mix with those ligand-to-metal transitions in which the final states involve the e_g and not the t_{2g} orbital of the metal. The relative intensities of the ligand-field transitions were correlated with the transition, ligand-to-e_g and correlations of intensities between corresponding transitions for analogous chloro- and bromo-complexes were also obtained. In another, essentially theoretical, paper, the use of Green functions in the description of the absorption spectrum of an orbitally-degenerate paramagnetic ion coupled to lattice phonons has been studied.[63]

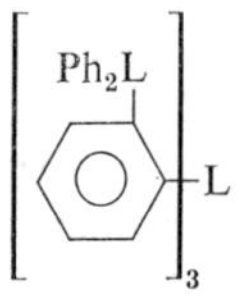

(QP L = P;
QAs L = As)

(2)

The ligand tris-(*o*-diphenylphosphinophenyl)phosphine [QP, (2)] and the corresponding arsenic compound, [QAs, (2)], are known to give five-co-ordinate trigonal bipyramidal complexes of the type $[MX(QL)]^{(n-1)+}$ with a number of transition metal ions, *e.g.* Fe^{II}, Co^{II}, Rh^{I}, Ir^{I}, Ni^{II}, Pd^{II}, and Pt^{II}. The electronic spectra of these complexes are of particular interest because they differ considerably from those of complexes of these metal ions in their more usual co-ordination numbers. The spectra of the iron(II), cobalt(II), and nickel(II) complexes of the type $[MCl(QP)]^+$ have now been interpreted using a ligand-field model.[2] Two-parameter ligand-field calculations for these d^6-, d^7-, and d^8- metal complexes were carried out using tensor operator theory, and the energy-level diagrams obtained were used for the assignment of the spectra. These complexes display intense absorption bands in the region 5,000–20,000 cm.$^{-1}$ ($\varepsilon_{max.}$ 500–5,000), but despite their high intensities, these bands appear to be due to predominantly ligand-field transitions. Several reasons are listed in support of this conclusion, and the high intensity is probably derived from the large admixture of higher configurations (both excited central-ion and charge-transfer states) by odd crystal-field components.

[62] R. F. Fenske, *J. Amer. Chem. Soc.*, 1967, **89**, 252.
[63] H. A. M. Van Eekelen and K. W. H. Stevens, *Proc. Phys. Soc.*, 1967, **90**, 199.

A very significant paper has appeared in which the utility of magnetic optical activity in the study of the *d–d* transitions of transition-metal complexes has been discussed.[1] The *d–d* transitions in the complexes $[Mn(H_2O)_6]^{2+}$, $[Co(H_2O)_6]^{2+}$, and $[Ni(H_2O)_6]^{2+}$ show large magnetic circular dichroism effects, whilst the *d–d* transitions in the complexes $[Co(NH_3)_6]^{3+}$ and $[Co(CN)_6]^{3-}$, and the spin-allowed transitions of $[Cr(NH_3)_6]^{3+}$, $[Cr(H_2O)_6]^{3+}$, and $[Cr(CN)_6]^{3-}$ exhibit only very small effects. The latter result is tentatively attributed to quenching of orbital angular momentum in the degenerate excited-states by vibronic interactions. A similar diminution was noted for $[Co(NH_3)_5Cl]^{2+}$, *cis*-$[Co(NH_3)_4(H_2O)_2]^{3+}$, and *cis*- and *trans*-$[Co\,en_2Cl_2]^{+}$, $[Co\,en_3]^{3+}$, $[Co\,ox_3]^{3-}$, and the spin-allowed transitions of $[Cr\,en_3]^{3+}$ and $[Cr\,ox_3]^{3-}$, with the exception of the $^1E \leftarrow {}^1A_1$ transition at 24,000 cm.$^{-1}$ in $[Co\,ox_3]^{3-}$ (en = ethylenediamine, ox = oxalato). Here a term characteristic of excited-state Zeeman splitting is found and the ^1E magnetic moment obtained is in good agreement with that predicted by ligand-field theory. Perhaps the most dramatic result is the huge effect found for spin-forbidden Cr^{III} transitions, which in some cases is greater than for the much more intense spin-allowed transitions. This observation greatly facilitates the assignment of these transitions.

Some elegant work on charge-transfer transitions has been reported by Day and his co-workers. Although the charge-transfer spectra of many metal halide complexes have been reported, the region above 45,000 cm.$^{-1}$ remains more or less unexplored. A method for measuring these spectra up to at least 60,000 cm.$^{-1}$ has now been devised, and the spectra of the anions MX_4^{2-} (M = Mn^{II}, Co^{II}, Ni^{III}, Cu^{II}, or Zn^{II}; X = Cl, Br, or I) are reported. The bands in the region 45,000–50,000 cm.$^{-1}$ are assigned as predominantly internal halide-transitions.[64] A study of the charge-transfer bands of iron (II), iron(III) and copper(I) phenanthroline complexes has also been reported.[65] The shifts produced by successive methyl substitution in the different positions of the phenanthroline molecule were shown to be additive, and the energy changes associated with the methyl groups in the different positions were extracted statistically. These values were interpreted by Hückel molecular orbital theory; it proved necessary to allow for an inductive effect of the ligand upon the energies of the orbitals of the metal, but the mixing between the donor- and acceptor-orbitals could be neglected. This treatment provides an assignment of the molecular orbital involved in charge transfer. The intensities of these charge-transfer bands were also considered, and it was suggested that the main source of intensity is the transition moment resulting from the transfer of charge itself, though borrowing from internal Fe^{II} and Cu^{I} (but not Fe^{III}) transitions, and internal ligand-transitions may be appreciable.[66]

64 B. D. Bird and P. Day, *Chem. Comm.*, 1967, 741.
65 P. Day and N. Sanders, *J. Chem. Soc.* (*A*), 1967, 1530.
66 P. Day and N. Sanders, *J. Chem. Soc.* (*A*), 1967, 1536.

Spectral and magnetic data have been presented for a new family of transition-metal chelates involving poly[(1-pyrazolyl)borate] ligands; neutral complexes with the ions Mn^{II}, Fe^{II}, Co^{II}, Ni^{II}, Cu^{II}, and Zn^{II} were studied. The optical data were analysed in terms of crystal-field theory, and the position of these ligands in the spectrochemical and nephelauxetic series was considered.[67]

Scandium.—It is now fairly clear that the electronic ground-state of ScF is $^1\Sigma^+$, whilst that for the isoelectronic TiO molecule is $^3\Delta_r$. This interesting inversion has been examined[68] by carrying out molecular orbital calculations based on the matrix Hartree–Fock expansion method of Roothaan.[69] It is found that $^3\Delta_r$ of configuration $\delta\sigma$ is the symmetry-restricted Hartree–Fock ground-state of both ScF and TiO, whilst a semi-empirical estimate of the correlation energy contribution shows that correlation depresses the $^1\Sigma^+$ state relatively more than $^3\Delta_r$, in both cases, and suggests that $^1\Sigma^+$ is the ground-state of ScF, in agreement with experiment.

A spectroscopic method for the determination of Sc^{III} based on a sensitive colour reaction with chromazurol-S has been described.[70]

Titanium, Zirconium, and Hafnium.—*Titanium*(IV), d^0. The difficulty of theoretical interpretation continues to restrict the number of publications dealing specifically with the spectra of d^0 systems. The electronic spectra of hexalogenotitanates(IV), ($[TiCl_6]^{2-}$ and $[TiBr_6]^{2-}$), and zirconates(IV), ($[ZrCl_6]^{2-}$ and $[ZrBr_6]^{2-}$), however, have been reported.[71] The diffuse reflectance and solution spectra in acetonitrile were recorded and in most cases were very similar; the solid-state spectra showed cation dependence. Bands associated with halogen $\pi \rightarrow$ M (t_{2g} and e_g) transitions are assigned and discussed in relation to other work. The diffuse reflectance spectra of several pseudo-octahedral complexes of the type *cis*-MX_4L_2 (X = Cl or Br) have also been recorded and the bands attributed to halogen $\pi \rightarrow$ M transitions.

Spectrophotometric measurements have shown[72] that the polymeric forms of Ti^{IV} in HCl solutions are not the same as in HCl–LiCl solutions. Evidence for the existence of tetrachloroperoxotitanate(IV) species in aqueous HCl and $HClO_4$ and also in anhydrous formic acid solutions has been presented.[73] A characteristic absorption band is obtained at 23,800 cm.$^{-1}$ The electronic spectrum of the organometallic compound, $Me_5C_5TiCl_3$, has also been recorded.[74]

[67] J. P. Jesson, S. Trofimenko, and D. R. Eaton, *J. Amer. Chem. Soc.*, 1967, **89**, 3148.
[68] K. D. Carlson and C. Moser, *J. Chem. Phys.*, 1967, **46**, 35.
[69] C. C. J. Roothaan, *Rev. Mol. Phys.*, 1951, **23**, 69.
[70] R. Ishida and N. Hasegawa, *Bull. Chem. Soc. Japan*, 1967, **40**, 1153.
[71] B. J. Brisdon, T. E. Lester, and R. A. Walton, *Spectrochim. Acta*, 1967, **23**, *A*, 1969.
[72] B. I. Nabivanets and L. N. Kudritskaya, *Russ. J. Inorg. Chem.*, 1967, **12**, 616.
[73] E. Wendling and J. de Lavillandre, *Bull. Soc. chim. France*, 1967, 2142.
[74] R. B. King and M. B. Bisnette, *J. Organometallic Chem.*, 1967, **8**, 287.

Titanium(III), d^1. The chemistry of titanium(III) continues to attract considerable interest, and the electronic spectra of a number of relatively simple systems have been studied. The polarised absorption spectra of octahedral α-$TiCl_3$ and α-$TiBr_3$ have been reported;[75] the $^2E_g \leftarrow {}^2T_{2g}$ transition is assigned at 14,300 and 12,500 cm.$^{-1}$, respectively. At higher wavenumbers, stronger absorption bands are observed which are undoubtedly due to charge-transfer transitions. In an elegant piece of work by Schläfer and Fritz,[76] the structures of the crystalline hexahydrates $TiCl_3,6H_2O$ and $TiBr_3,6H_2O$ were elucidated. Two ligand-field bands are observed in the reflectance spectra and are taken to indicate a symmetry lower than O_h. Taking the mean of the two observed bands, and using the rule of average environment, it was concluded that the chromophore is probably $[Ti(H_2O)_4X_2]^+$, although the species $[Ti(H_2O)_3X_3]$ could not be completely excluded. Far-i.r. measurements resolved this problem in favour of a *trans*-$[Ti(H_2O)_4X_2]^+$ structure. The observed electronic transitions may thus be assigned as $^2B_{1g} \leftarrow {}^2B_{2g}$ and $^2A_{1g} \leftarrow {}^2B_{2g}$ (D_{4h}).

Chloro-aquo-complexes of Ti^{III} in aqueous solution have also been studied.[77] The band assumed to be due to the $^2E_g \leftarrow {}^2T_{2g}$ transition shifts from about 20,100 cm.$^{-1}$ to about 14,500 cm.$^{-1}$ as the chloride ion concentration is increased. By applying the rule of average environment a 'formation curve' can be drawn for the number of chloride ions complexed to the central Ti^{III} ion and hence the equilibrium constants of the successive chloro-aquo-complexes can be determined.

Fowles and Russ[78] have isolated pyridinium salts of $[TiCl_6]^{3-}$ and $[TiBr_6]^{3-}$ in order to obtain accurate *Dq* values for both halide ions in these discrete entities. The observed band positions are recorded in the Table. Attempts to prepare the tetra-ethylammonium salts of these anions in methyl cyanide gave $[Et_4N]^+[TiX_4,2MeCN]^-$ and thermal decomposition of these gave $[Et_4N]^+[TiX_4]^-$, which appear to contain polymeric anions. The spectra of these salts are also reported and discussed. A theoretical treatment of the *d–d* spectrum of the $[TiF_6]^{3-}$ ion has appeared.[79] Assuming O_h symmetry, and using SCF radial wavefunctions for titanium and the crystal-field approximation, the energy of the $^2E_g \leftarrow {}^2T_{2g}$ transition was calculated to within 1% of the experimental value.

Evidence[80] for the existence of at least a small number of conduction electrons in the black oxide $ScTiO_{3\cdot003}$ is provided by its reflectance spectrum. A continuous absorption is observed from about 28,000–3,800 cm.$^{-1}$; it is probably superimposed on a Ti^{III} ligand-field band at 20,000 cm.$^{-1}$. The electronic spectra of the hexa-aquo-ions of Ti^{III} in host

[75] C. Dijkgraaf and J. P. G. Rousseau, *Spectrochim. Acta*, 1967, **23**, *A*, 1267.
[76] H. L. Schläfer and H. P. Fritz, *Spectrochim. Acta*, 1967, **23**, *A*, 1409.
[77] H. J. Gardner, *Austral. J. Chem.*, 1967, **20**, 2357.
[78] G. W. A. Fowles and N. J. Russ, *J. Chem. Soc.* (*A*), 1967, 517.
[79] E. Wendling and J. de Lavillandre, *Bull. Soc. chim. France*, 1967, 2738.
[80] A. F. Reid and M. J. Sienko, *Inorg. Chem.*, 1967, **6**, 521.

The absorption spectra of some Ti^{III} *complexes*

Compound	*Maxima* (cm.$^{-1}$)	*Assignment*	*Ref.*
α-$TiCl_3$ (s)	14,300	$^2E_g \leftarrow {}^2T_{2g}$	75
	18,900		
α-$TiBr_3$ (s)	12,500	$^2E_g \leftarrow {}^2T_{2g}$	75
	15,500		
	23,300		
$TiCl_3,6H_2O$ (s)	15,000	$^2B_{1g} \leftarrow {}^2B_{2g}$	76
	18,300	$^2A_{1g} \leftarrow {}^2B_{2g}$	
$TiBr_3,6H_2O$ (s)	13,900	$^2B_{1g} \leftarrow {}^2B_{2g}$	
	18,500	$^2A_{1g} \leftarrow {}^2B_{2g}$	
$(C_5H_6N)_3TiCl_6$ (s)	11,000 sh; 12,800	$^2E_g \leftarrow {}^2T_{2g}$	78
$(C_5H_6N)_3TiBr_6$ (s)	11,750	$^2E_g \leftarrow {}^2T_{2g}$	
$(C_5H_6N)_3TiCl_4Br_2$	10,500; 12,350	$^2E_g \leftarrow {}^2T_{2g}$	
[Ti $en_3]^{3+}$	~13,500	$^2E \leftarrow {}^2A_1$	82
	~37,500	$^2E \leftarrow {}^2A_1$	
[Ti $pn_3]^{3+}$	~14,000	$^2E \leftarrow {}^2A_1$	
	~37,500	$^2E \leftarrow {}^2A_1$	

s = Solid phase; pn = 1,2-diaminopropane

crystals of $C(NH_2)_3Al(SO_4)_2,6H_2O$ and of $(Tl,Al)Cl_3,6H_2O$ have been presented [81] and discussed.

The tris-bidentate cations [Ti $en_3]^{3+}$ and [Ti $pn_3]^{3+}$ have been characterised.[82] The diffuse reflectance spectra consist of a broad asymmetric band at about 13,000–15,000 cm.$^{-1}$ and a second band at about 37,500 cm.$^{-1}$. These bands are thought to arise from ligand-field transitions and are tentatively assigned to the two $^2E \leftarrow {}^2A_1$ transitions expected for a d^1 system in a trigonal field (D_3). Strong trigonal fields in Ti^{III} complexes have been noted previously.[83] Evidently the ligand-field strength of propylenediamine is virtually identical with that of ethylenediamine.

The u.v. and visible spectra of the complexes $TiBr_3$,3THF (THF = tetrahydrofuran) and $TiBr_3$,1·5(1,2-dimethoxyethane) have been reported and discussed.[84] The first allyl complexes of Ti^{III} have been prepared and the electronic spectra of the complexes, $(\pi\text{-}C_5H_5)_2TiR$ (R = allyl, methylallyl, or dimethylallyl), in cyclohexane have been recorded.[85] The titanium(III)–chromotropic acid system has been studied spectrophotometrically.[86]

Titanium(II), d^2. Basch and Gray [87] have developed semi-empirical approximations for the SCF-Hamiltonian matrix-elements in the LCAO–MO scheme. A limited application was made to TiO (and also CuO), where the empirical parameters were derived by fitting the

[81] I. M. Walker and R. L. Carlin, *J. Chem. Phys.*, 1967, **46**, 3931.
[82] R. J. H. Clark and M. L. Greenfield, *J. Chem. Soc.* (*A*), 1967, 409.
[83] B. R. McGarvey, *J. Chem. Phys.*, 1963, **38**, 388.
[84] G. W. A. Fowles, P. T. Greene, and T. E. Lester, *J. Inorg. Nuclear Chem.*, 1967, **29**, 2365.
[85] H. A. Martin and F. Jellinek, *J. Organometallic Chem.*, 1967, **8**, 115.
[86] M. M. Khan and N. Ahmad, *J. Inorg. Nuclear Chem.*, 1967, **6**, 2961.
[87] H. Basch and H. B. Gray, *Inorg. Chem.*, 1967, **6**, 639.

observed ground-state configuration for TiO. Other computed properties are then found to be in reasonable agreement with experiment. Titanium(II) complexes of the types $TiCl_2$,2L (L = C_4H_8O, $C_5H_{10}O$, py (pyridine), MeCN,½bipy (2,2′-bipyridyl), and ½phen (1,10-phenanthroline)) and $TiBr_2$,bipy have been prepared [88] (previously only $TiCl_2$,$2C_3H_7NO$ had been reported). The complexes are very dark brown in colour, apart from the bipyridyl and phenanthroline complexes which are dark blue. As expected they are readily oxidised by air. The visible spectra, measured on the solids and on solutions in the parent ligands, are described as 'disappointing' because of rising absorption above 10,000 cm.$^{-1}$, and only weak shoulders were observed.

Zirconium and Hafnium. The reflection spectra of the trihalides $ZrCl_3$, $ZrBr_3$, and $HfBr_3$ have been recorded,[89] and the spectra of $ZrCl_3$ and $ZrBr_3$ are compared with spectra presented in earlier work; the results do not agree. A preliminary report [90] on the first adducts of the zirconium(III) halides has only just appeared, in spite of the fact that the parent halides have been known for some time. Several complexes with nitrogen donor-ligands are reported. The electronic spectra of these complexes show at least one weak *d–d* transition in the 12,000–14,000 cm.$^{-1}$ region, together with much more intense charge-transfer bands between 20,000 and 25,000 cm.$^{-1}$.

Vanadium, Niobium, and Tantalum.—*Vanadium*(V), d^0. There is a paucity of spectral data on vanadium(V) complexes. Reports have mentioned data for a number of complexes formed by substituted aromatic hydroxamic acids [91] and the electronic spectrum of VS_4^{3-} has been discussed.[92]

Vanadium(IV), d^1. Although a great deal of optical and e.s.r. data are available for several oxovanadium(IV) complexes, there is as yet no totally satisfactory assignment of the observed bands for this unique d^1 system. In these complexes the vanadium atom never lies in the same plane as the four equatorial ligands, although all theoretical models, including the widely used Ballhausen and Gray scheme [93] have assumed an O=V—X angle of 90°. Selbin *et al.*[94] have investigated this problem by studying the spectra of six β-ketoenolate complexes in considerable detail. The spectra were obtained for several different phases, including the vapour phase for two of the compounds, and in some cases at two different temperatures. The data are analysed and a new totally empirical ordering of the vanadium *d* orbitals is presented (see Figure 2).

88 G. W. A. Fowles and T. E. Lester, *Chem. Comm.*, 1967, 47.
89 H. L. Schläfer and H. W. Wille, *Z. anorg. Chem.*, 1967, **351**, 279.
90 G. W. A. Fowles, B. J. Russ, and G. R. Willey, *Chem. Comm.*, 1967, 646.
91 R. L. Dutta and S. Ghosh, *J. Indian Chem. Soc.*, 1967, **44**, 369.
92 A. Müller, B. Krebs, W. Rittner, and M. Stockburger, *Ber. Bunsengesellschaft Phys. Chem.*, 1967, **71**, 182.
93 C. J. Ballhausen and H. B. Gray, *Inorg. Chem.*, 1962, **1**, 111.
94 J. Selbin, G. Maus, and D. L. Johnson, *J. Inorg. Nuclear Chem.*, 1967, **29**, 1735.

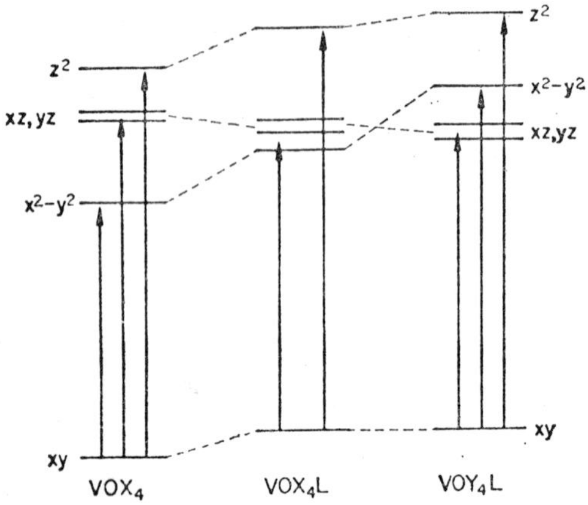

Figure 2 *Relative ordering of the* d *orbitals of vanadium for five-co-ordinate case (left) and six-co-ordinate with* Y *being a stronger ligand than* X *(centre and right)* (Reproduced by permission from *J. Inorg. Nuclear Chem.*, 1967, **29**, 1735)

A number of oxodichlorovanadium(IV) complexes of the type $VOCl_2,2L$ (L = Me_3N, Me_2NH, Me_2S, and Et_2S) have been reported.[95] These complexes are considered to be five-co-ordinate and their electronic spectra have been interpreted on the basis of the Ballhausen and Gray scheme.[93] Several other publications have included spectral data on oxovanadium(IV) complexes. These include reports on complexes with substituted pyridine-1-oxides,[96] aliphatic straight-chain 2-hydroxy-carboxylic acids,[97] tartrates,[98] citrates,[99] dithiocarbamates,[100] pyridine carboxylic acids,[101] a number of heterochelates,[102, 103] and a kojate.[104] The formation of a 1 : 1 complex between vanadium(IV) and chrome-azurol-S has been established using Job's method of continuous variations.[105]

Vanadium(III), d^2. For vanadium(III) complexes with O_h symmetry, three spin-allowed *d–d* transitions are expected as illustrated in Figure 3.

The magnetism and electronic spectra of several octahedral vanadium(III) complexes have been reported [106] and estimates of Dq and the Racah parameter, B, were obtained. The two $^3T_{1g}$ terms arising from the free-ion

[95] K. L. Baker, D. A. Edwards, G. W. A. Fowles, and R. G. Williams, *J. Inorg. Nuclear Chem.*, 1967, **29**, 1881.
[96] R. G. Garvey and R. O. Ragsdale, *J. Inorg. Nuclear Chem.*, 1967, **29**, 745.
[97] D. M. H. Bovey and E. R. Clark, *J. Inorg. Nuclear Chem.*, 1967, **29**, 755.
[98] R. E. Tapscott and R. L. Belford, *Inorg. Chem.*, 1967, **6**, 735.
[99] B. M. Nikolova and G. St. Nikolov, *J. Inorg. Nuclear Chem.*, 1967, **29**, 1013.
[100] B. J. McCormick, *Inorg. Nuclear Chem. Letters*, 1967, **3**, 293.
[101] R. L. Dutta and S. Ghosh, *J. Indian Chem. Soc.*, 1967, **44**, 273.
[102] R. L. Dutta and S. Ghosh, *J. Indian Chem. Soc.*, 1967, **44**, 296.
[103] R. L. Dutta and S. Ghosh, *J. Indian Chem. Soc.*, 1967, **44**, 306.
[104] R. L. Dutta and A. Syamal, *J. Indian Chem. Soc.*, 1967, **44**, 381.
[105] P. Sanyal, S. P. Sangal, and S. P. Mushran, *Bull. Chem. Soc. Japan*, 1967, **40**, 217.
[106] D. J. Machin and K. S. Murray, *J. Chem. Soc.* (*A*), 1967, 1498.

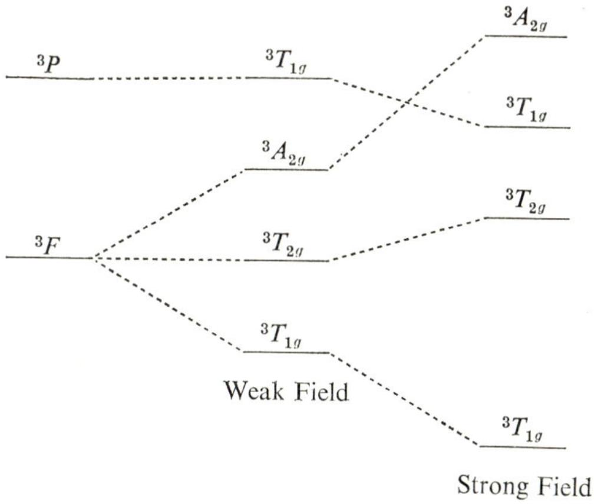

Figure 3 *Triplet terms of* $3d^2$ *in* O_h *field*

^3P and ^3F terms interact to give a ground-level wavefunction:

$$\psi = (1+c^2)^{-\frac{1}{2}}(\psi_{^3T_{1g}(F)} + c\psi_{^3T_{1g}(P)})$$

and a parameter A, related to the mixing coefficient, is defined as

$$A = \frac{1{\cdot}5-c^2}{1+c^2}$$

This parameter, A, was evaluated from the spectral measurements and found to agree well with the values derived from the magnetic data.

The electronic spectra of several six-co-ordinate vanadium trichloride and tribromide complexes with pyridine, 2,2′-bipyridyl, and 1,10-phenanthroline have been described.[107] One particularly interesting feature is that whilst the complex VBr_3,3py has the same spectrum in the solid state as in solution (either pyridine or benzene), the spectrum of VCl_3,3py changes significantly when dissolved in pyridine. It is suggested that the chloro-complex changes from a *trans* configuration in the solid state to a *cis* configuration in solution.

The positions of the band maxima and their extinction coefficients have been tabulated for the vanadium(III) thiocyanate complexes, $[V(NCS)_3(MeCN)_3]$,2MeCN, $K_3[V(NCS)_6]$,2MeCN, $[V(NCS)_3(quinoline)_2MeCN]$, and $[V(NCS)_3\,py_2MeCN]$ and assignments have been given.[108] A number of octahedral bromo-complexes of vanadium(III) have been detected spectrophotometrically in acetonitrile, trimethylphosphate, and propanediol-1,2-carbonate.[109]

[107] G. W. A. Fowles and P. T. Greene, *J. Chem. Soc.* (*A*), 1967, 1869.
[108] H. Böhland and P. Malitzke, *Z. anorg. Chem.*, 1967, **350**, 70.
[109] V. Gutmann and K. Fenkart, *Monatsh.*, 1967, **98**, 286.

Some interesting five-co-ordinate complexes VX_3,2L (X = Cl and Br; L = Me_3N, Me_2S, and C_4H_8S) and VCl_3,$2Et_2S$ have been prepared[110] and shown to be monomeric in non-donor solvents. Structural measurements indicate that the metal atoms have *trans*-trigonal-bipyramidal configurations and Jørgensen has predicted that in such an environment vanadium(III) should exhibit two electronic transitions in the near i.r. region. Two peaks were observed at around 5000 and 7000 cm.$^{-1}$ for solutions of the complexes in benzene or iso-octane, but these bands were absent for solutions in the free ligand when the spectra become typical of six-co-ordinate vanadium(III). On the basis of calculations by other workers, the peaks have been assigned for the VCl_3,$2Me_3N$ complex.

Vanadium-(I) *and* -(0). The electro-oxidation of π-cycloheptatrienyl-π-cyclopentadienylvanadium(0) to the corresponding vanadium(I) complex in acetonitrile solution has been studied and the electronic spectra of both complexes are reported.[111]

Niobium and Tantalum. Detailed spectral data for some new hexachloro-niobate(V), hexabromoniobate(V), oxopentachloro- and oxopentabromo-niobate(V) complexes in acetonitrile have been reported.[112] The observed bands are due to $\pi \rightarrow t_{2g}$ transitions, and the splitting of these bands is briefly discussed.

The study of metal-cluster compounds of the type $[Nb_6Cl_{12}]^{n+}$ has proved popular. The visible and u.v. spectra of aqueous solutions of $[Nb_6Cl_{12}]Cl_2$ and $[Nb_6Br_{12}]Br_2$ have been reported.[113] The results agree fairly well with earlier work, except that the extinction coefficients for the chloride are higher by 15–50%; the extinction coefficients for the bromide are reported for the first time. Incomplete spectral data for complexes containing $[Nb_6Cl_{12}]^{3+}$ and $[Nb_6Cl_{12}]^{4+}$ have also been presented.[114] An interesting trend is noted in that the two principal bands in the visible and u.v. region are shifted to lower energies by about 1500 cm.$^{-1}$ on each increase in the oxidation state of the cluster. The intensities of these bands scarcely change. A detailed study of these spectra is being undertaken. The electronic spectra of the compounds $(Ph_4N)_n(Nb_6Cl_{12})Cl_6$ (n = 2, 3, or 4) in ethanol and nitromethane are recorded.[115] The qualitative similarities in the spectra of these three compounds in a given solvent are noted, as are the differences between the two solvents.

Chromium, Molybdenum, and Tungsten.—*Chromium*(VI), d^0. The electronic spectra of chromium(VI) compounds have not been widely studied. U.v. and visible spectra are reported for hydrated and anhydrous chromates of

[110] M. W. Duckworth, G. W. A. Fowles, and P. T. Greene, *J. Chem. Soc.* (*A*), 1967, 1592.
[111] W. M. Gulick, jun., and D. H. Geske, *Inorg. Chem.*, 1967, **6**, 1320.
[112] C. Furlani and E. Zinato, *Z. anorg. Chem.*, 1967, **351**, 210.
[113] P. B. Fleming, L. A. Mueller, and R. E. McCarley, *Inorg. Chem.*, 1967, **6**, 1.
[114] P. B. Fleming, T. A. Dougherty, and R. E. McCarley, *J. Amer. Chem. Soc.*, 1967, **89**, 159.
[115] R. A. MacKay and R. F. Schneider, *Inorg. Chem.*, 1967, **6**, 549.

Mg, La, Nd, and Sm,[116] and the spectra in fused KSCN of potassium chromate, dichromate and trichromate and chromium(VI) oxide have also been recorded at 200°.[117]

Chromium(IV), d^2. The three new diperoxochromium(IV) complexes $[Cr\ en\ (NH_3)(O_2)_2],H_2O$, $[Cr\ pn\ (H_2O)(O_2)_2],2H_2O$, and $[Cr(isobutylenediamine)(H_2O)(O_2)_2],H_2O$ have been characterised, and their u.v. and visible spectra tabulated.[118, 119]

Chromium(III), d^3. A great deal of work on this oxidation state continues to be published. An empirical rule which predicts the type of luminescence to be expected in octahedral Cr^{III} complexes has been proposed:[120] complexes with ligands like Br^-, Cl^-, and F^-, which have small values of the ligand-field parameter Dq, exhibit fluorescence only, compounds where the co-ordinating atom is oxygen show both fluorescence and phosphorescence, or phosphorescence only, and complexes where nitrogen or carbon is co-ordinated around chromium, where greater values of Dq are obtained, show only phosphorescence. In agreement with this rule, the mixed complexes $[CrF_n(H_2O)_{6-n}]^{(3-n)+}$ (n = 1, 3, 5, or 6) fluorescence, whereas $[Cr(H_2O)_6]F_3$ and $NH_4[CrF_4\ en]$ show both types of emission, fluorescence and phosphorescence.[121] The ligand-field parameters, Dq, B_{55}, and B_{35}, have been obtained for the above compounds from reflectance spectra data.

The visible and u.v. spectra for the series of complexes $[Cr(CN)_n(H_2O)_{6-n}]^{(3-n)+}$ ($n = 1-5$) have been reported and the first systematic attempt made to interpret the absorption spectra of Cr^{III} complexes in non-cubic fields on the basis of complete ligand-field calculations in the limit of zero spin–orbit coupling.[122] The complex ions $[Cr(CN)(H_2O)_5]^{2+}$ and $[Cr(CN)_5(H_2O)]^{2-}$ are treated by the theory of tetragonal fields. Further assumptions based on the crystal-field model have been made to include the two *cis*-disubstituted complex ions under tetragonal fields. The complex *cis*-$[Cr(CN)_3(H_2O)_3]$ is treated by trigonal-field theory. The spectra of all the tetragonal systems were consistently explained with a Dt value of -300 cm.$^{-1}$ for substitution of a cyanide ion by a water molecule or $+300$ cm.$^{-1}$ for the reverse substitution. A crystal-field treatment leads to some interesting predictions which provide spectral criteria for distinguishing between the *cis* and *trans* isomers of d^3 disubstituted complexes. The spectra of the known *cis–trans* pairs of Cr^{III} complexes conform to these predictions.

116 R. G. Darrie, W. P. Doyle, and I. Kirkpatrick, *J. Inorg. Nuclear Chem.*, 1967, **29**, 979.
117 D. H. Kerridge and M. Mosley, *J. Chem. Soc.* (*A*), 1967, 1874.
118 E. A. V. Ebsworth, C. S. Garner, D. A. House, and R. G. Hughes, *Inorg. Nuclear Chem. Letters*, 1967, **3**, 61.
119 D. A. House, R. G. Hughes, and C. S. Garner, *Inorg. Chem.*, 1967, **6**, 1077.
120 H. L. Schläfer, H. Gausmann, and H. Witzke, *J. Chem. Phys.*, 1967, **46**, 1423.
121 H. L. Schläfer, H. Gausmann, and H. U. Zander, *Inorg. Chem.*, 1967, **6**, 1528.
122 R. Krishnamurthy, W. B. Schaap, and J. R. Perumareddi, *Inorg. Chem.*, 1967, **6**, 1338.

The multidentate ligands (o-$Ph_2P\cdot C_6H_4)_3P$ (QP) and (o-$Ph_2P\cdot C_6H_4)_2$-PPh (TP) form chromium(III) complexes of the type [CrX_3L] (X = Cl, Br, or NCS). The visible and u.v. spectra of the complexes CrX_3(QP) and CrX_3(TP) are practically identical,[123] and the former complexes are assigned an octahedral co-ordination. The spectra are typical of octahedral chromium(III) species; the high intensities of the bands assigned as ${}^4T_{2g} \leftarrow {}^4A_{2g}$ ($\varepsilon \sim 1000$) and ${}^4T_{1g} \leftarrow {}^4A_{2g}$ ($\varepsilon \sim 500$) are considered to be due to the mixing in of higher configurations (both excited central-atom and charge-transfer states) by the odd crystal-field components.[2]

The polarised crystal spectra of the hexa-aquochromium(III) ion in the host crystals of $C(NH_2)_3Al(SO_4)_2,6H_2O$ and several isomorphs and also in $AlCl_3,6H_2O$ have been discussed.[124] Trigonal-field parameters were evaluated from an analysis of the broad-band spectra and confirmed by a study of the fine-line spectra.

The ${}^2E \rightarrow {}^4A_2$ emission spectra of hexagonal $NaMgAl(C_2O_4)_3,9H_2O$ crystals doped with Cr^{3+} have been recorded.[125] The vibronic structure is very weak in emission and can be assigned to the ground-state intramolecular vibrations. A monoclinic modification exists in which the trigonal field splitting of ${}^2E$ is absent.

Where both nitro- ($M-NO_2$) and nitrito- ($M-ONO$) linkage isomers are known they can be readily distinguished on the basis of their i.r. bands and the energies of the *d–d* bands. Where only one linkage isomer is known electronic spectral measurements are usually inconclusive. Two new complexes *cis*-[Cr $en_2(ONO)_2]ClO_4$ and *trans*-[Cr $en_2(ONO)_2$]X (X = ClO_4 or NO_2) have been reported for which neither of the above criteria was conclusive, but an additional criterion has been developed which is applicable to these complexes.[126] In an appropriate dipolar aprotic solvent, nitrito-complexes give sharply resolved absorption bands in the region 25,000–33,000 cm.$^{-1}$, whereas the nitro-isomers do not.

The low-temperature spectrum of sodium thiochromite, $NaCrS_2$, has been studied to obtain the first reliable data for a transition-metal ion in a trigonally distorted sulphide-ligand field.[127] Assuming that the peaks at 13,350 and 14,600 cm.$^{-1}$ are the trigonal components of the ${}^4T_{2g}$ term, then a value of 2500 cm.$^{-1}$ is obtained for the first-order trigonal-field parameter. Dq is given as 1400 cm.$^{-1}$, and the nephelauxetic ratio, β, is estimated as 0·480; the latter value is in good agreement with that (0·484) reported earlier.[128]

Purple and green geometrical isomers of [Cr dienCl_3] have been isolated.[129] On the basis of a comparison of the i.r. and visible reflectance spectra with

[123] I. V. Howell, L. M. Venanzi, and D. C. Goodall, *J. Chem. Soc.* (*A*), 1967, 395.
[124] R. L. Carlin and I. M. Walker, *J. Chem. Phys.*, 1967, **46**, 3921.
[125] R. A. Condrate and L. S. Forster, *J. Mol. Spectroscopy*, 1967, **24**, 490.
[126] W. W. Fee, C. S. Garner, and J. N. MacB. Harrowfield, *Inorg. Chem.*, 1967, **6**, 87.
[127] S. L. Holt and A. Wold, *Inorg. Chem.*, 1967, **6**, 1594.
[128] A. L. Companion and M. Mackin, *J. Chem. Phys.*, 1965, **42**, 4219.
[129] D. A. House, *Inorg. Nuclear Chem. Letters*, 1967, **3**, 67.

the spectra of the known 1,2,6-[Co dienCl_3] complex, the green and purple isomers were assigned as *cis*- and *trans*-respectively. The complexes *cis*-[Cr pn_2X_2]X and *cis*-[Cr tm_2X_2]X (tm = 1,3-diaminopropane) have been made by the thermal matrix reaction.[130] The reflection spectra of the *cis*-diamine complexes contain two peaks in the visible region, whilst the tris-(1,3-diaminopropane) complex only displays a single peak. This observation made it relatively easy to identify reaction products.

The visible spectra of several new acido-tetraethylenepentamine complexes of chromium(III) have been recorded[131] and compared with the spectra of the related $[Cr(NH_3)_5X]^{n+}$ ions. Absorption maxima in 0·1N-perchloric acid solutions of the species $[Cr(MeNH_2)_6]^{3+}$ and $[Co(MeNH_2)_6]^{3+}$ have been recorded[132] and values for *Dq* are compared with those obtained for analogous systems with NH_3, propylenediamine, and ethylenediamine ligands. The maxima in the electronic spectra of several Cr^{III} ammine complexes have also been listed.[133]

A new magenta isomer of $[Cr\,en\,(H_2O)_3Cl]^{2+}$ has been reported[134] and, as expected, its visible absorption spectrum exhibits *d–d* absorption bands at wavelengths intermediate to those of the complexes $[Cr\,en\,(OH_2)_2Cl_2]^+$ and $[Cr\,en\,(H_2O)_4]^{3+}$. The spectrum is substantially different from that of the purple isomer of $[Cr\,en\,(H_2O)_3Cl]^{2+}$.

The electronic spectra of complexes $[Cr(NCS)_4L_2]^-$, where L is an amine, a tertiary phosphine, or (in one case) a tertiary arsine have been studied.[135] From the position of the ${}^4T_{2g} \leftarrow {}^4A_{2g}$ transition the relative positions of the ligands (L) in the spectrochemical series have been deduced.

The synthesis of a number of new diaquo- and triaquo-ethylenediamine-chromium(III) complexes from chromium(IV) diperoxoamines has been described and their visible absorption spectra are presented.[136] A number of complexes have also been reported with ammoniatriacetic acid (ataH_3). The electronic spectra indicate that the ammoniatriacetonato-ligand is functioning as tetradentate[137] in the complexes $[Cr\,ata\,acac]^+$, [Cr ata phen],4H_2O, [Cr ata bipy], $[Cr\,ata\,(NCS)_2]^{2+}$ and $[Cr\,ata\,(oxalate)]^{2+}$. Furthermore, the electronic spectra[138] indicate that in $[Cr\,ata_2]^{3-}$ and $[Cr\,ataH_2]^-$ the chromophore is $[CrN_2O_4]$, in $[Cr(OH)\,ata\,(H_2O)]^+$ it is $[CrNO_5]$, whilst in $[Cr(OH)\,ata\,(H_2O)_2]^+$ it is $[CrO_6]$.

[130] W. W. Wendlandt and L. K. Sveum, *J. Inorg. Nuclear Chem.*, 1967, **29**, 975.
[131] D. A. House and C. S. Garner, *Inorg. Chem.*, 1967, **6**, 272.
[132] M. Parris and N. F. Feiner, *Inorg. Nuclear Chem. Letters*, 1967, **3**, 337.
[133] E. Kyuno, M. Kamada, and N. Tanaka, *Bull. Chem. Soc. Japan*, 1967, **40**, 1848.
[134] D. M. Tully-Smith, R. K. Kurimoto, D. A. House, and C. S. Garner, *Inorg. Chem.*, 1967, **6**, 1524.
[135] M. A. Bennett, R. J. H. Clark, and A. D. J. Goodwin, *Inorg. Chem.*, 1967, **6**, 1625.
[136] R. G. Hughes and C. S. Garner, *Inorg. Chem.*, 1967, **6**, 1519.
[137] A. Uehara, E. Kyuno, and R. Tsuchiya, *Bull. Chem. Soc. Japan*, 1967, **40**, 2322.
[138] A. Uehara, E. Kyuno, and R. Tsuchiya, *Bull. Chem. Soc. Japan*, 1967, **40**, 2317.

I.r., visible, and u.v. spectral comparisons[139] suggest that the new compounds tris(diacetamido)chromium(III), $Cr(C_4H_6O_2N)_3$, and tris-(dibenzamido)chromium(III), $Cr(C_{14}H_{10}O_2N)_3$, are structurally analogous to Cr $acac_3$ and *tris*(dibenzoylmethano)chromium(III). The values of *Dq* for the β-diketo- and the diamido-ligands with chromium(III) are found to be essentially the same. It is concluded that the Cr—O π bonding cannot be extensive in these compounds.

Other N-donor complexes which have been studied by spectral means include *cis*-$[Cr\,pn_2Cl_2]^+$, *cis*-$[Cr\,pn_2(H_2O)Cl]^{2+}$, and *cis*-$[Cr\,pn_2(H_2O)_2]^{3+}$;[140] *cis*- and *trans*-$[CrCl_2\,en_2]^+$, and *cis*-$[CrCl(DMF)\,en_2]^{2+}$[141] (DMF = dimethylformamide) and also some 1 : 1 complexes of Cr^{III} with naphtholazopyrazolone dyes.[142]

The absorption of the Cr^{3+} ions in oxides containing (mainly) pentavalent ions as well has been reported.[143] The position of the intense absorption bands (assigned as charge-transfer transitions) in the short wavelength region of the visible and the crystal-field parameters *Dq* and *B* are tabulated. The results are discussed and applied to the problem of possible fluorescence of these materials.

Spectral data for predominantly oxygen-donor systems have also been described in studies concerned with the kinetics of the anation reaction of the *cis*-diaquobis(oxalato)chromium(III) ion with the oxalate ion,[144] the solvolysis of hexa-aquochromium(III) in dimethylsulphoxide,[145] and the isomerisation of *cis*- and *trans*-tetra-aquodichlorochromium(III).[146]

The photochemical preparation of chromium(III) mercaptides has been described according to the scheme:[147]

$$ArCr(CO)_3 \xrightarrow[h\nu]{Me_2S_2} Cr(SMe)_3 + Ar + 3CO$$

The electronic spectra, measured in KBr discs and also by the diffuse reflectance method, can be interpreted on the basis of an octahedral environment for Cr^{III}. The absorption spectra by diffuse reflectance for $[CrCl_3\,tu_3]$ (tu = thiourea) and trans-$[CrCl_2(H_2O)_4]Cl$ have also been reported.[148]

Chromium-(II), -(I), *and* -(0). Complexes of chromium(II) of the type $CrX_2,2Ph_3PO$ (X = Cl, Br, or I) have been studied.[149] The spectra of

139 C. S. Kraihanzel and D. N. Stehly, *Inorg. Chem.*, 1967, **6**, 277.
140 M. Esparza and C. S. Garner, *J. Inorg. Nuclear Chem.*, 1967, **29**, 2377.
141 D. A. Palmer and D. W. Watts, *Austral. J. Chem.*, 1967, **20**, 53.
142 M. Idelson, I. R. Karady, B. H. Mark, D. O. Rickter, and V. H. Hooper, *Inorg. Chem.*, 1967, **6**, 450.
143 G. Blasse, *J. Inorg. Nuclear Chem.*, 1967, **29**, 1817.
144 H. Kelm and G. M. Harris, *Inorg. Chem.*, 1967, **6**, 706.
145 K. R. Ashley, R. E. Hamm, and R. H. Magnuson, *Inorg. Chem.*, 1967, **6**, 413.
146 J. D. Salzman and E. L. King, *Inorg. Chem.*, 1967, **6**, 426.
147 D. A. Brown, W. K. Glass, and B. Kumar, *Chem. Comm.*, 1967, 736.
148 Y. Narusawa, K. Kanazawa, S. Takahashi, K. Morinaga, and K. Nakano, *J. Inorg. Nuclear Chem.*, 1967, **29**, 123.
149 D. E. Scaife, *Austral. J. Chem.*, 1967, **20**, 845.

$CrCl_2,2Ph_3PO$ and the green form of $CrBr_2,2Ph_3PO$ indicate tetragonally distorted six-co-ordination, with bridging through the halogen atoms. The spectra of the yellow form of $CrBr_2,2Ph_3PO$ and $CrI_2,2Ph_3PO$ suggest that these compounds have flattened tetrahedral co-ordination about chromium(II). Electronic spectra are also reported for the six-co-ordinate complexes $CrI_2,2Ph_3PO,2THF$, $CrBr_2,2THF$, and $CrBr_2,2MeCN$.

The absorption maxima for a number of chromium-(I) and -(O) phosphine-substituted carbonyls have been tabulated;[150] the phosphine ligands were tris-(*o*-diphenylphosphinophenyl)phosphine and bis-(*o*-diphenylphosphinophenyl)phenylphosphine. The u.v. spectra in dichloromethane have also been recorded[151] for $Me_2SnNCM(CO)_5$ and $HNCM(CO)_5$ (M = Cr, Mo, or W). The complexes formed from arenechromium tricarbonyl and 1,3,5-trinitrobenzene display a new band in the visible region of the spectrum; this band is shown to have its origin in a charge-transfer transition.[152] The ionisation potentials of several arenechromium tricarbonyls, as derived from charge-transfer spectra, are also reported.

Molybdenum(VI) *and Tungsten*(VI), d^0. The spectroscopic properties of molybdate(VI) solutions have been studied and the effect of the pH and Mo^{VI} concentration discussed.[153] Although it was impossible to define unambiguously by spectroscopic means the different Mo^{VI} species present, the entities suggested were: MoO_4^{2-} (or other monomolybdate forms); Mo—O—Mo; and MoO_2^{2+} (monomeric or polymeric).

The absorption spectra of aqueous solutions of $WS_2(SH)_2$, $(NH_4)_2WS_4$, and $(NH_4)_2WO_2S_2$ have been recorded[154] between 43,500 and 14,300 cm.$^{-1}$. The band at 25,500 cm.$^{-1}$ in WS_4^{2-} is 'split' in $WS_2(SH)_2$ owing to the lowering of the symmetry from T_d to C_{2v}. The electronic spectra of $MoO_2S_2^{2-}$ and $WO_2S_2^{2-}$ have also been reported.[155]

The electronic absorption spectra of $MoSe_4^{2-}$ and WSe_4^{2-} in aqueous solutions have been described.[156, 157] The band at 17,990 cm.$^{-1}$ in the spectrum of the violet aqueous solutions of $(NH_4)_2MoSe_4$ has been unambiguously assigned[157] as a charge-transfer transition from ligand to metal $^1T_2 \leftarrow {}^1A_1$. The existence of the ion $MoO_2Se_2^{2-}$ is reported for the first time[158] and it has been characterised by i.r. and u.v. spectra. The u.v. spectra in aqueous solution were reported and compared with related molybdenum and tungsten species. The bands are difficult to assign but

150 I. V. Howell and L. M. Venanzi, *J. Chem. Soc.* (*A*), 1967, 1007.
151 R. B. King, *Inorg. Chem.*, 1967, **6**, 25.
152 G. Huttner and E. O. Fischer, *J. Organometallic Chem.*, 1967, **8**, 299.
153 A. Bartecki and D. Dembicka, *J. Inorg. Nuclear Chem.*, 1967, **6**, 2907.
154 G. Gattow and A. Franke, *Z. anorg. Chem.*, 1967, **352**, 11.
155 A. Müller, B. Krebs, W. Rittner, and M. Stockburger, *Ber. Bunsengesellschaft. Phys. Chem.*, 1967, **71**, 182.
156 A. Müller, B. Krebs, O. Glemser, and E. Diemann, *Z. Naturforsch.*, 1967, **22b**, 1235.
157 A. Müller, B. Krebs, and E. Diemann, *Angew. Chem. Internat. Edn.*, 1967, **6**, 257.
158 A. Müller, B. Krebs, and E. Diemann, *Angew. Chem. Internat. Edn.*, 1967, **6**, 1081.

a band at 22,000 cm.$^{-1}$ is probably due to the transition $3b_2 \rightarrow 5a_1$, *i.e.* charge transfer Se $\rightarrow$ Mo.

The electronic spectra of the alkyl cyanide complexes $WOCl_4$,RCN and $WOBr_4$,MeCN (R = Me, Et, and Pr) have been discussed,[159] and the first charge-transfer band is assigned as the symmetry-forbidden $b_1 \rightarrow b_2$ transition. The u.v. solution spectra (in MeCN and CH_2Cl_2) of a number of WO_2Cl_2 complexes of the types WO_2Cl_2,2L and WO_2Cl_2B, where L is a mono- and B a bi-dentate ligand, are also reported.[160] The reflectance spectra of $WOCl_3$2L [L = $OPPh_3$ and $(Me_2N)_3PO$] are also recorded. No attempt has been made to assign the bands in the spectra.

Molybdenum(V) *and Tungsten*(V), d^1. The *d–d* bands have been assigned[161] for the ion $[MoOF_5]^{2-}$, assuming C_{4v} symmetry, as due to the transitions $^2E \leftarrow {}^2B_2$ (8300), $^2B_1 \leftarrow {}^2B_2$ (13,250) and $^2A_1 \leftarrow {}^2B_2$ ($\geqslant$27,600 cm.$^{-1}$). Good agreement with experiment was obtained by calculating the band positions using crystal-field theory and a SCF radial wavefunction in numerical tabular form; the use of Slater functions was unsatisfactory.

The electronic spectra of WCl_5, WBr_5, a number of Mo^V and W^V oxyhalide complexes of the type $[MOX_5]^{2-}$ and $[MOX_4]^-$ and halide complexes $[MX_6]^-$ have been described;[162] values of *Dq* are derived from the spectra. Complexes of the type $M^I[MoOCl_4]$ have also been reported:[163] data were obtained by diffuse reflectance, by absorption in single crystals, in solution in organic solvents, *e.g.* methanol, acetone, or acetonitrile, where the complexes probably achieve six co-ordination, and in aqueous hydrochloric acid where $[MoOCl_5]^{2-}$ is formed. It was concluded that the band order to oxygen and the degree of tetragonality in the Mo^V chromophores are to be regarded as substantially unchanged whatever the occupancy of the sixth octahedral position in the $[MoOCl_4]^-$ unit.

The electronic spectra of a number of W^V complexes formed by oxotetrahalides with 2,2′-bipyridyl, triphenylphosphine, dimethyl sulphide, and tetrahydrofuran have been reported,[159] although the precise nature of the products is obscure in some cases. The first two *d–d* transitions at ~14,000 ($b_2 \rightarrow e_\pi$*) and ~25,000 cm.$^{-1}$ ($b_2 \rightarrow b_1$*) may be picked out in certain cases. With the ligands 2,4,6-trimethylpyridine and benzonitrile stable complexes of the type $[MoCl_4(C_8H_{11}N)_2]Cl$ and $[WX_4L_2]X$ (X = Cl or Br; L = $C_8H_{11}N$ or C_7H_5N) have been obtained.[164] The u.v. and visible spectra of these complexes were very complicated, but solution spectra in nitromethane and acetonitrile were essentially identical with the diffuse

159 G. W. A. Fowles and J. L. Frost, *J. Chem. Soc.* (*A*), 1967, 671.
160 B. J. Brisdon, *Inorg. Chem.*, 1967, **6**, 1791.
161 E. Wendling and J. de Lavillandre, *Bull. Soc. chim. France*, 1967, 2743.
162 B. J. Brisden, D. A. Edwards, D. J. Machin, K. S. Murray, and R. A. Walton, *J. Chem. Soc.* (*A*), 1967, 1825.
163 O. Piovesana and C. Furlani, *Inorg. Nuclear Chem. Letters*, 1967, **3**, 535.
164 T. M. Brown and B. Ruble, *Inorg. Chem.*, 1967, **6**, 1335.

reflectance spectra. The u.v. and visible spectra of the octamethylpyrophosphoramide (L) complex, $MoOCl_3L$, has also been reported.[165]

The spectra of sulphur-donor complexes with dialkyldithiocarbamate[166] and toluene-3,4-dithiol (tdt)[167] have been investigated. The Mo^V dithiocarbamates contain a band[166] at about 19,500 cm.$^{-1}$ ($^2B_1 \leftarrow {}^2B_2$), whereas in the visible spectrum[167] of the red complex, Mo_2 tdt$_5$, a band is observed at 19,400 cm.$^{-1}$ (ε = 9400).

Molybdenum(IV) *and Tungsten*(IV), d^2. The electronic spectra of a number of eight-co-ordinate tungsten(IV) complexes have been studied. The compounds $W(CN)_4(CNR)_4$ (R = Et, Pr, and Bu) exhibit[168] two bands in the u.v. spectrum at 25,800 and 39,700 cm.$^{-1}$; these are at smaller wavenumbers than in $K_4[W(CN)_8]$ (27,200 and 40,650 cm.$^{-1}$). Tetrakis-(8-quinolinolato)tungsten(IV) has also been reported and has three intense absorption bands in the visible and near-u.v. region of the spectrum.[169] It has tentatively been suggested that these bands are too intense ($\varepsilon > 10^3$) for *d–d* transitions, but a band analysis is being carried out.

Compounds which have been isolated as photohydrolysis products of $[Mo(CN)_8]^{4-}$ or $[W(CN)_8]^{4-}$ are: red $Na_4[MoO_2(CN)_4],8H_2O$, red $K_4[MoO_2(CN)_4],6H_2O$, yellow-brown $K_4[WO_2(CN)_4],6H_2O$, blue $K_3[MoO(OH)(CN)_4],2H_2O$, and purple $K_3[WO(OH)(CN)_4]$. The electronic spectra of these diamagnetic complexes are reported and their visible bands assigned.[170]

Data have also appeared for compounds of the types $(\pi\text{-}C_5H_5)_2MX_2$ (M = Mo or W; X = Cl, Br, or I)[171] and $(\pi\text{-}C_5H_5)_2ML_2$ [M = Mo or W; $L_2 = (SPh_2)_2$, $(SH)_2$, $S_2C_2(CN)_2$, $o\text{-}S_2C_6H_3Me$, or $o\text{-}O_2C_6H_4$].[172]

Molybdenum(III) *and Tungsten*(III), d^3. The reflectance spectra of the complexes [bipy H]$^+$ [$MoCl_4$ bipy]$^-$ and [phen H]$^+$ [$MoCl_4$ phen]$^-$ have been measured.[173] Band assignments are given, and a detailed magnetic study is also reported.

Saillant *et al.* have described some reactions of $W_2Cl_9^{3-}$ with heterocyclic bases to form complexes of the type:

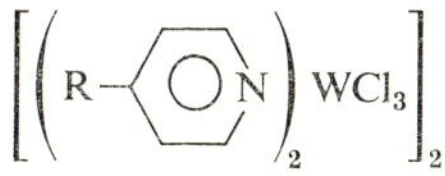

(R=H, Me, and Pr^i)

The electronic spectra of these complexes are given[174] and the spectrum of $W_2Cl_9^{3-}$ is also discussed. The assignments of the bands at 13,200 and

[165] M. D. Joesten, *Inorg. Chem.*, 1967, **6**, 1598.
[166] F. W. Moore and M. L. Larson, *Inorg. Chem.*, 1967, **6**, 998.
[167] A. Butcher and P. C. E. Mitchell, *Chem. Comm.*, 1967, 176.
[168] H. Latka, *Z. anorg. Chem.*, 1967, **353**, 243.
[169] R. D. Archer and W. D. Bonds, jun., *J. Amer. Chem. Soc.*, 1967, **89**, 2236.
[170] S. J. Lippard and B. J. Russ, *Inorg. Chem.*, 1967, **6**, 1943.
[171] R. L. Cooper and M. L. H. Green, *J. Chem. Soc.* (*A*), 1967, 1155.
[172] M. L. H. Green and W. E. Lindsell, *J. Chem. Soc.* (*A*), 1967, 1455.
[173] H. L. Schläfer and K. Christ, *Z. anorg. Chem.*, 1967, **349**, 289.
[174] R. Saillant, J. L. Hayden, and R. A. D. Wentworth, *Inorg. Chem.*, 1967, **6**, 1497.

15,900 cm.$^{-1}$ as being due to $^2E \leftarrow {}^4A_2$ and $^2T_2 \leftarrow {}^4A_2$ transitions respectively within the individual chromophores, as suggested by Jørgensen,[175] are considered incorrect.

Molybdenum(II), d^4. Molybdenum 'dichloride' has been shown[176] to react with various unidentate ligands to form complexes of the type $[(Mo_6Cl_8)Cl_4L_2]$. The close similarity between the electronic spectra of these complexes, the chloro-acid $H_2[(Mo_6Cl_8)Cl_6],8H_2O$, and the parent chloride confirms their structural relationship and the retention of the cluster. Of the three bands in the spectra, the highest (*ca.* 31,300 cm.$^{-1}$ in chlorides) is characteristic of the particular Mo_6X_8 cluster, and moves to a lower frequency for the heavier halogens. The spectra of complexes formed with a number of bidentate ligands and also terpyridyl are given.[176] The electronic spectra of complexes of the type $[(Mo_6Cl_8)Cl_4,2L]$ have also been reported for nine nitrogen-donor ligands.[177]

Quantitative solid potassium halide and hydrochloric acid solution absorption spectra of halogenotrimolybdates(II) have been examined.[178] The results indicate that the compounds are of a common structural type, both in solid and solution, but that at least two kinds of chloro- or bromo-molybdate(II) are present. Variations of the absorption intensities with hydrochloric acid molarity suggest the equilibrium:

$$[Mo_3Cl_{12}]^{6-} + Cl^- \rightleftharpoons [Mo_3Cl_{13}]^{7-}$$

and moreover similar intensity changes are observed for $K_6Mo_3Cl_{12}$ in KCl.

Lower Oxidation States of Molybdenum. The electronic spectrum of 1,4-diazobutadienetetracarbonylmolybdenum has been interpreted[179] in terms of charge-transfer transitions from the metal to the ligand on the basis of a qualitative MO scheme. Spectral data have also appeared, without detailed assignment, for *cis*-bicyclo[6,1,0]nona-2,4,6-trientricarbonylmolybdenum and the products of its thermal rearrangement,[180] and for a number of carbonyl-π-cyclopentadienyl complexes of molybdenum.[181] Spectral changes have been used to follow some substitution reactions[182] of tetracarbonylcyclo-octa-1,5-dienemolybdenum(0) and the kinetics of the reaction between trimethyl phosphite and some arenetricarbonylmolybdenum complexes.[183]

Manganese, Technetium, and Rhenium.—*Manganese*(VII), d^0. The absorption spectrum of $KMnO_4$ dissolved in $KClO_4$ has been remeasured at the

175 C. K. Jørgensen, *Acta Chem. Scand.*, 1959, **11**, 73.
176 J. E. Fergusson, B. H. Robinson, and C. J. Wilkins, *J. Chem. Soc.* (*A*), 1967, 486.
177 W. M. Carmichael and D. A. Edwards, *J. Inorg. Nuclear Chem.*, 1967, **29**, 1535.
178 W. van Bronswyk and J. C. Sheldon, *Austral. J. Chem.*, 1967, **20**, 2323.
179 H. Bock and H. tom Dieck, *Chem. Ber.*, 1967, **100**, 228.
180 W. Grimme, *Chem. Ber.*, 1967, **100**, 113.
181 R. J. Haines, R. S. Nyholm, and M. H. B. Stiddard, *J. Chem. Soc.* (*A*), 1967, 94.
182 F. Zingales, M. Graziani, and U. Belluco, *J. Amer. Chem. Soc.*, 1967, **89**, 256.
183 A. Pidcock, J. D. Smith, and B. W. Taylor, *J. Chem. Soc.* (*A*), 1967, 872.

temperatures of liquid hydrogen and helium using spectrographs with reasonably high dispersion.[184] Evidence is found for four spin-allowed electronic transitions. The first one is assigned as ${}^1T_2 \leftarrow {}^1A_1$, and it exhibits a great deal of vibrational structure starting with the O–O line located at 18,072 cm.$^{-1}$. The second transition, at 25,000–30,000 cm.$^{-1}$, is rather

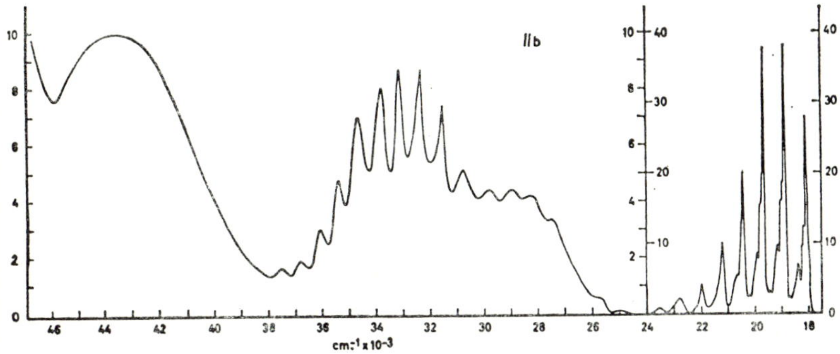

Figure 4 *Absorption spectrum of* $KMnO_4$ *dissolved in* $KClO_4$ *at liquid helium temperature. The electric vector is parallel to* b. *Relative extinction coefficient as ordinate*
(Reproduced by permission from *Theor. Chim. Acta*, 1967, 7, 314)

featureless, whilst the third, which has a maximum at 33,000 cm.$^{-1}$, shows a simple progression in quanta of 750 cm.$^{-1}$. The fourth band is completely featureless with maximum intensity at 43,500 cm.$^{-1}$. These results are compared with a simplified SCF–LCAO molecular orbital calculation.

The reactions of several manganese compounds with molten potassium thiocyanate have been investigated;[185] the absorption spectra of the green and blue solutions produced from Mn^{VII}, Mn^{VI}, Mn^{III}, and Mn^{II} were very like those of thiocyanate at 400° where partial decomposition to S_4 occurs.

Manganese(IV), d^3. The diffuse reflectance spectrum has been recorded[186] for the deep red-brown complex of stoicheiometry [MnO phen$_2$]$(ClO_4)_2$,-$O\cdot 5H_2O$. The antiferromagnetic behaviour of this complex has been interpreted in terms of a bridged structure. The electronic spectrum of the heteropoly-12-niobomanganate(IV) anion has also been examined[187] and shown to be very similar to that of the 9-molybdomanganate, $[MnMo_9O_{32}]^{6-}$.

Manganese(III), d^4. Few manganese(III) complexes have been studied, owing to the relative instability of this oxidation state. Recently, however, complexes with *trans*-1,2-diaminocyclohexanetetra-acetic acid and

[184] S. L. Holt and C. J. Ballhausen, *Theor. Chim. Acta*, 1967, **7**, 313.
[185] D. H. Kerridge and M. Mosley, *J. Chem. Soc.* (*A*), 1967, 352.
[186] H. A. Goodwin and R. N. Sylva, *Austral. J. Chem.*, 1967, **20**, 629.
[187] B. W. Dale and M. T. Pope, *Chem. Comm.*, 1967, 792.

hydroxyethylenediaminetriacetic acid have been reported and the effect of the pH on the electronic spectra has been studied.[188] The solid state spectra have also been recorded for the complexes $[MnF_3(H_2O)\,phen]$ and $[MnCl_3\,terpy]$;[186] these spectra are very similar to one another and to those previously reported for manganese(III) chelates of phenanthroline and bipyridyl.

The observed spectrum of the aquomanganese(III) ion in perchlorate media has been analysed into the spectra of the hexa-aquo-ion and the hydroxo-penta-aquo-ion.[189] Both have maxima at 21,000 and *ca.* 45,000 cm.$^{-1}$. The former band is identified with a *d–d* transition involving Jahn–Teller distortion from O_h symmetry and the latter with a partial ligand → metal electron transfer.

Manganese(II), d^5. The absorption spectrum of crystalline $RbMnF_3$ has been studied[190] at room temperature and at 77°K. The observed band-positions are fitted with four parameters *B*, *C*, *Dq*, and α, and the values of these parameters for the spectrum at 77°K are 840, 3080, 780, and 76 cm.$^{-1}$. The shift in band positions on reducing the temperature is attributed to an increase in *Dq*. Fine structure observed in the $^2T_{2g}(D)$ band is attributed to spin–orbit splitting, and the analysis yields a value of 320 cm.$^{-1}$.

The polarised absorption spectra of manganese(II) in Me_4NMnCl_3 (six-co-ordinate) and in $(Me_4N)_2MnCl_4$ (four-co-ordinate) have been studied[191] at 298, 77, and 2°K. Marked optical anisotropy was observed in both complexes. Details of the spectra indicate that the surroundings of the six-co-ordinate Mn^{II} are trigonally distorted from octahedral symmetry. In the four-co-ordinate species the symmetry is very nearly tetrahedral. The polarisation behaviour of the ion in both co-ordinations is discussed.

Although several electronic bands are expected for high-spin manganese(II), only one band was observed[192] for $Mn(NO_2)_4^{2-}$ below the strong absorption edge at *ca.* 25,000 cm.$^{-1}$. This band, however, is relatively intense (ε *ca.* 7) for manganese(II), and this is consistent with the presence of a non-centrosymmetric ligand field, as has been proposed on the basis of the i.r. results. However, this band may gain intensity from the nearby parity-allowed transition.

Spectral data have appeared for manganese(II) complexes with propone,[193] morpholine,[194] tri-2-pyridylamine,[195] 2-carboxyl-1,10-phenan-

188 R. E. Hamm and M. A. Suwyn, *Inorg. Chem.*, 1967, **6**, 139.
189 C. F. Wells and G. Davies, *J. Chem. Soc.* (*A*), 1967, 1858.
190 A. Mehra and P. Venkateswarlu, *J. Chem. Phys.*, 1967, **47**, 2334.
191 K. E. Lawson, *J. Chem. Phys.*, 1967, **47**, 3627.
192 D. M. L. Goodgame and M. A. Hitchman, *J. Chem. Soc.* (*A*), 1967, 612.
193 P. S. Gentile, P. Jargiello, T. A. Shankoff, and J. Carlotto, *J. Inorg. Nuclear Chem.*, 1967, **29**, 1685.
194 I. S. Ahuja, *J. Inorg. Nuclear Chem.*, 1967, **29**, 2091.
195 G. C. Kulasingam and W. R. McWhinnie, *J. Chem. Soc.* (*A*), 1967, 1253.

throline,[196] γ-butyrolactam and *N*-methyl-γ-butyrolactam.[197] Azido-complexes [198] and complexes of the types $Hg[Mn_xZn_{1-x}](NCS)_4$ [199] and $Mn(phthalimide)_2L_4$ [200] where L is a primary or secondary aliphatic amine have also been studied.

Manganese(I), d^6. A new polyhedral complex of manganese, $[B_6C_2H_8Mn(CO)_3]^-$ with the ligand, $B_6C_2H_8^{2-}$, has been reported and the electronic spectrum of the compound in acetonitrile was recorded.[201]

Technetium(VII) *and Rhenium*(VII), d^0. Pertechnetyl chloride, TcO_3Cl, can be prepared by adding small amounts of 12M-hydrochloric acid to a cold dilute solution of pertechnetate in 18M-sulphuric acid. The electronic spectrum has been recorded of this compound extracted into chloroform.[202] The red colour of concentrated aqueous $HTcO_4$ has been ascribed to a lower (VI)- or (V)-state. The reaction of TcO_4^- and ReO_4^- with hydrogen sulphide in aqueous solutions has been investigated by measuring the electronic absorption spectra.[203] Using a simple MO diagram the existence of ReS_4^- is proved and the electronic spectrum assigned. Small amounts of TcO_3S^- are possibly formed in the reaction of TcO_4^- with hydrogen sulphide.

Technetium(V) *and Rhenium*(V), d^2. The thiocyanato-compound $Cs[Re(SCN)_6]$ has been reported and its i.r. spectrum was interpreted in favour of S-bonding.[204] Only one peak is observed in the u.v.—visible region at 27,600 cm.$^{-1}$ and this band has been attributed to charge transfer on the basis of its extinction coefficient (ε $6{\cdot}12 \times 10^3$).

Technetium(IV) *and Rhenium*(IV), d^3. Cotton and Harris [205] have carried out extended Hückel calculations for the ions $ReCl_6^{2-}$, $OsCl_6^{2-}$, $IrCl_6^{2-}$, and $PtCl_6^{2-}$ using the Mulliken–Wolfsberg–Helmholz approximation for the off-diagonal elements of the Hamiltonian matrix. One of the key features of their procedure was to use orbital energies for the metal ions very close to those of the uncharged atoms and a very moderated dependence (1ev/unit charge) of the metal *d*-orbital energies on the effective charge of the metal atom. It is pointed out that this semi-empirical molecular orbital treatment is not well suited to giving a quantitative description of the electronic spectra because it does not explicitly include electron-repulsion integrals in the Hamiltonian. Nevertheless it can provide useful information about the expected trends in a given series of related molecules.

[196] H. A. Goodwin and R. N. Sylva, *Austral. J. Chem.*, 1967, **20**, 217.
[197] S. K. Madan and J. A. Sturr, *J. Inorg. Nuclear Chem.*, 1967, **29**, 1669.
[198] V. Gutmann and W. K. Lux, *J. Inorg. Nuclear Chem.*, 1967, **29**, 2391.
[199] D. E. Scaife, *Inorg. Chem.*, 1967, **6**, 625.
[200] G. Narain and P. Shukla, *Austral. J. Chem.*, 1967, **20**, 227.
[201] M. F. Hawthorne and A. D. Pitts, *J. Amer. Chem. Soc.*, 1967, **89**, 7115.
[202] C. L. Rulfs, R. A. Pacer, and R. F. Hirsch, *J. Inorg. Nuclear Chem.*, 1967, **29**, 681.
[203] A. Müller, B. Krebs, and E. Diemann, *Z. anorg. Chem.*, 1967, **353**, 259.
[204] R. A. Bailey and S. L. Kozak, *Inorg. Chem.*, 1967, **6**, 2155.
[205] F. A. Cotton and C. B. Harris, *Inorg. Chem.*, 1967, **6**, 379.

A new black crystalline form of rhenium(IV) chloride has been reported [206] and the spectrum of a solution of this chloride dissolved in acidified methanol suggests the formation of $[Re_2Cl_9]^{2-}$. The diffuse reflectance spectra of $[Re_2Cl_9]^{2-}$ and $[Re_2Cl_8]^{2-}$ are shown to be different, although both complex anions show a sharp band at 13,900 cm.$^{-1}$, characteristic of a dimeric rhenium halide species. In acetonitrile the $[Re_2Cl_9]^{2-}$ species show some decomposition to $[Re_2Cl_8]^{2-}$. In a later publication,[207] both the anions $[Re_2Cl_9]^-$ and $[Re_2Br_9]^-$ are reported and their electronic spectra, recorded in dichloromethane, are tabulated. The electronic spectrum of the complex anion $[Re(NCS)_6]^{2-}$ has also been discussed.[208] The electronic spectra of the red $K_2Tc(OH)Cl_5$ and the yellow K_2TcCl_6 have been studied in the solid state. Their spectra are very similar and assignments are discussed.[209] The u.v. and visible spectra of three hexathiocyanatorhenium(IV) compounds have been recorded.[210] The spectra display two intense charge-transfer bands and a series of low-intensity peaks which appear to be ligand-field transitions. A distinct similarity is noted between the more intense bands and those observed for octahedral rhenium(IV) complexes, such as $[ReCl_6]^{2-}$.

Technetium(III) *and Rhenium*(III), d^4. The metal–metal and metal–halogen bonding in $[Re_2Cl_8]^{2-}$ has been investigated by performing an extended Hückel MO calculation.[211] The calculations suggest that the π-bonding contribution to the Re—Re bond is five times that of the δ bonding and almost three times that of the σ bonding. The ordering of the molecular orbitals is discussed with respect to the magnetic properties and the observed and calculated spectral properties. The reactions of the $[Re_2Cl_8]^{2-}$ and $[Re_2Br_8]^{2-}$ ions with several carbonylic acids and sulphur ligands have also been studied.[212] The electronic spectra of the carboxylate complexes are discussed and related to axial substitution. The reflectance spectrum of the thiourea complex, $[ReX_3\,tu_3]$, differs from that of the polymeric tetramethylthiourea (tmtu) complex, $[ReX_3\,tmtu]_n$, and it is concluded that thiourea readily cleaves the Re–Re bond. The spectra of 2,5-dithiahexane complexes are also discussed. The electronic spectrum of the isothiocyanate complex $[Re_2(NCS)_8]^{2-}$ has been recorded by the diffuse reflectance technique and also in solution in acetonitrile. The spectrum is assigned on the basis of an energy-level scheme described earlier.

A red crystalline compound of analytical composition $CsReCl_3(NO_3)$ has been formulated as containing the trinuclear anion $[Re_3Cl_9(NO_3)_3]^{3-}$

[206] F. A. Cotton, W. R. Robinson, and R. A. Walton, *Inorg. Chem.*, 1967, **6**, 223.
[207] F. Bonati and F. A. Cotton, *Inorg. Chem.*, 1967, **6**, 1353.
[208] F. A. Cotton, W. R. Robinson, R. A. Walton, and R. Whyman, *Inorg. Chem.*, 1967, **6**, 929.
[209] M. Elder, J. E. Fergusson, G. J. Gainsford, J. H. Hickford, and B. R. Penfold, *J. Chem. Soc.* (*A*), 1967, 1423.
[210] R. A. Bailey and S. L. Kozak, *Inorg. Chem.*, 1967, **6**, 419.
[211] F. A. Cotton and C. B. Harris, *Inorg. Chem.*, 1967, **6**, 924.
[212] F. A. Cotton, C. Oldham, and R. A. Walton, *Inorg. Chem.*, 1967, **6**, 214.

on the basis of its absorption spectrum.[213] The maxima in the visible spectra of the complexes $ReX_3(MeCN)(PPh_3)_2$ (X = Cl or Br) in dichloromethane have been recorded.[214]

Iron, Ruthenium, and Osmium.—*Iron Dithiolate Complexes.* Monomeric five-co-ordinate dithiolate complexes of the type $[LFeS_4C_4(CF_3)_4]$ (L = Ph_3P, Ph_3As, Ph_3Sb, and $(PhO)_3P$) have been described.[215] They exhibit similar electronic spectra in dichloromethane and the addition of further ligand L, to the solution does not perturb the spectrum. A series of five-co-ordinate mononitrosyl dithiolates has also been reported by McCleverty and his co-workers.[216] The complexes are formulated as $[Fe(NO)(—S_4)]^z$ ($z = +1$, 0, -1, -2, or -3 and the ligands include disubstituted *cis*-ethylene-1,2-dithiolates, tetrachlorobenzene-1,2-dithiolate, and toluene-3,4-dithiolate; their electronic spectra are recorded but no spectral assignments are made. Some analogous compounds of cobalt are also reported in the latter paper.

Iron(III), d^5. The vapour-phase absorption spectra of $FeCl_3$ and $FeBr_3$ in the interval from 13,300–50,000 cm.$^{-1}$ at temperatures between 200° and 450° have been observed.[217] Dimers appear to be the principal absorbing species in this temperature range, and two peaks are observed.

Assignments have been made for the first three low-intensity spin-forbidden *d–d* transitions by comparing both the diffuse reflectance spectra of the complexes $[FeCl_n(H_2O)_{6-n}]^{(3-n)+}$ (n = 0, 1, 3, 5, and 6) and the solution spectra of the lower complexes.[218] Ligand-field calculations reveal a very sharp decrease in the parameters Dq and B on increasing n. The hydrolysis of concentrated solutions of $FeCl_3$ has also been investigated by studies of the *d–d* transitions in the visible region of the spectrum.[219]

The spectrum of the tetrachloroferrate(III) anion is well known, but its interpretation has proved difficult. The ligand-field approach to this problem has recently been critically reviewed.[220] The influence of differences in the delocalisation of the metal t_2 and e orbitals has been considered. This covalency influence could be assessed either by using different nephelauxetic parameters or by introducing different crystal-field parameters. Both approaches gave a satisfactory explanation of the relative positions and the number of bands in the spectrum, but the agreement between the calculated and observed energies was not very good.

213 J. H. Hickford and J. E. Fergusson, *J. Chem. Soc.* (*A*), 1967, 113.
214 G. Rouschias and G. Wilkinson, *J. Chem. Soc.* (*A*), 1967, 993.
215 A. L. Balch, *Inorg. Chem.*, 1967, **6**, 2158.
216 J. A. McCleverty, N. M. Atherton, J. Locke, E. J. Wharton, and C. J. Winscom, *J. Amer. Chem. Soc.*, 1967, **89**, 6082.
217 J. D. Christian and N. W. Gregory, *J. Phys. Chem.*, 1967, **71**, 1579.
218 S. Balt and A. M. A. Verwey, *Spectrochim. Acta*, 1967, **23**, *A*, 2069.
219 S. Balt, *J. Inorg. Nuclear Chem.*, 1967, **29**, 2307.
220 S. Balt, *Rec. Trav. chim.*, 1967, **86**, 1025.

A novel series of five-co-ordinate iron(III) complexes of the type $[FeX(S_2CNR_2)_2]$ (X = halogen, R = alkyl) has been reported.[221] These complexes have been characterised chemically and their five-co-ordinate nature established by the methods of molecular-weight determination, magnetochemistry, and crystal-structure determination. The molecular structure is approximately square pyramidal with the iron atom lying above the basal plane formed by the four sulphur atoms (see Figure 5). The optical spectra of these compounds are rich in absorption

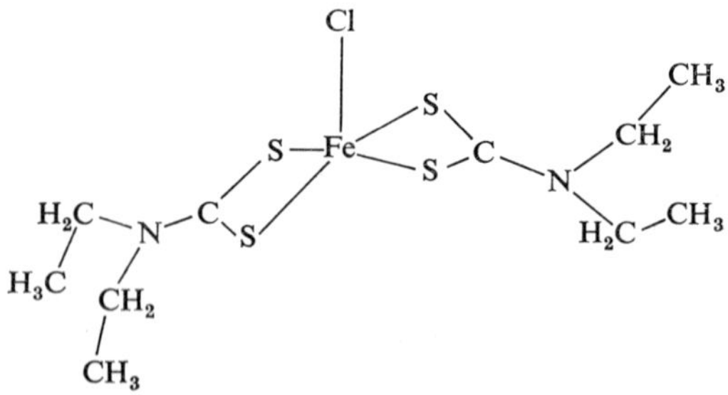

Figure 5 *Molecular structure of* $[FeCl(S_2CNEt_2)_2]$
(Reproduced by permission from *Inorg. Chem.*, 1967, **6**, 714)

bands, the majority of which appear to be charge-transfer transitions. However, two weak bands at 6500 and 8400 cm.$^{-1}$ are probably due to spin-allowed *d–d* transitions. The tentative assignments $xy \rightarrow xz,yz$ at 6500 cm.$^{-1}$ and $xy \rightarrow z^2$ at 8400 cm.$^{-1}$ are given, or alternatively both transitions may terminate in x^2-y^2 from z^2 and xz,yz respectively. Attempts to assign these bands are complicated by the likelihood that interaction between the metal and ligand π-systems would lift the degeneracy of the $3d_{xz}$, $3d_{yz}$ pair of orbitals.

Diffuse reflectance spectra have been recorded for a number of hydroxo-bridged iron(III) complexes.[222] The electronic spectra have also been given[223] for the two intermediates $[(EDTA)Co^{III}\text{-}NC\text{-}Fe^{II}(CN)_5]^{5-}$ and $[(EDTA)Co^{III}\text{-}NC\text{-}Fe^{III}(CN)_5]^{4-}$ formed in the redox reaction between $[Co(EDTA)]^{2-}$ and $[Fe(CN)_6]^{3-}$. Spectrophotometric studies of the Fe^{III}-5,7-dibromo-oxine-*N*-oxide system indicate the presence of at least two complexes.[224] The dissociation of the monothiocyanate complex of iron(III) has also been followed spetrophotometrically.[225]

Iron(II), d^6. Depending upon the strength of the ligand-field, octahedral d^6 complexes may exist in either a $^5T_{2g}$ or a $^1A_{1g}$ electronic ground state. At a field strength close to the cross-over point, the energy separation between

[221] R. L. Martin and A. H. White, *Inorg. Chem.*, 1967, **6**, 712.
[222] R. Driver and W. R. Walker, *Austral. J. Chem.*, 1967, **20**, 1375.
[223] D. H. Huchital and R. G. Wilkins, *Inorg. Chem.*, 1967, **6**, 1022.
[224] A. N. Bhat, R. D. Gupta, and B. D. Jain, *J. Indian Chem. Soc.*, 1967, **44**, 187.
[225] T. J. Conocchioli and N. Sutin, *J. Amer. Chem. Soc.*, 1967, **89**, 282.

the two states of different multiplicity would be expected to attain values within the thermally accessible range. Equilibria between the two states should then be possible, but in spite of considerable effort few definite examples of compounds displaying this sort of behaviour are known. The magnetic moments of [Fe phen$_2$(NCS)$_2$] and [Fe phen$_2$(NCSe)$_2$] have been measured [226] in the temperature range between 77 and 440°K; the magnetic moments are about 5·20 B.M. at 440°K, but they show a pronounced decrease at critical temperatures, T_c, of 174°K and 232°K respectively. These results agree qualitatively with the predictions made on the basis of the van Vleck equation for a $^1A_{1g}$ ground-state slightly separated from a higher lying $^5T_{2g}$ state. Mössbauer spectra confirm these conclusions. Thus it is concluded that the two compounds exist above T_c in a $^5T_{2g}$ state and below T_c in a $^1A_{1g}$ state. These conclusions are also supported by the temperature dependence of the electronic (and vibrational) spectra. The values of the spectral parameters 10*Dq* 11,900 and *B ca.* 640 for the $^5T_{2g}$ state as well as 10*Dq ca.* 16,300 and *B ca.* 580 cm.$^{-1}$ for the $^1A_{1g}$ state of both complexes were estimated from the *d–d* transitions.

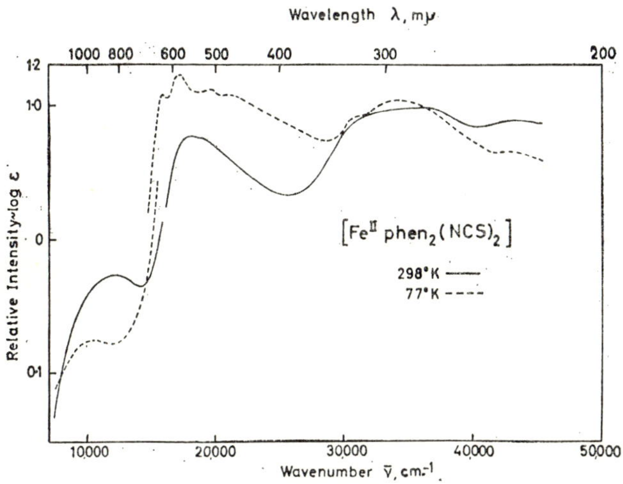

Figure 6 *Electronic spectrum of* [Fe phen$_2$(NCS)$_2$] *at* 298°K (*full line*) *and at* 80°K (*broken line*). *Left part of the spectrum: undiluted sample; right part of the spectrum: sample diluted with* LiF (1 : 50)
(Reproduced by permission from *Inorg. Chem.*, 1967, **6**, 53)

The Mössbauer quadrupole splittings for the tetragonal complexes FeL$_4$X$_2$ (L = isoquinoline or γ-picoline; X = Cl, Br, I, or NCS) and Fepy$_4$X$_2$ (X = Cl or NCS) have been measured and related to δt_{2g}, the splitting of the t_{2g} orbitals by the tetragonal component of the ligand field.[227] It is concluded that δt_{2g} is small for all these complexes and that

[226] E. König and K. Madeja, *Inorg. Chem.*, 1967, **6**, 48.
[227] C. D. Burbridge, D. M. L. Goodgame, and M. Goodgame, *J. Chem. Soc.* (*A*), 1967, 349.

the d_{xy} orbital lies lowest for the halides, whereas the orbital doublet, (d_{xz},d_{yz}), may lie lowest in energy in the isothiocyanates. Support for the assignment of d_{xy} as the lowest-energy orbital in the halogeno-complexes is obtained by comparing the electronic spectra of the Fe^{II} complexes with those of the corresponding Ni^{II} compounds, for which δt_{2g} was previously evaluated by completely independent methods. Spectral data for the corresponding Co^{II} complexes are also reported and it appears that, in general, for an octahedral halogeno-complex ML_4X_2 where L is a heterocyclic amine and M is Fe^{II}, Co^{II}, or Ni^{II}, the orbital splitting pattern varies little on changing the identity of the metal ion.

Venanzi and his co-workers [228] have reported some interesting five-co-ordinate iron(II) complexes of the type $[FeX(QP)]^+$ (X is Cl, Br, or I). These compounds provide the first examples of low-spin five-co-ordinate complexes of iron(II) and they have been assigned trigonal bipyramidal structures. In the visible and u.v. spectra there are two low-frequency absorption regions at *ca.* 18,000 and at *ca.* 9000 cm.$^{-1}$. These are attributed to ligand-field transitions despite their fairly high intensities.[2] The six-co-ordinate complexes $[FeX_2(QP)]$ have similar spectra to one another but they are very different from those of the five-co-ordinate complexes. Spectral data have also been reported for the five-co-ordinate complexes FeX_2L (X = Cl or Br; L is the tridentate ligand bis-[2-dimethylamino-ethyl] oxide).[229] (Note. QP = 2.)

The electronic absorption spectrum of ferrocene in the vapour state, in liquid solutions, and in glassy matrices has been investigated [230] at 77–420°K, and the spectrum was shown to contain at least eleven distinct electronic absorption bands. Three of these appear to be triplet ← singlet transitions, and to be made allowed by spin-vibronic perturbations. Absorption data have also been presented for two ferrocenylphosphine oxides and it is concluded from these data that there is only a very weak d_π–p_π interaction between the phosphoryl and ferrocenyl groups.[231]

Measurements of the Jahn–Teller splittings of the $^5E_g(^5D) \leftarrow {}^5T_{2g}(^5D)$ transitions are reported [232] for FeF_2, $KFeF_3$, $KMgF_3{:}Fe^{2+}$, $FeSiF_66H_2O$, $FeCl_2$, and $MgO{:}Fe^{2+}$. The effect of the site symmetry of the ferrous ion on the magnitude of the splittings observed is discussed.

The spectra of several di-2-pyridylamine complexes of iron(II) have been reported.[233, 234] This ligand lies between ethylenediamine and 1,10-phenthroline in the spectrochemical series, and the value of Dq for the tris-complex, FeL_3^{2+}, seems to be close to the position of spin-pairing [226] for

228 M. T. Halfpenny, J. G. Hartley, and L. M. Venanzi, *J. Chem. Soc.* (*A*), 1967, 627.
229 M. Ciampolini and N. Nardi, *Inorg. Chem.*, 1967, **6**, 445.
230 A. T. Armstrong, F. Smith, E. Elder, and S. P. McGlynn, *J. Chem. Phys.*, 1967, **46**, 4321.
231 E. W. Neuse, *J. Organometallic Chem.*, 1967, **7**, 349.
232 G. D. Jones, *Phys. Rev.*, 1967, **155**, 259.
233 W. R. McWhinnie, R. C. Poller, and M. Thevarasa, *J. Chem. Soc.* (*A*), 1967, 1671.
234 C. D. Burbridge and D. M. L. Goodgame, *J. Chem. Soc.* (*A*), 1967, 694.

iron(II). Complexes of tri-2-pyridylamine with ferrous perchlorate, chloride, and bromide are also reported[233] and shown to contain the spin-paired bis-(tri-2-pyridylamine)iron(II) cation. The first spin-allowed *d–d* transition ($^1T_{1g} \leftarrow {}^1A_{1g}$) is assigned at 19,000 cm.$^{-1}$.

Complexes of the sexadentate ligand 1,8-bis-(8-quinolylmethyleneimino)-3,6-diazo-octane, A, with iron(II), cobalt(III), and nickel(II) have been described.[235] The mode of co-ordination of this base is such that its complexes must be closely related structurally to octahedral complexes of 1-amino-2-(8-quinolylmethyleneimino)ethane, B. The absorption spectra of $[FeA](ClO_4)_2$ and $[FeB_2]ClO_4$ in nitromethane, and of $[CoA]Cl_3$ and $[CoB_2]Cl_3$ in water, are illustrated. The reflectance spectrum of the 5-aminoquinoxaline complex, FeL_2Cl_2, has been reported,[236] but since the spectrum was complicated by an intense band a structural assignment was not given (cf. octahedral Co^{II} and Ni^{II} complexes).

The magnetic properties of salts of the bis-2(2-pyridylamino)-4-(2-pyridyl)thiazoleiron(II) ion are anomalous and are markedly dependent upon the anion of the salt.[237] However, the diffuse reflectance spectra did not show very significant differences between complexes which were essentially spin-paired or spin-free. It was concluded that the striking visual differences in colour result from differences in band intensities rather than in band positions.

Buckingham and his co-workers have reported the electronic spectra of several β-diketone complexes of iron(II). The reaction of [Fe $acac_2$] with pyridine was followed spectrophotometrically in benzene solutions, and it was concluded that the following species were formed: [Fe $acac_2]_2$ py, [Fe $acac_2$ py$]_2$ and [Fe $acac_2$ py_2], although only the latter compound was isolated in the solid state.[238] In another study,[239] the polymeric nature of several β-diketone complexes was investigated by means of a number of physical techniques. Spectral data were obtained from benzene solutions over the concentration range 0·005–0·06 M. Striking deviations from Beer's law at intermediate concentrations were noted, which agree qualitatively with the conclusion drawn from molecular-weight data that there is a concentration-dependent molecular association process occurring in this concentration range. A limiting spectrum was approached at higher concentrations.

Some interesting new bridged addition compounds formed by dicyanobis-(1,10-phenanthroline)iron(II) and the fluorides and chlorides of silicon, germanium, and tin have been reported.[240] A comparison of the charge-transfer bands indicates that for the 1 : 1 adducts these bands

[235] J. Dekkers and H. A. Goodwin, *Austral. J. Chem.*, 1967, **20**, 69.
[236] O. Dahl and P. H. Nielsen, *Acta Chem. Scand.*, 1967, **21**, 1153.
[237] R. N. Sylva and H. A. Goodwin, *Austral. J. Chem.*, 1967, **20**, 479.
[238] D. A. Buckingham, R. C. Gorges, and J. T. Henry, *Austral. J. Chem.*, 1967, **20**, 497.
[239] D. A. Buckingham, R. C. Gorges, and J. T. Henry, *Austral. J. Chem.*, 1967, **20**, 281.
[240] J. J. Rupp and D. F. Shriver, *Inorg. Chem.*, 1967, **6**, 755.

are shifted to higher energy in the series $SiF_4 < GeF_4 < SnF_4$. This order parallels the generally accepted order of acidity. Also, except for the SnF_4 adducts, the metal-to-ligand charge-transfer band is shifted to higher energies as the proportion of Lewis acid in the complex is increased.

Electronic spectral data have also been reported for iron(II) complexes derived from substituted hydrazones of 2,6-diacetylpyridine, (3),[241] substituted 1,10-phenanthrolineiron(II) cyanide complexes of the type $[Fe\,phen_2(CN)_2]$,[242] the thiourea complexes $[Fetu_2(SCN)_2]$, $[Fetu_4Cl_2]$, and $[FeL_2Cl_2]$ (L = methylthiourea, 1,3-dimethylthiourea, and 1,3-diethylthiourea),[243] complexes of *NNN'N'*-tetramethylated diamines,[244] and the charge-transfer spectra of $K_2BaFe(NO_2)_6$.[245]

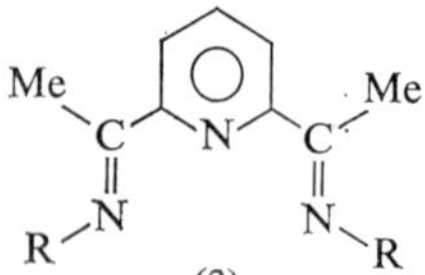

(3)

Iron(−I), d^9. The absorption spectrum of salmon coloured solutions of sodium tris-(2,2′-bipyridyl)ferrate(−I) in 1,2-dimethoxyethane has been recorded;[246] this spectrum is similar to those of Fe $bipy_3$ and Na^+ $bipy^-$.

Ruthenium(VIII) *and Osmium*(VIII), d^0. The electronic spectrum of osmium tetroxide, OsO_4, has been measured in the vapour state and in solution.[247] The irregularity of the main intervals between peaks, and their sensitivity to temperature was indicative of a perturbed spectrum, but the spectrum was too diffuse to analyse convincingly. Three fragmentary systems of quite different appearance were observed in all samples and may be due to weakly forbidden transitions. The spectrum of ruthenium tetroxide, RuO_4, in the vapour phase was also investigated, but the resolution was no better. The reproducible features of both spectra are tabulated.

Ruthenium(III), d^5. Several 6 : 1 azide complexes of ruthenium(III), rhodium(III), iridium(III), and platinum(IV) and 4 : 1 complexes of palladium(II), platinum(II), and gold(III) have been prepared with large organic cations which do not absorb in the near-u.v. region.[248] The ligand-field and charge-transfer bands in the spectra were assigned. Resulting from this study, the azide ligand was placed between sulphur-bonded SCN^- and diethyldithiophosphate in the spectrochemical series, and between bromide and diethyldithiophosphate in the nephelauxetic series.

241 J. D. Curry, M. A. Robinson, and D. H. Busch, *Inorg. Chem.*, 1967, **6**, 1570.
242 A. A. Schilt and T. W. Leman, *J. Amer. Chem. Soc.*, 1967, **89**, 2012.
243 R. A. Bailey and T. R. Peterson, *Canad. J. Chem.*, 1967, **45**, 1135.
244 I. Bertini and F. Mani, *Inorg. Chem.*, 1967, **6**, 2032.
245 K. G. Caulton and R. F. Fenske, *Inorg. Chem.*, 1967, **6**, 562.
246 C. Mahon and W. L. Reynolds, *Inorg. Chem.*, 1967, **6**, 1927.
247 E. J. Wells, A. D. Jordan, D. S. Alderdice, and I. G. Ross, *Austral. J. Chem.*, 1967, **20**, 2315.
248 H. H. Schmidtke and D. Garthoff, *J. Amer. Chem. Soc.*, 1967, **89**, 1317.

The visible and u.v. spectra for a number of ethylenediamineruthenium(III) complexes have also been recorded.[249] The spectra support the formulation $[RuX_2\,en_2][RuX_4\,en]$ (X = Cl or Br) rather than distribution isomers such as $[Ru\,en_3][RuX_6]$, $[Ru\,en_3][RuX_4\,en]_3$, or $[RuX_2\,en_2]_3[RuX_6]$.

Ruthenium(II), d^6. The visible and u.v. spectra have been reported[250] for a number of bis-(2,2′-bipyridyl) and mixed 2,2′-bipyridyl, 2,2′,2″-terpyridyl complexes of ruthenium(II) in aqueous solutions and in nitromethane. Two bands were observed in the visible range for tris-(2-picolylamine)-ruthenium(II) chloride; in this case identical spectra were obtained both in the solid state and in solution (H_2O and $MeNO_2$).[251] A band is observed[252] at 19,700 cm.$^{-1}$ in the electronic spectrum of H_2RuCl_5NO and an intense absorption beyond 40,000 cm.$^{-1}$. Evidence for Ru^{II} chloro-complexes in acqueous hydrochloric acid solutions has also appeared.[253]

Cobalt, Rhodium, and Iridium.—*Cobalt*(III), d^6. The electronic spectra have been recorded[254] for some new polyhedral complexes $[Co(B_7C_2H_9)_2]^-$ and $[(C_5H_5)Co(B_7C_2H_9)]$. Several reports have appeared dealing essentially with the kinetics of reactions of cyano-cobalt(III) complexes.[255–258]

It appears that *cis*-$[Co(NH_3)_4Cl_2]IO_3$ is the product formed when solid *trans*-$[Co(NH_3)_4Cl_2]IO_3,2H_2O$ is heated.[259] Although there are several reports in the literature on the visible spectrum of *cis*-$[Co(NH_3)_4Cl_2]Cl$ there is considerable disagreement between them. The absorption maximum at 18,700 cm.$^{-1}$ exhibited by the product is in the region expected for the *cis*-$[Co(NH_3)_4Cl_2]^+$ cation and certainly the analogous *cis*-$[Co\,en_2Cl_2]^+$ cation also displays a maximum at about 18,700 cm.$^{-1}$. The compound originally formulated as $[Co(NH_3)_4(NO_2)_2]NO_3,HNO_3$ has been reformulated as *cis*-$[Co(NH_3)_4(HNO_3)(NO_2)](NO_3)_2$. Washing or warming this compound gives a product which is shown by electronic spectral measurements to be *cis*-$[Co(NH_3)_4(NO_2)_2](NO_3)$.[260] Studies of the competition between water and other species as entering groups in the replacement of the ligand, L, from the complexes $[Co^{III}(NH_3)_5L]^{n+}$ [L = H_2O, Cl^-, Br^-, NO_3^-, $-NCS^-$, or $(MeO)_3PO$] have been extended.[261] The reactions were followed spectrophotometrically and the absorption spectra of the species involved are tabulated. The ${}^1T_{1g} \leftarrow {}^1A_{1g}$ transitions

249 J. A. Broomhead and L. A. P. Kane-Maguire, *J. Chem. Soc.* (*A*), 1967, 546.
250 N. R. Davies and T. L. Mullins, *Austral. J. Chem.*, 1967, **20**, 657.
251 F. G. Nasouri, M. W. Blackmore, and R. J. Magee, *Austral. J. Chem.*, 1967, **20**, 1291.
252 R. Bramley, B. N. Figgis, and R. S. Nyholm, *J. Chem. Soc.* (*A*), 1967, 861.
253 M. G. Adamson, *Austral. J. Chem.*, 1967, **20**, 2517.
254 M. F. Hawthorne and T. A. George, *J. Amer. Chem. Soc.*, 1967, **89**, 7114.
255 R. Grassi, A. Haim, and W. K. Wilmarth, *Inorg. Chem.*, 1967, **6**, 237.
256 R. Barca, J. Ellis, Maak-Sand Tsao, and W. K. Wilmarth, *Inorg. Chem.*, 1967, **6**, 243.
257 P. H. Tewari, R. W. Gaver, H. K. Wilcox, and W. K. Wilmarth, *Inorg. Chem.*, 1967, **6**, 611.
258 M. D. Johnson, M. L. Tobe, and Lai-Yoong Wong, *J. Chem. Soc.* (*A*), 1967, 491.
259 H. E. Le May and J. C. Bailar, *J. Amer. Chem. Soc.*, 1967, **89**, 5577.
260 R. Ugo and R. D. Gillard, *J. Chem. Soc.* (*A*), 1967, 2078.
261 G. E. Dolbear and H. Taube, *Inorg. Chem.*, 1967, **6**, 60.

have been tabulated for some 28 octahedral cobalt(III) complexes, many of which contain *N*-donor ligands.[262] Spectral data have also been presented for the ions $[Co(MeNH_2)_5Cl]^{2+}$ and $[Co(MeNH_2)_5H_2O]^{3+}$,[263] a series of new complexes of the type $[Co\,bipy_2X_2]^+$ (X = Cl, Br, I, or NO_2^-),[264] and several tris-(*N*-alkyl-2-hydroxypropiophenimine)cobalt(III) complexes.[265]

The visible absorption spectra for a series of salts of the type $[Co\,en_2LCl]^{2+}$ (L = ethanolamine, 2- and 3-hydroxypropylamine,[266] and hydroxylamine[267]) have been described and it is concluded from these data that the complexes have *cis*-configurations. On the other hand, the complexes $[Co(\pm)\text{-}pn_2NCSX]SCN$ (X = Cl or Br) have been assigned *trans*-configurations by comparing their *d–d* spectra with those of *cis*- and *trans*-bis(ethylenediamine)cobalt(III) systems.[268] The electronic absorption spectra of *cis*-$[Co\,en_2ClN_3]^+$, *cis*- and *trans*-$[Co\,en_2N_3(H_2O)]^{2+}$ ions,[269] and several cobalt(III) hexamines have been reported[270] and the kinetics of octahedral cobalt(III) complexes in dipolar aprotic solvents have been studied.[271, 272] The inter-relations between Werner's green and red μ-amido-μ-peroxo-bis{bis(ethylenediamine)cobalt} salts $[en_2Co(NH_2)(O_2)Co\,en_2](NO_3)_4$ and $[en_2Co(NH_2)(O_2)Co\,en_2]H(NO_3)_4,2H_2O$ respectively have been examined, and the electronic spectra of several other dibridged dicobalt ethylenediamine complexes reported.[273]

The electronic spectra of three isomers of the type $[Co(CN)_2\,trien]^+$ have been recorded and compared with that of *cis*-$[Co(CN)_2\,trien]ClO_4$. Structures (4), (5), and (6) were suggested for the isomers consistent with their spectra.[274] The spectra of other triethylenetetramine complexes of the types *cis*- and *trans*-$[Co\,trienX_2]^{n+}$ (X = Cl, H_2O, or NO_2; 2X = CO_3) have also been described and some assignments given.[275]

A number of cobalt(III) complexes, containing the tetradentate macrocyclic Schiff base, $C_{16}H_{32}N_4$, (7) have been described; the complexes were of the type $[Co(C_{16}H_{32}N_4)L_2]$ (L = Cl^-, Br^-, OH^-, NCS^-, N_3^-, CN^-, NO_2^-, or H_2O). It was concluded that the ligands L are *trans* to each other, and an analysis of the visible spectra indicates that the macrocyclic Schiff base has an appreciably higher crystal-field strength

262 E. Lazzarini, *J. Inorg. Nuclear Chem.*, 1967, **29**, 7.
263 M. Parris, *J. Chem. Soc.* (*A*), 1967, 583.
264 A. A. Vlček, *Inorg. Chem.*, 1967, **6**, 1425.
265 A. Chakrovorty and B. Behera, *Inorg. Chem.*, 1967, **6**, 1812.
266 S. C. Chan and F. Leh, *J. Chem. Soc.* (*A*), 1967, 1730.
267 S. C. Chan and F. Leh, *J. Chem. Soc.* (*A*), 1967, 573.
268 S. C. Chan, C. L. Chik, and B. Hui, *J. Chem. Soc.* (*A*), 1967, 607.
269 D. A. Buckingham, I. I. Olsen, and A. M. Sargeson, *Inorg. Chem.*, 1967, **6**, 1807.
270 D. M. Palade, *Russ. J. Inorg. Chem.*, 1967, **12**, 520.
271 I. R. Lantzyke and D. W. Watts, *J. Amer. Chem. Soc.*, 1967, **89**, 815.
272 W. R. Fitzgerald and D. W. Watts, *J. Amer. Chem. Soc.*, 1967, **89**, 821.
273 M. Mori and J. A. Weil, *J. Amer. Chem. Soc.*, 1967, **89**, 3732.
274 K. Kuroda and P. S. Gentile, *Inorg. Nuclear Chem. Letters*, 1967, **3**, 151.
275 A. M. Sargeson and G. H. Searle, *Inorg. Chem.*, 1967, **6**, 787.

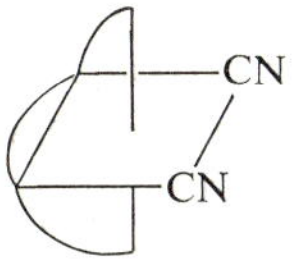

(4)

cis-α-cyano
(light yellow)

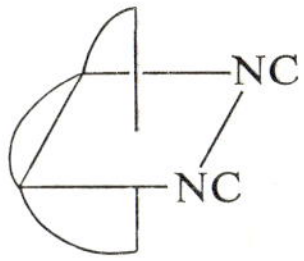

(5)

cis-α-isocyano
(deep yellow)

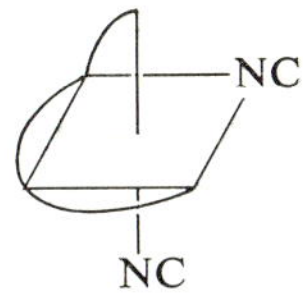

(6)

cis-β-isocyano
(deep yellow)

than is found for related primary and secondary amine ligands.[276] The electronic spectra of cobalt(III) complexes of the tetradentate ligand β,β′,β″-triaminotriethylamine (tren) of the type *cis*-[Co tren X_2]$^{n+}$ (X = Cl, Br, or H_2O) and *cis*-[Co tren BrH_2O]$^{2+}$ have been tabulated and compared with the spectra of related tetramine complexes taken from the literature.[277]

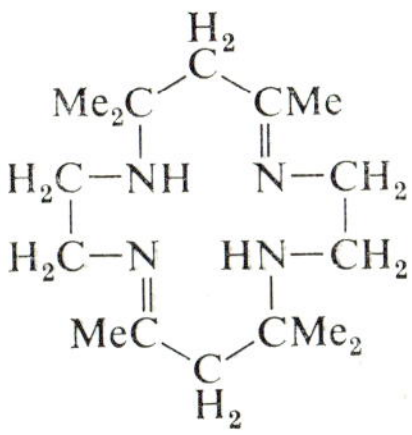

(7)

The charge-transfer spectra of the hexanitro-complexes of Co^{III}, Rh^{III}, Fe^{II}, Co^{II}, Ni^{II}, and Cu^{II} have been reported and interpreted on the basis of a molecular model with T_h symmetry.[245] A molecular-orbital energy-level diagram which accounts for the charge-transfer spectra is presented. By means of this diagram the weak band at 21,500 cm.$^{-1}$ in $[Co(NO_2)_6]^{3-}$ is shown to be an orbitally forbidden charge-transfer transition from a non-bonding oxygen orbital to the predominantly metal e_g orbital. This assignment obviates anomalous spectroscopic properties previously attributed to the nitro-group.

The electronic absorption spectra have been investigated in the visible region for the two geometrical isomers (red and violet) of the tris-(β-alaninato)cobalt(III) complex.[278] The main difference between the spectra appears in the region of the first absorption band; the violet isomer exhibits a distinct inflection in this region, while the red isomer has a symmetrical absorption band. The inflection in the spectrum of the violet isomer is attributed to a splitting of the $^1T_{1g}$ state due to a rhombic crystal-field component in the 1,2,6-isomer. The red isomer, which has a sharp single band at 19,011 cm.$^{-1}$, belongs to the cubic crystal-field and is assigned the facial 1,2,3-configuration.

[276] N. Sadasivan, J. A. Kernohan, and J. F. Endicott, *Inorg. Chem.*, 1967, **6**, 770.
[277] S. K. Madan and J. Peone, jun., *Inorg. Chem.*, 1967, **6**, 463.
[278] M. B. Ćelap, S. R. Niketić, T. J. Janjić, and V. N. Nikolić, *Inorg. Chem.*, 1967, **6**, 2063.

Reactions have been reported in which metal ions promote the formation of peptide bonds through the co-ordination of ester carbonyl groups.[279] The visible spectra of the dipeptide and glycinamide complexes of cobalt(III) are listed, along with the spectra of other relevant cobalt(III) complexes. Two bands corresponding to the two spin-allowed *d–d* transitions for octahedral cobalt(III) are observed in the visible spectra of each complex. Spectral data have been recorded for *cis*- and *trans*-diacetatobis-(ethylenediamine)cobalt(III) ions,[280] and the visible and near-u.v. absorption spectra for sixteen complexes of the type [Co ox_x gly_y en_z] have been measured and discussed.[281]

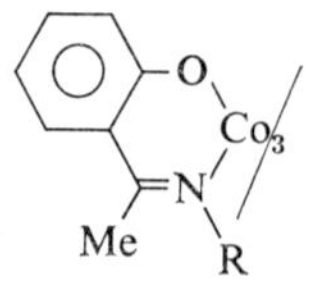

(8)

(R = Me, Et, Pr^n, Bu^n)

The electronic spectra of the *trans*-complexes (8) of the type tris-(*N*-alkyl-2-hydroxyacetophenoneimine)cobalt(III) have been studied.[282] Two bands are observed in the visible range at about 15,900 cm.$^{-1}$ for the complex with R = Me, whilst for the remaining chelates only one asymmetric band is observed. The two bands in the R = Me complex are assigned as $^1B_2 \leftarrow {}^1A_1$ (17,000 cm.$^{-1}$) and $^1B_1 \leftarrow {}^1A_1$ (15,630 cm.$^{-1}$) on the basis of a C_{2v} model. The u.v. and visible spectra have been reported and briefly discussed[283] for the organo-cobalt chelates of bis(salicylaldehyde)ethylenedi-imine, (salen) of the types RCo(salen)L and RCo(salen) (R = Me, Et, Pr, Bu, and Ph; L = H_2O, NH_3, py, or benzimidazole).

Spectral data have appeared for several diketone complexes of cobalt(III). Boucher[284] has reported the spectra of the complexes [$CoLNO_2$ py] (L = acac, propionylaceto, and methoxyacetylaceto) and [CoL_2 en]I. A broad band at 18,700–18,900 cm.$^{-1}$ has been assigned to the $^1T_{1g} \leftarrow {}^1A_{1g}$ transition (pseudo O_h microsymmetry). The $^1T_{2g} \leftarrow {}^1A_{1g}$ transition, however, is obscured by the tail of an intense charge-transfer absorption in the u.v. region. The charge-transfer absorptions $t_{2g} \rightarrow \pi^*$ and two higher energy $\pi \rightarrow \pi^*$ absorptions are assigned to the enolate anion. The metal–ligand charge-transfer band at 29,700–30,900 cm.$^{-1}$ appears to be independent of the β-diketone. In methanol solution tris(isonitrosoacetylacetonato)cobalt(III) has two very intense bands at 40,000 and

279 J. P. Collman and E. Kimura, *J. Amer. Chem. Soc.*, 1967, **89**, 6096.
280 V. Carunchio, G. Illuminati, and G. Ortaggi, *Inorg. Chem.*, 1967, **6**, 2168.
281 N. Matsuoka, J. Hidaka, and Y. Shimura, *Bull. Chem. Soc. Japan*, 1967, **40**, 1868.
282 A. Chakravorty and K. C. Kalia, *Inorg. Chem.*, 1967, **6**, 690.
283 G. Costa, G. Mestroni, and L. Stefani, *J. Organometallic Chem.*, 1967, **7**, 493.
284 L. J. Boucher, *Inorg. Chem.*, 1967, **6**, 2162.

37,030 cm.$^{-1}$ and a slightly weaker band at 28,110 cm.$^{-1}$. The ligand has a band at 43,470 cm.$^{-1}$ and the appearance of two very intense bands in the complex in this region has been taken to indicate that the energy states of the π-electron system of the ligand suffer substantial alterations on complex formation.[285] Other reports have included spectral data on the solvolysis products of the *trans*-dinitrobis(acetylacetonato)cobaltate(III) ion[286] and on tris(hexafluoroacetylacetonato)cobalt(III).[287]

The optical spectra of single crystals of high-purity garnets $Y_3Fe_5O_{12}$, $Y_3Al_5O_{12}$, and $Y_3Ga_5O_{12}$ doped with cobalt ions have been studied in quite some detail.[288] The spectra for all three crystals are essentially the same, and the spectrum has been attributed to the rarely observed tetrahedral Co^{3+}. The ligand-field spectra have been discussed[289] for heteropolyanions of the type $[XZW_{11}O_{40}H_2]^{n-}$ (X, Z = Si, Co^{III}; Ge, Co^{III}; and Ge, Co^{II}).

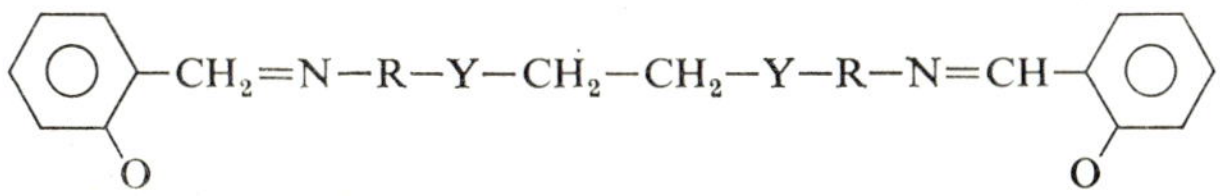

(9) R = o-C_6H_4

The electronic spectra of the complexes $[Co(o\text{-}NH_2C_6H_4YCH_2)_2Cl_2]Cl$ (Y = O, S, or NH) and [CoL]Cl [L = (9) with Y = O, S, or NH] have been investigated.[290] Comparison of the spectra shows that the ligand-field strength of the ether-oxygen in these cobalt(III) complexes is similar to that of the sulphide-sulphur and lower than that of the imino-nitrogen. The ligand-field strengths of ethylxanthate and ethylthioxanthate in cobalt(III) compounds have been investigated using the Dharmatti–Kanekar n.m.r. method.[291] These results, together with those obtained earlier and confirmed in this investigation, indicate that the ligand-field strength increases in the order: diethyldithiophosphate $[(EtO)_2PS_2^-]$ < diethyldithiocarbamate $(Et_2NCS_2^-)$ < ethyl trithiocarbonate $(EtSCS_2^-)$ < O – ethyl dithiocarbonate $(EtOCS_2^-)$.

Cobalt(II), d^7. Cobalt(II) complexes containing water and isocyanides (MeNC and PhNC) as the only ligands have been investigated.[292] The previously reported blue (phenyl isocyanide)cobalt(II) complex has now been shown to be the sesquihydrate and to contain the six-co-ordinate complex ion $[Co(PhNC)_5(H_2O)]^{2+}$. The similarity of the electronic spectra of the blue phenyl isocyanide complex in solution and in the solid state and of the methyl isocyanide complex in aqueous solution shows that they

285 N. J. Patel and B. C. Haldar, *J. Inorg. Nuclear Chem.*, 1967, **29**, 1037.
286 B. P. Cotsoradis and R. D. Archer, *Inorg. Chem.*, 1967, **6**, 800.
287 M. Kilner, F. A. Hartman, and A. Wojcicki, *Inorg. Chem.*, 1967, **6**, 406.
288 D. L. Wood and J. P. Remeika, *J. Chem. Phys.*, 1967, **46**, 3595.
289 T. J. R. Weakley and S. A. Malik, *J. Inorg. Nuclear Chem.*, 1967, **6**, 2935.
290 R. D. Cannon, B. Chiswell, and L. M. Venanzi, *J. Chem. Soc.* (*A*), 1967, 1277.
291 C. R. Kanekar, M. M. Dhingra, V. R. Marathe, and R. Nagarajan, *J. Chem. Phys.*, 1967, **46**, 2009.
292 J. M. Pratt and P. R. Silverman, *J. Chem. Soc.* (*A*), 1967, 1286.

have the same structure. The blue solid can be reversibly dehydrated to a yellow anhydrous compound, which is shown to contain the five-co-ordinate ion $[Co(PhNC)_5]^{2+}$. The complex present in freshly prepared solutions containing cobalt(II) and cyanide ions has been identified[293] as being predominantly the six-co-ordinate species, $[Co(CN)_5H_2O]^{3-}$, by comparison of its electronic spectrum with those of the analogous isocyanide complexes. No evidence was found for the hexacyanide $[Co(CN)_6]^{4-}$ under equilibrium conditions, but another complex, probably the tetracyanide $[Co(CN)_4(H_2O)_2]^{2-}$, is formed in very dilute solutions. In contrast with the above work, Alexander and Gray[294] assumed a five-co-ordinate structure for the cyanide and interpreted the e.s.r .and optical spectra in terms of a square pyramidal structure for $Co(CN)_5{}^{3-}$ in solution. The u.v. and visible spectrum[295] of the product of the reaction of the pentacyanocobalt(II) ion and hydrogen is shown to be identical with that of hydropentacyanocobaltate(II).

The electronic spectra of the complexes $[CoN(CH_2CH_2NH_2)_3X]X$ (X = I or NCS) suggest five-co-ordinate structures. The spectra are shown to be very different from those of cobalt(II) complexes with tetrahedral or octahedral stereochemistries.[296] They are, instead, very similar to the spectrum of the compound $[CoN(CH_2CH_2NMe_2)_3Br]Br$, which has been shown to be five-co-ordinate by an *X*-ray study. The electronic spectra of the five-co-ordinate complexes $[CoLX_2]$ (X = Cl or Br; L = tridentate *N*-donor Schiff-base formed from *N*-methyl-*o*-amino-benzaldehyde and *NN*-diethylethylenediamine) have also been reported.[297] The spectra are very similar to those of other high-spin cobalt complexes, such as $[Co\{MeN(CH_2{\cdot}CH_2{\cdot}NMe_2)_2\}Cl_2]$, for which a distorted five-co-ordinate structure has been established by *X*-ray crystallography. The first three bands in the spectra can be assigned to transitions between levels originating from splitting of the ^4F term. As for other five-co-ordinate complexes, the first band is assignable to a $^4E''(F) \leftarrow {}^4A_2'(F)$ transition and the other two to the two components of the $^4E'(F) \leftarrow {}^4A_2'(F)$ transition in D_{3h} symmetry. The fact that the latter transition is split into two components with an energy difference of about 1500 cm.$^{-1}$ is indicative of a high distortion in these complexes. The final band, which is due to a $^4A_2'(P) \leftarrow {}^4A_2'(F)$ transition is expected to be almost independent of the ligand-field strength and the geometry of the complex and is, as expected, around 16,000 cm.$^{-1}$.

The Schiff-base complexes $[Co(X\text{-sal}{\cdot}R)_2]$ (X = 5-chloro-, 5-bromo-, 5-nitro-, 3,5-dibromo-, or 5,6-benzo; R = 2,6-dimethylphenyl- or 2,6-diethylphenyl-) have been shown to be tetrahedral in non-donor solvents

293 J. M. Pratt and R. J. P. Williams, *J. Chem. Soc.* (*A*), 1967, 1291.
294 J. J. Alexander and H. B. Gray. *J. Amer. Chem. Soc.*, 1967, **89**, 3356.
295 M. G. Burnett, P. J. Connolly, and C. Kemball, *J. Chem. Soc.* (*A*), 1967, 800.
296 M. Ciampolini and P. Paoletti, *Inorg. Chem.*, 1967, **6**, 1261.
297 L. Sacconi, I. Bertini, and R. Morassi, *Inorg. Chem.*, 1967, **6**, 1548.

and in the solid state on the basis of their electronic spectra and magnetic properties.[298] However, electronic spectra indicate that the five-co-ordinate species, [Co(X-sal R)$_2$ py], are present in pyridine solutions; some of the pyridine solvates were isolated in the solid state. These authors[299] have also used the electronic spectra of cobalt(III) complexes formed by Schiff bases, derived from 3-methoxysalicylaldehyde and amines, as a diagnostic test in assigning configurations. The electronic spectra have also been reported for complexes with quadridentate Schiff bases.[300]

Spectral data have appeared for several cobalt(II) complexes containing substituted pyridines.[301, 302] Wasson *et al.*[301] have given the ligand-field parameters *Dq*, *B*, and β for the pseudo-tetrahedral complexes [Co(4-X-py)$_2$Cl$_2$] (X = OH, SPh, CH$_2$·OH, or —CH(OH)—), and have shown that the ligands studied lie close to benzimidazole in the spectrochemical series. The maxima in the reflection spectrum have also been tabulated and assigned[302] for the pseudohalide complexes [CoL$_2$X$_2$] [L = 3- or 4-methylpyridine, 2,3- or 2,4-dimethylpyridine, quinoline, or isoquinoline; X = N(CN)$_2$ or C(CN)$_3$].

A series of planar cobalt(II) complexes with *O*-alkyl-1-amidinoureas (10) have been reported.[303] Neutral complexes are formed by proton loss from

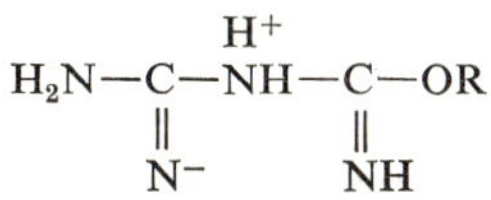

(10)

the ligand, and the electronic spectra of these complexes have been used to obtain the ordering of the *d* orbitals. The electronic spectra of a number of tetrahedral and octahedral complexes of the type [M(diamine)X$_2$] (M = Co or Ni; X = Cl, Br, I, NO$_3$, or NCS; diamine = *NNN'N'*-tetramethylated ethylenediamine, 1,2-diaminopropane, or trimethylenediamine) have been reported.[304] The spectral and magnetic studies indicate that all the cobalt halide complexes and [Co{Me$_2$N·[CH$_2$]$_3$·NMe$_2$}(NCS)$_2$] are tetrahedral.

Three high-spin approximately square pyramidal complexes β-[Co paphy X$_2$] [X = Cl or Br; paphy = pyridine-2-aldehyde-2'-pyridylhydrazone (11)] and Co terpy Cl$_2$ have been reported and shown to have

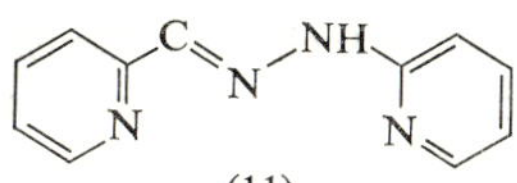

(11)

[298] S. Yamada and E. Yoshida, *Bull. Chem. Soc. Japan.*, 1967, **40**, 1298.
[299] E. Yoshida and S. Yamada, *Bull. Chem. Soc. Japan*, 1967, **40**, 1395.
[300] W. Weigold and B. O. West, *J. Chem. Soc.* (*A*), 1967, 1310.
[301] J. R. Wasson, K. K. Gauguli, and L. Theriot, *J. Inorg. Nuclear Chem.*, 1967, **29**, 2807.
[302] H. Köhler and B. Seifert, *Z. anorg. Chem.*, 1967, **352**, 265.
[303] V. Rasmussen and W. A. Baker, *J. Chem. Soc.* (*A*), 1967, 1712.
[304] L. Sacconi, I. Bertini, and F. Mani, *Inorg. Chem.*, 1967, **6**, 262.

similar electronic reflectance spectra.[305] The spectra have been compared with those of other complexes definitely known to contain high-spin five-co-ordinate cobalt(II), and it is apparent that the spectra of square pyramidal cobalt(II) compounds are distinguished from those of four- and six-co-ordinate complexes with cubic symmetry by an absorption maximum (ε *ca.* 20) in the 11,000–14,000 cm.$^{-1}$ range. A novel metal–metal bonded cobalt(II) complex formulated as $[\{(H_2O)CoB—\}_2](ClO_4)_4,H_2O$ has been reported,[306] where the ligand, B, is hexamethyl-1,4,8,11-tetra-azacyclotetradecadiene (12). The electronic spectrum in aqueous solution consists of

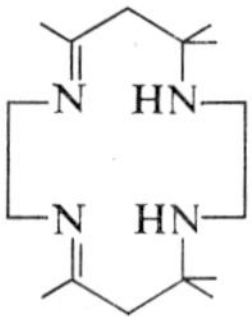

(12)

intense bands at 45,500 and 30,300 cm.$^{-1}$, and these are assigned as B → M and M → B charge-transfer transitions respectively. A charge-transfer band has also been located for a salicylideneimine complex of cobalt(II) in the region 25,000–27,000 cm.$^{-1}$ and assigned to a metal-to-ligand electron transfer.[307] The high energy of the charge-transfer band for complexes derived from *N*-phenylsalicylideneimines bearing *ortho*-substituents on the phenyl ring has been discussed in terms of steric strain. Weak bonding between the *o*-methoxy-groups and the metal is suggested to explain the low energy of the charge-transfer band for the chelate bis-(*N*-*o*-anisylsalicylideneiminato)cobalt(II). Spectral data have also been reported for cobalt(II) complexes with the nitrogen-donor ligands imidazole, thiazole,[308] furfurylamine derivatives,[309] azopyridines,[310] azo-dyestuffs,[311] 2,2′-biquinoline,[312] and 1,2-dipyridylethylene isomers.[313]

Crystalline complexes containing the five-co-ordinate $[CoL_2X]^+$ species {X = Cl, Br, or I; L = bidentate ligand [diphenyl(*o*-methylthiophenyl)-phosphine (13) diphenyl(*o*-methylselenophenyl)phosphine (14), or diphenyl-(*o*-diphenylarsinophenyl)phosphine (15)]} have been isolated.[3] The magnetism and electronic spectra can be interpreted in terms of a square pyramidal arrangement of donor atoms. Three distinct bands appear below 22,000 cm.$^{-1}$ and despite their high intensity they are thought to be predominantly due to ligand-field transitions. Thus, changes in the anionic

[305] I. G. Dance, J. Lewis, and F. Lions, *J. Chem. Soc.* (*A*), 1967, 565.
[306] J. L. Love and H. K. J. Powell, *Inorg. Nuclear Chem. Letters*, 1967, **3**, 113.
[307] A. A. Diamantis, H. Weigold, and B. O. West, *J. Chem. Soc.* (*A*), 1967, 1281.
[308] W. J. Eilbeck, F. Holmes, and A. E. Underhill, *J. Chem. Soc.* (*A*), 1967, 757.
[309] M. D. Joesten, K. G. Claus, and K. P. Lannert, *J. Inorg. Nuclear Chem.*, 1967, **29**, 1421.
[310] P. J. Beadle and R. Grzestowiak, *Inorg. Nuclear Chem. Letters*, 1967, **3**, 245.
[311] R. Price, *J. Chem. Soc.* (*A*), 1967, 521.
[312] G. H. Faye and J. L. Horwood, *Canad. J. Chem.*, 1967, **45**, 2335.
[313] H. Brierley and W. J. Geary, *J. Chem. Soc.* (*A*), 1967, 963.

ligand cause the bands to shift as predicted by the spectrochemical series and not as expected for charge-transfer transitions.

(13)

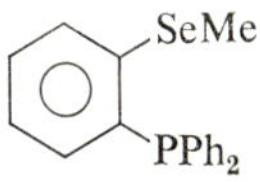
(14)

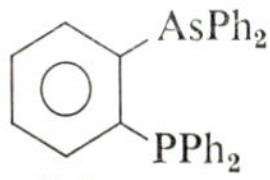
(15)

The electronic structures of the complexes $Co(Ph_2P[CH_2]_2 \cdot PPh_2]_2X_2$ (X = Cl, Br, or I) have been investigated by studying their e.s.r. and optical electronic spectra in detail.[314] It is concluded that the electronic ground-state is $|(\overset{+}{xy})(\overset{-}{xy})(x^2 \overset{+}{-} y^2)\rangle$, and this is consistent with a five-co-ordinate structure. The energies of the one-electron *d* orbitals were estimated using the observed optical data, and the interelectronic interaction energies were calculated for the low-spin d^7 system.

Few reports on arsenic-donor complexes of cobalt(II) have appeared during the year, but the positions of the absorption maxima in solution and in the solid state have been recorded[315] for the diarsine complexes *trans*-Co $diars_2X_2$ [diars = *o*-phenylenebis(dimethylarsine); X = ClO_4, NO_3, or Cl].

A number of cobalt(II) complexes of β-ketoamines (16) derived from aromatic amines and β-diketones, such as acetylacetone and benzoyl-acetone, have been prepared.[316] The electronic spectra are indicative of tetrahedral stereochemistry in non-donor solvents, but in pyridine the complexes may be tetrahedral, five-co-ordinate, or octahedral depending upon the ligand. The reaction of *NN'*-ethylenebis(salicylideneiminato)-cobalt(II) and its conversion into a methyl derivative containing a direct cobalt–carbon bond has been studied and the electronic absorption spectra have been recorded for the compounds involved.[317]

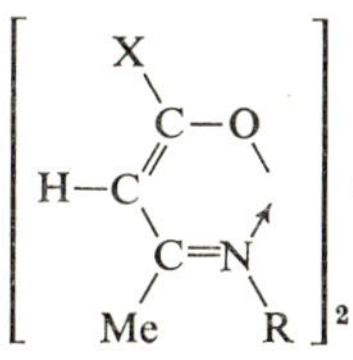
(16)

The electronic spectra for complexes of the general formula $[M\,py_2(CX_xH_{3-x}CO_2)_2]$ (x = 1, 2, or 3; M = Co^{II}, Ni^{II}, or Cu^{II}; X = F or Cl) have been reported,[318] along with a number of analogous tetrapyridine derivatives of Co^{II} and Ni^{II}. The interpretation of the splitting of the main

[314] W. DeW. Harrocks, jun., G. R. Van Hecke, and D. DeW. Hall, *Inorg. Chem.*, 1967, **6**, 694.
[315] G. A. Rodley and P. W. Smith, *J. Chem. Soc.* (*A*), 1967, 1580.
[316] S. Yamada and E. Yoshida, *Bull. Chem. Soc. Japan*, 1967, **40**, 1854.
[317] F. Calderazzo and C. Floriani, *Chem. Comm.*, 1967, 139.
[318] A. B. P. Lever and D. Ogden, *J. Chem. Soc.* (*A*), 1967, 2041.

band in the visible region, usually near 18,000–20,000 cm.$^{-1}$ in the electronic spectra of the Co^{II} complexes, presents something of a problem and is the major theme of this report. These complexes exhibit a broad band near 9000 cm.$^{-1}$ and this may safely be assigned to the $^4T_{2g} \leftarrow {}^4T_{1g}$ transition. Furthermore, certain of the complexes exhibit a weak band or shoulder between 16,000 and 17,500 cm.$^{-1}$. This is approximately twice the energy of the first band, and in accordance with a rule proposed in this paper, this may well be due to the $^4A_{2g} \leftarrow {}^4T_{1g}$ transition. Finally, it is concluded that the peak and shoulder in the main band are due to transitions to the $^4A_{2g}$ and ^4E_g components of the $^4T_{1g}(P)$ level in D_{4h} symmetry.

The remarkable band-splitting effects in the spectra of the spinel-type phases $Ni_{2-x}Mg_xGeO_4$ ($0 \geqslant x \geqslant 1{\cdot}2$) and $Co_{2-x}Mg_xGeO_4$ ($0 \geqslant x \geqslant 1$) have been explained by assuming that the co-ordination octahedra of Ni^{II} and Co^{II} are compressed along their trigonal axes. A quantitative crystal-field treatment is given.[319] The absorption of Co^{II} in both 2,5-spinels [320] and 2,6-spinels [321] has also been reported.

I.r. and electronic spectral data for complexes of the type $CoL_2(NO_2)_2$ (L = substituted pyridine-*N*-oxide or a substituted quinoline-*N*-oxide) have been interpreted in terms of a six-co-ordinate structure with chelating nitrite groups.[322] The electronic spectra for a number of Co^{II}, Cu^{II}, and Ni^{II} complexes of diphenylcyclopropenone have been reported.[323] In all cases the co-ordination occurs via the carbonyl oxygen and the position of this ligand in the spectrochemical series is close to that of water but greater than that of pyridine-*N*-oxide. The spectrum of Co(OMe)Cl has been interpreted on the basis of octahedral stereochemistry, whilst the spectra of Co(OMe)Br and its solvate Co(OMe)Br,2MeOH are assigned to tetrahedral and octahedral Co^{II} respectively.[324] The spectrum of Co(OMe)I proved more difficult to interpret due to an intense charge-transfer band obscuring the visible region.

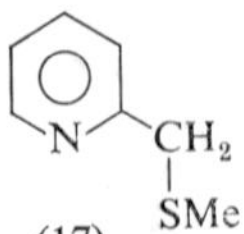

(17)

Magnetic and electronic spectral data have been used to elucidate the structures of the tetrahedral complexes $Co(R_2PS_2)_2$ (R = Et or Pr) and octahedral adducts with pyridine and thiophen.[325] The diffuse reflectance spectra of the complexes $CoLX_2$ [X = Cl, Br, or NCS and L = bis-(2-dimethylaminoethyl)sulphide] are clearly diagnostic of a five-co-ordinate

319 D. Reinen, *Theor. Chim. Acta*, 1967, **8**, 260.
320 H. Kasper, *J. Inorg. Nuclear Chem.*, 1967, **29**, 2693.
321 H. Kasper, *Z. anorg. Chem.*, 1967, **354**, 78.
322 L. El-Sayed and R. O. Ragsdale, *Inorg. Chem.*, 1967, **6**, 1644.
323 C. W. Bird and E. M. Briggs (née Hollins), *J. Chem. Soc. (A)*, 1967, 1004.
324 G. A. Kakos and G. Winter, *Austral. J. Chem.*, 1967, **20**, 2343.
325 W. Kuchen and A. Judat, *Chem. Ber.*, 1967, **100**, 991.

structure; tentative assignments are given.[326] Cobalt(II) complexes with the ligand 2-methylthiomethylpyridine, (mmp), (17), have been described.[327] The reflectance spectra for the solid compounds $CoX_2\,mmp_2$ (X = Cl, Br, or SCN) and $[Co\,mmp_3](ClO_4)_2$ are indicative of six-co-ordination. The electronic spectra have also been reported for Co^{II} complexes of thiourea containing oxyanions[328] and octahedral complexes of the type CoX_2,L (X = Cl, Br, or I; L = 1,2-dithiocyanatoethane).[329]

An elegant study of the effect of temperature on the equilibria:

$$X^- + [MX_2,4\text{solvent}]_{\text{octahedral}} \rightleftharpoons [MX_3,\text{solvent}]^-_{\text{tetrahedral}}$$

has been reported for some cobalt and nickel halides in water and alcohols.[330] High temperatures and high halide-ion concentrations favour the tetrahedral species which were identified by characteristic intense absorptions. The absorption spectra of Co^{II} have been reported in molten mixtures of $ZnCl_2$ and $AlCl_3$, both as a function of melt composition and of temperature.[331] The spectral changes were interpreted in terms of an octahedral–tetrahedral equilibrium involving chloro-complexes of Co^{II}, analogous to those observed earlier for Ni^{II} in mixtures of KCl and $ZnCl_2$.

The pink colour of an aqueous solution of $CoCl_2$ changes to a reddish violet with the addition of a large amount of sodium perchlorate.[332] Spectral data indicate the formation of chloro-complexes presumably due to the decrease in water activity caused by the added inert salt. A similar effect was observed in aqueous solutions of Cu^{II} and Ni^{II} chlorides. Two complete series of the type $[Li\,da_2]_2[MCl_nBr_{4-n}]$ (M = Co^{II} or Cu^{II}; da = diacetamide) have been reported; the co-ordination of diacetamide to lithium provides a unique method of preparing and stabilising mixed halogeno-complexes of these transition elements.[333] The electronic spectra indicate tetrahedral co-ordination for the Co^{II} series, whereas the evidence indicates that Cu^{II} is involved in either planar or octahedral co-ordination. A note has appeared in which the absorption spectra of gaseous $CoCl_2$ and $CoBr_2$ were reported and briefly discussed,[334] and the absorption spectra of aqueous and organic phases obtained in the extraction of Co^{II} from HCl solution by tri-*n*-octylamine in benzene have been illustrated.[335]

Lower Oxidation States of Cobalt. All possible substitution products of the type $Co(NO)(CO)_x(PF_3)_{3-x}$ have been isolated.[336] They are all reddish liquids and their visible and u.v. spectra are similar. The absorption spectra

[326] M. Ciampolini and J. Gelsomini, *Inorg. Chem.*, 1967, **6**, 1821.
[327] P. S. K. Chia, S. E. Livingstone, and T. N. Lockyer, *Austral. J. Chem.*, 1967, **20**, 239.
[328] C. D. Flint and M. Goodgame, *J. Chem. Soc.* (*A*), 1967, 1718.
[329] D. C. Goodall, *J. Chem. Soc.* (*A*), 1967, 203.
[330] D. E. Scaife and K. P. Wood, *Inorg. Chem.*, 1967, **6**, 358.
[331] C. A. Angell and D. M. Gruen, *J. Inorg. Nuclear Chem.*, 1967, **29**, 2243.
[332] K. Mizutani and K. Sone, *Z. anorg. Chem.*, 1967, **350**, 216.
[333] P. S. Gentile, T. A. Shankoff, and J. Carlotto, *J. Inorg. Nuclear Chem.*, 1967, **29**, 1393.
[334] A. Trutia and M. Musa, *Spectrochim. Acta*, 1967, **23**, *A*, 1165.
[335] T. Sato, *J. Inorg. Nuclear Chem.*, 1967, **29**, 547.
[336] R. J. Clark, *Inorg. Chem.*, 1967, **6**, 299.

of sodium tetracarbonylcobaltate in water and of cobalt tricarbonyl nitrosyl in n-heptane have been reported for the first time,[337] and the u.v. spectrum of hexacarbonyl(decafluorotolan)dicobalt(0) (18) has also been reported.[338]

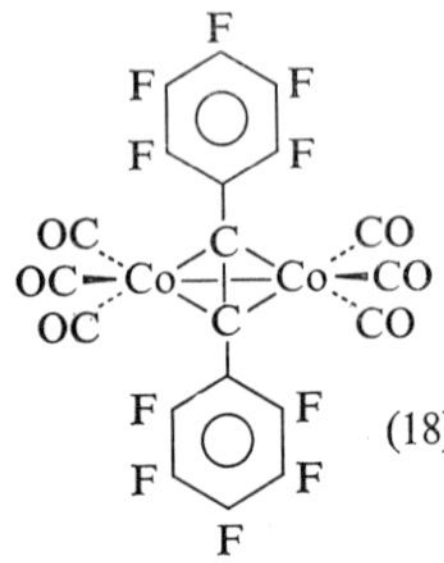

(18)

Rhodium and Iridium. The absorption spectra of K_3RhCl_6 and K_3IrCl_6 dissolved in LiCl–KCl eutectic mixtures have been studied.[339] The Rh^{III} and Ir^{III} species are shown to exist as octahedral MCl_6^{3-} complexes with lower values of *Dq* and higher values of β than in aqueous solution. The spectra of the *cis*- and *trans*-isomers of $[RhCl_4(H_2O)_2]^-$ have been assigned,[340] and in a separate publication the spectra of $[RhCl_4(H_2O)_2]^-$ and $[RhCl_5(H_2O)]^{2-}$ are illustrated.[341]

Several new Rh^{III} complexes with *N*-methyl-substituted ethylenediamines have been reported and from their electronic spectra the spectrochemical series for these ligands is deduced.[342] Electronic spectral data have also been presented for *trans*-$[Rh\,en_2(N_2H_4)Cl]^{2+}$, *trans*-$[Rh\,en_2(N_2H_4)_2]^{3+}$,[343] *trans*-$[Rh\,en_2L(H_2O)]^{2+}$, *trans*-$[Rh\,en_2L(OH)]^+$ (L = Cl, Br, or I),[344] a neptunium(v)–rhodium(III) complex,[345] and complexes of $RhCl_3$ with 1,4-thioxan and thioether ligands of the type $RS\cdot CH_2\cdot CH_2\cdot SR$.[346]

The u.v. and visible spectra of the complexes $[Ir\,bipy_3]^{3+}$ and $[Ir\,phen_3]^{3+}$ have been reported and assigned.[347] The luminescence in these complexes is attributed to a $\pi^* \rightarrow d$ transition. The electronic spectra have also been presented[348] for the complexes *cis*- and *trans*-$[Ir(H_2O)_2Cl_4]^-$, 1,2,6- and 1,2,3-$[Ir(H_2O)_3Cl_3]$, together with the spectra for the analogous iridium(IV) complexes.

337 E. A. C. Lucken, K. Noack, and D. F. Williams, *J. Chem. Soc. (A)*, 1967, 148.
338 J. M. Birchall, F. L. Bowden, R. N. Haszeldine, and A. B. P. Lever, *J. Chem. Soc. (A)*, 1967, 747.
339 J. R. Dickinson and K. E. Johnson, *Canad. J. Chem.*, 1967, **45**, 1631.
340 K. L. Bridges and J. C. Chang, *Inorg. Chem.*, 1967, **6**, 619.
341 W. Robb and M. M. de V. Steyn, *Inorg. Chem.*, 1967, **6**, 616.
342 G. W. Watt and P. W. Alexander, *J. Amer. Chem. Soc.*, 1967, **89**, 1814.
343 D. J. Baker and R. D. Gillard, *Chem. Comm.*, 1967, 520.
344 H. L. Bott and A. J. Pöe, *J. Chem. Soc. (A)*, 1967, 205.
345 R. K. Murmann and J. C. Sullivan, *Inorg. Chem.*, 1967, **6**, 892.
346 R. A. Walton, *J. Chem. Soc. (A)*, 1967, 1852.
347 K. R. Wunschel and W. E. Ohnesorge, *J. Amer. Chem. Soc.*, 1967, **89**, 2777.
348 A. A. El-Awady, E. J. Bounsall, and C. S. Garner, *Inorg. Chem.*, 1967, **6**, 79.

Nickel, Palladium, and Platinum.—*Nickel*(IV), d^6 *and Nickel*(III), d^7. The electronic spectra of the complexes $[Ni^{IV}(B_9C_2H_{11})_2]$ and $[Ni^{III}(B_9C_2H_{11})_2]^-$ have been reported; their colours are yellow and brown respectively.[349]

Nickel(II), d^8. The electronic spectra of species containing Ni^{II} d^8, were by far the most extensively investigated of all compounds during the year. The next most popular species were those of the +2 oxidation state of the neighbouring elements Co^{II} d^7 and Cu^{II} d^9. Indeed these three, together with Co^{III} d^6, comprise almost half of the references on the electronic spectra of the transition elements.

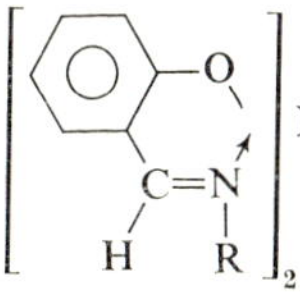

(19)

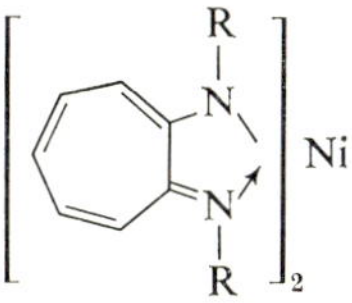

(20)

The anomalous magnetic and spectroscopic behaviour of a number of *N*-substituted nickel(II) salicylaldimine (19) and aminotroponeimine (20) complexes in solution has been examined over a range of pressures.[350] For bis-(*N*-phenylsalicylaldimine)nickel(II), the very broad peak between 8300 and 10,000 cm.$^{-1}$, attributed to the associated species of this complex, is seen to increase in intensity with pressure, and it is thus concluded that pressure enhances association. This is confirmed by the decrease in the intensity of the peak at 16,400 cm.$^{-1}$, attributed to the monomeric species. These spectral data are consistent with the fact that the magnetic susceptibility of the salicylaldimine complexes increases with pressure. The susceptibility of the aminotroponeimineates, on the other hand, is found to decrease with pressure, indicating that the volume of the planar, diamagnetic form of the complex is smaller in solution than that of the tetrahedral form. In nickel(II) *N,N'*-di-2-naphthylaminotroponeimineate the intensity of the peak at 12,200 cm.$^{-1}$ increases with increasing pressure and this peak has been attributed to an absorption by the planar form of the complex.

Complete interpretations of the spectra of tetragonal Ni^{II} complexes have attracted very little attention compared with Co^{III} complexes. A crystal-field analysis has now been carried out for the tetragonal complexes $Ni\,py_4Cl_2$ and $Ni\,py_4Br_2$, and the splitting of the absorption bands due to the strong axial-field is discussed.[351] Detailed assignments have been made and the agreement between experiment and theory is excellent. The difference between the solution and solid-state spectra is also discussed. The new pyridine complexes $Ni\,py_6X_2$ ($X = NCO^-$, NCS^-, or $NCSe^-$) have also been prepared and compared with the previously known

[349] L. F. Warren, jun., and M. F. Hawthorne, *J. Amer. Chem. Soc.*, 1967, **89**, 470.
[350] A. H. Ewald and E. Sinn, *Inorg. Chem.*, 1967, **6**, 40.
[351] D. A. Rowley and R. S. Drago, *Inorg. Chem.*, 1967, **6**, 1092.

tetrakispyridine complexes.[352] The diffuse reflectance spectra of these six compounds indicate a very close similarity in the environment of the Ni^{II} ion in every case, and it is concluded that each pseudo-halide ion is co-ordinated through nitrogen to give a *trans*-octahedral structure regardless of whether six or four molecules are present in the complex.

Polarised single-crystal, liquid-nitrogen temperature colloid, and room-temperature solution spectra have been obtained for the bis-dimethylglyoxime, -ethylmethylglyoxime, and -heptoxime complexes of Ni^{II}. These data have led to the first precise characterisation of the striking spectral changes that typically accompany solid formation in the nickel glyoximates.[353] Assignments for the 'new' band in the solid state and for the other transitions observed have been made in terms of an energy-level diagram suitable for planar complexes with intraligand π-bonds. Insight is thus obtained into the electronic structures of both the isolated molecules and the crystals.

The tridentate Schiff-base formed from *N*-methyl-*o*-aminobenzaldehyde and *NN*-diethylethylenediamine (MABen—NEt_2, donor atoms *NNN*) forms high-spin nickel(II) complexes with the general formula Ni(MABen—NEt_2)X_2 (X = Cl, Br, I, NCS, or NO_3).[297] Spectral data for the chloride and bromide complexes suggest a five-co-ordinate structure both in the solid state and in solution. For the bromide complex, an increase in temperature causes a decrease in intensity of the band at 13,100 cm.$^{-1}$, and the appearance of two new bands indicative of a pseudo-tetrahedral environment. The spectrum of the iodide complex shows differences in both the form and frequency which cannot be attributed to the difference in field-strength between iodine and the other halogens; it is thought to have a five-co-ordinate structure but with lower symmetry. The spectra of the nickel nitrate and nickel thiocyanate complexes, although they are almost identical, are completely different from those of the halide complexes and are assigned O_h symmetry.

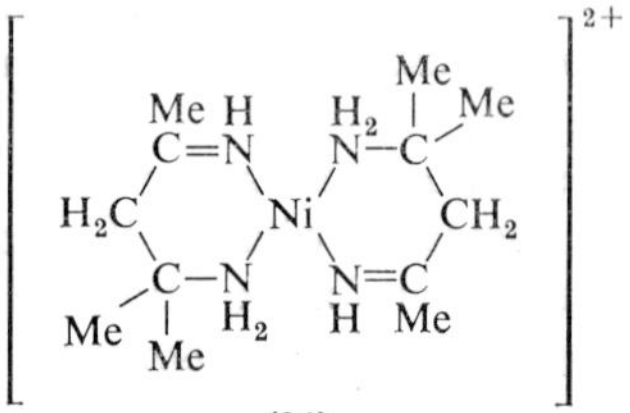

(21)

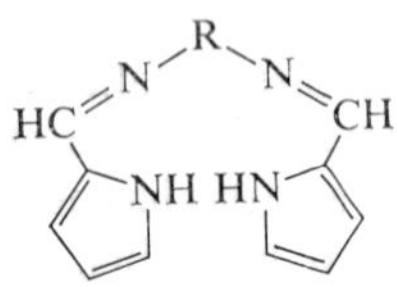

(22)

Some extremely interesting complexes have been isolated from the reaction of $[Ni(NH_3)_6]^{2+}$ with acetone.[354] The complexes $[NiL_2]^{2+}$ (21),

[352] A. H. Norbury, E. A. Ryder, and R. F. Williams, *J. Chem. Soc.* (*A*), 1967, 1439.
[353] B. G. Anex and F. K. Krist, *J. Amer. Chem. Soc.*, 1967, **89**, 6114.
[354] N. J. Rose, M. S. Elder, and D. H. Busch, *Inorg. Chem.*, 1967, **6**, 1924.

with the co-anions Cl^-, Br^-, I^-, $ZnCl_4^{2-}$, and PF_6^-, all have similar electronic spectra with a single band in the 23,000 cm.$^{-1}$ region. These data, coupled with the spin-paired nature of the chloride salt, are consistent with a square planar structure. The compounds $[NiL_2(NCS)_2]$ and $[NiL_3](PF_6)_2$ are spin-free six-co-ordinate complexes as indicated both by their effective magnetic moments and their electronic spectra. The spectra of square planar complexes of the type 2,2′-R-bis(nitrilomethylidyne)-dipyrrolnickel(II) (22) (R = *o*-aryl or a saturated chain of 2–5 carbons) have also been reported.[355] The presumably $\pi \rightarrow \pi^*$ ligand-transitions undergo a red shift upon complexation. The single band observed as a shoulder for the tri-, tetra-, and penta-methylene complexes can be attributed to the only spin-allowed transition in square planar complexes. The order of ligand-field strength with size of R in (22) is: trimethylene > tetramethylene > pentamethylene. A yellow, four-co-ordinate diamagnetic nickel(II) complex $[Ni\,en_2](Ph_4B)_2$ has been prepared by dehydrating the blue complex $[Ni\,en_2(H_2O)_2](Ph_4B)_2$. Since $[Ni\,en_2(H_2O)_2](ClO_4)$, $[Ni\,en_2(H_2O)_2](Ph_4B)_2$, and $[Ni\,en_3]^{2+}$ all exhibit similar absorption maxima it is concluded that the blue complex $[Ni\,en_2(H_2O)_2](Ph_4B)_2$ has a tetragonally distorted octahedral structure.[356] The visible spectrum of the yellow complex is typical of square planar nickel(II).

The electronic spectra of complexes of the type $NiL_2(NO_2)_2$ [L = 2-(aminomethyl)pyridine, 2-(methylaminomethyl)pyridine, 2-(aminomethyl)-6-methylpyridine, and their saturated derivatives] and their analogous isothiocyanate compounds have been studied.[357] The nitro- and nitrito-complexes were differentiated on the basis that the nitro-group is a strong-field and the nitrito-group a weak-field ligand. The spectra of several Ni^{II} and Cu^{II} complexes with 2-(aminomethyl)-6-methylpyridine and *N*-methyl-2-aminomethylpyridine have also been reported.[358]

A series of orange-yellow square planar complexes of *O*-alkyl-1-amidinourea [ROC(:NH)·NH·C(:NH)·NH_2 (= L)] of the type NiL_2X_2 have been reported.[359] A complex in which the ligand itself becomes anionic is also reported, and a spectral comparison of these two types of complexes permits an assignment of relative energies for the *d* orbitals. Complexes with the tridentate ligands di-(2-pyridyl-β-ethyl)amine and di-(2-pyridyl-β-ethyl)sulphide of the general formula $[NiLX_2]$ (X = Cl, Br, I, or NCS) have been isolated.[360] Physical measurements, including the electronic spectra, indicate that the halide complexes have high-spin five-co-ordinate configurations both in the solid state and in solution.

[355] J. H. Weber, *Inorg. Chem.*, 1967, **6**, 258.
[356] J. H. Nelson and R. O. Ragsdale, *Inorg. Nuclear Chem. Letters*, 1967, **3**, 585.
[357] L. El-Sayed and R. O. Ragsdale, *Inorg. Chem.*, 1967, **6**, 1640.
[358] S. Utsuno and K. Sone, *Bull. Chem. Soc. Japan*, 1967, **40**, 105.
[359] V. Rasmussen and W. A. Baker, jun., *J. Chem. Soc.* (*A*), 1967, 580.
[360] S. M. Nelson and J. Rodgers, *Inorg. Chem.*, 1967, **6**, 1390.

Benzimidazole (23) complexes of the type NiL_4X_2 (X = Cl, Br, I, or ClO_4) and a number of acetone solvates have been described,[361] and their electronic spectra obtained at room temperature and at *ca.* 80°K. The diamagnetic complexes are shown to contain planar NiL_4 units, whereas the paramagnetic compounds are six-co-ordinate with pronounced distortions. All these compounds dissociate in solution giving complexes of stoicheiometry NiL_2X_2. The electronic spectra of the solids, NiL_2X_2 (X = Br or I), indicated essentially tetrahedral structures, but the low temperature spectra indicate pronounced departures from pure T_d symmetry.[362]

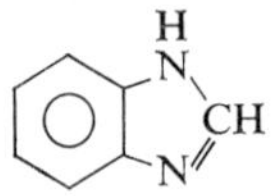

(23)

The absorption spectra of square planar complexes of Ni^{II} and Pd^{II} with the tetradentate nitrogen-donor ligand ethylenebisbiguanide, $C_6H_{16}N_{10}$, have been reported.[363] From *ca.* 6200–33000 cm.$^{-1}$ only a single peak, was detected for $NiLCl_2$ at 20,900 cm.$^{-1}$, whilst a similar peak was observed for $PdLCl_2$ at 32,500 cm.$^{-1}$; this is consistent with the expectation of a substantial increase in the magnitude of the ligand-field splitting in going from the first to the second transition series. The position of this band was found to be independent of the solvent for the Ni^{II} complex, and in consequence the transition was assigned as $d_{xy} \rightarrow d_{x^2-y^2}$ ($^1A_{2g} \leftarrow {}^1A_{1g}$) rather than $d_{z^2} \rightarrow d_{x^2-y^2}$.

The series of complexes [Ni en_2(ONO)]X (X = Cl, Br, or I) have been examined.[364] The electronic spectra suggest a *trans*-octahedral configuration for the nickel ion with no evidence that the halide ions are co-ordinated. The electronic and i.r. spectra are consistent with a symmetrically bridging nitrito-group. Complexes containing 2, 3, 4, or 6 molecules of *o*-phenylenediamine and some new complexes of 4-methyl-*o*-phenylenediamine with 3, 4, or 6 diamines present have been characterised by several techniques including electronic spectra.[365] Criteria for detecting unidentate diamines are discussed.

Electronic spectral data have also been reported for the following systems, which involve *N*-donor atoms: complexes with 1,10-phenanthroline, 2,2′-bipyridyl and analogous ligands;[366] *N*-donor complexes of the types NiL_4X_2, NiL_2X_2, and $NiLX_2$;[367] heterocyclic complexes formed

361 D. M. L. Goodgame, M. Goodgame, and M. J. Weeks, *J. Chem. Soc.* (*A*), 1967, 1125.
362 D. M. L. Goodgame, M. Goodgame, and M. J. Weeks, *J. Chem. Soc.* (*A*), 1967, 1676.
363 D. J. MacDonald, *Inorg. Chem.*, 1967, **6**, 2269.
364 B. J. Hathaway and R. C. Slade, *J. Chem. Soc.* (*A*), 1967, 952.
365 D. R. Marks, D. J. Phillips, and J. P. Redfern, *J. Chem. Soc.* (*A*), 1967, 1464.
366 C. M. Harris and E. D. McKenzie, *J. Inorg. Nuclear Chem.*, 1967, **29**, 1047.
367 A. K. Majumdar, A. K. Mukherjee, and A. K. Mukherjee, *J. Indian Chem. Soc.*, 1967, **44**, 211.

by reaction of mixtures of diaminoethane and 1,2- or 1,3-diaminopropanenickel complexes with acetone;[368] cationic amine complexes adsorbed on silica gel;[369] thiourea complexes of the types $Ni(NCS)_2,2L$ and $Ni(NCS)_2,4L$;[370] complexes of Ni^{II} with ethylenediamine and perchlorate or tetraphenylborate;[371] benzylamine complexes;[372] complexes of *N*-thiocarbamoyl-*N'*-carbamoyl hydrazine and its derivatives;[373] adducts of bis-(2-guanidiniumbenzimidazole)nickel(II) with NH_3 or pyridine;[374] complexes derived from 1,3-di-iminoisoindoline-containing ligands;[375, 376] sulphinamide complexes;[377] bis-(*N*-hydroxyalkyl-3-methoxysalicylideneiminato)- and bis-(*N*-hydroxyalkylsalicylideneiminato)-nickel(II) complexes ($R = CH_2{\cdot}CH_2{\cdot}OH$, $CH_2{\cdot}CH(Me){\cdot}OH$, $Me_2C{\cdot}CH_2{\cdot}OH$, or $CH_2{\cdot}CH_2{\cdot}CH_2{\cdot}OMe$);[378] complexes of furfurylamine derivatives;[309] imidazole and thiazole complexes;[308] complexes of *NNN'N'*-tetramethyldiamines[304] and 2,2'-biquinoline;[312] complexes of 2,4,6-tri-(2'-pyridyl)-1,3,5-triazine (24);[379] complexes of the type $NiLX_2$ and $(NiL_2)X_2$ where X is a univalent anion and L is 2,6-bis-(1,3-di-iminoisoindolin-1-yl)pyridine (25);[380] bis-(*N*-substituted-2-acetiminodimedonato)nickel(II) complexes;[381] isomers of hexamethyl-1,4,8,11-tetra-azacyclotetradecadienenickel(II).[382]

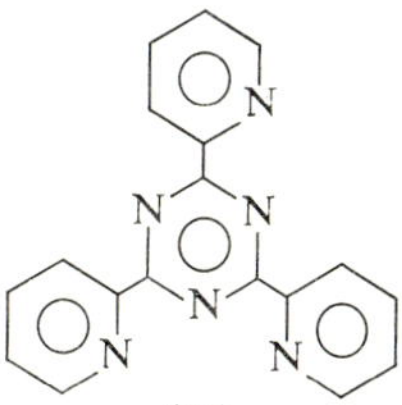

(24)

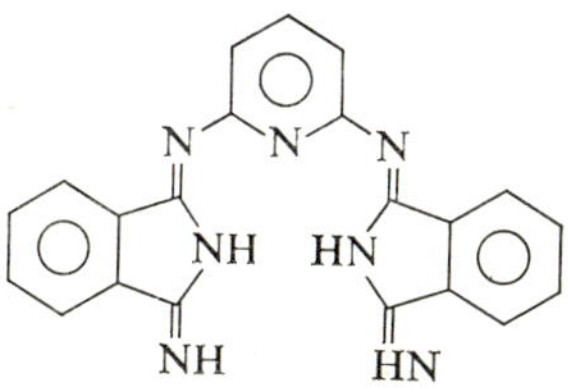

(25)

Several five-co-ordinate Ni^{II} complexes, containing phosphorus-donor atoms, have been reported particularly by Meek and his co-workers.[383, 384, 385]

[368] N. F. Curtis and D. A. House, *J. Chem. Soc.* (*A*), 1967, 537.
[369] B. J. Hathaway and C. E. Lewis, *Chem. Comm.*, 1967, 504.
[370] C. Puglisi and R. Levitus, *J. Inorg. Nuclear Chem.*, 1967, **29**, 1069.
[371] M. E. Farago, J. M. James, and V. C. G. Trew, *J. Chem. Soc.* (*A*), 1967, 820.
[372] R. J. Fitzgerald and R. S. Drago, *J. Amer. Chem. Soc.*, 1967, **89**, 2879.
[373] A. D. Ahmed and P. K. Mandal, *J. Inorg. Nuclear Chem.*, 1967, **29**, 2759.
[374] S. P. Ghosh and A. K. Banerjee, *J. Indian Chem. Soc.*, 1967, **44**, 589.
[375] T. J. Hurley, M. A. Robinson, and S. I. Trotz, *Inorg. Chem.*, 1967, **6**, 389.
[376] M. A. Robinson, S. I. Trotz, and T. J. Hurley, *Inorg. Chem.*, 1967, **6**, 392.
[377] K. M. Nykerk, D. P. Eyman, and R. L. Smith, *Inorg. Chem.*, 1967, **6**, 2262.
[378] S. Yamada, Y. Kuge, and K. Yamanouchi, *Bull. Chem. Soc. Japan*, 1967, **40**, 1864.
[379] R. S. Vagg, R. N. Warrener, and E. C. Watton, *Austral. J. Chem.*, 1967, **20**, 1841.
[380] J. A. Scruggs and M. A. Robinson, *Inorg. Chem.*, 1967, **6**, 1007.
[381] S. McKinley and E. P. Dudek, *Inorg. Nuclear Chem. Letters*, 1967, **3**, 561.
[382] L. G. Warner, N. J. Rose, and D. H. Busch, *J. Amer. Chem. Soc.*, 1967, **89**, 703.
[383] G. Dyer and D. W. Meek, *Inorg. Chem.*, 1967, **6**, 149.
[384] T. D. DuBois and D. W. Meek, *Inorg. Chem.*, 1967, **6**, 1395.
[385] M. O. Workman, G. Dyer, and D. W. Meek, *Inorg. Chem.*, 1967, **6**, 1543.

Using the tetradentate ligand tris-(*o*-methylselenophenyl)phosphine (TSeP) (26), a series of five-co-ordinate, trigonal bipyramidal nickel(II) complexes

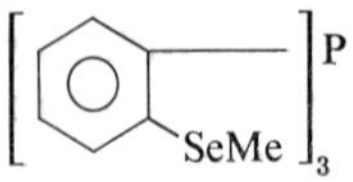

(26)

were characterised.[383] The complexes are diamagnetic and uni-univalent electrolytes consistent with the formulation $[Ni(TSeP)X]ClO_4$ when the fifth ligand is an anion, and di-univalent electrolytes when the fifth ligand is a neutral molecule. The extremely intense green to blue colours of these complexes, due to a strong absorption band at about 16,000 cm.$^{-1}$, appear to be characteristic of diamagnetic five-co-ordinate nickel(II) compounds.

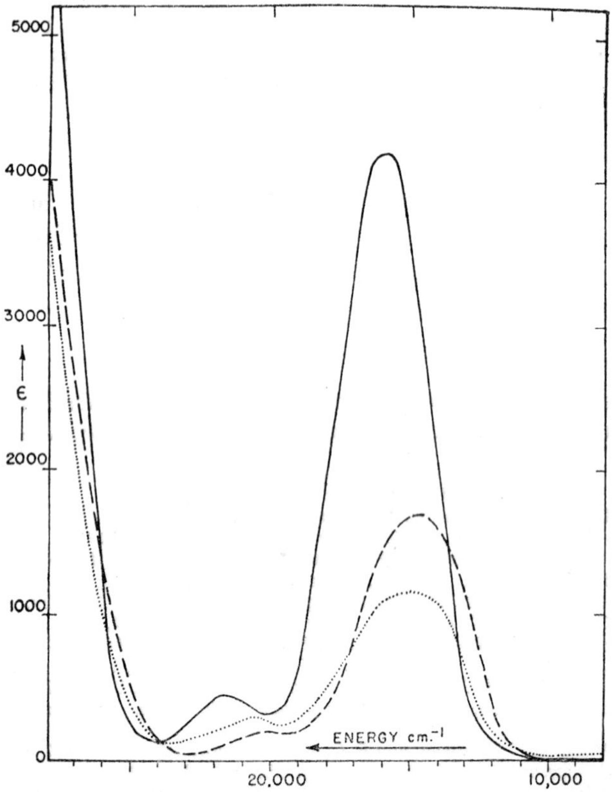

Figure 7 *Electronic absorption spectra of the five-co-ordinate complexes* (——) $[NiAs\text{-}(o\text{-}C_6H_4AsPh_2)_3Br]^+$, (----) $[NiP\text{-}(o\text{-}C_6H_4SeCH_3)_3Br]$, *and* (····) $[NiP\text{-}(o\text{-}C_6H_4SCH_3)_3Br]^+$
(Reproduced by permission from *Inorg. Chem.*, 1967, **6**, 152)

A trigonal bipyramidal structure is indicated for these complexes by a comparison of the electronic absorption spectra of the corresponding TSeP, P(*o*-$C_6H_4SMe)_3$, and As(*o*-$C_6H_4AsPh_2)_3$ bromide complexes (see Figure 7). Detailed comparisons indicate that the spectrochemical series for the donor groups is $R_2Se < R_2S < R_3As < R_3P$. The intensity of the spectral band corresponding to the $d_{x^2-y^2}$, $d_{xy} \rightarrow d_{z^2}$ transition is consistently greater for the selenium complexes than for sulphur complexes, but less than for the arsenic or phosphorus complexes. The five-co-ordinate complexes $[NiL_2X]ClO_4$ [X = Cl, Br, I, NO_2, NCS, or NCSe; L = QAS, (2)] have also been reported.[384] Their electronic spectra differ significantly from those of trigonal bipyramidal nickel(II) complexes with ligands containing similar donor atoms and may be interpreted in terms of a square pyramidal structure. The polydentate phosphine ligands, bis-(*o*-methylthiophenyl)phenylphosphine (DSP) (27), diphenyl-(*o*-methylthiophenyl)-phosphine (SP) (28), and diphenyl-(*o*-methylselenophenyl)phosphine (SeP) (29), give five-co-ordinate complexes of the types $[Ni(DSP)X_2]$, $[Ni(DSP)_2](ClO_4)_2$, $[Ni(DSP)(bidentate)](ClO_4)_2$, $[Ni(SP)_2Br]ClO_4$, and $[Ni(SeP)_2Br]ClO_4$.[385] The absorption spectra do not permit a definitive structural assignment for these complexes. The spectra are not characteristic of trigonal bipyramidal complexes nor the one known distorted square pyramidal complex, *i.e.* $[Ni(triarsine)Br_2]$. The spectra are considered to be consistent with a square pyramidal structure, but this must be considered tentative until one or two of these structures have been determined by *X*-ray crystallography.

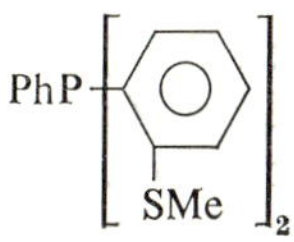

(27)

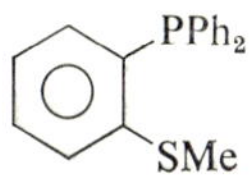

(28)

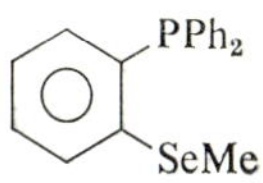

(29)

A series of five-co-ordinate diamagnetic complexes of the general formula $[Ni(DPES)_2X]ClO_4$ (X = Cl, Br, or I; DPES = the bidentate ligand 2-diethylphosphosphinoethyl ethyl sulphide) have been reported.[386] A square pyramidal structure has been suggested, since the electronic spectra differ from those of known trigonal bipyramidal adducts. The electronic spectra of the complexes $Ni(TPN)X_2$, $Ni(TPN)XPPh_4$ [X = Cl, Br, or I; TPN = $N(CH_2{\cdot}CH_2{\cdot}PPh_2)_3$] and $Ni(TAN)XBPh_4$ [X = Br or I; TAN = $N(CH_2{\cdot}CH_2{\cdot}AsPh_2)_3$] are very similar to those of known trigonal bipyramidal nickel(II) complexes; the spectra in the solid state and in CH_2Cl_2 or CH_3NO_2 are identical.[387] Spectral data indicate the formation of five-co-ordinate intermediates in the reaction between the complexes *trans*-$Ni(PR_3)_2(CN)_2$ with the corresponding phosphine in solution.[388]

[386] J. F. Sieckhaus and T. Layloff, *Inorg. Chem.*, 1967, **6**, 2185.
[387] L. Sacconi and I. Bertini, *J. Amer. Chem. Soc.*, 1967, **89**, 2236.
[388] P. Rigo, C. Pecile, and A. Turco, *Inorg. Chem.*, 1967, **6**, 1636.

Five-co-ordinate complexes of the types $[NiLX]^+$ and $[NiLH_2O]^{2+}$ (X = Cl, Br, I, or CN) are formed with the tetradentate arsenic ligand tris-(3-dimethylarsinopropyl)arsine.[389] These complexes are assigned a trigonal bipyramidal structure on the basis of the close similarity between their electronic spectra and the spectrum of the compound $[Ni\{P(CH_2 \cdot CH_2 \cdot CH_2 \cdot AsMe_2)_3\}CN]ClO_4$, which is known to have a trigonal bipyramidal structure. The absorption maxima have been recorded for solids and acetone solutions of several complexes of the types $NiLX_2$ and NiL_2X_2 {L = 2-[β-(diphenylarsino)-ethyl]pyridine (30a)}; these complexes are compared with some 2-[β-(diphenylphosphino)ethyl]pyridine (30b) complexes.[390]

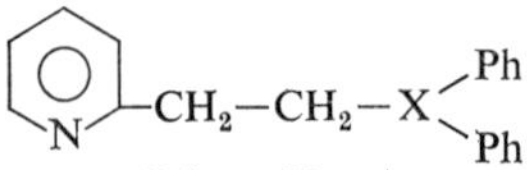

(30) a; X = As
b; X = P.

Goodgame *et al.*[391] have shown that the orange-brown compound previously formulated as $K_4Ni(NO_2)_6$ is a monohydrate. Upon dehydration its spectral properties suggest the formation of $[Ni(NO_2)_4(ONO)_2]^{4-}$; the reduction in the ligand field is in accord with the conversion of some nitro-groups to nitrito, since the latter occupy a much lower position in the spectrochemical series. The reflectance spectra of the complexes Ni(X-salen·NEt_2)(catechol) (X = 3,4-dibenzo-, 5,6-dibenzo, H, 5-chloro-, or 5-bromo- and salen = salicylaldimine) are similar to the previously known five-co-ordinate complex Ni(5-chlorosalen·NEt_2)$_2$. Five-co-ordination is confirmed for Ni(salen·NEt_2)(catechol) by *X*-ray crystallography.[392] The reflectance spectrum of the isomorphous cobalt(II) complex Co(salen·NEt_2)(catechol) is similar to the spectra of other five-co-ordinate Co^{II} complexes with Schiff bases.

Nickel(II) complexes with monoethanolamine, $NH_2C_2H_4OH$, have been investigated spectrophotometrically.[393, 394] The absorption spectra of anhydrous nickel bromide and its complexes with 2, 3, 4, and 5 moles of monoethanolamine have been recorded in methanol, and the position of monoethanolamine in the spectrochemical series determined. Spectral data have also appeared for cyclohexylenediaminetetra-acetate,[395] diglycine, triglycine, tetraglycine,[396] and heterocyclic dienol [397] complexes.

389 G. S. Benner and D. W. Meek, *Inorg. Chem.*, 1967, **6**, 1399.
390 E. Uhlig and M. Maaser, *Z. anorg. Chem.*, 1967, **349**, 300.
391 D. M. L. Goodgame and M. A. Hitchman, *Inorg. Chem.*, 1967, **6**, 813.
392 L. Sacconi, P. L. Orioli, and M. DiVaira, *Chem. Comm.*, 1967, 849.
393 V. V. Udovenko and Yu. S. Duchinskii, *Russ. J. Inorg. Chem.*, 1967, **12**, 510
394 V. V. Udovenko and Yu. S. Duchinskii, *Russ. J. Inorg. Chem.*, 1967, **12**, 517.
395 D. W. Margerum, P. J. Menardi, and D. L. Janes, *Inorg. Chem.*, 1967, **6**, 283.
396 M. K. Kim and A. E. Martell, *J. Amer. Chem. Soc.*, 1967, **89**, 5138.
397 M. Brierley and W. J. Geary, *J. Chem. Soc.* (*A*), 1967, 321.

The reflectance spectra of the compounds $NiX_2\,mmp_2$ (X = Cl, Br, SCN, or I) [327] and NiX_2(*o*-methylthioaniline)$_2$ (X = Cl, Br, or I) [398] are indicative of an octahedral environment for the Ni^{II} ion, although there is evidence of some tetragonal distortion in the bromo- and iodo-complexes of the latter type. Fivefold co-ordination is indicated for the complexes Ni(Me_4 daes)X_2 [X = Cl, Br, or NCS; Me_4 daes = bis-(2-dimethylamino-ethyl)sulphide], since the electronic spectra are distinctly similar to those of known high-spin five-co-ordinate complexes.[326] The electronic spectra of $Ni(R^1R^2N{\cdot}CH_2{\cdot}CH_2{\cdot}S{\cdot}CH_2{\cdot}CO_2)_2$ complexes have been recorded and the non-cubic-field components are considered to be small since the bands can be assigned on the basis of an octahedral model.[399] A series of γ-thiobutyrolactam and *N*-methyl-γ-thiobutyrolactam complexes have also been characterised by their electronic absorption spectra.[400]

The structural environment of nickel(II) in Na_2O–B_2O_3 glasses has been studied by optical (and e.s.r.) techniques.[401] The optical spectrum indicates an octahedral environment. A value of -304 cm.$^{-1}$ is obtained for the spin–orbit coupling constant and a comparison with the free-ion value (λ -335 cm.$^{-1}$) suggests a certain amount of covalency, but less than is usual for nickel(II) complexes. Some oscillator strengths have been calculated from absorption data measurements for nickel(II) ions in aqueous solution.[402] The tetragonal field was found to be the cause of the red band splitting rather than the first-order *L–S* coupling or second-order intermixing.

The preparation of several complexes formed by biuret [(bt), (31)], and transition-metal ions in both neutral and alkaline solution have been described.[403] The reflectance spectrum of the nickel bromide derivative,

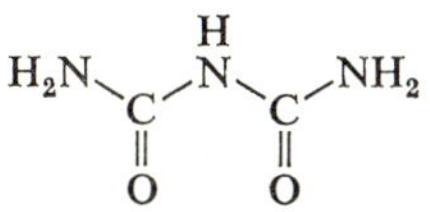

(31)

Ni bt_2Br_2, is typical of octahedrally co-ordinated nickel(II), and a mean value for 10*Dq* of 8460 cm.$^{-1}$ is indicated for four oxygen-donor atoms and two bromide ions. Applying the rule of average environment, gives a value of 9190 cm.$^{-1}$ for 10*Dq* for biuret as a neutral ligand co-ordinating through the oxygen atoms. Attempts to check this value with the chloride derivative were unsuccessful due to poor resolution of the spectrum. The reflectance spectrum of the potassium bis(biuret)nickelate(II) dihydrate

[398] L. F. Lindoy, S. E. Livingstone, and T. N. Lockyer, *Austral. J. Chem.*, 1967, **20**, 471.
[399] N. J. Rose, C. A. Root and D. E. Busch, *Inorg. Chem.*, 1967, **6**, 1431.
[400] S. K. Madan and M. Sulich, *J. Inorg. Nuclear Chem.*, 1967, **29**, 2765.
[401] H. G. Hecht, *J. Chem. Phys.*, 1967, **47**, 1840.
[402] N. S. Chhonkar, *Indian J. Phys.*, 1967, **41**, 21.
[403] A. W. McLellan and G. A. Melson, *J. Chem. Soc.* (*A*), 1967, 137.

complex is characteristic of a square planar environment about the nickel atom. The data indicate that biuret as a di-anionic ligand produces a much stronger ligand field than the neutral ligand.

Tricyclic 1 : 1 chelates containing both co-ordinated and free ester groups are formed by the condensation products from 1,2-diamines and hydroxymethylene malonic acid with Ni^{II} and Cu^{II}. Contrary to comparable N—O co-ordinating six-membered ring chelates, with acetyl or benzoyl oxygen as the ligand atom, the ester group co-ordination is shown to bring about a decrease of the ligand-field strength.[404] The electronic spectra of some *o*-hydroxyarylcarbonyl nickel(II) complexes have also been discussed,[405] and the spectra of chloro- and fluoro-sulphate salts have been compared;[406] the latter study shows that the chloro- and fluoro-sulphate ions can be positioned closer to fluoride than to chloride in both the spectrochemical and nephelauxetic series. The spectra of several nickel(II) complexes with monothio-β-diketones have been recorded, and the bands assigned as metal *d–d*, metal–metal, ligand–metal, and ligand–ligand transitions.[407] The preparation of mercaptides of the type $Ni(SR)_2$ has been described.[408] Although $Ni(SMe)_2$ was insoluble in organic solvents, increasing the size of the R group increased the solubility and some solution spectra were obtained in cyclohexane. Three characteristic bands were observed at about 24,100, 29,400, and 40,000 $cm.^{-1}$, and these were considered to be too intense for *d–d* transitions. The electronic spectrum of $[Ni\{1,2\text{-bis(mercapto)-}o\text{-carborane}\}_2]^{2-}$ has been studied to ascertain the type of metal–sulphur interaction.[409] An empirical comparison with square planar dithio-containing compounds in which the Ni—S bonding (*a*) is largely of the σ type, and (*b*) has a large amount of π character, indicated that the carborane complex was of the former type. It was concluded that the carborane molecule does not provide an effective network for π delocalisation.

The electronic spectra of base adducts of xanthates and dithiocarbamates with nickel(II) both in solution and in the solid state are characteristic of nickel(II) complexes with distorted octahedral structures.[410] The spectrum of the tris(xanthato)nickel(II) complex is similar to those of the base adducts except that the ligand-field transitions for the anion are more intense. This suggests that the base adducts may be *trans* and thus nearly centrosymmetric. Certain 1,1-dithiolate complexes have been found to react with elemental sulphur to form new complexes containing one additional sulphur atom per ligand molecule.[411] The visible and u.v. spectra show a

404 E. G. Jäger, *Z. anorg. Chem.*, 1967, **349**, 139.
405 D. P. Graddon and G. M. Mockler, *Austral. J. Chem.*, 1967, **20**, 21.
406 D. A. Edwards, M. J. Stiff, and A. A. Woolf, *Inorg. Nuclear Chem. Letters*, 1967, **3**, 427.
407 S. H. H. Chaston and S. E. Livingstone, *Austral. J. Chem.*, 1967, **20**, 1079.
408 D. C. Bradley and C. H. Marsh, *Chem. and Ind.*, 1967, 361.
409 H. D. Smith, jun., M. A. Robinson, and S. Papetti, *Inorg. Chem.*, 1967, **6**, 1014.
410 D. Coucouvanis and J. P. Fackler, *Inorg. Chem.*, 1967, **6**, 2047.
411 D. Coucouvanis and J. P. Fackler, *J. Amer. Chem. Soc.*, 1967, **89**, 1346.

characteristic change accompanying sulphur addition. The visible spectra have also been given for a number of *N*-cyanocarbamate complexes of nickel(II) but a detailed interpretation was not given.[412]

An elegant study of the electronic absorption spectra of gaseous $NiCl_2$, $NiBr_2$, and NiI_2 has been reported.[413] The low-intensity *d–d* transitions are interpreted on the basis of a ligand-field calculation (including spin–orbit coupling), for a $3d^8$ configuration in an axial field. The main features of the spectra are due to transitions to $^3P_\pi$ and $^1D_{\Delta,\pi}$ excited-states which are strongly intermixed in the bromide and iodide. An analysis of the nephelauxetic effect in these levels leads to the conclusion that the F_2 parameter decreases more rapidly than the F_4 parameter in the halide series from chloride to iodide.

It has been known for some time that nickel(II) in certain binary molten chloride salt solvents may occur as an equilibrium mixture of two co-ordination forms. One of these forms has a distribution of co-ordination geometries which on the average is more or less tetrahedral, while the other is more or less octahedral. By following changes in the electronic spectra with composition it has recently been shown[414] that Ni^{II} in molten mixtures of $ZnCl_2$ and CsCl has two such tetrahedral–octahedral co-ordination equilibria, one of which occurs in melts containing 50–70 mole % $ZnCl_2$ and the other in about 92–100 mole % $ZnCl_2$. The explanation of the electronic spectra of $NiCl_2$ in molten LiCl–KCl mixtures proved difficult.[415] It was found essential to regard the nickel centres as having a distribution of co-ordination geometries that was strongly influenced by the outer-shell cations. A fairly detailed model was developed to explain the results. The spectra of nickel centres in $MgCl_2$–KCl mixtures at temperatures up to 1060° has also been studied and the observed behaviour was found to be qualitatively similar to that reported for LiCl–KCl mixtures.[416]

A method has been described for measuring the optical absorption spectra of the crystalline and liquid phases of a transition-metal salt close to its melting point.[417] By supercooling the liquid, the spectra of both phases can be measured at the same temperature. This technique has been applied to Cs_3NiCl_5 (m.p. 547°) and $CsNiCl_3$ (m.p. 758°). It was shown that when Cs_3NiCl_5 melts, the approximately tetrahedral arrangement of the chloride ions about the nickel is only slightly affected, whereas when $CsNiCl_3$ melts, the octahedral arrangement in the crystal is completely destroyed and replaced by a different distribution of geometries.

The blue colour typical of tetrahedral $NiCl_4^{2-}$ has been obtained by warming a solution containing 0·015M-nickel perchlorate hexahydrate and

[412] F. A. Cotton and J. A. McCleverty, *Inorg. Chem.*, 1967, **6**, 229.
[413] C. W. De Kock and D. M. Gruen, *J. Chem. Phys.*, 1967, **46**, 1096.
[414] W. E. Smith, J. Brynestad, and G. P. Smith, *J. Amer. Chem. Soc.*, 1967, **89**, 5983.
[415] J. Brynestad, C. R. Boston, and G. P. Smith, *J. Chem. Phys.*, 1967, **47**, 3179.
[416] J. Brynestad and G. P. Smith, *J. Chem. Phys.*, 1967, **47**, 3190.
[417] C. R. Boston, J. Brynestad, and G. P. Smith, *J. Chem. Phys.*, 1967, **47**, 3193.

2·5M-tetramethylammonium chloride.[418] The spectrum obtained closely resembles that established for tetrahedral or very slightly distorted tetrahedral $NiCl_4^{2-}$.

Lower Oxidation States of Nickel. Extremely large Faraday effects have been noted [419] for $Ni(CO)_4$ [and $Fe(CO)_5$], $Ni(PCl_3)_4$, $Ni[P(OMe)_3]_4$, and for three mixed derivatives of the type $Ni(CO)_n[P(OMe)_3]_{4-n}$. Further, it was suggested that such measurements might be used to detect multiple bonding via π electrons and perhaps even be used to analyse its detailed character. Contributions to the molecular magnetic rotation from the nickel–phosphorus bond in the complexes $Ni[P(OEt)Cl_2]_4$, $Ni[P(OPr)Cl_2]_4$, and $Ni[P(OBu)Cl_2]_4$ have also been evaluated by assuming that the ligand contributions are the same as in the free compound.[420] The values do not vary greatly and a mean value of 857 ± 5 μr was obtained.

Palladium and Platinum. Most of the recent spectral data for palladium and platinum complexes have been obtained with the metal in the formal oxidation state of +2. The u.v. and visible spectra have, however, been presented [421] for the platinum(IV) complexes *trans*-$[Pt(NH_3)_4(SCN)_2]^{2+}$, *trans*-$[Pt(NH_3)_4(SCN)Cl]^{2+}$, and *trans*-$[Pt(NH_3)_4Cl_2]^{2+}$.

The u.v. spectra of several platinum–olefin complexes of the type $[ol\ PtCl_3]^-$ (ol = C_2H_4, C_3H_5OH, $C_3H_5NH_3^+$, $C_4H_7NEt_3^+$, and $C_3H_5PEt_3^+$) have been studied to elucidate the relative energies of the *d* orbitals.[422] The three main bands observed in the region 31,000–45,000 cm.$^{-1}$, with extinction coefficients between 750 and 3325, are assigned to metal-$d \rightarrow$ olefin-π^* transitions. On the basis of this assignment, the relative energies of the *d* orbitals in these complexes are given as $d_{x^2-y^2} > d_{xy} > d_{z^2} > d_{yz} > d_{xz}$. The dimeric cyclopentadienyl complex, $Pt_2(\pi\text{-}C_5H_5)_4$, has also been reported, and its electronic spectrum was taken to indicate both σ and π contributions to the bonding of the cyclopentadienyl rings.[423]

Little spectral data have appeared for nitrogen-donor ligand complexes of palladium or platinum. The aqueous solution spectra of the ethylenediamine-*NN'*-diacetic acid (H_2EDDA) complexes $Pt(H_2EDDA)Cl_2$, Pt(HEDDA)Cl, and Pt(EDDA) have been illustrated,[424] and the spectra of a number of palladium(II) complexes with morpholine biguanide have been reported.[425] Spectrophotometric studies have been carried out on the equilibria between several choroammine complexes of palladium(II) in ammonium salt solutions [426] and it is reported that measurements of the

418 R. K. Scarrow and T. R. Griffiths, *Chem. Comm.*, 1967, 425.
419 F. Gallais and H. Haraldsen, *Compt. rend.*, 1967, **264**, *C*, 1.
420 P. Cassoux and J.-F. Labarre, *Compt. rend.*, 1967, **264**, *C*, 736.
421 W. R. Mason, E. G. Berger, and R. C. Johnson, *Inorg. Chem.*, 1967, **6**, 248.
422 R. G. Denning, F. R. Hartley, and L. M. Venanzi, *J. Chem. Soc.* (*A*), 1967, 1322.
423 E. O. Fischer and H. Schuster-Woldam, *Chem. Ber.*, 1967, **100**, 705.
424 S. P. Tanner, F. Basolo, and R. G. Pearson, *Inorg. Chem.*, 1967, **6**, 1089.
425 P. Spacu and C. Gheorghiu, *Z. anorg. Chem.*, 1967, **351**, 201.
426 R. A. Reinhardt, N. L. Brenner, and R. K. Sparkes, *Inorg. Chem.*, 1967, **6**, 254.

u.v. spectra of the *cis*- and *trans*-isomers of dichloro- and dibromo-diamminepalladium(II) yield little information.[427]

The polarised single-crystal reflection spectra of Magnus' green salt, $[Pt(NH_3)_4][PtCl_4]$, and two of its tetra-alkylamine analogues have been measured, and the corresponding absorption spectra were obtained through Kramers–Krönig analyses. It is found that a strong u.v. transition exists with the proper polarisation for the visible out-of-plane bands to gain their intensity from it through vibronic mixing of zero-order electronic wavefunctions. Possible assignments of the u.v. band are also discussed.[428]

Diphenyl(*o*-diphenylarsinophenyl)phosphine sulphide (32) has been shown to form stable chelate complexes with palladium(II), platinum(II), and copper(I).[429] By comparing the electronic absorption spectra of the palladium(II) complexes with the analogous complexes of analogous complexes of diphenyl(*o*-diphenylarsinophenyl)phosphine (33), it is apparent that the phosphine sulphide (32) exerts a weaker ligand field than the corresponding phosphine (33). A number of polydentate phosphorus–sulphur ligands have been investigated with palladium(II); only four-co-ordinate complexes could be isolated, whereas five-co-ordinate complexes were previously obtained with nickel(II). The electronic spectra of these four-co-ordinate complexes are recorded.[430]

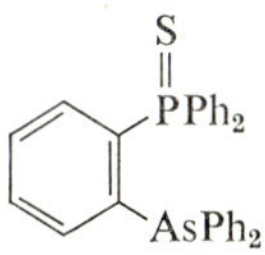

(32)

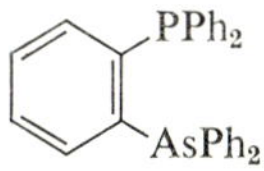

(33)

The bromination of a number of platinum(II) complexes with olefinic tertiary arsines has been studied spectrophotometrically, and the u.v. spectra of both the initial complexes and the bromination products are tabulated.[431] The use of sodium *p*-(mercaptoacetamido)benzenesulphonate (34) as a spectrophotometric reagent for palladium has been suggested,[432]

(34)

and the electronic spectra of the dimeric palladium and platinum complexes $(MX_2,L)_2$ (X = Cl or Br; L = bidentate 1,2-bis(isopropylseleno)ethane), in both ethanol and chloroform solutions have been recorded.[433]

[427] J. C. Coe and A. A. Malik, *Inorg. Nuclear Chem. Letters*, 1967, **3**, 99.
[428] B. G. Anex, M. E. Ross, and M. W. Hedgcock, *J. Chem. Phys.*, 1967, **46**, 1090.
[429] P. Nicpon and D. W. Meek, *Inorg. Chem.*, 1967, **6**, 145.
[430] G. Dyer, M. O. Workman, and D. W. Meek, *Inorg. Chem.*, 1967, **6**, 1404.
[431] M. A. Bennett, J. Chatt, G. J. Erskine, J. Lewis, R. F. Long, and R. S. Nyholm, *J. Chem. Soc.* (*A*), 1967, 501.
[432] H. K. L. Gupta, M. R. Bhandari, and N. C. Sogani, *Bull. Chem. Soc. Japan*, 1967, **40**, 215.
[433] N. N. Greenwood and G. Hunter, *J. Chem. Soc.* (*A*), 1967, 1520.

The intriguing question of the ordering of the metal *d* orbitals in the d^8 tetrahalide square planar complexes has been investigated. Basch and Gray[434] have reported SCCC–MO calculations and compared them with detailed assignments for both the $d \rightarrow d$ and charge-transfer spectra of $PtCl_4^{2-}$ and $PdCl_4^{2-}$. The *d*-orbital ordering of $z^2 < xz,yz < xy < x^2 - y^2$ was found for a wide variation in analytical orbital functions, VOIP's and *F* factors, with and without ligand–ligand overlap. It is, however, stressed that this ordering need not be correct for all square planar complexes, and in fact these halides probably represent the case in which the d_{z^2} orbital is at its lowest relative position in the ligand-field-level scheme. Cotton and Harris[435] obtained the same order for the *d* orbitals in $PtCl_4^{2-}$ by performing a semi-empirical MO calculation of the extended Hückel type. This result was obtained over a wide range of variations in the Pt valence-state ionisation potentials, the Pt wavefunctions, and the Wolfsberg–Helmholz factor in both the Mulliken–Wolfsberg–Helmholz and the Ballhausen–Gray approximations.

A number of spectrophotometric studies on halide systems have been reported. Spectrophotometric evidence has appeared[436] for the species $[Pt_2Br_6]^{2-}$, formed in the acid hydrolysis of $[PtBr_4]^{2-}$, and mixed ligand complexes of palladium(II) with bromide and iodide have been noted.[437] The tetraiodo- and tri-iodoaquo-platinate(II) ions have also been identified spectrophotometrically, but all attempts to isolate a solid salt of the tetraiodoplatinate(II) ion failed.[438]

Copper, Silver, and Gold.—*Copper*(II), d^9. The e.s.r. and optical spectra of the unstable purple species produced by the reaction of copper(II) and the cyanide ion have been studied at low temperatures in a number of different solvents.[439] The optical spectra depend very markedly upon the solvent and all the experimental facts can be satisfactorily accounted for in terms of an essentially square planar complex $[Cu(CN)_4]^{2-}$.

Optical and magnetic measurements have been carried out on single crystals of copper-doped dichloro-(1,10-phenanthroline)zinc.[440] The superhyperfine structure in the e.s.r. spectrum has been resolved and its directional dependence used to establish that the local environment of the cupric ion is nearly tetrahedral. Furthermore, the ground-state metal orbital was found to contain less than 3% of the 4*p* state. Two *d–d* transitions were observed at 11,630 and 13,800 cm.$^{-1}$ with extinction coefficients of the order of 100, a value which also suggests relatively small 4*p* admixture.

[434] H. Basch and H. B. Gray, *Inorg. Chem.*, 1967, **6**, 365.
[435] F. A. Cotton and C. B. Harris, *Inorg. Chem.*, 1967, **6**, 369.
[436] J. E. Teggins, D. R. Gano, M. A. Tucker, and D. S. Martin, jun., *Inorg. Chem.*, 1967, **6**, 69.
[437] S. C. Srivastava and L. Newman, *Inorg. Chem.*, 1967, **6**, 762.
[438] B. Corain and A. J. Poë, *J. Chem. Soc.* (*A*), 1967, 1318.
[439] A. Longo and T. Buch, *Inorg. Chem.*, 1967, **6**, 556.
[440] G. F. Kokoszka, C. W. Reimann, and H. C. Allen, *J. Phys. Chem.*, 1967, **71**, 121.

Similar studies have been carried out for single crystals of copper-doped tris(phenanthroline)zinc(II) nitrate dihydrate.[441] In this case the e.s.r. data were interpreted in terms of a Jahn–Teller effect for the Cu^{II} ion. The optical *d–d* transitions and their polarisation properties are also reported; the optical axes do not coincide with the principal magnetic axis.

A semi-empirical SCF–MO calculation with zero-differential overlap has been carried out in an investigation of the ground- and excited-states of copper dimethylglyoxime.[442] The calculations confirm the existence of a strong metal–ligand π bond. The calculated electronic spectrum agrees well with experiment and assignments were made. The usual classification of excited states into ligand-field and charge-transfer states was found not to be valid. Spectral measurements in solution have shown that the complexes $[Ph_4As]_2[Cu\{N(CN)_2\}_4]$ and $[Me_4N]_2[Cu\{N(CN)_2\}_4]$ have distorted tetrahedral structures.[443] In the solid state the Ph_4As^+ salt has the same structure, but the Me_4N^+ salt is polymeric and consists of distorted octahedral units. The electronic spectra of bidentate and tetradentate pyrrole-2-aldimine complexes of copper(II) in co-ordinating (pyridine) and non-co-ordinating (chloroform, benzene, and hexane) solvents and also in Nujol mulls have been reported.[444] Certain trends in the positions and intensities of the bands are indicated, but assignments to specific ligand-field transitions could not be achieved.

The five-co-ordinate complexes $[Cu\,dach_2Cl](ClO_4)$ and $[Cu\,dach_2Br](ClO_4)$ (dach = 1,4-diazacycloheptane) have been reported.[445] The visible absorption maxima are given for the solid state and for nitromethane and aqueous solutions. These data are interpreted in terms of a square pyramidal geometry. A five-co-ordinate antipyridine (A), (35), complex $[CuA_5](ClO_4)_2$ has also been reported.[446] In this case the *d–d* spectrum again suggests C_{4v} symmetry.

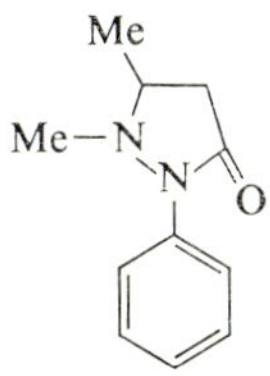

(35)

The electronic spectra of the mono- and bis-2,2′-biquinolyl (biq) complexes $[Cu\,biq\,X_2]$ (X = Cl, Br, or NO_3) and $[Cu\,biq_2(ClO_4)_2]$ have

[441] G. F. Kokoszka, C. W. Reimann, H. C. Allen, and G. Gordon, *Inorg. Chem.*, 1967, **6**, 1657.
[442] B. Roos, *Acta Chem. Scand.*, 1967, **21**, 1855.
[443] H. Köhler and B. Seifert, *Z. Naturforsch*, 1967, **22b**, 238.
[444] A. Chakravorty and T. S. Kannan, *J. Inorg. Nuclear Chem.*, 1967, **29**, 1691.
[445] W. K. Musker and M. S. Hussain, *Inorg. Nuclear Chem. Letters*, 1967, **3**, 271.
[446] J. Gopalakrishnan, A. Ravi, and C. C. Patel, *Bull. Chem. Soc. Japan*, 1967, **40**, 791.

been shown to be consistent with a pseudo-tetrahedral arrangement about the copper atoms.[447] The absorption spectra for the system $Cu(NO_3)_2$–2-methylbenzimidazole–methanol have been recorded for both constant concentration of $Cu(NO_3)_2$ and also for constant total concentrations.[448] Analysis of the spectra indicates that $Cu(NO_3)_2$ in methanol and its complexes with 2-methylbenzimidazole have tetragonally distorted octahedral configurations. Spectral data for copper(II) complexes with di-2-pyridyl ketone, 2-benzoylpyridine,[449] and *NNN′N′*-tetramethylated diamines [244] have also been reported.

An interesting new type of isomerism has been discovered for co-ordination compounds.[450] The copper(II) chelates of the Schiff-bases derived from α-alanine and salicylaldehyde (36) and from pyruvic acid and salicylamine (37) have been studied by a number of physical methods. Compound (36) is blue-green or dark green, whilst (37) is yellow-green or light green. The ligand-field bands are similar, but large differences are observed in the u.v. region. The authors propose to call this form of isomerism 'fused chelate ring isomerism'.

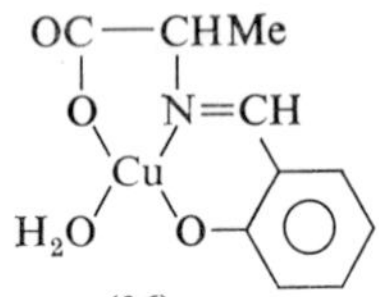

(36)

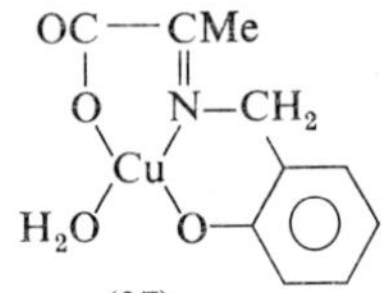

(37)

The spectra of several amino-acid complexes of copper(II) have been reported.[451, 452] The first examples of stereoselective isomerism in the solid state of bis(amino-acidato)copper(II) complexes using the ligands phenylalanine and tyrosine have been noted. The absorption maxima have been recorded for each isomer, and some tentative assignments are given, based upon the fact that *trans*-bis(amino-acidato)copper(II) complexes generally absorb around 16,100 $cm.^{-1}$. The 1 : 1 and 1 : 2 complexes of copper(II) with *meso* and racemic 2,3-diaminosuccinic acids have also been described and their reflectance spectra discussed.[453] Single-crystal spectra of copper oxinate have been studied.[454] The copper atom in this complex is surrounded by two nitrogen and two oxygen atoms in a square planar geometry, and a weak broad band in the i.r. region (*ca.* 10,000 $cm.^{-1}$) is regarded as a *d–d* transition.

[447] C. M. Harris, H. R. H. Patil, and E. Sinn, *Inorg. Chem.*, 1967, **6**, 1102.
[448] M. V. Artemenko and K. F. Slyusarenko, *Russ. J. Inorg. Chem.*, 1967, **12**, 513.
[449] R. R. Osborne and W. R. McWhinnie, *J. Chem. Soc.* (*A*), 1967, 2075.
[450] Y. Nakao, S. Sasaki, K. Sakurai, and A. Nakahara, *Bull. Chem. Soc. Japan*, 1967, **40**, 241.
[451] S. H. Laurie, *Chem. Comm.*, 1967, 155.
[452] S. H. Laurie, *Austral. J. Chem.*, 1967, **20**, 2609.
[453] Y. Yoshikawa and K. Yamasaki, *Bull. Chem. Soc. Japan*, 1967, **40**, 813.
[454] G. Basu, *J. Inorg. Nuclear Chem.*, 1967, **29**, 1544.

Copper(II) complexes of thiosemicarbazide (tsc) of the types [Cu tscX_2] (X = Cl, Br, or $\frac{1}{2}SO_4$) and [Cu tsc$_2X_2$] (X = Cl, Br, NO_3, ClO_4, or $\frac{1}{2}SO_4$) are reported.[455] Reflectance spectra for these complexes are given, and it is concluded that thiosemicarbazide produces a greater ligand-field and nephelauxetic effect than does semicarbazide in the corresponding copper(II) complexes.

A study of the visible spectrum, including polarisation measurements on single crystals at 25°, of bis(dipivaloylmethanido)copper(II), Cu dpm$_2$, has been reported:[456] it is concluded that all four *d–d* transitions occur within a range of a few thousand wavenumbers as predicted earlier by an extended Hückel MO calculation.[457] This result is also consistent with the e.s.r. data. The observed polarisations are sufficient to rule out a magnetic dipole mechanism as the main source of intensity, but they do not in themselves provide a basis for assignment of the transitions if the intensity mechanism is taken to be a vibronic one. Assignments previously proposed are confirmed except that the band at 48,600 cm.$^{-1}$ is now postulated to be a metal-to-ligand charge-transfer band instead of a $\pi \rightarrow \pi^*$ band.

The polarised spectra of (001) and (100) faces of bis-(3-phenyl-2,4-pentanedionato)copper exhibit four *d–d* bands all principally *y*-polarised.[458] The transitions are of electric-dipole, vibronically allowed, single-molecule origin. It is inferred that several vibrations are effective, but that the bulk of the intensity is borrowed from a charge-transfer transition, $B_{3u} \leftarrow B_{1g}$ in D_{2h}.

In the optical absorption spectrum of copper monochloroacetate 2·5 hydrate a very intense band was found at 14,000 cm.$^{-1}$ and two other bands were centred at 11,000 and 26,000 cm.$^{-1}$. The bands at 11,000 and 14,000 cm.$^{-1}$ are tentatively assigned as $d_{x^2-y^2} \rightarrow d_{xy}$ and $d_{x^2-y^2} \rightarrow d_{xz}, d_{yz}$ transitions respectively; this assignment is consistent with the magnetic data. The origin of the higher-energy band, however, is still in doubt.[459] The dimeric copper(II) levulinate complex of formula [Cu($O_2C{\cdot}CH_2{\cdot}CH_2{\cdot}COMe)_2,H_2O$] exhibits absorption maxima at 14,700 and 26,700 cm.$^{-1}$ in its reflectance spectrum.[460] The band at 26,700 cm.$^{-1}$ is considered to be characteristic of the dimeric molecule. The same spectrum is obtained in an acetone solution, and thus a dimeric structure is also implied in solution. The appearance of a shoulder near 27,000 cm.$^{-1}$ in the reflectance spectrum of the phthalic acid complex [Cu{$C_6H_4(CO_2)_2$}]$_2$, is thought to confirm a dimeric structure.[461] Some alkali-metal copper acetates and their acetic acid solvates have been described.[462] They show

[455] M. J. Campbell and R. Grzeskowiak, *J. Chem. Soc.* (*A*), 1967, 396.
[456] F. A. Cotton and J. J. Wise, *Inorg. Chem.*, 1967, **6**, 917.
[457] F. A. Cotton, C. B. Harris, and J. J. Wise, *Inorg. Chem.*, 1967, **6**, 909.
[458] R. L. Belford and J. W. Carmichael, *J. Chem. Phys.*, 1967, **46**, 4515.
[459] G. F. Kokoska, H. C. Allen, and G. Gordon, *J. Chem. Phys.*, 1967, **47**, 10.
[460] J. Gopalakrishnan and C. C. Patel, *Inorg. Chem.*, 1967, **6**, 2111.
[461] I. Shimizu, R. Tsuchiya, and E. Kyuno, *Bull. Chem. Soc. Japan*, 1967, **40**, 1162.
[462] H. D. Hardt and G. Streit, *Z. anorg. Chem.*, 1967, **350**, 84.

nearly identical absorptions in the 15,400 cm.$^{-1}$ region due to the acetato-cuprate ions.

The absorption spectrum of 'Egyptian Blue', $CaCuSi_4O_{10}$, which contains Cu^{II} in a square planar environment, has been reported.[463] The broad band at about 16,000 cm.$^{-1}$ was resolved into three bands at 12,900, 15,900, and 18,800 cm.$^{-1}$. An independent study on $BaCuSi_4O_{10}$ gave an almost identical spectrum [464] with bands at 12,900, 15,800, and 18,800 cm.$^{-1}$. The spectra were interpreted in each case by supposing that the order of increasing energy of the 3*d* orbitals of the copper is $d_{z^2} < d_{xz}, d_{yz} < d_{xy} < d_{x^2-y^2}$.

A study of two binuclear copper complexes with pyridine-*N*-oxide has been carried out to obtain experimental data on the nature of the metal–metal interaction.[465] The compounds studied were dichlorobis(pyridine-*N*-oxide)copper(II), $[(C_5H_5NO)_2CuCl_2]$, and dichloroaquopyridine-*N*-oxide-copper(II), $[(C_5H_5NO)CuCl_2(H_2O)]$. The results are discussed on the basis that to a first approximation the bands in the 8000–15,000 cm.$^{-1}$ region can be assigned as the usual monomeric *d*–*d* transitions. The diffuse reflectance spectra of several substituted heterocyclic *N*-oxide complexes have also been reported.[466]

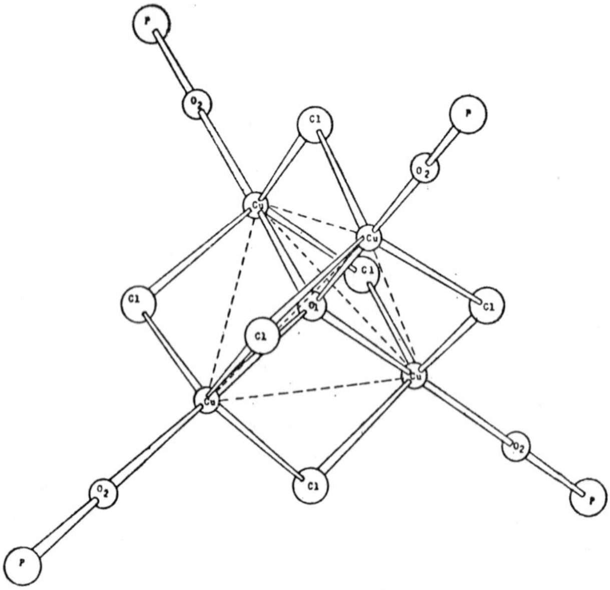

Figure 8 *Perspective drawing of the structure of μ_4-oxo-hexa-μ-chloro-tetrakis[(triphenylphosphine oxide)copper(II)]. The phenyl groups are omitted*
(Reproduced by permission from *Inorg. Chem.*, 1967, **6**, 496)

463 A. Ludi and R. Giovanoli, *Naturwiss.*, 1967, **54**, 88.
464 M. G. Clark and R. G. Burns, *J. Chem. Soc.* (*A*), 1967, 1034.
465 G. F. Kokosyka and H. C. Allen, *J. Chem. Phys.*, 1967, **46**, 3013.
466 R. Whyman, D. B. Copley, and W. E. Hatfield, *J. Amer. Chem. Soc.*, 1967, **89**, 3135.

The preparation and properties of μ_4-oxo-hexa-μ-chloro-tetrakis-[(triphenylphosphine oxide)copper(II)], $Cu_4OCl_6(Ph_3PO)_4$, have been reported.[467] It is one of the few molecular compounds in which oxygen is four-co-ordinate. Furthermore, it is an example of a trigonal bipyramidal complex of copper(II), although it is somewhat distorted. The visible and near-i.r. spectrum of this complex is similar to that reported for the trigonal bipyramidal $CuCl_5^{3-}$ ion. The absorption bands in this complex at 9900 and 11,200 cm.$^{-1}$ occur at higher energy than those of $CuCl_5^{3-}$ (8200 and 10,400 cm.$^{-1}$), as would be expected from a comparison of the ligand-field strength of oxygen and chlorine atoms.

Two independent reports have dealt with the polarised absorption spectra of the tetrachlorocuprate ion with the copper atom in a slightly distorted tetrahedral environment (D_{2d}).[468, 469] The polarised absorption spectrum of single crystals of tetrachlorobistrimethylbenzylammonium-cuprate(II) has been studied, with the *d–d* bands of primary interest.[468] The bands at about 6000 and 9000 cm.$^{-1}$ are assigned to the $^2E \leftarrow {}^2B_2$ and $^2A_1 \leftarrow {}^2B_2$ transitions (Figure 9) respectively. No experimental evidence

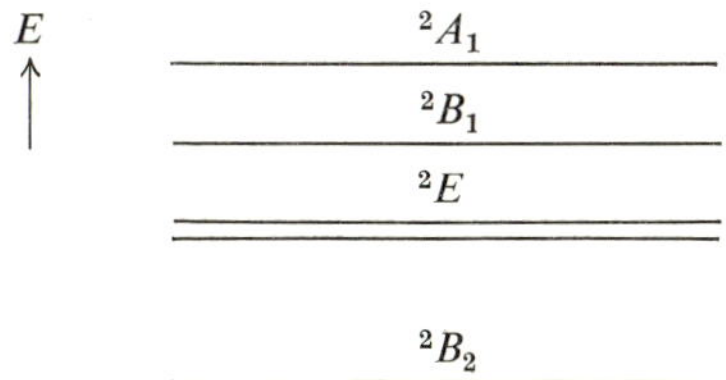

Figure 9 *Proposed electronic energy-level diagram for* $CuCl_4^{2-}$

for a third transition, $^2B_1 \leftarrow {}^2B_2$, was obtained. The charge-transfer transitions have been investigated [469] by studying the polarised u.v. absorption spectra of the $CuCl_4^{2-}$ ion oriented in single crystals of Cs_2CuCl_4 and Cs_2ZnCl_4. The spectral bands were assigned by means of an analysis which takes into account the observed variations with lattice of the band intensities and positions and which allows for the effects upon the spectra of terms of rhombic symmetry known to be present in the ionic Hamiltonian. The resulting band assignments differ from those proposed earlier by Ferguson.

The tetrachlorocuprate ion, $CuCl_4^{2-}$, can also adopt a square planar geometry, and the e.s.r. and electronic absorption spectra of this arrangement have been studied by Willett *et al.*[470] In both the square planar and tetrahedral species, the positions of the charge-transfer bands are quite similar. The *d–d* transitions are found to be critical in distinguishing between these two geometries. There is a ligand-field band at 12,800 cm.$^{-1}$

[467] J. A. Bertrand, *Inorg. Chem.*, 1967, **6**, 495.
[468] C. Furlani, E. Cervone, F. Calzona, and B. Baldanza, *Theor. Chim. Acta*, 1967, **7**, 375.
[469] M. Scharnoff and C. W. Reimann, *J. Chem. Phys.*, 1967, **46**, 2634.
[470] R. D. Willett, O. L. Liles, and C. Michelson, *Inorg. Chem.*, 1967, **6**, 1885.

in the square planar species whilst it is located in the 5500–9000 cm.$^{-1}$ range in the tetrahedral form, the exact value depending upon the amount of distortion in the latter. At low temperatures a phase-transition occurs in the square planar complexes, and a colour change from a light yellow to a pale green takes place. Spectral studies, however, indicate only slight changes (< 500 cm.$^{-1}$) in the positions of the major absorption bands. The ligand-field absorption spectrum of the complex $(MeNH_3)_2CuCl_4$, which contains elongated octahedra $[CuCl_6]$, consists of three bands at 10,800, 12,200, and 13,300 cm.$^{-1}$. Both the empirical ordering of the energy levels of the partly filled *d* shell and the e.s.r. spectrum are consistent with a high tetragonality of the chromophore.[471]

The electronic spectra have been reported for copper solutions in glacial acetic acid in the presence of an excess of lithium chloride or bromide.[472] The species responsible for the principal peaks are identified as Li_2CuCl_4 (planar), 26,700 cm.$^{-1}$; Li_2CuCl_4 (distorted tetrahedral), 22,200 cm.$^{-1}$; $LiCuBr_3$, 16,000 cm.$^{-1}$; and Li_2CuBr_4 (distorted tetrahedral), 19,600 cm.$^{-1}$. The equilibria in the lithium chloride solutions have been studied in detail over a range of water and lithium acetate concentrations, and four principal copper-containing species, $CuAc_2$, $CuCl_2$, and the two forms of Li_2CuCl_4, are detected. The spectra of the copper halide species obtained when small amounts of ethanol (or water) are added to a solution of anhydrous copper(II) chloride dissolved in anhydrous ethylacetate saturated with hydrogen chloride have been examined.[473] In the original anhydrous solution, the low extinction is indicative of a planar species, and the most reasonable choice is a planar trichloromonoethylacetate species. With added alcohol (or water) the square planar $CuCl_4{}^{2-}$ ion is indicated.

Salts of the types $[M^{III}(NH_3)_6]CuCl_5$ (M = Co, Cr, Rh, or Ru) and $[dienH_3]CuCl_5$, which contain the trigonal bipyramidal complex ion $CuCl_5{}^{3-}$, have been studied by diffuse reflectance spectroscopy.[474] Near-i.r. bands in the 4000–6000 cm.$^{-1}$ region are assigned to vibrational combinations of the NH_3 ligand, since they are found in practically the same positions in the complex and in $[M^{III}(NH_3)_6]Cl_3$. Two prominent bands in the near-i.r. region from 8000–10,000 cm.$^{-1}$ at room temperature and from 9000–11,000 cm.$^{-1}$ at liquid nitrogen temperature are assigned as ${}^2E' \leftarrow {}^2A_1$ and ${}^2E'' \leftarrow {}^2A_1$ ligand-field transitions of the Cu^{II} ion in D_{3h} symmetry. Three charge-transfer bands are observed in the u.v. region at 24,200, 27,200, and 37,800 cm.$^{-1}$.

The electronic absorption spectra of several red copper(II) chloride compounds containing planar $[Cu_2Cl_6]^{2-}$ dimers have been studied.[475] In

[471] C. Furlani, A. Sgamellotti, F. Magrini, and D. Cordischi, *J. Mol. Spectroscopy*, 1967, **24**, 270.
[472] R. P. Eswein, E. S. Howald, R. A. Howald, and D. P. Keeton, *J. Inorg. Nuclear Chem.*, 1967, **29**, 437.
[473] R. Myers and R. D. Willet, *J. Inorg. Nuclear Chem.*, 1967, **29**, 1546.
[474] G. C. Allen and N. S. Hush, *Inorg. Chem.*, 1967, **6**, 4.
[475] R. D. Willett and O. L. Liles, *Inorg. Chem.*, 1967, **6**, 1666.

addition to the normal charge-transfer and *d–d* transitions typical of square-planar $CuCl_4^{2-}$ ions, a new band characteristic of the dimeric species is observed at 19,000 cm.$^{-1}$. This band is polarised parallel to the Cu–Cu direction of the dimer. A molecular orbital description of the bonding in the dimer is presented and the origin of the 19,000 cm.$^{-1}$ band rationalised. Several compounds containing a di-bridged structure similar to that of the $[Cu_2Cl_6]^{2-}$ dimer have also been studied.[476] When examined with polarised light, each compound studied exhibited pleochroism with the absorption much stronger in one direction than in the other. Thus dimers of this type are characterised by a strongly polarised absorption in the visible region which may be used to identify the existence of such dimeric species.

Quantitative intensity data have been presented[477] for the mixed-valence absorption in single crystals of hexamminecobalt(III) chlorocuprates(I, II) as a function of the mole fraction of copper(I). The low intensity of the band agrees with its assignment as an intermolecular charge-transfer from a chlorocuprate(I) anion to a chlorocuprate(II).

Copper(I), d^{10}. Copper(I) complexes of the type [Cu dmp_2X], (dmp = 2,9-dimethylphenanthroline; X = phenylalanine, tryptophan, or tyrosine) have been reported.[478] These three complexes are soluble in most polar solvents and have identical absorption spectra in a given solvent. The spectra are also identical to those of similar complexes of the type [Cu dmp_2X] where X is a monovalent anion. It is concluded that in solution the amino-acid is not co-ordinated to the Cu^I ion. The diffuse reflectance spectra of several Cu^I complexes of bis(pyrazyl)alkane ligands (38) have been studied.[479] These complexes exhibit absorption bands near 24,000 and 27,000 cm.$^{-1}$ which are almost certainly charge-transfer in origin. The most probable assignment of these bands is thought to involve transitions from the copper(I) atom to an empty π^* antibonding pyrazine molecular orbital.

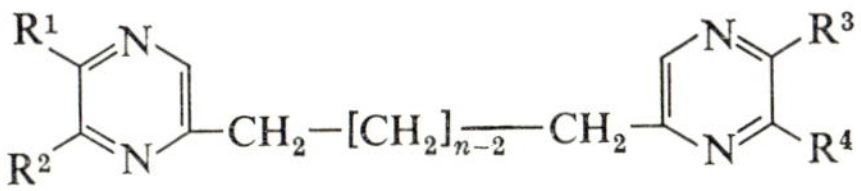

(38)

Silver and Gold. The e.s.r. and diffuse reflectance spectra of [Ag py_4]S_2O_8, both as a pure compound and diluted in [Cd py_4]S_2O_8, have been recorded.[480] The data are consistent with a model having a square planar geometry, although it is somewhat distorted in the matrix of the pure

[476] R. D. Willett, *J. Inorg. Nuclear Chem.*, 1967, **29**, 2482.
[477] P. Day and D. W. Smith, *J. Chem. Soc.* (*A*), 1967, 1045.
[478] S. H. Laurie, *Austral. J. Chem.*, 1967, **20**, 2597.
[479] B. J. Edmondson and A. B. P. Lever, *J. Inorg. Nuclear Chem.*, 1967, **29**, 2777.
[480] H. G. Hecht and J. P. Frazier, *J. Inorg. Nuclear Chem.*, 1967, **29**, 613.

silver compound. The remaining electronic spectral data, which were collected for compounds of silver, appeared in publications primarily concerned with reaction kinetics. For example, the kinetics of oxidation–reduction reactions [481] involving Ag^{II} and the reactions of Ag^{II} with dithionate [482] have been studied spectrophotometrically.

Spectral data on gold compounds have also been scarce. However, the reactivity of amines towards Au^{III} complexes [483] and the hydrolysis of tetrachlorogold(III) [484] have been studied spectrophotometrically.

Zinc, Cadmium, and Mercury.—Studies in the u.v. region show that only the methyl and ethyl homologues (39) of the four-co-ordinate bis-(*N*-alkylsalicylaldiminato)zinc(II) complex form five-co-ordinate pyridine adducts, even in pyridine solutions. The reaction between pyridoxamine (40), potassium α-ketoisovalerate (41), and zinc acetate in methanol solution is found to yield the zinc chelate of pyridoxylidenevaline (42). Spectrophotometric studies [486] of this non-enzymatic transamination reaction show that it proceeds smoothly and completely at room temperature and involves two slow steps.

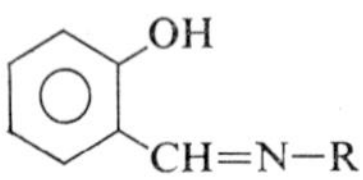

(39)

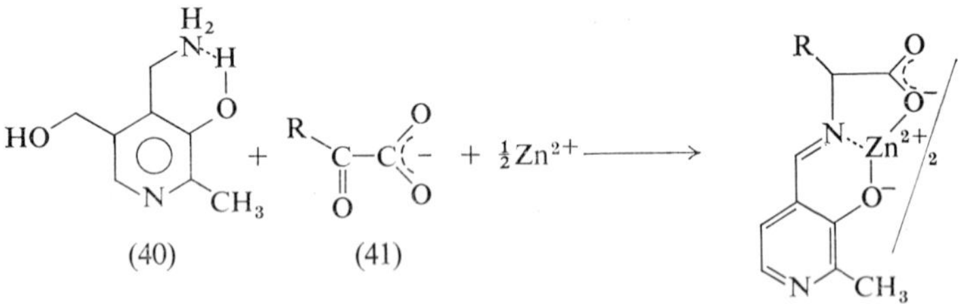

(40) (41) (42)

A very interesting spectral study [487] of the gaseous $Zn–ZnCl_2$ system in the range 700–950° reveals that ZnCl is the only detectable new species. The two sequences observed in the electronic spectrum of ZnCl, previously attributed to the two components of a $^2\Pi \leftarrow ^2\Sigma$ transition, are centred at 33,850 and 34,120 cm.$^{-1}$, compared with 33,830 and 34,080 cm.$^{-1}$ reported earlier (1929). The spectra of several 2,2′-bipyridyl complexes of zinc

[481] D. H. Huchital, N. Sutin, and B. Warnqvist, *Inorg. Chem.*, 1967, **6**, 838.
[482] G. Veith, E. Guthals, and A. Viste, *Inorg. Chem.*, 1967, **6**, 639.
[483] L. Cattalini, A. Doni, and A. Orio, *Inorg. Chem.*, 1967, **6**, 280.
[484] W. Robb, *Inorg. Chem.*, 1967, **6**, 382.
[485] G. E. Batley and D. P. Graddon, *Austral. J. Chem.*, 1967, **20**, 877.
[486] Y. Matsushima and A. E. Martell, *J. Amer. Chem. Soc.*, 1967, **89**, 1331.
[487] J. D. Corbett and R. A. Lynde, *Inorg. Chem.*, 1967, **6**, 2199.

alkyls have been reported[488] and these intensely coloured compounds are said to show typical charge-transfer absorption bands.

The complexes formed by HgI_2 with KI in dimethyl sulphoxide (DMSO) and also with the iodides of post-transition metals in DMSO and dimethylformamide (DMF) have been studied. It is concluded[489] that HgI_3^- is the predominant species in DMSO solutions containing a 1 : 1 mole ratio of KI and HgI_2; this conclusion is based upon the appearance of a band at 32,900 cm.$^{-1}$ which had previously been shown to characterise the above species. The further addition of KI reduces this peak and gives a new absorption band at about 29,950 cm.$^{-1}$; this agrees closely with the average value of 29,850 cm.$^{-1}$ found for the HgI_4^{2-} ion. Similarly it is concluded[490] that in dilute solutions HgI_3^- is formed by the reaction of HgI_2 with iodides of post-transition metals in both DMF and DMSO. In more concentrated solutions the formation of heteropolynuclear complexes is assumed.

3 The Lanthanides and Actinides

The Lanthanides.—*General Considerations.* Assuming an icosahedral crystal field symmetry, K_h, the major groupings of the crystal-field levels in the optical spectra of lanthanide double nitrates may be understood.[491] Structure within these groups may be explained by assuming an approximate symmetry of T_h with a small distortion to C_3. Tentative crystal-field parameters for T_h symmetry are reported for $Ce_2Mg_3(NO_3)_{12},24H_2O$ and $Nd_2Mg_3(NO_3)_{12},24H_2O$. Tripositive lanthanide ions in the electric-field environment of an ethyl sulphate crystal give rise to sharp-line optical absorption spectra. The Zeeman-effect theory is developed for the case in which the crystal-field and Zeeman splittings are of comparable magnitudes;[492] only the case in which the magnetic field is perpendicular to the *c* axis is treated. The u.v. spectra of mixed chelates of La, Pr, Nd, Sm, and Y with propionylacetones and several *N*-co-ordinating ligands have been recorded in ethanol.[493] The crystal-field parameters of lanthanum sulphate have been calculated by studying the absorption spectrum of Pr^{3+}- and Nd^{3+}-doped $La_2(SO_4)_3,9H_2O$.[494] These values are compared with previously obtained parameters for lanthanum ethyl sulphate. The lattice parameter $A_2^0\langle r^2\rangle$ is shown to be quite large in the sulphate in comparison with the ethyl sulphate.

The hypersensitive transitions, *i.e.* bands which show abnormal variations in intensity and fine structure, in the absorption spectra of 6-, 7-, and 8-co-ordinate Nd, Ho, and Er β-diketonates in non-aqueous

488 K. H. Thiele and H. Rau, *Z. anorg. Chem.*, 1967, **353**, 127.
489 A. Buckingham and R. P. H. Gasser, *J. Chem. Soc.* (*A*), 1967, 1964.
490 F. Gaizer and M. T. Beck, *J. Inorg. Nuclear Chem.*, 1967, **29**, 21.
491 S. D. Devine, *J. Chem. Phys.*, 1967, **47**, 1844.
492 T. Murao, W. J. Haas, R. W. G. Syme, F. H. Spedding, and R. H. Good, *J. Chem. Phys.*, 1967, **47**, 1572.
493 N. K. Dutt and S. Upadhyaya, *J. Inorg. Nuclear Chem.*, 1967, **29**, 1368.
494 C. R. Viswanathan, K. Radisavljevis, and E. Y. Wong, *J. Chem. Phys.*, 1967, **46**, 4231.

solvents have been studied.[495] These transitions are shown to be characteristic of the co-ordination and symmetry of the lanthanide ion. Changes in the absorption spectra with changes in the co-ordination of the lanthanide ion were also observed for Pr, Sm, Eu, Dy, Tm, and Yb chelates.

Neodymium. Gluconate complexes of Nd have been investigated[496] by studying the absorption spectra in the 23,300 cm.$^{-1}$ region ($^2P_{\frac{1}{2}} \leftarrow {}^4I_{9/2}$ transition). Solutions containing different component ratios at various pH values were studied and the sublevels of the Nd splitting in the fields of the ligands recorded.

The electronic spectra of Nd^{3+} in single crystals of Nd_2O_3 and neodymium-doped La_2O_3 have been observed[497] at 10, 80, and 290°K. Over 300 absorption lines were assigned from the five crystalline Stark-levels of the $^4I_{9/2}$ ground-state to over 20 excited J manifolds in each crystal. Spectroscopic evidence has also been presented for a volatile complex formed by $NdCl_3$ and $AlCl_3$;[498] the stoicheiometry and structure of the complex is unknown.

Europium. The polarised emission and absorption spectra of single crystals of europium-doped $YVO_4{}^{3+}$ have been measured and analysed.[499] The spectra are consistent with D_{2d} site symmetry and no significant violation of the appropriate electric- and magnetic-dipole selection rules was found. Most of the observed transitions could be identified and the complete energy-level structure of the ion has been defined for all states below the $^5D_3$. A crystal-field model has been used to describe the structure of the ^7F multiplet. Although the simple electrostatic-field approximation gives an adequate description of the experimental data, the necessity for a more refined treatment is indicated.

The fluorescent and absorption spectra of Eu^{3+} in the naphthalene sulphonate complex, $EuL_3,2H_2O$, have been studied;[500] the energy levels ^7F_n up to $^7F_5$ and the levels ^5D_n up to $^5D_3$ have been determined. The fluorescent emission spectrum of the closely related tris-(9,10-anthraquinone-2-sulphonate) of europium has also been reported.[501] A comparison of this spectrum with those of other salts and chelates of europium indicate that this compound should be regarded as a simple salt rather than a chelate.

Emission spectra of various adducts of europium–trisdibenzoylmethide in powder form have been measured at 77°K at high resolution.[502] Analysis

[495] D. G. Karraker, *Inorg. Chem.*, 1967, **6**, 1863.
[496] N. A. Kostromina and T. V. Ternovaya, *Zhur. neorg. Khim.*, 1967, **12**, 511.
[497] J. R. Henderson, M. Muramoto, and J. B. Gruber, *J. Chem. Phys.*, 1967, **46**, 2515.
[498] D. M. Gruen and H. A. Øye, *Inorg. Nuclear Chem. Letters*, 1967, **3**, 453.
[499] C. Brecher, H. Samelson, A. Lempicki, R. Riley, and T. Peters, *Phys. Rev.*, 1967, **155**, 178.
[500] B. Blanzat and J. Loriers, *Compt. rend.*, 1967, **265**, *C*, 849.
[501] B. Blanzat and J. Loriers, *Compt. rend.*, 1967, **264**, *C*, 4.
[502] K. Kreher, E. Butter, and W. Seifert, *Z. Naturforsch.*, 1967, **22b**, 242.

of the spectra leads to the conclusion that the point symmetry of the europium ion is C_{2v}, but in some cases a higher symmetry such as D_{4d} or D_{2d} is found as a first approximation. The proposed structure is a face-centred trigonal prism. Spectral data have also been presented for some europium tetrakisbenzoyltrifluoroacetonates,[503] ten chelates of the type [piperidinium]$^+$ [Eu(RCO·CH·CO·CF$_3$)$_4$]$^-$ [504] and Eu^{II} in M-perchloric acid.[505]

Gadolinium. A preliminary report of the spectra of Gd^{3+} in SrF_2 and BaF_2 has appeared.[506] The observed lines, comprising the groups of transitions between the ground state $^8S_{\frac{7}{2}}$ and the Stark sublevels of $^6P_{\frac{7}{2}}$ and $^6P_{\frac{5}{2}}$, are tabulated and discussed.

Holmium. Free-ion and crystal-field parameters have been obtained[507] for Ho^{3+} in $LaCl_3$, and these parameters reproduce the observed energy-level structure quite well.

Thulium. The absorption spectrum of tricyclopentadienylthulium has been measured at room temperature.[508] According to present plausible assignments, it appears that this is only the second example of a lanthanide complex exhibiting an unusually large ligand-field splitting of its $4f^n$-multiplet terms. An estimate of the Racah parameter E^3 indicates a remarkably large nephelauxetic effect for the $4f$-electrons. Above 20,000 cm.$^{-1}$ rather broad absorption bands of medium intensity appear in the spectrum; these bands are ascribed to electron transitions from non-bonding ligand orbitals into appropriate $4f$ orbitals. Several of the f–f transitions show hypersensitive behaviour (see page 317).

Ytterbium. Two very elegant papers concerned with the electronic spectra of ytterbium compounds have appeared. Pappalardo and Jørgensen[509] measured the spectra at liquid helium temperatures of ytterbium tricyclopentadienide, Yb(C$_5$H$_5$)$_3$, and its solution in a 2-methyltetrahydrofuran glass. The $^2F_{\frac{5}{2}} \leftarrow {}^2F_{\frac{7}{2}}$ transitions of the green Yb(C$_5$H$_5$)$_3$ and certain of its adducts with Lewis bases indicate a fairly unperturbed $4f^{13}$ configuration. The green colour is due to a weak band which shows a very pronounced fine structure at low temperatures. This absorption is probably due to an electron transfer from a linear combination of cyclopentadienide orbitals having particularly low ionisation energy because of strong ligand–ligand repulsion. A molecular orbital treatment of M(C$_5$H$_5$)$_3$ in trigonal symmetry is discussed.

The spectra of the divalent lanthanide ions in crystals arise from two or three types of transitions, namely $4f \rightarrow 4f$, $4f \rightarrow 5d$, and possibly charge

[503] T. M. Shepherd, *J. Inorg. Nuclear Chem.*, 1967, **29**, 2551.
[504] R. G. Charles and E. P. Riedel, *J. Inorg. Nuclear Chem.*, 1967, **29**, 715.
[505] R. T. M. Fraser, R. N. Lee, and (in part) K. Hayden, *J. Chem. Soc.* (*A*), 1967, 741.
[506] J. Makovsky, *J. Chem. Phys.*, 1967, **46**, 390.
[507] K. Rajnak and W. F. Krupke, *J. Chem. Phys.*, 1967, **46**, 3532.
[508] R. D. Fischer and H. Fischer, *J. Organometallic Chem.*, 1967, **8**, 155.
[509] R. Pappalardo and C. K. Jørgensen, *J. Chem. Phys.*, 1967, **46**, 632.

transfer. To understand transitions of the second type a knowledge of the energy levels of the $4f^{n-1}\,5d$ configurations in a crystal field is required. With this objective, the late T. S. Piper *et al.*[510] calculated the energy levels of the configurations *fd* and $f^{13}\,d$ in cubic crystal-fields as a function of the *f*- and *d*-electron crystal-field parameters. The spectrum of Yb^{2+}, with the ground-state configuration of $4f^{14}$, has been recorded in $SrCl_2$ for comparison with these calculations. Both the calculated energies and calculated intensities agree well with experiment.

The Actinides.—*Thorium.* *o*-Carboxy-[511] and *o*-hydroxyphenylazochromotropic acid[512] have been used as colorimetric reagents for the determination of thorium. In each case 1 : 1 complexes are formed in acidic solutions with Th^{IV}.

Protactinium. Two publications have appeared dealing with the absorption spectra of Pa^{IV} in aqueous solution.[513, 514] The spectrum of Pa^{IV} in hydrochloric acid solutions changes little as the acid concentration is increased from 1–4M and three well resolved bands, probably due to $5f \rightarrow 6d$ transitions, are observed. Bagnall and Brown[514] noted spectral changes with further increase in the chloride ion concentrations (hydrochloric acid or sodium chloride added). However, solutions produced by the zinc-amalgam reduction of Pa^{V} in hydrochloric acid gave relatively uncomplexed Pa^{IV} species, since the Zn^{2+} produced in the reaction complexes preferentially with the available chloride ions. On the other hand Mitsuji,[513] who also studied Pa^{IV} solutions produced by the zinc-amalgam method, concluded that Pa^{IV} was reluctant to form chloro-complexes and apparently overlooked the possibility of competition from Zn^{2+} ions.

The absorption spectra of Pa^{IV} in ND_4F–D_2O solutions and also in various fluorides have been presented.[515] For $7RbF,6PaF_4$, $3NaF,PaF_4$, and PaF_4, an intense i.r. absorption is observed at 5485 cm.$^{-1}$. This is assigned as a $^2F_{\frac{7}{2}} \leftarrow {}^2F_{\frac{5}{2}}$ transition (which equals $\frac{7}{2}\xi_{5f}$) and leads to a value of the spin–orbit coupling constant for Pa^{IV}, ξ_{5f}, equal to 1567 cm.$^{-1}$.

Spectral data have also been recorded and discussed for protactinium(IV) halides[516] and the oxalate, $PaO(OH)C_2O_4,xH_2O$.[517]

Uranium. The absorption spectrum of the uranyl ion in perchlorate media was found to become more complex in the visible region when the ion

[510] T. S. Piper, J. P. Brown, and D. S. McClure, *J. Chem. Phys.*, 1967, **46**, 1353.
[511] K. Tôei, H. Miyata, and T. Harada, *Bull. Chem. Soc. Japan*, 1967, **40**, 1141.
[512] K. Tôei, H. Miyata, S. Nakashima, and S. Kiguchi, *Bull. Chem. Soc. Japan*, 1967, **40**, 1145.
[513] T. Mitsuji, *Bull. Chem. Soc. Japan*, 1967, **40**, 2091.
[514] K. W. Bagnall and D. Brown, *J. Chem. Soc.* (*A*), 1967, 275.
[515] L. B. Asprey, F. H. Kruse, and R. A. Penneman, *Inorg. Chem.*, 1967, **6**, 544.
[516] D. Brown and P. J. Jones, *J. Chem. Soc.* (*A*), 1967, 719.
[517] R. Muxart, J. Vernois, R. Guillaumont, and H. Arapaki-Strapelias, *Bull. Soc. chim. France*, 1967, 2890.

hydrolyses.[518] However, the resolved band positions do not shift as hydrolysis proceeds and it was concluded that the hydrolysis affects neither the uranyl structure nor the O—U—O bonding. The extraction of U^{VI} from sulphuric acid solutions by means of solutions of tri-n-octylamine in benzene and in carbon tetrachloride has been studied and the relevant spectra are presented;[519] some spectral data have also appeared for uranyl complexes with anions of carboxylic acids.[520]

The spectra of gaseous UCl_4 between 4000 and 23,000 cm.$^{-1}$ have been obtained[521] from 950–1250°K. These spectra were compared with those of single crystals of UCl_4 at 77°K and at room temperature; the spectra of solid UCl_4 were also obtained over the range 300–827°K. The energy levels of the gaseous state do not shift appreciably with temperature but, in general, the spectrum of the gas is shifted to the low-energy side of the high-temperature solid spectra. With the exception of a general shift, the spectrum of the vapour is remarkably similar to that of the solid. The local symmetry of U^{IV} in the solid is D_{2d}, and although one might expect the molecule to be tetrahedral in the gas phase, a Jahn–Teller distortion may lower the symmetry towards D_{2d}.

The electronic spectrum of a single-domain crystal of $CsUO_2(NO_3)$ at 20°K has been investigated.[522] Selection rules are given for a trigonally co-ordinated uranyl complex in a crystal having two molecules per unit cell. Lower-lying levels, whose excitation gives rise to the fluorescent and magnetic series, are assigned as the A_2 and E species of the D_3 point group. A number of U^{IV} complexes, which are thought to be eight-co-ordinate, are reported along with their electronic spectra in a number of media.[523] Detailed assignments were not given. The absorption spectra for U^{IV} in 1–12M-hydrochloric acid solution have also been reported, and the effect of added zinc chloride described.[514]

The electronic spectra of both U^{IV} and U^{III} in molten $LiF–BeF_2$, $LiF–BeF_2–ZrF_4$, and LiF–NaF–KF have been obtained[524] at temperatures up to 540°. Based on a comparison of these spectra with those obtained in other molten salt systems and aqueous systems, it is suggested that the co-ordination number of the uranium species in the molten fluoride media is possibly 8 or 9.

Few U^{III} compounds are known; the halides and the hydride have been prepared, and the visible and u.v. spectra of aqueous solutions of U^{III} in perchloric acid and hydrochloric acid have been recorded. The isolation of the first hydrated U^{III} salt, $U_2(SO_4)_3,8H_2O$, has now been reported.[525]

518 J. T. Bell and R. E. Biggers, *J. Mol. Spectroscopy*, 1967, **22**, 262.
519 C. Deptula and S. Minc, *J. Inorg. Nuclear Chem.*, 1967, **29**, 221.
520 C. Miyake and H. W. Nürnberg, *J. Inorg. Nuclear Chem.*, 1967, **29**, 2411.
521 J. R. Morrey, D. G. Carter, and J. B. Gruber, *J. Chem. Phys.*, 1967, **46**, 804.
522 J. B. Newman, *J. Chem. Phys.*, 1967, **47**, 85.
523 J. Selbin and J. D. Ortego, *J. Inorg. Nuclear Chem.*, 1967, **29**, 1449.
524 J. P. Young, *Inorg. Chem.*, 1967, **6**, 1486.
525 R. Barnard, J. I. Bullock, and L. F. Larkworthy, *Chem. Comm.*, 1967, 1270.

A comparison of its diffuse reflectance spectrum with those of the U^{IV} compounds $U(SO_4)_2,8H_2O$ and $U(SO_3)_2,8H_2O$ confirms that little if any oxidation of the U^{III} had occurred during the preparation.

Visible spectra are presented which indicate that, in anhydrous acetic acid, the species M^{III}, M^{IV}, and M^{VI} are reasonably stable for uranium and plutonium.[526]

Transuranium Elements. Fourteen absorption groups were identified[527] as levels of the ground electronic configuration $5f^4$ of neptunium(III) in the polarised absorption and emission spectra of Np^{III} in $LaBr_3$. The assumption of a pure $5f^4$ electronic configuration does not lead to a satisfactory interpretation of the spectrum, but the inclusion of an electrostatic configuration interaction allows a satisfactory fitting of the observed levels.

The electronic absorption spectra of gaseous plutonium trichloride, tetrachloride, and tribromide were measured[528] in the range 4000–30,000 cm.$^{-1}$ at temperatures of about 1000°. Transitions of the types $5f^n \rightarrow 5f^n$ and $5f^n \rightarrow 5f^{n-1}\,6d$ were identified.

The first chloro-complexes of americium(V) and americium(VI) have been reported[529] and they are formulated as $Cs_8(AmO_2)_3Cl_{11}$ and $Cs_2AmO_2Cl_4$. The oxidation states of these two compounds were confirmed by recording the visible spectra of the compounds in dilute perchloric, nitric, or hydrochloric acid solutions.

Sixteen absorption bands in the range from 6667–31,250 cm.$^{-1}$ have been observed for berkelium(III) in a variety of aqueous acid solutions at berkelium concentrations of $3{\cdot}6 \times 10^{-3}$ M (or less).[530] The weaker bands were identified by the use of time-averaging techniques followed by computer processing of the data. Bk^{III} was oxidised to the quadrivalent state and part of its absorption spectrum was recorded.

The first observations of the absorption spectrum of einsteinium(III) in 3–6M-hydrochloric acid solutions in the region from 31,250–15,384 cm.$^{-1}$ have been reported.[531]

4 Optically Active Co-ordination Compounds

Introduction.—The resurgence of interest in optically active co-ordination compounds has continued. The Proceedings of the NATO Summer School held in Bonn during 1965 have been reported in a book,[532] and the papers presented at a Discussion Meeting at the Royal Society in London in 1966

526 M. Alei, Q. C. Johnson, H. D. Cowan, and J. F. Lemons, *J. Inorg. Nuclear Chem.*, 1967, **29**, 2327.
527 W. F. Krupke and J. B. Grubers, *J. Chem. Phys.*, 1967, **46**, 542.
528 D. M. Gruen and C. W. DeKock, *J. Inorg. Nuclear Chem.*, 1967, **29**, 2569.
529 K. W. Bagnall, J. B. Laidler, and M. A. A. Stewart, *Chem. Comm.*, 1967, 24.
530 R. G. Gutmacher, E. K. Hulet, R. Lougheed, J. G. Conway, W. T. Carnall, D. Cohen, T. K. Keenan, and R. D. Baybarz, *J. Inorg. Nuclear Chem.*, 1967, **29**, 2341.
531 B. B. Cunningham, J. R. Peterson, R. D. Baybarz, and T. C. Parsons, *Inorg. Nuclear Chem. Letters*, 1967, **3**, 519.
532 'Optical Rotatory Dispersion and Circular Dichroism in Organic Chemistry,' ed. G. Snatzke, Heyden & Son Ltd., London, 1967.

have also been published. At the latter meeting, topics of particular relevance to this Report included the general principles of circular dichroism,[533] the interpretation of natural and magnetically induced optical activity,[534] the effect of ring size and and conformation on the rotatory strength of tris(bidentate) complexes,[535] stereospecific induction,[536] the angular overlap model applied to chiral chromophores and the parentage interrelation of absolute configurations,[537] and the circular dichroism and configuration of complexes of amino-acids.[538] A summary [539] of the 1965 Meldola Lecture delivered by R. D. Gillard on 'Optically Active Co-ordination Compounds' was published, to be followed shortly afterwards by a review article [540] based upon this lecture.

The application of the Cotton effect in the determination of absolute and relative configurations in co-ordination compounds still constitutes a major field of research, where the configuration is traditionally related to the long-wavelength Cotton effect. For Co^{III} complexes, containing two or three rings with five or less members, there is a reliable rule which states that if the Cotton effect of the longest wavelength *d–d* absorption is dominantly positive, then the enantiomer has the configuration corresponding to that of the D-(+)-$[Co\,en_3]^{3+}$ ion. At the present time no exceptions to this empirical rule are known. Two apparent exceptions have

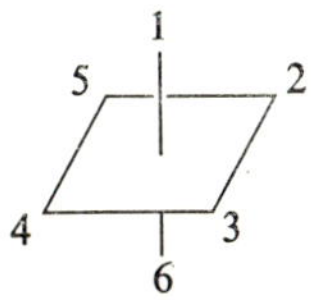

been discussed [541] for complexes of the type (+)-$[Co\,en_2(AA)]^{n+}$ and 1,2,6$[Co(AA)_3]$, where AA is an amino-acid anion. The apparent discrepancy of this rule when applied to certain Co^{II} complexes with polydentate ligands has also been resolved,[542] and it was pointed out that for complexes of C_2 symmetry the Cotton effect corresponding to the $^1A \leftarrow {}^1A$ transition should be considered.

The empirical and non-empirical [543] optical methods for the determination of absolute configurations of dissymmetric co-ordination compounds have been shown [544] to be mutually consistent and to be supported by the *X*-ray diffraction work on the absolute configuration of (−)-$[Fe\,phen_3]^{2+}$.

533 S. F. Mason, *Proc. Roy. Soc.*, 1967, *A*, **297**, 3.
534 A. Moscowitz, *Proc. Roy. Soc.*, 1967, *A*, **297**, 16.
535 F. Woldbye, *Proc. Roy. Soc.*, 1967, *A*, **297**, 79.
536 B. Bosnich, *Proc. Roy. Soc.*, 1967, *A*, **297**, 88.
537 C. E. Schäffer, *Proc. Roy. Soc.*, 1967, *A*, **297**, 96.
538 R. D. Gillard, *Proc. Roy. Soc.*, 1967, *A*, **297**, 134.
539 R. D. Gillard, *Chem. in Britain*, 1967, **3**, 1.
540 R. D. Gillard, *Chem. in Britain*, 1967, **3**, 205.
541 R. D. Gillard, *Nature*, 1967, **214**, 168.
542 R. D. Gillard, *Nature*, 1967, **214**, 1007.
543 A. J. McCaffery, S. F. Mason, and B. J. Norman, *Proc. Chem. Soc.*, 1964, 259.
544 S. F. Mason and B. J. Norman, *Inorg. Nuclear Chem. Letters*, 1967, **3**, 285.

The use of n.m.r. spectra has been suggested [545] as a rapid means of assigning the absolute configurations of amino-acid complexes of the type D-cis-α and L-cis-β-[Co(L,L-$\alpha\alpha'$-dimethyltriethylenetetramine)AA]$^+$, where AA is either glycinate, L-alaninate, or D-alaninate. The correlation was noted by first assigning configurations on the basis of the sign of the Cotton effect in the 18,200–22,250 cm.$^{-1}$ region.

Amino-acid Complexes.—An elegant study of the stereochemistry of complexes of the type [Co trienAA]$^{2+}$, where AA is the anion of glycine or sarcosine, *i.e.* *N*-methylglycine, has appeared.[546] Three geometrical isomers are possible, α (43), β_1 (44), and β_2 (45).

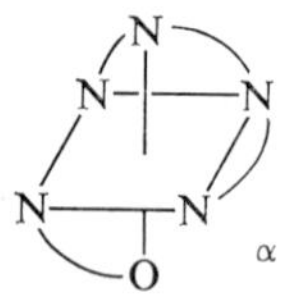

(43)

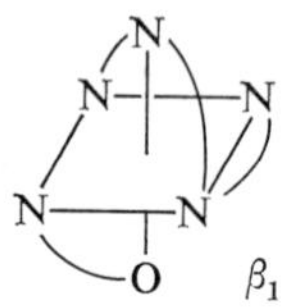

(44)

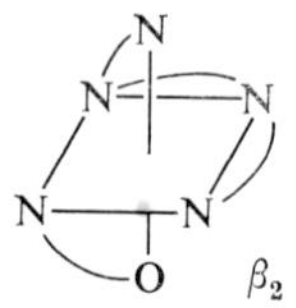

(45)

Only the complexes β_1- and β_2-[Co trien (gly)]$^{2+}$ and β_2-[Co trien sar]$^{2+}$ could be prepared, and each was resolved. Both the synthetic methods used and the rotatory dispersion, circular dichroism, and i.r. spectra suggest that these complexes contain the β-triethylenetetramine configuration, and visible and proton n.m.r. spectra are consistent with this assignment. Repulsive energies were calculated for three different ligand conformations in the sarcosine complex and this analysis suggested the stereospecific co-ordination of sarcosine in the β_2-[Co trien sar]$^{2+}$ ion. From the known absolute configuration of D-(+)-[Co en$_2$ sar]$^{2+}$, the absolute configurations of the D-(+)-β_2-*SS*-[Co trien gly]$^{2+}$ and L-(−)β_2-*RRS*-[Co trien sar]$^{2+}$ ions are deduced.

The steric requirements of D- (or L-) aspartate, functioning as a tridentate ligand, dictate the formation of both geometric and optical isomers in the complex [Co(D- or L-aspartate)(diethylenetriamine)]$^+$. The circular dichroism spectra and the chirality described by the chelate rings allow a tentative assignment of configuration to be made for the three isomers found.[547] The assignments are consistent with visible absorption data and

[545] R. G. Asperger and C. F. Liu, *J. Amer. Chem. Soc.*, 1967, **89**, 708.
[546] L. G. Marzilli and D. A. Buckingham, *Inorg. Chem.*, 1967, **6**, 1042.
[547] J. I. Legg and D. W. Cooke, *J. Amer. Chem. Soc.*, 1967, **89**, 6854.

chromatographic behaviour. The three complexes *trans*-*N*-[Co(oxalato)L_2]$^-$ (L = glycine, β-alanine, or L-α-alanine) have been resolved and their absolute configurations inferred [548] from their circular dichroism spectra.

Using the empirical rule,[540] the isomer of the bis(glycylglycinato)-cobaltate(III) anion, which is more strongly absorbed on starch, is assigned a D-configuration (see Figure 10).[549] The optical configurations of this ion and of the analogous complexes of glycyl-L-leucine, L-alanylglycine, and L-alanyl-L-alanine are also correlated.

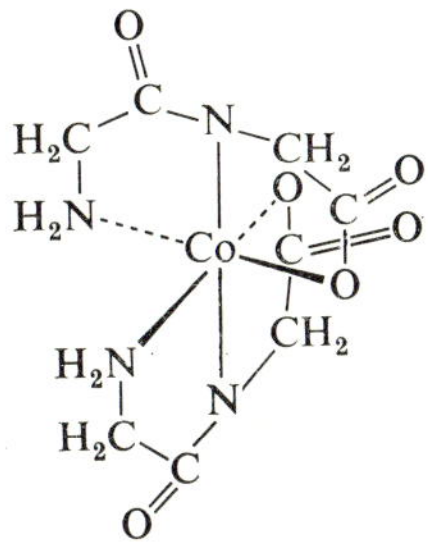

Figure 10 *The configuration of* D-(+)-[Co glygly$_2$]$^-$

Cobalt–Nitrogen Donor Complexes.—Vibronic structure in circular dichroism has been reported for the carbonyl chromophore, but it had not been observed in transition-element complexes. Vibrational structure is now reported [550] for the complex (+)-[Co en_3]$^{3+}$. Absorption and circular dichroism spectra were obtained using a crystal of (±)-[Rh en_3Cl_3]$_2$,NaCl,-6H_2O doped with *ca.* 1 mole % (+)-[Co en_3]$^{3+}$. At 77 and *ca.* 5°K the low-energy side of the E_a circular dichroism band is lost, just as in the absorption spectrum, and vibrational structure appears (see Figure 11). The implications of these observations are still being considered, but it is anticipated that the two circular dichroism components of the $^1T_{1g}(O_h)$ band arise not from the trigonal component, as assumed previously, but from Jahn–Teller components.

The polarised crystal spectra have been reported [551] for the *trans*-[Co en_2Cl_2]$^+$ ion and the optically active 1,2-diaminopropane complex, *trans*-[Co pn_2Cl_2]$^+$, at temperatures down to 25°K. The temperature dependence of the spectral intensities suggests a vibronic intensity gaining mechanism, and since the spectra of the two complexes are so similar it is concluded that the conformation of the chelate rings is unimportant in the determination of the absorption spectra. The spectra are discussed in detail and assigned in a vibronic D_{4h} model. The D_{4h} model, however, cannot be correct for *trans*-[Co pn_2Cl_2]$^+$ since a centre of inversion is

548 J. Hidaka and Y. Shimura, *Bull. Chem. Soc. Japan*, 1967, **40**, 2312.
549 R. D. Gillard, (Miss) P. M. Harrison, and E. D. McKenzie, *J. Chem. Soc.* (*A*), 1967, 618.
550 R. G. Denning, *Chem. Comm.*, 1967, 120.
551 R. Dingle, *J. Chem. Phys.*, 1967, **46**, 1.

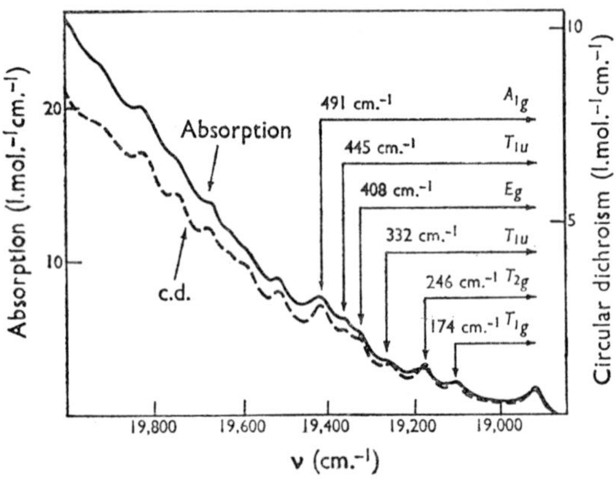

Figure 11 *The single-crystal circular dichroism* (– – –) *and absorption spectra* (——) *of the low-energy portion of the* $^1T_{1g}(O_h)$ *band in* $[(+)\text{-Co en}_3]^{3+}$ *ion at* 5°K

incompatible with optical activity. Models with D_2, *i.e.* ignoring equatorial methyl groups, or C_2 symmetry are possibilities and it is proposed that the spectrum is only very weakly allowed under the appropriate electronic selection rules and that the major part of the intensity still arises by a vibronic mechanism. However, the results presented in this paper are in good accord with the hypothesis that there is no vibronic contribution to the circular dichroism.

Highly stereoselective reactions, involving L-glutamic and D-tartaric acids, to produce bisethylenediaminecobalt(III) derivatives have been reported.[552] The stereoselectivity is kinetic in origin in the case of glutamic acid, though there is some evidence that the formation of L-[Co en_2{(+)-tartrate}] is favoured thermodynamically as well as kinetically. Absolute configurations of the complexes mentioned are presented. The bisdiamine complex (+)-*trans,trans*-[Co$(NO_2)_2$(*N*-methylethylenediamine)$_2$] has also been studied.[553] The visible and u.v. absorption spectra, rotatory dispersion, and circular dichroism spectra are presented.

The co-ordination of L-3,8-dimethyltriethylenetetramine (L-3,8-diMetrien) to cobalt(III) leads to the formation of $(+)_{589}$-*trans*-[Co(L-3,8-diMetrien)Cl_2]Cl, [Figure 12(a)], ion in preference to the possible *cis*-isomers.[554] The circular dichroism spectrum of the $(+)_{589}$-*trans*-[Co trienCl_2]$^+$ ion is similar in form and sign [555] to that of the

552 J. H. Dunlop, R. D. Gillard, and N. C. Payne, *J. Chem. Soc.* (*A*), 1967, 1469.
553 D. A. Buckingham, L. G. Marzilli, and A. M. Sargeson, *J. Amer. Chem. Soc.*, 1967, **89**, 3428.
554 S. Yoshikawa, T. Sekihara, and M. Goto, *Inorg. Chem.*, 1967, **6**, 169.
555 D. A. Buckingham, P. A. Marzilli, A. M. Sargeson, S. F. Mason, and P. G. Beddoe, *Chem. Comm.*, 1967, 433.

compound shown in Figure 12(a), which suggests that these complex ions have the same absolute configuration and that the optical rotatory power has an analogous origin in the two cases. Steric considerations suggest the configuration shown in Figure 12(a) for the $(+)_{589}$-*trans*-[Co(L-3,8-diMetrien)Cl_2]$^+$ ion, and by analogy the structure shown in Figure 12(b) for the $(+)_{589}$-*trans*-[Co trien Cl_2]$^+$. These assignments are supported by further comparative circular dichroism studies.

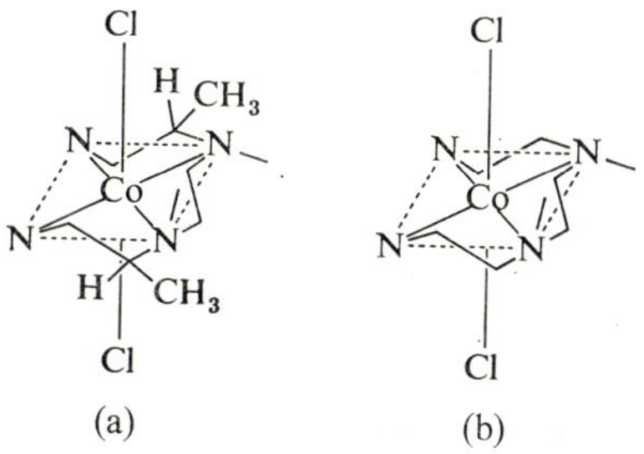

Figure 12

Rotatory dispersion, circular dichroism, visible and u.v. spectra for the isomer (+)-[Co$(NH_3)_4$ (*N*-methylethylenediaminetetrammine)]I_3,H_2O have been recorded.[556] The maximum optical rotation is at 21,000 cm.$^{-1}$ and this wavelength was used to follow the rates of racemisation. The relationship between the absolute configurations of a number of acid-alcohols and the Cotton effect has been studied [557] by investigating a number of Co^{III} complexes of the types *cis*-[Co$(NH_3)_4$(H_2O)Ac](ClO_4)$_2$, *cis*-[Co$(NH_3)_4$$Ac_2$]$ClO_4$, and [Co$(NH_3)_4$lactate]$SO_4$. These complexes absorb in the visible region whereas the parent-acid alcohols absorb at *ca.* 45,460 cm.$^{-1}$, a region inaccessible on most spectropolarimeters.

The circular dichroism spectra of a series of optically active Co^{III} chelate complexes having in common the ligand ethylenediamine-*NN'*-diacetic acid in a *trans*-configuration have also been reported.[558] These spectra were used to determine the absolute configurations of the complexes. The absolute configuration of (−)-β-[Co tetrenCl] (tetren = tetraethylenepentamine) has been assigned [131] as L on the basis of optical rotatory dispersion and circular dichroism evidence.

Some interesting asymmetric syntheses, asymmetric transformations, and asymmetric inductions take place in L-2,3-butanediol.[559] For example, it was shown that when a solution of the racemic *cis*-dichlorobis(ethylenediamine)cobalt(III) ion is heated in the optically active solvent, the ion slowly anti-racemises to give an optically active product. The asymmetric

[556] D. A. Buckingham, L. G. Marzilli, and A. M. Sargeson, *J. Amer. Chem. Soc.*, 1967, **89**, 825.
[557] J. Bolard, *Compt. rend.*, 1967, **264**, *C*, 73.
[558] J. I. Less, D. W. Cooke, and B. E. Doublas, *Inorg. Chem.*, 1967, **6**, 700.
[559] B. Bosnich, *J. Amer. Chem. Soc.*, 1967, **89**, 6143.

synthesis of alanine via the template action of a dissymmetric cobalt(III) complex has also been reported.[560]

Polynuclear Cobalt Complexes.—The circular dichroism of the hexa-μ-hydroxododeca-amminetetracobalt(III) ion, (46), which was the first purely inorganic complex to be resolved, has been examined[561] together with its hexaethylenediamine analogue (47). The chirality rule[562] connecting the sign of the E_a circular dichroism with the stereochemistry of the complex suggests that the oxychelate rings in the (−)-isomers of (46) and (47) have an absolute configuration around the central cobalt(III) ion similar to that of the ethylenediamine rings in the (+)-[Co en$_3$]$^{3+}$ ion. The circular dichroism of (47) has also been discussed by Kern and Wentworth,[563] who concluded that it was impossible to assign uniquely the absolute configurations of the central and peripheral octahedra on the

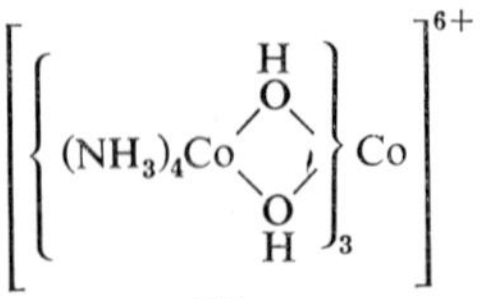

(46)

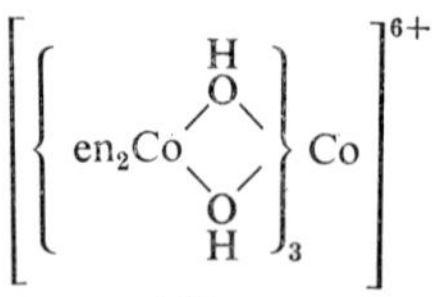

(47)

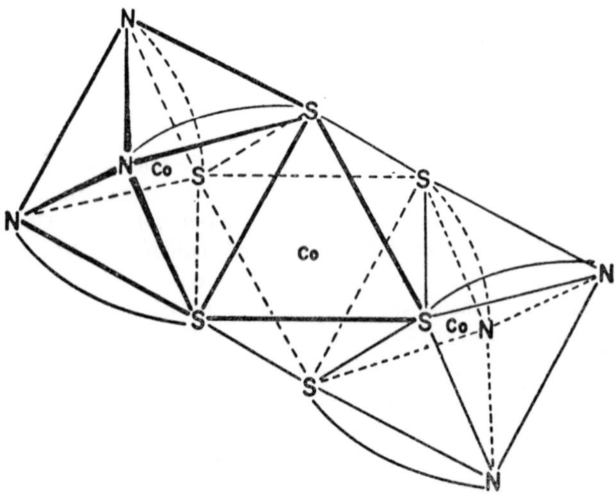

Figure 13 *The structure of the* [Co$_3$L$_6$]$^{3+}$ *cation* (L = 2-*aminoethanethiol*) (Reproduced by permission from *Inorg. Chem.*, 1967, **6**, 1563)

[560] R. G. Asperger and C. F. Liu, *Inorg. Chem.*, 1967, **6**, 796.
[561] S. F. Mason and J. W. Wood, *Chem. Comm.*, 1967, 209.
[562] R. E. Ballard, A. J. McCaffery, and S. F. Mason, *J. Chem. Soc.*, 1965, 2883.
[563] R. D. Kern and R. A. D. Wentworth, *Inorg. Chem.*, 1967, **6**, 1018.
[564] Y. Sasaki, J. Fujita, and K. Saito, *Bull. Chem. Soc. Japan*, 1967, **40**, 2206.

basis of the signs of the rotational strengths from the dominant dichroism maxima.

The circular dichroism of the paramagnetic μ-peroxo-μ-amido-tetrakis-(1,2-diaminopropane)dicobalt ion has also been studied[564] and a Λ(L)-type of configuration suggested. The electronic and circular dichroism spectra of the optical isomers of the hexakis-(2-aminoethanethiolo)tricobalt(III) (see Figure 13) and hexakis-(2-aminoethanethiolo)dicobalt(III)zinc(II) cations have been reported.[565] Only the terminal Co(III) ions contribute to the visible circular dichroism spectrum of the heterometallic complex ion, and this spectrum was subtracted from the tricobalt(III) complex to obtain the circular dichroism spectra of the central cobalt atom. Opposite configurations about the central and terminal cobalt ions are assigned from the circular dichroism.

From polarimetric measurements of the association between (+)-$[Co\ en_3]^{3+}$ and $[Fe(CN)_6]^{4-}$ ions in solutions of total sodium ion concentration equal to 2M, the stability constants for the two first 'outer-sphere' complexes have been determined.[566] The absorption and circular dichroism spectra are reported for the species $[(+)\text{-}Co\ en_3\{Fe(CN)_6\}_n]^{(3-4n)+}$ (n = 0, 1, and 2). An interpretation of the circular dichroism spectra was given earlier.[567]

Complexes of Chromium, Iron, Nickel, and Copper.—The stereospecific formation of a novel optically active binuclear chromium(II) complex containing two L-tartrate bridges has been described.[568] The complex is formulated as $Ba[Cr_2(L\text{-}tart)_2H\ bipy_2]_2$ and its absorption, circular dichroism and rotatory dispersion spectra are all reported. A $[\Lambda-\Lambda]$ absolute configuration is suggested for this ion.

An optical rotatory dispersion study has been reported for iron(II) and iron(III) complexes with L-cysteine at several pH values;[569] also included is a report of an investigation into a new type of optically active complex containing carbonyl or nitrosyl and optically active L-cysteine.

The optical activity of nickel(II) complexes has attracted rather little attention. Several new series of tetrahedral bis(salicylaldimino)- and bis(β-ketoamino)-nickel(II) complexes have been reported.[570] The formation of true enantiomeric complexes is demonstrated by using the (+)- and (−)-amines, when exact mirror-image optical rotatory dispersion curves are obtained for the (+)(+)- and (−)(−)-forms of a typical complex, *e.g.* $Ni(5\text{-}Me\text{-}PhEt\text{-}sal)_2$. The electronic spectra are also recorded for all these complexes in solution and are used in configurational discussions.

565 G. R. Brubaker and B. E. Douglas, *Inorg. Chem.*, 1967, **6**, 1562.
566 R. Larsson, *Acta Chem. Scand.*, 1967, **21**, 257.
567 R. Larsson, S. F. Mason, and B. J. Norman, *J. Chem. Soc.* (*A*), 1966, 301.
568 S. Kaizaki, J. Hidaka, and Y. Shimura, *Bull. Chem. Soc. Japan*, 1967, **40**, 2207.
569 A. Tomita, H. Hirai, and S. Makishima, *Inorg. Chem.*, 1967, **6**, 1746.
570 R. E. Ernst, M. J. O'Connor, and R. H. Holm, *J. Amer. Chem. Soc.*, 1967, **89**, 6104.

The resolution has been accomplished[571] of a Ni^{II} complex containing tribenzo[*b*, *f*, *j*][1,5,9]triazacycloduodecine (tri). The complex, [Ni tri $(H_2O)_3]^{2+}$, comprises only the second class of dissymmetric six-co-ordinate Ni^{II} complexes to be resolved. Circular dichroism spectra and optical rotatory dispersion curves are reported and discussed.

Data[572] have been presented for an amino-acid–copper(II) complex which demonstrate that cross-plane, steric, and electronic interactions between ligands and ligand-field changes have small or negligible effects on the Cotton-effect amplitudes (measures of rotational strength) of the *d*–*d* transition in the complex. On the other hand, the Cotton-effect amplitudes were found to be very sensitive to the substituted glycine, $RCH{\cdot}NH_2{\cdot}CO_2H$, and the nature of this substituent effect has been further studied[573] with particular emphasis on amino-acids which could be tridentate.

Platinum and Palladium Complexes.—Reliable circular dichroism data for optically active olefins are still difficult to obtain. However, in certain cases the formation of Pt^{II} complexes has been shown to shift the circular dichroism bands into a more convenient instrumental range, and this technique may be used to study relative and absolute configurations.[574] The complex, *cis*-[Pt $en_2Cl_2]Cl_2$, has been resolved[575] into its optical isomers and optical rotatory dispersion curves are presented.[575] The reaction of $(+)_{450}$-[Pt $en_2Cl_2]Cl_2$ with a stoicheiometric amount of ethylenediamine in aqueous solution at room temperature yields optically pure D-[Pt $en_3]Cl_4$.

The nature of the absorption bands in square planar metal complexes of the d^8 type has been investigated by measuring the circular dichroism, rotatory dispersion, and absorption spectra of a number of Pt^{II}, Pd^{II}, and Au^{III} complexes containing 1-propylenediamine and 1-cyclohexanediamine.[576]

[571] L. T. Taylor and D. H. Busch, *J. Amer. Chem. Soc.*, 1967, **89**, 5372.
[572] K. M. Wellman, T. G. Mecca, W. Mungall, and C. R. Hare, *J. Amer. Chem. Soc.*, 1967, **89**, 3646.
[573] K. M. Wellman, W. Mungall, T. G. Mecca, and C. R. Hare, *J. Amer. Chem. Soc.*, 1967, **89**, 3647.
[574] E. Premuzic and A. I. Scott, *Chem. Comm.*, 1967, 1078.
[575] C. F. Liu and J. Doyle, *Chem. Comm.*, 1967, 412.
[576] H. Ito, J. Fujita, and K. Saito, *Bull. Chem. Soc. Japan*, 1967, **40**, 2584.

7
Mössbauer Spectroscopy

1 Introduction

Nuclear resonant absorption or Mössbauer spectroscopy is the most recently developed of the spectroscopic techniques described in this volume and for this reason many readers will be least familiar with its use. The fundamental physical principles governing the Mössbauer effect restrict its observation to solid compounds or to species frozen in solid matrices containing specific resonant isotopes. It therefore has no application to liquid or gaseous phases. Although this immediately restricts chemical application, there is still considerable scope in environmental studies of atoms in crystalline solids and powders, especially as many other techniques give little information about this phase.

Since this is the first annual specialist review to appear in the field, it was thought appropriate to include a rather broad selection of subject matter in the chapter so that a fuller appreciation of the capabilities of the new technique could be more easily realised. Over 300 relevant papers were published during 1967. Over half of these refer to practical measurements using ^{57}Fe and about one-third to all other isotopes. Chemical compounds of iron and oxide phases containing iron, each account for about one-sixth of the total. Much of the data have been obtained in physics laboratories, but are frequently of considerable chemical significance.

A number of reviews have appeared during the year, including several at an elementary level on chemical applications.[1-7] A book on 'Hyperfine Interactions' includes some edited general lectures by R. L. Mössbauer,[8] as well as some specialist reviews. The latter cover such subjects as magnetic relaxation effects,[9] the more particular case of diluted ferric alum,[10]

[1] J. S. van Wieringen, *Philips Tech. Rev.*, 1967, **28**, 33.
[2] G. K. Wertheim, *Phys. Today*, 1967, **20**, 31.
[3] N. N. Greenwood, *Chem. in Britain*, 1967, **3**, 56.
[4] N. N. Greenwood, *Rec. Chem. Progr.*, 1967, **28**, 181.
[5] N. N. Greenwood, *Proc. Roy. Austral. Chem. Inst.*, 1967, 273.
[6] U. Zahn, *Chimica*, **21**, 145.
[7] V. I. Goldanskii, *Angew. Chem.*, 1967, **79**, 844.
[8] R. L. Mössbauer, 'Hyperfine Interactions', ed. by A. F. Freeman and R. B. Frankel, Academic Press, New York, 1967, p. 497.
[9] Ref. 8, A. J. Dekker, p. 679.
[10] Ref. 8, H. de Waard and R. M. Housley, p. 691.

rare-earth experiments involving Coulomb excitation techniques,[11] and the use of polarised γ rays.[12] Other reviews on hyperfine effects have included a general text,[13] and a survey of available data and interpretations on haemoglobin derivatives.[14] A long review by Herber[15] contains much useful information but is disappointingly out of date as evinced by the lack of references post 1965.

In the present review the sequence of sections is:

Theoretical; instrumentation; iron-57; tin-119; rare-earth elements; and other elements.

New Mössbauer transitions on which information was published for the first time during 1967 include ^{57}Fe (136 kev); ^{152}Sm (121·8 kev); ^{160}Gd (75·3 kev); ^{162}Dy (81 kev); ^{164}Er (91·5 kev); ^{172}Yb (79 kev); ^{176}Yb (82 kev); ^{189}Os (69·6 kev); ^{191}Ir (82·4 kev); ^{193}Ir (139 kev); and ^{238}U (45 kev).

In addition, experimental results were published on the known Mössbauer resonances: ^{57}Fe (14·4 kev); ^{73}Ge (67 kev); ^{83}Kr (9·3 kev); ^{99}Ru (90 kev); ^{121}Sb (37·2 kev); ^{125}Te (35·5 kev); ^{129}I (27·8 kev); ^{129}Xe (39·6 kev); ^{151}Eu (21·7 kev); ^{153}Eu (97, 103 kev); ^{155}Gd (86·5 kev); ^{156}Gd (89·0 kev); ^{158}Gd (79·5 kev); ^{161}Dy (25·6, 74·6 kev); ^{164}Dy (73·4 kev); ^{166}Er (80·6 kev); ^{168}Er (79·8 kev); ^{169}Tm (8·4 kev); ^{170}Yb (84·3 kev); ^{171}Yb (75·9 kev); ^{174}Yb (77 kev); ^{182}W (100 kev); ^{183}W (47, 99 kev); ^{186}W (122·5 kev); ^{191}Ir (129 kev); ^{193}Ir (73, kev); ^{195}Pt (98·8 kev); ^{197}Au (77·3 kev); ^{237}Np (59·5 kev).

2 Theoretical

Although considerable theoretical developments in the interpretation of spectra have taken place during the year, it is convenient to leave such discussions to the later sections so that they can be treated together with the actual results. However, several papers have appeared concerning more general theoretical aspects.

Relaxation effects produced by time-fluctuating electronic spins have been observed in a large number of systems, and mathematical formulation of the principles involved is receiving considerable attention from several workers.[16–19] Nussbaum and Dash have predicted emission lines narrower than the natural linewidth:[20] if a localised lattice-mode is stimulated during the preceding nuclear transitions, it is possible that the relaxation time of this mode and the nuclear lifetime are comparable; the resulting amplitude

[11] Ref. 8, J. C. Walker, p. 650.
[12] Ref. 8, p. 696.
[13] J. J. van Loef, *Physica*, 1967, **33**, 188.
[14] G. Lang, *J. Appl. Phys.*, 1967, **38**, 915.
[15] R. H. Herber, *Progr. Inorg. Chem.*, 1967, **8**, 1.
[16] A. J. Dekker and F. van der Woude, *Physica*, 1967, **33**, 195.
[17] H. Gabriel, *Phys. Stat. Sol.*, 1967, **23**, 195.
[18] I. Nowik, *Phys. Letters*, 1967, **24**, *A*, 487.
[19] L. Pal, *Acta Phys. Acad. Sci. Hung.*, 1967, **23**, 161.
[20] R. H. Nussbaum and J. G. Dash, *Phys. Stat. Sol.*, 1967, **19**, *K*31.

modulation of the exponential time-dependence is predicted to result in an emission line being potentially narrower than the Heisenberg width. This has not so far been observed.

A new method for measuring long spin–lattice relaxation times when the spin–spin relaxation is fast has been described.[21] The practical measurements required involve determination of the transmission at a particular point in the spectrum, *i.e.* constant Doppler velocity, with a modulating magnetic field applied. The counting rate varies with time in a predictable manner.

A new theoretical treatment of the pressure dependence of the chemical isomer shift has been given.[22]

Interest has also been shown in the automatic analysis of data, especially where complex functions are involved. The numerical methods available for optimising the fit of theoretical functions to practical data have been critically discussed.[23, 24] The problem of deriving the magnetic and quadrupole hyperfine parameters from ^{57}Fe spectra in the difficult case where the asymmetry parameter is non-zero has been treated in full, and the appropriate computer methods described.[25, 26]

An 'equivalent width' for a resonance line has been defined and is the ratio of the measured area to measured maximum absorption.[27] It is independent of background corrections and the recoil-less fraction of the source, and its use is proposed for quantitative analysis of spectra.

Apparent velocity shifts in the positions of resonance lines have been shown to be simply related to the experimental geometry if the absorbers are thin.[28] The shape of the absorption line has been calculated for absorbers consisting of particles randomly dispersed in a non-resonant carrier medium.[29] Large granules can cause a significant reduction in the observed effect. Continued reduction in particle size eventually results in a reversion to the uniform absorber model.

3 Instrumentation

A surprisingly large number of papers are still appearing on the instrumentation of Mössbauer spectrometers. Perhaps the simplest to be described this year is a demonstration spectrometer using an intensity modulated oscilloscope scan to show a resonance peak position.[30] The basic instrument in use is the constant-acceleration velocity-scan system

21 I. Nowik, *Phys. Letters*, 1967, **24**, *A*, 300.
22 K. Mahesh, *Nuclear Instr. Methods*, 1967, **54**, 212.
23 B. J. Duke and T. C. Gibb, *J. Chem. Soc.* (*A*), 1967, 1478.
24 G. M. Bancroft, A. G. Maddock, W. K. Ong, R. H. Prince, and A. J. Stone, *J. Chem. Soc.* (*A*), 1967, 1966.
25 W. Kundig, *Nuclear Instr. Methods*, 1967, **48**, 219.
26 G. R. Hoy and S. Chandra, *J. Chem. Phys.*, 1967, **47**, 961.
27 T. Yoshioko and H. Kohno, *Nuclear Instr. Methods*, 1967, **48**, 193.
28 N. Hershkowitz, *Nuclear Instr. Methods*, 1967, **53**, 172.
29 J. D. Bowman, E. Kankeleit, E. N. Kaufmann, and B. Persson, *Nuclear Instr. Methods*, 1967, **50**, 13.
30 V. Beltrán-López and G. Loria, *Rev. Sci. Instr.*, 1967, **38**, 707.

with data storage in a multichannel analyser operated in the 'time mode'. Several transducer drives for this have been described.[31-36] Variations on the theme include a time-delay system to prevent all but a predetermined portion of the velocity spectrum (not necessarily symmetrical about zero velocity) to be rejected.[37] This has application if high-resolution work is required when only a small capacity analyser is available. Forward-backward address scaling systems to accumulate the positive and negative acceleration sweeps in the same channel groups have been described.[38-39]

Constant-velocity drives continue to be detailed occasionally,[40] and a method of measuring the transmission at an arbitrary set of velocity values which are not necessarily equispaced has been described;[41] the distance between holes on a velocity programme tape is sensed optically.

The modulation techniques well known in e.s.r. and other branches of spectroscopy have been used in an instrument designed to produce the Mössbauer spectrum in a derivative form.[42] A novel means of Doppler-shifting the γ-ray energy when both source and absorber are stationary is to use a Bragg reflection from a moving crystal of aluminium.[43] A method for detecting resonance absorption in a 2π-counting geometry has been developed in which the absorber is placed in contact with the detector surface and the conversion X-rays are counted;[44] the 136 kev resonance in ^{57}Fe was detected in this manner. Another type of resonance detector employs a gas proportional-counter with a thin film of the resonant element in the window.[45]

A new method of polarising spectra has been described.[46] Polarised sources giving multiple emission lines are not monochromatic, but it is now shown that a polariser such as a magnetised-iron metal foil can be used to linearly polarise a monochromatic single-line source. The polariser is moved with a constant Doppler velocity such that one line in its spectrum coincides with the source line. Examples are shown for an iron foil absorber which is Doppler scanning conventionally. Application is forecast

31 P. E. Clark, A. W. Nichol, and J. S. Carlow, *J. Sci. Instr.*, 1967, **44**, 1001.
32 J. Pahor, D. Kelsin, A. Kodre, D. Hanzel, and A. Moljk, *Nuclear Instr. Methods*, 1967, **46**, 289.
33 F. W. D. Woodhams, *J. Sci. Instr.*, 1967, **44**, 285.
34 K. P. Aleshin, I. I. Lukashevich, V. V. Sklyarevskii, E. P. Stepanov, and N. I. Filippov, *Prib. Tekh. Eksp.*, 1967, 53.
35 B. N. Veits, G. G. Gurevich, and Yu. D. Lisin, *Prib. Tekh. Eksp.*, 1967, 56.
36 Y. Hazony, *Rev. Sci. Instr.*, 1967, **38**, 1760.
37 M. Michalski, J. Piekoszewski, and A. Sawicki, *Nuclear Instr. Methods*, 1967, **48**, 349.
38 Y. Reggev, S. Bukshpan, M. Pasternak, and D. Segal, *Nuclear. Instr. Methods*, 1967, **52**, 193.
39 E. Nadav and M. Palmai, *Nuclear Instr. Methods*, 1967, **56**, 165.
40 K. Cassell and A. H. Jiggins, *J. Sci. Instr.*, 1967, **44**, 212.
41 R. C. Knauer and J. G. Mullen, *Rev. Sci. Instr.*, 1967, **38**, 1624.
42 T. Bressani, P. Brovetto, and E. Chiavassa, *Nuclear Instr. Methods*, 1967, **47**, 164.
43 A. Nielsen and J. Olsen, *Nuclear Instr. Methods*, 1967, **52**, 173.
44 N. Hershkowitz and J. C. Walker, *Nuclear Instr. Methods*, 1967, **53**, 273.
45 J. J. Muray and B. W. Worster, *IEEE, Trans. Nuclear Sci.*, 1967, **14**, 11.
46 S. Shtrikman, *Solid State Comm.*, 1967, **5**, 701.

in, for example, the study of antiferromagnets and non-axial electric-field gradients. The selection of suitable materials for absorber mounts has been discussed,[47] and a cryostat for use between 0 and $-180°$ with a stability of 0·01° has been described.[48] Use of a demagnetisation cryostat using chromium potassium alum as the refrigerant allows temperatures down to 0·08°K to be achieved. In this way the magnetic structure of $FeCl_2,4H_2O$, which has a Néel point of about 1°K, has been observed.[49] Similar low-temperature work on iron metal was carried out using a ^3He–^4He dilution refrigeration technique.[50]

There has also been considerable experimental advance in the use of *in situ* nuclear bombardment processes, but these will be referred to with the appropriate isotopes.

4 Iron-57

^{57}Fe is the most important of the known Mössbauer isotopes. It is one of the simplest resonances to generate and detect and this, coupled with the vast extent of iron technology and the wide range of chemical compounds of iron, provides a never-ending supply of interesting problems which can be investigated by Mössbauer spectroscopy. As will be seen in the next few sections, the results obtained are frequently extremely rewarding. This introductory section deals with sources, both conventional and novel, with charge-states following electron capture and with general interpretation of iron spectra. This will be followed by sections dealing with the individual compounds of iron, with oxide and sulphide systems containing iron, with interstitial systems, and with alloys of iron.

The standard preparation is by electron capture in ^{57}Co and various metallic source matrices have now become universally popular. Detailed preparations of ^{57}Co–Pd sources[51] and ^{57}Co in stainless steel, copper, iron, and palladium matrices[52] have been given.

Although electron capture in ^{57}Co is the best method of populating the ^{57}Fe 14·4 kev level, other nuclear processes have been investigated. Coulomb excitation by bombardment with 3·0 Mev α-particles is one possibility. A target foil of α-iron gives a magnetic spectrum in which the internal magnetic field and the recoil-free fraction have values essentially the same as seen in the normal ^{57}Co decay process.[53] An Fe_2O_3 target shows a 50% reduction in recoil-free fraction but otherwise behaves normally. The high energy of the Coulomb excitation process will displace the excited atom from its original site, but the observation of a normal

[47] L. May and D. K. Snediker, *Nuclear Instr. Methods*, 1967, **55**, 183.
[48] A. J. Nozik and M. Kaplan, *Analyt. Chem.*, 1967, **39**, 854.
[49] M. Shinohara, A. Ishigaki, and K. Ono, *Japan J. Appl. Phys.*, 1967, **6**, 982.
[50] G. J. Ehnholm, T. E. Katila, O. V. Lounasmaa, and P. Reivari, *Phys. Letters*, 1967, 25, *A*, 758.
[51] D. Dubrev, T. Ruskov, and T. Tomov, *Compt. rend. Acad. bulg. Sci.*, 1967, **20**, 537.
[52] I. Dezsi and B. Molnar, *Nuclear Instr. Methods*, 1967, **54**, 105.
[53] E. T. Ritter, P. W. Keaton, jun., Y. K. Lee, R. R. Stevens, jun., and J. C. Walker, *Phys. Rev.*, 1967, **154**, 287.

spectrum indicates that the recoiling atom must come to rest on a normal lattice site by a replacement collision in a timescale less than the 14·4 kev excited-state lifetime ($t_{\frac{1}{2}} = 10^{-7}$ sec.). In the case of iron metal, nearly all the atoms must do this, but in the oxide it appears that there is an appreciable probability of the displaced atom coming to rest on an oxygen site, in which case its contribution to the resonant spectrum is either slight or broadened into the background.

An improved technique, using recoil implantation through vacuum, renders coincidence counting unnecessary by giving a substantial reduction in radioactive background.[54] ^{57}Fe atoms in a target are excited by 36 Mev ^{16}O ions and are thereby displaced from the target matrix. They then travel through high vacuum to a catcher material which is shielded from the direct ion-beam. Copper, aluminium, gold, and iron catcher foils gave spectra which showed that the displaced atoms came to rest on a normal impurity site within 10^{-7} of a second, whereas the silicate mineral olivine did not give a detectable spectrum.

Neutron capture experiments have been reported for ^{56}Fe in several target matrices.[55] Iron foil gives the normal ^{57}Fe Mössbauer spectrum, but Fe_2O_3 shows a slight reduction in internal magnetic field. $FeSO_4,7H_2O$ appears to suffer considerable oxidation to ferric during the neutron capture process.

The 136·4 kev resonance of ^{57}Fe has finally been detected in a special scattering experiment.[56] The resonance is very weak because of a low recoil-free fraction and small resonance cross-section. Some hyperfine structure was partially resolved. Detection of scattered 14·4 kev γ rays has also proved to be a feasible method of recording nuclear resonance.[57]

Considerable interest has centred for several years on the production of unstable charge-states during the electron capture process in ^{57}Co. CoO, NiO (doped with ^{57}Co), $(NH_4)_2Fe(SO_4)_2,6H_2O$ (doped with ^{57}Co), $CoSO_4,7H_2O$, and $CoCl_2,4H_2O$ have all been stated to show evidence for both iron-(II) and -(III) charge states in the Mössbauer spectrum, it being presumed that the iron(III) state is still decaying. The logical method of proof is to use delayed coincidence techniques to observe the spectrum after a longer time interval than usual, and this method has proved that none of the spectra of the compounds mentioned show time dependence other than 'time filtering' effects.[58] It is clear that the co-existing iron-(II) and -(III) charge-states are in equilibrium rather than metastable. Decay of ^{57}Co in bulk NiO shows the co-existence of both Fe^{2+} and Fe^{3+} species,[59]

[54] G. D. Sprouse, G. M. Kalvius, and S. S. Hanna, *Phys. Rev. Letters*, 1967, **18**, 1041.
[55] W. G. Berger, J. Fink, and F. E. Obenshain, *Phys. Letters*, 1967, **25**, *A*, 466.
[56] N. Hershkowitz and J. C. Walker, *Phys. Rev.*, 1967, **156**, 391.
[57] A. N. Artemev, V. V. Sklyarevskii, G. V. Smirnov, and E. P. Stepanov, *Soviet Phys.—J.E.T.P.*, 1967, **25**, 768 (*Zhur. eksp. teor. Fiz.*, 1967, **52**, 1157).
[58] W. Triftshauser and P. P. Craig, *Phys. Rev.*, 1967, **162**, 274.
[59] J. D. Siegwarth, *Phys. Rev.*, 1967, **155**, 285.

but ultrafine particles show only the iron(III) spectrum (with superparamagnetic relaxation effects), indicating a change in the equilibrium condition.[60]

$CoCl_2$, $CoCl_2,2H_2O$, and $CoCl_2,4H_2O$ show only Fe^{2+} charge states after ^{57}Co decay, but $CoSO_4$, $CoSO_4,H_2O$, $CoSO_4,4H_2O$, $Co_2(SO_4)_3,18H_2O$, CoF_2, and CoF_3 show both Fe^{2+} and Fe^{3+} states.[61]

An ultrasonic calibration of the internal magnetic field in iron metal has enabled the g values to be determined accurately.[62] The value of g_0 obtained (45·47 mHz) is for the bulk material, whereas the previously assumed best value from n.m.r. methods (45·46 MHz) applies only to the domain walls. The two values do, however, agree to within 0·1%. The γ-ray transition-energy is found to be $14{\cdot}39 \pm 0{\cdot}02$ kev and g_0/g_1 is $1{\cdot}749 \pm 0{\cdot}003$.

The Walker–Wertheim–Jaccarino (WWJ) treatment of the chemical isomer shift in iron compounds remained mathematically uncontested for six years, although its authors would be amongst the first to agree that it conflicts with many semi-empirical chemical estimates of electron densities in actual compounds made by Danon and others. A new treatment has appeared which gives a recalibration of the chemical isomer shift scale.[63] The s-electron density at the iron nucleus in oxides is calculated as a function of the Fe—O distance when the overlap between the iron inner shells and the oxygen $2p$-wavefunctions is included. These results are correlated with the pressure dependence of the chemical isomer shift of the Fe^{2+} impurity in the CoO and give an upper limit of $-4{\cdot}0 \times 10^{-4}$ for $\delta r/r$, the fractional change in nuclear radius on excitation. This is much smaller than the WWJ value of $-1{\cdot}8 \times 10^{-3}$. The large shift between Fe^{2+} and Fe^{3+} salts is easily interpreted on the new model as an increase in the amount of $4s$ bonding in the ferric case, as previously suggested on chemical grounds.

Compounds of Iron.—In this section the topics dealt with are: high-spin Fe^{II} compounds; high-spin Fe^{III} compounds; spin-crossover, and intermediate (spin $\frac{3}{2}$) states; low-spin complexes. Compounds of iron in oxide and sulphide systems and interstitial compounds and alloys of iron are dealt with in later sections.

High-spin Iron(II) *Compounds.* Undoubtedly more progress has been made to date in the interpretation of the spectra of iron(II) compounds than in any other subdivision. This is largely because (*a*) antiferromagnetic interactions are frequently observed in these compounds at low temperatures, and (*b*) there is a significant orbital contribution from the $3d^6$ configuration, which affects both the internal magnetic field and the

[60] K. J. Ando, W. Kündig, G. Constabaris, and R. H. Linquist, *J. Phys. and Chem. Solids*, 1967, **28**, 2291.
[61] J. M. Friedt and J. P. Adloff, *Compt. rend.*, 1967, **264**, *C*, 1356.
[62] T. E. Cranshaw and P. Reivari, *Proc. Phys. Soc.*, 1967, **90**, 1059.
[63] E. Simanek and Z. Sroubek, *Phys. Rev.*, 1967, **163**, 275.

quadrupole splitting. Temperature-dependence studies can provide considerable data regarding the energy-level spacings in iron(II) complexes.

One significant problem which is still unresolved is estimation of the importance of a lattice contribution to the observed electric-field gradient. Ingalls, in 1964, deduced that the lattice term was always opposite in sign to, and smaller in magnitude than, the valence-electron term in near-octahedral complexes. Some new calculations using direct lattice-sum calculations for $FeSiF_6,6H_2O$, $FeSO_4,7H_2O$, $FeCl_2,4H_2O$, and $FeCl_2,2H_2O$ suggest that this may not be so.[64] Restriction of the sum to the first co-ordination sphere only can give errors not only of magnitude but also of sign. The value for the fluorosilicate is very small despite the high distortion from octahedral symmetry in the cation environment. A new value of +0·206 barn was deduced for the quadrupole moment of the excited $I = \frac{3}{2}$ state, but this is unlikely to be more accurate than many of the previous estimates.

The interesting phenomenon of magnetically induced quadrupole interactions has been described. $RbFeF_3$ has a cubic perovskite structure and, above the Néel point of *ca.* 102°K, shows only a single-line resonance in accord with iron(II) ion in cubic symmetry.[65, 66] Below 87°K, however, $RbFeF_3$ shows two internal magnetic fields and is presumably in a ferrimagnetically ordered state. It is antiferromagnetic in the intermediate region and although still crystallographically cubic shows a small, but significant, quadrupole splitting because introduction of the magnetic axis lowers the symmetry. This produces exchange splitting of the low-lying spin–orbit triplet. The spin axis is seen to be in the $\langle 111 \rangle$ direction. $KFeF_3$ shows a similar lowering in symmetry below the Néel point of 115°K.[67] In this case spin-relaxation effects are recorded in the region 95–115°K.

A detailed study of the antiferromagnetic spin-ordering process in FeF_2 has been made with particular emphasis on the region just below the critical point of 78·12°K.[68, 69] The critical exponents of the sub-lattice magnetisation of MnF_2 and FeF_2 do not differ significantly despite the much larger anisotropy in FeF_2 which is isostructural with the manganous compound.

The antiferromagnetic phase of $FeCl_2,4H_2O$ ($T_N = 1{\cdot}1$°K) has been studied down to 0·4°K using a ^3He cryostat and both single-crystal and powdered absorbers.[70] At this temperature the internal magnetic field is 260 kOe and the spin direction is angled at about 30° to the *b* axis and 15° to the electric-field gradient tensor which is positive in sign. The

[64] A. J. Nozik and M. Kaplan, *Phys. Rev.*, 1967, **159**, 273.
[65] U. Ganiel, M. Kestigian, and S. Shtrikman, *Phys. Letters*, 1967, **24**, *A*, 577.
[66] G. K. Wertheim, H. J. Guggenheim, H. J. Williams, and D. N. E. Buchanan, *Phys. Rev.*, 1967, **158**, 446.
[67] R. Fatehally, G. K. Shenoy, N. P. Sastry, and R. Nagarajan, *Phys. Letters*, 1967, **25**, *A*, 453.
[68] G. K. Wertheim and D. N. E. Buchanan, *Phys. Rev.*, 1967, **161**, 478.
[69] G. K. Wertheim, *J. Appl. Phys.*, 1967, **38**, 971.
[70] C. E. Johnson and M. S. Ridout, *J. Appl. Phys.*, 1967, **38**, 1272.

temperature dependence of the internal field is anomalous, and line-broadening effects suggest that the electron-spin relaxation times are comparable to the nuclear precession times.

Iron(II) fluorosilicate shows significant anisotropic effects in an applied external magnetic field at or below 4·2°K (see Figure 1).[71] Single-crystal

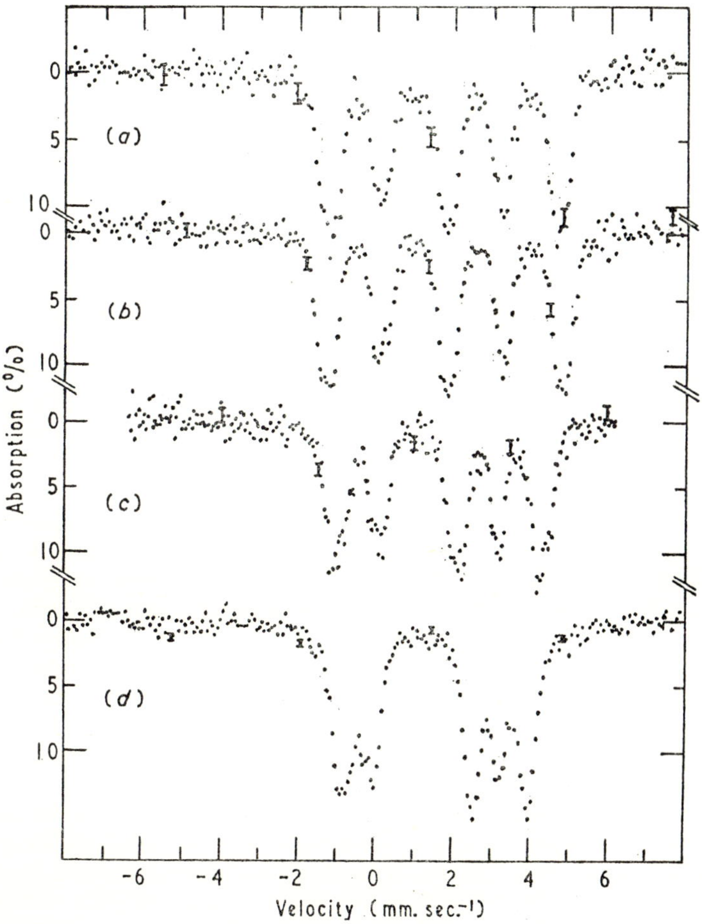

Figure 1 *Mössbauer spectra of* $FeSiF_6,6H_2O$ *single crystal at* 4·2°K *in an external field of* 30 kG *applied at angles of* (*a*) 90, (*b*) 70, (*c*) 45, *and* (*d*) 20° *to the trigonal axis*
(Reproduced by permission from *Proc. Phys. Soc.*, 1967, **92**, 748)

samples with $H_{ext.}$ parallel to the trigonal axis show only a small hyperfine effect similar to that seen for polycrystalline samples. $H_{ext.}$ perpendicular

[71] C. E. Johnson, *Proc. Phys. Soc.*, 1967, **92**, 748.

to the trigonal axis produces a large effective field, and a detailed theoretical interpretation is given. A new value for $\langle r^{-3}\rangle$ of a $3d$ electron of 3·5 atomic units (1 a.u. = 0·529Å) is calculated, being 20% less than for the free-ion value probably as a result of covalency, and a new value for the quadrupole of +0·18 barn is deduced.

A number of phases exist in the boracites $Fe_3B_7O_{13}Cl$, $Fe_3B_7O_{13}Br$, and $Fe_3B_7O_{13}I$.[72] The high-temperature cubic phases show only a small quadrupole effect, but the low-temperature orthorhombic phase contains two distinct site symmetries. A hitherto unsuspected trigonal phase, with only one site symmetry, was found at lower temperatures. The trigonal-to-orthorhombic phase change takes place over a range of temperature, and hysteresis effects occur for the iodide. There is a sharp drop in recoil-free fraction on assuming the cubic structure.

The iron(II) phosphate, vivianite, $Fe_3(PO_4)_2,8H_2O$, contains two iron sites in the ratios 1 : 2.[73, 74] Single-crystal paramagnetic data [74] show the electric-field gradient tensor to be positive and in the *ac* plane for both sites. The asymmetry parameter, η, is small for the Fe_{II} site and nearly zero for Fe_I indicating an $|xy\rangle$ ground-state. Low-temperature data is somewhat contradictory in interpretation. The more detailed paper [74] suggests that the spin direction is in the *ac* plane with adjacent pairs of Fe_{II} spins ferromagnetically coupled. Fe_I and Fe_{II} spin-axes are nearly colinear but not necessarily so. The two-field-gradient directions are perpendicular.

Several series of octahedral iron(II) complexes with ligands bonding through nitrogen have been referred to. The yellow and black forms of $Fe(NCS)_2,4py$ (py = pyridine) examined at 80 and 300°K show no differences, and it is concluded that the black form is probably an impure preparation of the yellow.[75] On the assumption of axial symmetry, the quadrupole splitting value, Δ, of 1·97 mm./sec. at 80°K is said to indicate a doublet ground-state. However, the strong temperature dependence of Δ would seem to indicate a large spin–orbit effect and the conclusion should be treated with caution. Room-temperature spectra for complexes of the type FeL_4X_2 (X = Cl, Br, I, or NCS; L = isoquinoline, γ-picoline, or pyridine) were used to deduce orbital splittings, assuming axial symmetry.[76] A wide range of quadrupole splitting values was found (0·19–3·18 mm./sec.) and although some of the complexes clearly have a singlet ground-state, this may not be true in all cases. Two of the compounds, $Fe\,py_4Cl_2$ and $Fe(\gamma\text{-picoline})_4Cl_2$, and a third similar one, $Fe(\beta\text{-picoline})_4Cl_2$, have also been studied at low temperature.[77]

72 H. Schmid and J. M. Trooster, *Solid State Comm.*, 1967, **5**, 31.
73 S. Chandra and G. R. Hoy, *Phys. Letters*, 1967, 24, *A*, 377.
74 U. Gonser and R. W. Grant, *Phys. State. Sol.*, 1967, **21**, 331.
75 A. V. Ablov, V. I. Goldanskii, E. F. Makarov, and R. A. Stukan, *Doklady Akad. Nauk S.S.S.R.*, 1967, **173**, 595.
76 C. D. Burbridge, D. M. L. Goodgame, and M. Goodgame, *J. Chem. Soc.* (*A*), 1967, 349.
77 A. Martin, G. T. Szabo, I. Dezsi, and A. I. Kiss, *Acta Chim. Acad. Sci. Hung.*, 1967, **52**, 215.

Several complexes with di- and tri-2-pyridylamine (dipyam and tripyam) have been studied.[78, 79] $[Fe\ tripyam_2]^{2+}[FeCl_4]^{2-}$ contains a low-spin cation and a high-spin anion.[79] Complexes of the type Fe dipyam$_2$X$_2$ appear to be chelated six-co-ordinate high-spin iron(II) with large deviations from O_h symmetry. Fe dipyam$_2$(NCS)$_2$ is anomalous in that two distinct iron(II) environments are found. Fe dipyam X$_2$ compounds show smaller chemical isomer shifts which is compatible with the proposed distorted tetrahedral co-ordination.

A detailed temperature-dependence analysis of the quadrupole effects in complexes containing $[FeCl_4]^{2-}$, $[FeBr_4]^{2-}$, $[Fe(NCSe)_4]^{2-}$, and $[Fe(NCS)_4]^{2-}$ anions gives values for the static distortions of the tetrahedra.[80] Small deviations from the simple theoretical treatment are seen in $(NMe_4)_2FeCl_4$. Application of an external magnetic field shows the sign of the electric-field gradient tensor to be negative, and the ground state to be d_{z^2}. The large second-order Doppler shift and the rapid decrease in line intensity with rising temperature reflect the low effective Debye temperatures of these complexes.

The effect of pressure on high-spin iron(II) compounds is discussed on page 343.

The temperature dependence of the Mössbauer parameters of frozen aqueous solutions of iron(II) chloride and sulphate is complex and a function of the thermal history of the sample.[81] It is found that the hydrated iron(II) ion can be used both to influence and to probe the structure of the frozen solvent matrix. Quenching the solution to $-196°$ traps the $Fe(H_2O)_6{}^{2+}$ ion in a cubic ice lattice. The orbital state in this form indicates an axial distortion with an $|xy\rangle$ ground-state. Warming to $-80°$ causes a non-reversible transition to hexagonal ice with some reduction in symmetry of the iron(II) environment. The resonance disappears in iron(II) chloride solutions at $-40°$ because of eutectic formation. The hydrate $FeCl_2,6H_2O$ was produced but is unstable below $-10°$.

High-spin Iron(III) *Compounds.* Very few high-spin iron(III) compounds have been studied this year. Iron(III) fluoride, FeF_3, has been shown to be antiferromagnetic below 362°K. The iron environment is very nearly cubic and the quadrupole interaction is only 0·04 mm./sec.[82, 83] No discontinuity in the chemical isomer shift was detected at the Néel point, although signs of a possible spin-relaxation process were noted below it. $FeCl_3$ is antiferromagnetic below 15°K. The temperature dependence of

[78] C. D. Burbridge and D. M. L. Goodgame, *J. Chem. Soc.* (*A*), 1967, 694.
[79] W. R. McWhinnie, R. C. Poller, and M. Thevarasa, *J. Chem. Soc.* (*A*), 1967, 1671.
[80] P. R. Edwards, C. E. Johnson, and R. J. P. Williams, *J. Chem. Phys.*, 1967, **47**, 2074.
[81] A. J. Nozik and M. Kaplan, *J. Chem. Phys.*, 1967, **47**, 2960.
[82] V. Bertelsen, J. M. Knudsen, and H. Krogh, *Phys. Stat. Sol.*, 1967, **22**, 59.
[83] G. K. Wertheim, H. J. Guggenheim, and D. N. E. Buchanan, *Solid State. Comm.*, 1967, **5**, 537.

the internal magnetic field is as expected, but a discontinuity of 0·04 mm./sec. in the chemical isomer shift at the Néel point is as yet unexplained.[84]

The interlaminar $FeCl_3$–graphite compound, which is analysed as $C_{7\cdot11}FeCl_3$, gives a single-line spectrum similar to that of pure $FeCl_3$. Reduction in hydrogen produces an intercalated species giving a spectrum similar to that of $FeCl_2$ but a minor second species could not be identified. The results are considered to refute several earlier theories as to the nature of these phases.[85]

Room-temperature spectra of $Fe(NO_3)_3,9H_2O$ and $NH_4Fe(SO_4)_2,12H_2O$ in applied magnetic fields of up to 50 kOe show interesting relaxation effects.[86] The earlier observation of line narrowing in the sulphate in fields of up to 2·5 kOe is confirmed, but the linewidth is unaffected above this value. It is now shown that this sharp line is superimposed on a very diffuse background. The broad component is attributed to the $|\pm\frac{3}{2}\rangle \rightarrow |\pm\frac{1}{2}\rangle$ transitions and the narrow component to $|\pm\frac{1}{2}\rangle \rightarrow |\mp\frac{1}{2}\rangle$. Above about 1 kOe the Zeeman splitting of the spin states is larger than the crystal-field terms so that spectra are field-independent. The nitrate is similar but narrowing is slower. No satisfactory explanation of the low-field data has yet been formulated. Earlier work on the iron(III) alum at low temperatures has been extended to higher applied fields.[87] A maximum internal field of 598 kOe and two atomic spin correlation times are required to simulate the observed spectra.

Spectra at 78°K of a series of iron(III) diketone complexes show broad asymmetric quadrupole doublets indicative of electronic relaxation processes.[24] Spectra at room temperature of ferric citrate, benzoate, and malate are very broad.[88] At 78°K, extra satellite lines from magnetic interactions are seen, indicating that the broadening is due to spin relaxation as already observed in many other ferric compounds.

Frozen solutions have been used to identify the chemical species involved in solvent extraction processes using non-aqueous solvents.[89] Iron(III) enriched in ^{57}Fe was extracted with nitrobenzene from solutions containing Cl^-, Br^-, and SCN^-. The bromide and chloride extracts contain FeX_4^- tetrahedral species, but it is suggested that a quadrupole split spectrum from the thiocyanate extract indicates an $Fe(SCN)_4L_2^-$ species, where L is nitrobenzene or water. No spin-relaxation broadening was observed with nitrobenzene as solvent, but chloride and bromide solutions extracted with trioctylphosphine oxide dissolved in benzene show the typical six-line spectrum of the $|\pm\frac{5}{2}\rangle$ Kramer's doublet on a broader unresolved

84 C. W. Kocher, *Phys. Letters*, 1967, **24**, *A*, 93.
85 B. V. Liengme, M. W. Bartlett, J. G. Hooley, and J. R. Sams, *Phys. Letters*, 1967, **25**, *A*, 127.
86 R. M. Housley, *J. Appl. Phys.*, 1967, **38**, 1287.
87 W. Bruckner, G. Ritter, and H. Wegener, *Z. Physik*, 1967, **200**, 421.
88 V. G. Bhide, S. K. Data, G. K. Shenoy, and P. H. Umadikar, *Phys. Letters*, 1967, **24**, *A*, 91.
89 A. G. Maddock, L. O. Medeiros, and G. M. Bancroft, *Chem. Comm.*, 1967, 1067.

background. The spectrum obtained is concentration dependent because of the effect this has on the spin–spin relaxation time.

The effects of pressures of up to 200 kbar on iron compounds are proving extremely interesting. In a series of 12 iron-(II) and -(III) compounds the chemical isomer shift was found to generally decrease with increasing pressure corresponding to an increase in the *s*-electron density at the nucleus.[90] The change is larger for Fe^{2+} because of greater changes in the $3d$–$3s$ shielding. The compounds FeF_2, $FeSO_4,7H_2O$, $KFeF_3$, $FeC_2O_42H_2O$, K_3FeF_6, and $FeF_3,3H_2O$ all show well behaved reversible pressure shifts. The quadrupole splitting often increases with pressure. $KFeF_3$, which is nominally cubic, gives a large splitting of 2·02 mm./sec. at 200 kbar. Significant distortion is induced by pressures below 25 kbar. Iron(III) sulphate, phosphate, acetyl acetonate, citrate, dichromate, and basic acetate all show substantial reduction (reversibly) to iron(II) with increasing pressure. Iron(III) acetyl acetonate is anomalous in that the chemical isomer shift for Fe^{3+} increases with pressure up to about 40 kbar before decreasing again. Iron(III) compounds often reduce photochemically, suggesting only a moderate energy barrier for the process. Compression apparently reduces this barrier to the point where thermal excitation can cause electron transfer from the ligands.

Spin-crossover and Intermediate (Spin $\frac{3}{2}$) States. Several examples of $^5T_2 \rightarrow {}^1A_1$ ligand-field crossover have been investigated. Thiocyanatobis-(1,10-phenanthroline)iron(II) and the corresponding selenocyanate, Fe $phen_2(NCS)_2$ and Fe $phen_2(NCSe)_2$, both give a low-spin Fe^{II} spectrum at liquid-nitrogen temperature and a high-spin Fe^{2+} spectrum at room temperature.[91] The magnetic susceptibility data are inconsistent with a straightforward thermal excitation equilibrium. The thiocyanate shows the co-existence of both spin states within the excited-state lifetime in the transition region, 163–173°K.[92] An iron(II) poly-(1-pyrazolyl)borate (see Figures 2 and 3) chelate shows a similar behaviour at 192°K,[93] as do tris-(2-aminomethylpyridine)iron(II) chloride and iodide at 77°K.[94] Magnetic dilution in the latter case with up to 85% of the corresponding zinc complex has no effect, thereby eliminating the possibility of anti-ferromagnetic interactions. The thermal relaxation time between the singlet and quintet states may be long because of the difference in multiplicity.

A study of pressure effects on phthalocyanineiron(II) showed that a mixture of both high- and low-spin states was present at room temperature, the proportion of the high-spin state increasing (reversibly) with increase

[90] A. R. Champion, R. W. Vaughan, and H. G. Drickamer, *J. Chem. Phys.*, 1967, **47**, 2583.
[91] E. Konig and K. Madeja, *Inorg. Chem.*, 1967, **6**, 48.
[92] I. Dezsi, B. Molnar, T. Tarnoczi, and K. Tompa, *J. Inorg. Nuclear Chem.*, 1967, **29**, 2486.
[93] J. P. Jesson and J. F. Weiher, *J. Chem. Phys.*, 1967, **46**, 1995.
[94] G. A. Renovitch and W. A. Baker, jun., *J. Amer. Chem. Soc.*, 1967, **89**, 6377.

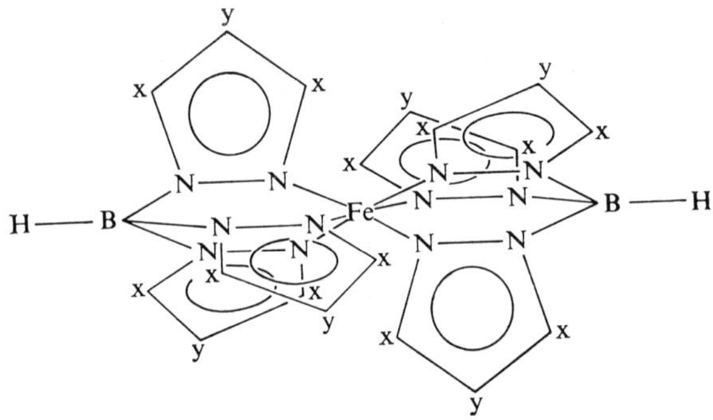

Figure 2 *Basic chelate structure* (x = Me, y = H)
(Reproduced by permission from *J. Chem. Phys.*, 1967, **46**, 1995)

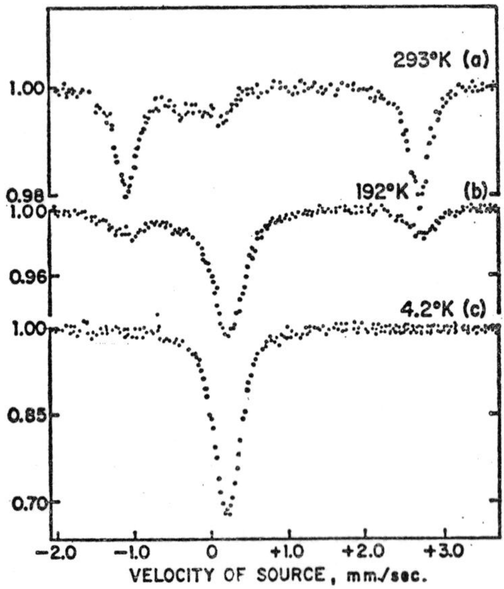

Figure 3 *Temperature-dependence of Mössbauer spectrum of iron* (II) *poly*-(1-*pyrazolyl*)*borate*
(Reproduced by permission from *J. Chem. Phys.*, 1967, **46**, 1995)

in pressure.[95] The energy levels presumably alter so that the $|d_{z^2}\rangle$, $|d_{xz}\rangle$ and $|d_{yz}\rangle$ levels are raised in energy, making excitation to a $|d_{x^2-y^2}\rangle$ state feasible.

The spin-$\frac{3}{2}$ complex bis-(*NN*-diethyldithiocarbamato)iron(III) chloride has proved unusual in that it is ferromagnetic below 2·5°K.[96] The magnetic

[95] A. R. Champion and H. G. Drickamer, *Proc. Nat. Acad. Sci. U.S.A.*, 1967, **58**, 876.
[96] H. H. Wickman, A. M. Trozzolo, H. J. Williams, G. W. Hull, and F. R. Merritt, *Phys. Rev.*, 1967, **155**, 563 (erratum *loc. cit.*, 1967, **163**, 526).

spectra are somewhat complicated just below the Curie point, possibly due to the relaxation rate being nearly as fast as in the paramagnetic state (see Figure 4). The internal magnetic field of 329 kOe at 1·6°K is in good agreement with the empirical rule of $H = 220\langle Sz\rangle$ kOe (*i.e.* 110 kOe per unpaired electron). There is little or no orbital contribution to *H*. The quadrupole splitting is 2·68 mm./sec. at 4·2°K and shows little dependence on temperature. The corresponding isopropyl complex is paramagnetic but shows a magnetic hyperfine effect at 1·2°K because of long electron-spin relaxation times.[97]

Low-spin Complexes. Low-spin complexes of iron are more difficult to study by Mössbauer spectroscopy because of the reduction in the importance of the valence electron contribution to the electric-field gradient and the wide variation in covalent bonding which can occur. Generally speaking the observed data are often consistent with the Pauling neutrality principle in that the charge density at the central atom is largely independent of environment. This is seen, for instance, in a series of ferrocyanides and ferricyanides with different cations.[98] The difference in chemical isomer shift between corresponding pairs of compounds is only of the order of 0·1 mm./sec. A more significant difference is the presence of a small quadrupole splitting in the ferricyanides arising from ligand-field splitting of the t_{2g}^5 configuration. It appears that a spherically symmetric cation such as K^+, Mn^{2+}, or Zn^{2+} produces a smaller distortion in the $Fe(CN)_6^{3-}$ octahedron than for example Co^{2+}, Ni^{2+}, or Cu^{2+}. The Mn, Cu, and Ni ferricyanides all become ferromagnetic below 15°K with observed internal magnetic fields at 4·2°K of 195, 266, and 269 kOe respectively.[99] This is the first observation of internal fields in $S = \frac{1}{2}$ iron compounds. The values do not agree very well with the $220\langle Sz\rangle$ rule and it is proposed that this is due to alterations in the value of $\langle r^{-3}\rangle$ in the Fermi term as a result of covalent bonding, but the orbital contributions have not been estimated. The Zn and Cd ferricyanides are not ferromagnetic, so a paramagnetic cation may be necessary for the appropriate exchange processes. Analysis of the temperature dependence of the quadrupole splitting in ferricyanides will give data on the energy-level separations. A general theoretical treatment has been developed for this,[100] and applied qualitatively to some experimental data for

$$K_3Fe(CN)_6,\ Na_3Fe(CN)_6,H_2O,\ \text{and}\ Cu_3[Fe(CN)_6]_2,14H_2O.$$

An accurate study of $K_3Fe(CN)_6$ suggests that this compound is somewhat anomalous.[101] The data deviate at low temperatures from an attempted

[97] H. H. Wickman and F. R. Merritt, *Chem. Phys. Letters*, 1967, **1**, 117.
[98] K. Chandra, D. Raj, and S. P. Puri, *J. Chem. Phys.*, 1967, **46**, 1466.
[99] B. Sawicki, J. Sawicki, and A. Z. Hynkiewicz, *Soviet Phys.—Solid State*, 1967, **9**, 1100 (*Fiz. Tverd. Tela*, 1967, **9**, 1410).
[100] R. M. Golding, *Mol. Phys.*, 1967, **12**, 13.
[101] W. T. Oosterhuis, G. Lang, and S. Debenedetti, *Phys. Letters*, 1967, **24**, *A*, 346.

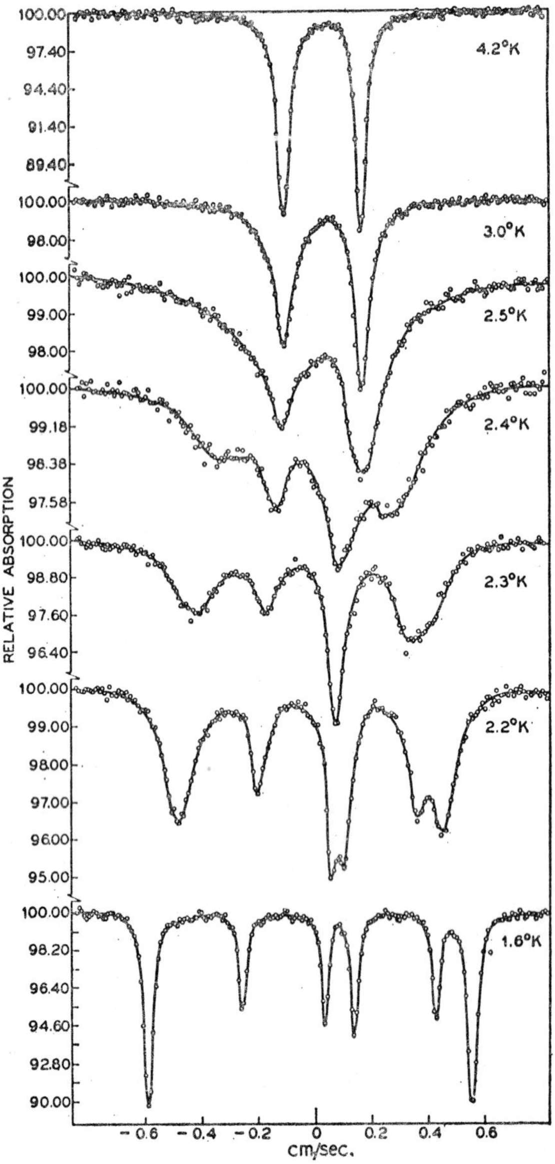

Figure 4 *Mössbauer spectra of bis-*(NN-*diethyldithiocarbamato*)*iron*(II) *chloride as a function of temperature,* $1{\cdot}6 \leqslant T \leqslant 4{\cdot}2$
(Reproduced by permission from *Phys. Rev.*, 1967, **155**, 563)

theoretical fit, and observed line-broadening may be due to a lengthening of the electronic spin relaxation time.

Pressure effects on low-spin complexes are as interesting as those already detailed for high-spin compounds. $K_4Fe(CN)_6$ shows a decrease in chemical isomer shift with increasing pressure.[102] Although the resonance line broadens, no quadrupole split is actually resolved. The initial pressure coefficient for $K_3Fe(CN)_6$ is very high and indicates considerable changes in the bonding of the ferricyanide. The quadrupole splitting increases substantially and a third peak, probably from an $Fe(CN)_6^{4-}$ species, is seen. At 50 kbar there is a phase transition which produces pronounced discontinuities in the pressure dependence of all the Mössbauer parameters. The proportion of pressure-induced ferrocyanide decreases substantially at this point. The reduction of the iron, which is reversible, is obviously related to the overlap of the metal and ligand orbitals. Insoluble Prussian Blue $Fe_4[Fe(CN)_6]_3$ shows reduction of Fe^{3+} to Fe^{2+} with increasing pressure.

The spectrum of the unusual hexa-azideiron(III) complex, formula $Na(Me_4N)_2[Fe(N_3)_6]$, is a single line with a shift intermediate between those of the ferro- and ferri-cyanides.[103] The azido-ligand therefore has a π-bond strength similar to the cyanides, and the complex is octahedral. Mössbauer data on a series of pentacyano-complexes are correlated with other available data to formulate the following order of π-bonding strength:

$$NO^+ > CO > CN^- > Ph_3P > SO_3^{2-} > NO_2^- \approx NH_3 \approx Ph_3As \approx Ph_3Sb.$$

A detailed single-crystal study of sodium nitroprusside shows some evidence for anisotropy of the recoil-free fraction.[104] Correlation of the electric-field gradient and chemical isomer shift with available MO calculations shows the high degree of electron delocalisation between the Fe atom and the NO ligand.

Spectra of *cis*- and *trans*-isomers of alkyl and aryl isocyanide complexes of the type $[Fe(CNR)_4(CN)_2]$, and some related species, are consistent with the proposed stereochemistry.[105] The ratio of quadrupole splitting in the two isomers is predicted to be 1 : 2, and this is found to be approximately correct. The two forms of dicyanobis-(1,10-phenanthroline)iron(II) appear to be identical and the assumed *cis–trans*-isomerism of this complex may be incorrect.

Fe tripyam$_2$(ClO$_4$)$_2$ is a low-spin complex without significant quadrupole splitting.[79] No valence-electron term is expected for a t_{2g}^6 configuration and the ligand symmetry is presumably close to octahedral. The related [Fe tripyam$_2$]$^{2+}$[FeCl$_4$]$^{2-}$ shows the additional lines from $FeCl_4^{2-}$.

Preliminary data are available [106] on some iron 1,2-dithiolene complexes belonging to the general classes $[FeS_4C_4R_4]_n^z$ ($z = 0$, -1, or -2; $n = 1$

[102] A. R. Champion and H. G. Drickamer, *J. Chem. Phys.*, 1967, **47**, 2591.
[103] E. Fluck and P. Kuhn, *Z. anorg. chem.*, 1967, **350**, 263.
[104] J. Danon and L. Iannarella, *J. Chem. Phys.*, 1967, **47**, 382.
[105] R. R. Berrett and B. W. Fitzsimmons, *J. Chem. Soc.* (*A*), 1967, 525.
[106] T. Birchall, N. N. Greenwood, and J. A. McCleverty, *Nature*, 1967, **215**, 625.

or 2), $[Fe(NO)S_4C_4R_4]^z$ ($z = +1, 0, -1, -2$, or -3), and $[FeS_6C_6(CN)_6]^z$ ($z = -2$ or -3). The electron density at the central cation is found to be affected by the overall charge as shown, for example, by an increase in the chemical isomer shift of +0·15 mm./sec. from $[FeS_6C_6(CN)_6]^{2-}$ to $[FeS_6C_6(CN)_6]^{3-}$ (formal iron oxidation states of +4 and +3 respectively). $Et_4N[FeS_4C_4(CN)_4]$ is shown to be similar to the known dimeric species $Bu_4{}^nN[FeS_4C_4(CN)_4]$ with a square pyramidal co-ordination to sulphur. The related $Ph_4P[Fe\,pyS_4C_4(CN)_4]$ is five-co-ordinate but monomeric. $[Fe_2(NO)_2S_6C_6Ph_6]$ contains two different iron environments, but no other structural data are available.

Pressures of up to 200 kbar applied to ferrocene [107] reduce the chemical isomer shift, *i.e.* increase the *s*-electron density at the nucleus. The change is greater than predicted by a simple second-order Doppler shift, and can be interpreted as a decreased shielding of the 3*s* electrons by the 3*d* shell. MO calculations to estimate the occupation and radial function of the 3*d* orbitals gave good agreement with experiment, and also explained the observed decrease in the quadrupole splitting.

The Mössbauer spectrum of the ferrocene–tetracyanoethylene charge-transfer complex is very similar to that of ferrocene and indicates that there is unlikely to be any direct bonding between $(CN)_4C_2$ and the iron;[108] this was subsequently confirmed by the full *X*-ray structure.[109] The spectrum of a ferrocenylcarbonium ion, $(C_5H_5)Fe(C_5H_5CH_2{}^+)$, has been discussed but no definite conclusions were stated as regards the possible bonding structures of the compounds.[110] Other diagnostic applications have been to show that the two iron atoms in the compound formulated as $(PhNCO)_2Fe_2(CO)_6$ [111] and $(PhN)_2Fe_2(CO)_6$ [112] are in equivalent environments.

Data have been given for a series of phosphorus- and sulphur-bridged derivatives of $Fe_2(CO)_9$.[113] Substitution of a terminal carbonyl group by triethylphosphine has only a small effect on the Mössbauer parameters which are also insensitive to substitution of an alkyl by an aryl group in the bridging units. Much larger effects are observed when terminal carbonyl groups are replaced by a π-cyclopentadienyl group or when the bridging atom is changed from phosphorus to sulphur. The sulphur atom is a less effective σ-electron donor than phosphorus in these compounds.

Mössbauer and n.m.r. spectra of a series of tricarbonyl iron cyclo-octatetraene complexes have been used in conjunction to distinguish

[107] R. W. Vaughan and H. G. Drickamer, *J. Chem. Phys.*, 1967, **47**, 468.
[108] R. L. Collins and R. Pettit, *J. Inorg. Nuclear Chem.*, 1967, **29**, 503.
[109] E. Adam, M. Rosenblum, S. Sullivan, and T. N. Margulis, *J. Amer. Chem. Soc.*, 1967, **89**, 4540.
[110] J. J. Dannenberg and J. H. Richards, *Tetrahedron Letters*, 1967, 4747.
[111] W. T. Flannigan, G. R. Knox, and P. L. Pauson, *Chem. and Ind.*, 1967, 1094.
[112] M. Dekker and G. R. Knox, *Chem. Comm.*, 1967, 1243.
[113] T. C. Gibb, R. Greatrex, N. N. Greenwood, and D. T. Thompson, *J. Chem. Soc.* (*A*), 1967, 1663.

between 1,3- and 1,5-diene bonding arrangements.[114] The two types show significant differences in chemical isomer shift and quadrupole splittings.

Phthalocyanineiron(II), which has a square planar cation environment, and a series of its bis-complexes with monodentate organic bases, such as pyridine, which contain six-co-ordinate iron, all show large quadrupole splittings as a result of the high anisotropy in the covalent bonding.[115] The quadrupole splitting of some of the bis-complexes increases with increasing temperature. This unusual observation may be a result of significant contributions from lattice terms rather than a valence-electron effect. It was also shown that the Mössbauer spectrum of chlorophthalocyanineiron(III) gave no definite evidence of the suggested $S = \frac{3}{2}$ spin state of this complex.

Imidazole, pyridine, and piperidine adducts of α, β, γ, δ-tetraphenylporphyrin- and protoporphyrin-iron(II) have been investigated to study the π-bonding characteristics of the ligand with the central metal cation.[116] New measurements have been reported on some previously studied haemoglobin derivatives.[117] Existing Mössbauer data and MO calculations for five haemoglobin derivatives have been compared.[118] The MO approach proves to be more successful than ligand-field theory in interpreting the quadrupole data, but neither is successful in accounting for the chemical isomer shifts observed.

An analysis of some unpublished data by Schulman and Wertheim for iron(III) haemin involves temperature-dependent spin–spin relaxation times.[119] The compound gives a doublet spectrum at 4°K but only a broadened singlet at 298°K. The $|\pm\frac{1}{2}\rangle$ Kramer's doublet is about 12 cm.$^{-1}$ below the $|\pm\frac{3}{2}\rangle$ doublet. There is a faster dipole transition probability in the $|\pm\frac{1}{2}\rangle$ levels than the $|\pm\frac{3}{2}\rangle$, $|\pm\frac{5}{2}\rangle$ levels, so that at higher temperatures, when all are equally occupied, the relaxation time is longer and magnetic broadening occurs.

Oxide and Sulphide Systems containing Iron.—(i) *Binary oxides and sulphides.* The comparatively large differences in the Mössbauer parameters of high-spin iron-(II) and -(III) cations makes Mössbauer spectroscopy particularly valuable in the study of non-stoicheiometric oxide phases and ionic solid-state systems in general.

Although the better known oxide systems have already been examined in previous years, several have been re-investigated in more detail. FeO shows very unusual pressure effects.[120] The chemical isomer shift of the

[114] R. Grubbs, R. Breslow, R. Herber, and S. J. Lippard, *J. Amer. Chem. Soc.*, 1967, **89**, 6864.
[115] A. Hudson and H. J. Whitfield, *Inorg. Chem.*, 1967, **6**, 1120.
[116] L. M. Epstein, D. K. Straub, and C. Maricondi, *Inorg. Chem.*, 1967, **6**, 1720.
[117] H. M. Kappler, A. Trautwein, A. Mayer, and H. Vogel, *Nuclear Instr. Methods*, 1967, **53**, 157.
[118] M. Weissbluth and J. E. Maling, *J. Chem. Phys.*, 1967, **47**, 4166.
[119] M. Blume, *Phys. Rev. Letters*, 1967, **18**, 305.
[120] R. W. Vaughan and H. G. Drickamer, *J. Chem. Phys.*, 1967, **47**, 1530.

iron(II) cation shows strong pressure-dependence similar to that of other iron(II) compounds mentioned in the preceding section. Gross magnetic broadening commences at 50 kbar (see Figure 5) and can be attributed directly to the influence of the defect structure.

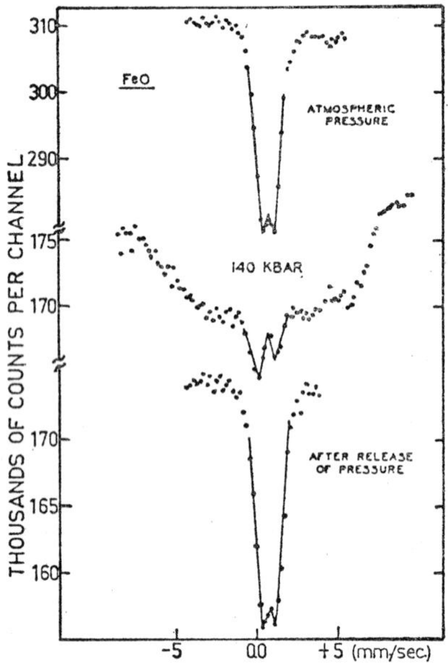

Figure 5 *Mössbauer spectra of* FeO
(Reproduced by permission from *J. Chem. Phys.*, 1967, **47**, 1530)

The Morin transition of haematite, α-Fe_2O_3, in which the electron-spin axis changes direction has gained yet more attention. A magnetic field of up to 9 kOe applied perpendicular to the [111] axis lowers the Morin temperature and the field dependence gives an estimate of the temperature dependence of the antiferromagnetic anisotropy energy.[121] Similar experiments with fields of up to 20 kOe depressed the transition temperature by as much as 10°, and the actual transition was found to be abrupt.[122] Small particles of α-Fe_2O_3 have a larger lattice-spacing than the bulk material because of surface effects, and the reduction in the Morin transition thereby induced [123] can be related to the more conventional pressure-dependence of the bulk.[120] The transition is still sharp in the small particles, emphasising that the change in lattice spacing is homogeneous throughout the whole microcrystal. Calcinated $FePO_4$ shows very similar magnetic properties to

121 D. J. Simkin and R. A. Bernheim, *Phys. Rev.*, 1967, **153**, 621.
122 G. Cinader, P. J. Flanders, and S. Shtrikman, *Phys. Rev.*, 1967, **162**, 419.
123 D. Schroeer and R. C. Nininger, jun., *Phys. Rev. Letters*, 1967, **19**, 632.

those of α-Fe_2O_3.[124] Below 25°K it is antiferromagnetic with a change in the sign of the quadrupole effect at 19°K.

The suspected transition to an antiferromagnetic state below 34°K in $Fe(OH)_2$ has been established by the observation of a magnetic spectrum.[125] The electric-field gradient is parallel to the *c* axis, and the spin axis is in the *c* plane. The ligand-field splitting of the $t_{2g}{}^4$ levels is probably large.

The internal magnetic fields observed in α-FeOOD (301 and 355 kOe at room temperature) are very similar to those in α-FeOOH, implying that deuteriation does not affect the electronic structure.[126] The temperature dependence of the spectra of β- and γ-FeOOH has been followed up to and beyond their ultimate decomposition to α-Fe_2O_3.[127] Both oxides show electronic spin relaxation as evinced by partial collapse of the hyperfine spectrum below the magnetic transition-temperature. Studies of rusts on iron metal in wire and sheet form show the presence of Fe_2O_3, β-FeOOH, and γ-FeOOH in various proportions. Corrosion can be studied over a period of time.[128]

The quadrupole splitting in FeS_2 increases with pressure but the change can be adequately explained on a lattice-term basis only.[107] A more accurate measurement of the spectrum of FeS has shown that earlier reports of values of $-0{\cdot}32$ and $-0{\cdot}36$ mm./sec. for e^2qQ taken from the combined Zeeman quadrupole interaction by first-order calculations are incorrect.[129] Inclusion of second-order terms and allowing e^2qQ and H to be non-collinear shows that $|e^2qQ|$ is of the order of 1·8 mm./sec., although a unique solution was not obtained.

The selenides $Fe_{1-x}Se$ ($0 \leqslant x \leqslant 0{\cdot}165$) are more complex than the oxides and sulphides.[130] At least two overlapping magnetic patterns are seen, and the spectra are somewhat dependent on sample preparation. The chemical isomer shifts are lower than in FeO because of the increased covalent character.

(ii) *Spinel structures.* The spinel oxides are still a popular area of research. Two papers have discussed the normal spinels, $CdFe_2O_4$ and $ZnFe_2O_4$, which contain octahedral iron(III), $M^{2+}[Fe^{3+}Fe^{3+}]O_4$. Point-charge summations have been used in an attempt to correlate the electric-field gradients and observed quadrupole-splittings of the trigonally distorted iron sites,[131] and their low-temperature (<10°K) antiferromagnetic forms have a limiting internal magnetic field of about 490 kOe at 0°K.[132]

[124] W. Bruckner, W. Fuchs, and G. Ritter, *Phys. Letters*, 1967, **26**, *A*, 32.
[125] H. Miyamoto, T. Shinjo, Y. Bando, and T. Takada, *J. Phys. Soc. Japan*, 1967, **23**, 1421.
[126] A. Burewicz, D. Kulgawczuk, and A. Szytuza, *Phys. Stat. Sol.*, 1967, **21**, *K*85.
[127] I. Dezsi, L. Keszthelyi, D. Kulgawczuk, B. Moinar, and N. A. Eissa, *Phys. Stat. Sol.*, 1967, **22**, 617.
[128] I. Dezsi, A. Vertes, and L. Kiss, *Magyar Kém. Folyóirat*, 1967, **73**, 421.
[129] S. S. Hafner, B. J. Evans, and G. M. Kalvius, *Solid State Comm.*, 1967, **5**, 17.
[130] G. A. Fatseas, *Compt. rend.*, 1967, **265**, *B*, 1073.
[131] A. Hudson and H. J. Whitfield, *Mol. Phys.*, 1967, **12**, 165.
[132] J. Sawicki, *Czech. J. Phys.*, 1967, **17**, 371 (*Chem. Abs.*, 1967, **67**, No. 6792*q*).

The inverse iron(III) spinel $NiFe_2O_4$ has been the subject of considerable controversy. The Yafet–Kittel model of ferrites embodies a pair of antiparallel spin sub-lattices in each of the tetrahedral *A* and octahedral *B* sites. Kedem and Rothem [133] resolved the very similar hyperfine fields of these two sites by octahedral chromium substitution to give $NiFeCrO_4$, and interpret the data in detail as being best fitted by the Yafet–Kittel model. Chappert and Frankel [134] applied a 70 kOe external field to $NiFe_2O_4$ which separated out only two antiparallel spin sub-lattices with internal fields at 4°K of −506 (tetrahedral) and −548 kOe (octahedral). This is consistent with the simpler Néel two-sub-lattice model with *A* and *B* spins opposed. $NiFe_{0.3}Cr_{1.7}O_4$ was found to have at least one-third of the iron *B* sites whereas it was originally considered to be all on *A* sites, and this system does correspond to a type of Yafet–Kittel arrangement. The temperature-dependence of the two internal fields can only be explained satisfactorily if strong positive Ni^{2+}–Fe^{3+} interactions are assumed at the *B* sites.[135]

The site occupancy in $MnFe_2O_4$ has been redetermined by applying a 17 kOe field to separate the fields at the *A*- and *B*-site spin sub-lattices by a further 34 kOe.[136] The improved resolution gives an iron ratio $n_A : n_B$ of 0·12 in excellent agreement with neutron diffraction data but which differs from the previous Mössbauer estimate of this ratio (0·3 ± 0·1). The *A* and *B* sites have internal fields of 483 and 430 kOe at room temperature, and there is some suggestion of structure in the *B*-site spectrum. Combined n.m.r. and Mössbauer data have been used to establish the existence of two forms of $MnFe_2O_4$ which are formulated as:
$Mn_{0.8}{}^{2+}Fe_{0.2}{}^{3+}[Mn_{0.2}{}^{3+}Fe_{1.6}{}^{3+}Fe_{0.2}{}^{2+}]O_4$ and
$Mn_{0.48}{}^{2+}Fe_{0.52}{}^{3+}[Mn_{0.06}{}^{2+}Mn_{0.46}{}^{3+}Fe_{1.02}{}^{3+}Fe_{0.46}{}^{2+}]O_4$. Annealing the latter at 450° for 5 hours gave the former by cation diffusion and mutual redox interaction between Mn^{3+} and Fe^{2+}.[137]

The oxide $Fe_{0.3}Mn_{2.7}O_4$ shows a quadrupole splitting produced by a tetragonal distortion below 1373°K.[138] Above this temperature the splitting is smaller because of a change to trigonal symmetry. The changes in site symmetry are interpreted on the basis of a Jahn–Teller effect in Mn^{3+} at the octahedral site.

Although the Mössbauer spectra of iron(III) cations in disordered spinel oxides are relatively insensitive to minor changes in environment, this has not been found to be so with iron(II). The inverse spinel Fe_2TiO_4, and its solid solutions with Fe_3O_4, shows gross broadening of the spectra below

133 D. Kedem and T. Rothem, *Phys. Rev. Letters*, 1967, **18**, 165.
134 J. Chappert and R. B. Frankel, *Phys. Rev. Letters*, 1967, **19**, 570.
135 J. P. Morel, *J. Phys. and Chem. Solids*, 1967, **28**, 629.
136 G. A. Sawatzky, F. Van der Woude, and A. H. Morrish, *Phys. Letters*, 1967, **25**, *A*, 147.
137 H. Yasuoka, A. Hirai, T. Shinjo, M. Kiyama, Y. Bando, and T. Takada, *J. Phys. Soc. Japan*, 1967, **22**, 174.
138 M. Bornaz, G. Filoti, A. Gelberg, and M. Rosenberg, *Phys. Letters*, 1967, **24**, *A*, 449.

the Néel point.[139, 140] This is because the local variations in site symmetry cause large changes in the orbital and dipolar contributions to the internal magnetic field. The cation on the nominally cubic tetrahedral site shows a quadrupole effect of the same order as that at the trigonally distorted octahedral site for similar reasons. $Fe_{2\cdot4}Ti_{0\cdot6}O_4$ shows a broad iron(II) spectrum at 77°K but a sharp ferric pattern. Broadening of both components at 300°K may indicate a rapid charge-transfer mechanism. Site occupancies previously proposed for Fe_2VO_4 are questioned.[140]

The orbital state of iron(II) ions as an impurity on tetrahedral sites in natural $MgAl_2O_4$ crystals has been found to be significantly different from that in the synthetic material.[141] Two types of site distortion are seen, neither of which appears to be a Jahn–Teller effect.

$FeCr_2S_4$ is a normal cubic spinel which is magnetic below 4·2°K. Above the ordering temperature there is no quadrupole splitting because of the cubic symmetry at the *A*-site iron(II) cation,[142] but when the spins are ordered the orbital degeneracy is removed via the spin–orbit coupling to produce a small quadrupole effect. The distortion is quite small, being only about 5 cm.$^{-1}$, and the large positive orbital and dipolar terms cause the effective internal field to decrease again slightly at low temperatures.[143] $FeIn_2S_4$ is an inverted spinel and the trigonal site-symmetry around the ferrous ion causes a large electric-field gradient.[142, 144] The chemical isomer shifts are much lower for the sulphur spinels than for the oxide spinels. These are probably the most covalent $S = 2$ compounds of iron yet studied, and show that the supposed discontinuity between the ranges of observed shifts for high-spin iron-(II) and -(III) complexes was merely a result of insufficient data.

(iii) *Other systems.* Chemical isomer shift and quadrupole data for Cu_2FeSnS_4 from both ^{57}Fe and ^{119}Sn prove that the cation distribution is unambiguously $(Cu^+)_2Fe^{2+}Sn^{4+}S_4$.[145]

Temperature-dependence studies on synthetic and natural $CuFeO_2$ verify the $Cu^+Fe^{3+}O_2$ cation distribution.[146] Both the chemical isomer shifts and the internal fields below the Néel temperature of ~19°K are consistent with iron(III). The spin direction is along the *c* axis and the electric-field gradient tensor is positive as predicted from point-charge lattice summations. $FeNb_2O_6$ contains octahedrally co-ordinated iron(II) with a small tetragonal distortion. The temperature dependence of the quadrupole splitting gives a t_{2g} level splitting of 360 cm.$^{-1}$.[147]

[139] S. K. Banerjee, W. O'Reilly, and C. E. Johnson, *J. Appl. Phys.*, 1967, **38**, 1289.
[140] S. K. Banerjee, W. O'Reilly, T. C. Gibb, and N. N. Greenwood, *J. Phys. and Chem. Solids*, 1967, **28**, 1323.
[141] K. Ono, L. Chandler, and A. Ito, *Phys. Letters*, 1967, **24**, *A*, 273.
[142] C. M. Yagnik and H. B. Mathur, *Solid State Comm.*, 1967, **5**, 841.
[143] M. Eibschütz, S. Shtrikman, and Y. Tenenbaum, *Phys. Letters*, 1967, **24**, *A*, 563.
[144] M. Eibschütz, E. Hermon, and S. Shtrikman, *Solid State Comm.*, 1967, **5**, 529.
[145] M. Eibschütz, E. Hermon, and S. Shtrikman, *J. Phys. and Chem. Solids*, 1967, **28**, 1633.
[146] A. H. Muir, jun., and H. Wiedersich, *J. Phys. and Chem. Solids*, 1967, **28**, 65.
[147] M. Eibschütz, U. Ganiel, and S. Shtrikman, *Phys. Rev.*, 1967, **156**, 259.

Fayalite, Fe_2SiO_4, has the olivine structure with two different octahedral sites containing iron(II). Two quadrupole doublets are seen.[148, 149] $CaFeSiO_4$ has one of the two sites replaced by calcium, but since the observed quadrupole splitting is smaller than either in Fe_2SiO_4, no positive identification of sites in the latter could be made.[148] Below 66°K it is weakly ferromagnetic, becoming antiferromagnetic below 20°K. Although the two sites are barely resolved in the quadrupole spectrum below room temperature, the magnetic spectra give two widely differing internal-field values (323 and 120 kOe at 9°K). Earlier proposals of a change in the spin directions are refuted.[149]

Calcium–iron oxide systems have proved popular. High-temperature quadrupole splitting data have been correlated with point-charge lattice summation for $CaFe_2O_4$, $Ca_2Fe_2O_5$, $Ca_2Fe_{1\cdot5}Al_{0\cdot5}O_5$, and Ca_2FeAlO_5.[150, 151] The antiferromagnetic form of $Ca_2Fe_2O_5$ shows preferential substitution of Sc^{3+} at octahedral sites, but of Ga^{3+} at tetrahedral sites.[152] Aluminium also substitutes at tetrahedral sites [153] but the claim to have observed two Néel temperatures in $Ca_2Fe_2O_5$ was shown to be in error,[154] and it was also pointed out that $Ca_2Fe_2O_5$ and Ca_2FeAlO_5 do not have the same crystal structure. The observed quadrupole spectra of $CaFe_2O_4$, FeOF, and $Fe_xMn_{1-x}MO_3$ (M = rare-earth metal or Y) have been correlated with the crystal structures.[155] FeOF was shown to have anionic disorder and to be magnetically ordered at low temperatures. A proposed structure was confirmed by neutron diffraction.

Ilmenite, $FeTiO_3$, shows significant changes in the electron density at the nucleus with increasing pressure as a result of the orbital contributions in iron(II).[120] The quadrupole splitting is increased by pressure whereas the chemical isomer shift decreases. Four samples of natural ilmenite have been studied and the quadrupole splitting observed and analysed on a crystal-field model.[156]

The spectra of the orthorhombic oxides $Al_{2-x}Fe_xO_3$ $(0\cdot6<x<1\cdot0)$ at 4·2°K show large hyperfine magnetic fields.[157] Application of an external field shows that there are two iron sites and that these have antiparallel alignment. Rhombohedral $Al_{1\cdot6}FeO_{0\cdot4}O_3$ under the same conditions shows a non-unique magnetic field which seems to indicate a canted-spin structure.

148 M. Eibschütz and U. Ganiel, *Solid State Comm.*, 1967, **5**, 267.
149 W. Kündig, J. A. Cape, R. H. Lindquist, and G. Constabaris, *J. Appl. Phys.*, 1967, **38**, 947.
150 A. Hudson and H. J. Whitfield, *J. Chem. Soc.* (*A*), 1967, 376.
151 F. Wittmann, *Phys. Letters*, 1967, **24**, *A*, 252.
152 R. W. Grant, H. Wiedersich, S. Geller, U. Gonser, and G. P. Espinosa, *J. Appl. Phys.*, 1967, **38**, 1455.
153 H. J. Whitfield, *Austral. J. Chem.*, 1967, **20**, 859.
154 S. Geller, R. W. Grant, U. Gonser, H. Wiedersich, and G. P. Espinosa, *Phys. Letters*, 1967, **25**, *A*, 722.
155 J. Chappert, *J. Phys. Radium*, 1967, **28**, 555.
156 Sh. Sh. Bashkirov, G. D. Kurbatov, R. A. Manapov, I. N. Penkov, E. K. Sadykov, and V. A. Chistyakov, *Doklady Akad. Nauk S.S.S.R.*, 1967, **173**, 407.
157 M. Schieber, R. B. Frankel, N. A. Blum, and S. Foner, *J. Appl. Phys.*, 1967, **38**, 1282.

The $Ga_{2-x}Fe_xO_3$ and $AlFeO_3$ systems have been confirmed to be ferrimagnetic.[158] The oxides $Rh_{2-x}Fe_xO_3$ are similar to α-Fe_2O_3 but show relaxation phenomena below the Néel points and a Morin transition which occurs over a wide range of temperature (50–100°) because of the cation randomisation.[159] The oxide phase $Mn_{2-x}Fe_xO_3$ ($0{\cdot}04 \leqslant x \leqslant 0{\cdot}5$) contains two distinct iron cation sites and shows a complex antiferromagnetic behaviour at low temperatures.[160, 161]

The magnetic interactions of the two iron sites in dysprosium–iron garnet have been studied in detail below the Néel point.[162] The yttrium ferrite garnets $Y_3Fe_{5-x}Al_xO_{12}$ are interesting in that increasing the aluminium content on the *B* sites gradually destroys the strong *A–B* site exchange coupling and causes broadening and eventual collapse (at $x = 1{\cdot}2$) of the internal magnetic-field effects on the spectrum.[163] Site occupancy and the magnetic structure have been studied in the $Y_{3-x}Ca_xFe_{5-x}Sn_xO_{12}$ system.[164]

The influence of the rare-earth cations on the magnetic properties of the oxides $MFeO_3$ (M = Y, La, Pr, Nd, Sm, Eu, Gd, Tb, Dy, Ho, Er, Tm, Yb, or Lu) has been studied in overwhelming detail.[165] $TbFeO_3$ shows three types of magnetic structure.[166] The iron spins are coupled antiferromagnetically below 681°K; the Fe and Tb spins are coupled together below 8·4°K, but the Fe is decoupled from the Tb below 3·1°K. The Mössbauer spectrum is barely affected by the two transitions because, although the magnetic axis changes direction, the quadrupole effect is so small in all cases as to be almost negligible. The ferroelectric quaternary oxide phases $(Fe_3Mn_{1-x})MO_3$ ($0 < x \leqslant 1$; M = Y, Ho, Er, Tm, Yb, or Lu) show a change from orthorhombic to hexagonal structure as x decreases and several of these phases have been investigated.[167]

The large difference in chemical isomer shift between iron-(II) and -(III) cations makes the Mössbauer spectrum extremely useful for defining formal oxidation states in natural minerals. Neptunite, formula $[LiNa_2K(Fe, Mn, Mg)_2Ti_2O_2(Si_8O_{22})]$, contains at least 95% of the iron as Fe^{2+} so that earlier postulates of considerable Ti^{3+}–Fe^{3+} content have been disproved.[168] Glauconite has the ideal formula $K(Al_2Fe_2{}^{3+}Fe_3{}^{3+}Mg_3)$ $(OH)_2[Si_3AlO_{10}]$, but the Mössbauer spectrum shows only iron(III) to be present;[169] the green colour is not due primarily to iron(II). Site

158 J. M. Trooster and A. Dymanus, *Phys. Stat. Sol.*, 1967, **24**, 487.
159 I. Dezsi, Gy. Erlaki, and L. Keszthelyi, *Phys. Stat. Sol.*, 1967, **21**, *K*121.
160 R. R. Chevalier, G. Roult, and E. F. Bertaut, *Solid State Comm.*, 1967, **5**, 7.
161 S. Geller, R. W. Grant, J. A. Cape, and G. P. Espinosa, *J. Appl. Phys.*, 1967, **38**, 1457.
162 G. Crecelius, D. Quitman, and S. Hüfner, *Solid State Comm.*, 1967, **5**, 817.
163 V. F. Belov and L. A. Aliev, *Soviet Phys.—Solid State*, 1967, **8**, 2229 (*Fiz. Tverd. Tela*, 1966, **8**, 2791).
164 I. S. Lyubutin, E. F. Makarov, and V. A. Povitskii, *Zhur. eksp. teor. Fiz.*, 1967, **53**, 65.
165 M. Eibschütz, S. Shtrikman, and D. Treves, *Phys. Rev.*, 1967, **156**, 562.
166 E. F. Bertant, J. Chappert, J. Mareschal, J. P. Rebouillat, and J. Sivardiere, *Solid State Comm.*, 1967, **5**, 293.
167 J. Chappert, *J. Phys. Radium*, 1967, **28**, 81.
168 G. M. Bancroft, R. W. Burns, and A. G. Maddock, *Acta Cryst.*, 1967, **22**, 934.
169 V. Hoffmann, E. Fluck, and P. Kuhn, *Angew. Chem.*, 1967, **79**, 581.

occupations have also been determined in five orthopyroxenes of the enstatite–orthoferrosilite series and a manganiferous orthopyroxene.[170] Earlier work on oxidation and dehydration products from amosite and crocidolite has been confirmed and extended.[171] Resonance saturation effects have been used to determine site populations in the amphibole mineral cummingtonite.[172] The values agree with *X*-ray data.

A detailed study of high-spin Fe^{2+} in a square planar environment in the mineral gillespite, $BaFeSi_4O_{10}$, has been published.[173] The small observed quadrupole effect is found to be due to a near cancellation of large valence-electron and lattice contributions. Temperature-dependence data for the spectrum are given with orientation and anisotropy data. It has been found generally possible to distinguish between tetrahedral and octahedral site-symmetries in clay minerals and micas.[174] Analysis of mixtures of layer-silicates may be feasible.

Further analytical applications of Mössbauer spectroscopy include a study of the effects of ammonia adsorption on a supported iron catalyst. The catalyst itself shows evidence of iron-(II) and -(III); the latter is reduced by ammonia but can be regenerated by out-gassing at elevated temperature.[175] Adsorption of simple iron-(II) and -(III) salts onto various ion-exchange resins is found to involve only weak bonding to the resin;[176] the wet resins give no Mössbauer effect unless frozen, and the adsorbed species seem virtually identical to the parent compound, with the exception of ferric sulphate which may be partially hydrolysed. The precipitation of magnetite from solutions containing magnesium has also been followed by Mössbauer spectroscopy;[177] ultrafine particles of iron(III) oxides prepared in high-area silica gels have been studied and often show relaxation effects.[178–180] Amorphous glasses in the systems Fe_2O_3–$NaPO_3$, Fe_3O_4–$NaPO_3$, and $Fe_2(OH)(PO_3)_5$ are also reported on,[181] as are sodium trisilicate glasses containing iron.[182]

Impurity effects in oxides are beyond the scope of this review but two papers of more general interest should be briefly mentioned. The Fe^+ oxidation state has been seen in MgO, and analysis of the crystal-field terms and the hyperfine field suggests that the core-polarisation field per

170 G. M. Bancroft, R. G. Burns, and R. A. Howie, *Nature*, 1967, **213**, 1221.
171 H. J. Whitfield and A. G. Freeman, *J. Inorg. Nuclear Chem.*, 1967, **29**, 903.
172 G. M. Bancroft, *Phys. Letters*, 1967, **26**, *A*, 17.
173 M. G. Clark, G. M. Bancroft, and A. J. Stone, *J. Chem. Phys.*, 1967, **47**, 4250.
174 C. E. Weaver, J. M. Wampler, and T. D. Pecuil, *Science*, 1967, **156**, 504.
175 M. C. Hobson, jun., *Nature*, 1967, **214**, 79.
176 J. L. Mackey and R. L. Collins, *J. Inorg. Nuclear Chem.*, 1967, **29**, 655.
177 G. J. Kakabadse, J. Riddoch, and D. St. P. Bunbury, *J. Chem. Soc.* (*A*), 1967, 576.
178 A. A. Van der Giessen, *J. Phys. and Chem. Solids*, 1967, **28**, 343.
179 W. Kundig, K. J. Ando, R. H. Lindquist, and G. Constabaris, *Czech. J. Phys.*, 1967, **17**, 467.
180 M. C. Hobson, jun., and A. D. Campbell, *J. Catalysis*, 1967, **8**, 294.
181 M. Kamo, Y. Takashima, and S. Ohashi, *Bull. Chem. Soc. Japan*, 1967, **40**, 2812.
182 J. P. Gosselin, O. Shimony, L. Grodzins, and A. R. Cooper, *Phys. and Chem. Glasses*, 1967, **8**, 56.

spin moment is still of the order of −127 kOe for this cation.[183] Although a quadrupole splitting has been observed for an Fe^{2+} impurity atom at a cubic site in MgO below 14°K, previous suggestions of a Jahn–Teller effect are countered by a theory that random strains in the crystal are of sufficient magnitude to lift the degeneracy of the ground-state triplet.[184]

Interstitial Compounds and Alloys.—The Mössbauer spectrum of Fe_2B is virtually unaffected by the insertion of up to 10% of manganese instead of iron.[185] No satellite lines were observed of the type often found in iron alloys. It is postulated that the internal field arises solely from the *d*-electron core polarisation term and that the conduction polarisation term is negligible because the Fermi surface contains no iron 4*s*-electron density.

Several groups of workers have studied the martensite solid solutions of carbon in iron metal. In samples containing about 5 atoms % carbon it is possible to detect hyperfine fields from iron atoms with 0, 1, and 2 carbon neighbours.[186–188] There is some confusion as to whether martensite is body-centred tetragonal or cubic in structure, but the presence of three fields can only be correlated with a body-centred cubic structure in which there is a small local distortion at each carbon so that the *X*-ray pattern is microscopically tetragonal.[188] Some paramagnetic austenite is also usually present in the samples,[187] and in one case was reported to show some fine structure.[186] Tempering destroys the weak satellite lines of the martensite and gives cementite and iron metal. The observed increase in the value of the principal hyperfine field with increasing carbon content can be ascribed to alterations in the exchange integral with increasing interatomic distance.[189]

Fe_5C_2, cementite Fe_3C, and the boron-substituted phase $Fe_3(C_{1-x}B_x)$ are all magnetic at room temperature. Boron substitution causes an increase in internal field, and also a broadening of the lines which shows the presence of some short-range effects of the interstitial atoms.[190] Similar systems with metal substitution, $(Fe_{1-x}Mn_x)_3C$ and $(Fe_{1\cdot1}Mn_{3\cdot9})C_2$ have also been studied.[191] The iron–nitrogen austenite and martensite phases are similar to the carbides, and the work on Fe_4N of Shirane and co-workers has been repeated and confirmed.[186] Nitrogen (0·01%) in an alloy of composition $Fe_{0\cdot75}Cr_{0\cdot25}$ produces a new non-magnetic phase in quantities dependent on the quenching temperature, but its constitution is not known.[192]

183 J. Chappert, R. B. Frankel, and N. A. Blum, *Phys. Letters*, 1967, **25**, *A*, 149.
184 F. S. Ham, *Phys. Rev.*, 1967, **160**, 328.
185 H. Bernas and I. A. Campbell, *Phys. Letters*, 1967, **24**, *A*, 74.
186 P. M. Gielen and R. Kaplow, *Acta Metallurgica*, 1967, **15**, 49.
187 H. Ino, T. Moriya, F. E. Fujita, and Y. Maeda, *J. Phys. Soc. Japan*, 1967, **22**, 346.
188 M. Ron, A. Kidron, H. Schechter, and S. Niedzwiedz, *J. Appl. Phys.*, 1967, **38**, 590.
189 T. Zemcik, *Phys. Letters*, 1967, **24**, *A*, 148.
190 H. Bernas, I. A. Campbell, and R. Fruchart, *J. Phys. and Chem. Solids*, 1967, **28**, 17.
191 G. P. Huffman, P. R. Errington, and R. M. Fisher, *Phys. Stat. Sol.*, 1967, **22**, 473.
192 R. B. Roy, B. Solly, and R. Wappling, *Phys. Letters*, 1967, **24**, *A*, 583.

The phosphides FeP, Fe_2P, and Fe_3P are similar in behaviour to the boride and carbide compounds, and the decreasing phosphorus content in the above sequence causes a decrease in the number of electrons donated by the phosphorus into the 3*d*-band, thereby resulting in a progressive increase in the magnetic moment and the internal magnetic field.[193] Thus FeP is paramagnetic, even at 30°K, and shows a quadrupole-split spectrum due to the co-ordination of iron by a distorted octahedron of phosphorus atoms. Fe_2P is more complex, having two iron sites with tetrahedral and square pyramidal co-ordination, respectively, by phosphorus; the spectrum is broad and unresolved at 195°K but at 90°K shows magnetic interaction with hyperfine fields of 140 and 115 kOe. In Fe_3P there are three iron environments and these give rise to three hyperfine fields having magnitudes of 185, 265, and 295 kOe at 90°K.

Although a considerable number of papers have been written on the subject of metallic systems, nearly all fall outside the present scope. (See Bibliography at end of Chapter for a list of papers not individually discussed in this review.) It is perhaps sufficient to say that the major application at the present time seems to be in the study of order–disorder phenomena in binary alloys and in permanent magnet materials.

However, three or four papers are of more general interest. A ^{3}He–^{4}He dilution refrigerator has been used to cool a source of ^{57}Co in iron metal to 0·08°K.[50] Unequal thermal population of the cobalt magnetic levels produces a non-uniform population by decay of the ^{57}Fe magnetic levels and consequently an asymmetry is observed in the hyperfine spectrum. The new data extend results obtained in earlier work to much lower temperatures and are an interesting demonstration of retrospect study of the state of parent atoms prior to their decay.

A new attempt at calculating the chemical isomer shift in iron metal has been made,[194] but it is still not possible to give a unique electron configuration to the iron; the band-structure model suggests $3d^{7}4s^{1}$, the density of states $3d^{7}4s^{0.5}4p^{0.5}$, and the Mössbauer data about $3d^{6.5}4s^{1.5}$.

The chemical isomer shifts of ^{57}Co impurity atoms in 32 metals are related to the electronic structure of the host.[195] Plots of shift *vs.* number of outer electrons show similar trends for the first, second, and third transition-series, with a change in direction of the shift at the half- and completely-filled *d*-shell (assuming a $d^{n}s^{2}$ configuration).

The electrolytic hydrogenation of nickel doped with ^{57}Co has been shown by Mössbauer spectroscopy to produce a well defined new non-magnetic hydride phase.[196] It co-exists with the nickel in the early stages of the process, and there are no phases of intermediate composition. ^{57}Fe–$NiH_{0.6}$ has a larger chemical isomer shift than ^{57}Fe–Ni and an

[193] R. E. Bailey and J. F. Duncan, *Inorg. Chem.*, 1967, **6**, 1444.
[194] R. Ingalls, *Phys. Rev.*, 1967, **155**, 157 (erratum *loc. cit.*, 1967, **162**, 518).
[195] S. M. Quaim, *Proc. Phys. Soc.*, 1967, **90**, 1065.
[196] G. K. Wertheim and D. N. E. Buchanan, *J. Phys. and Chem. Solids*, 1967, **28**, 225.

increase in both the 3*d* and 4*s* density on the iron, by acceptance of electrons from the hydrogen, is suggested. Introduction of hydrogen into stainless steel produces similar effects.[196] The Fe–Pd–H alloys, on the other hand, are a single phase and there is no evidence that the electrons from the hydrogen atoms interact significantly with the iron 3*d*-electrons.[197] The change in chemical isomer shift in this instance is attributed to expansion of the lattice.

5 Tin-119

Although the effort which has been invested in ^{119}Sn Mössbauer studies is quite extensive, this isotope is still very much less important than ^{57}Fe; the results are also often less rewarding. Although the range of velocities covered by hyperfine effects is about the same for the two elements, the much larger natural linewidth of the tin resonance considerably reduces the resolution which can be achieved in practice. Further, the comparative rarity of internal hyperfine magnetic fields in tin compounds reduces the number of effective diagnostic parameters which are obtainable unless external fields can be applied. These practical difficulties are not ameliorated by the gross lack of information concerning appropriate molecular structures and electronic wavefunctions.

The, as yet, limited and difficult field of tin(II) chemistry has few investigators. A review of the early and often inaccurate data on tin(II) inorganic compounds has appeared and has included some redetermined Mössbauer parameters.[198] The gradual tendency of the tin(II) compound $[Sn_3O(OH)_2SO_4]$ to decompose to tin(IV) oxide at temperatures above 230° has been followed in the Mössbauer spectrum.[199] The similarity of the spectra of $Sr_2SnNO_3F_7,2H_2O$ and $Pb_2SnNO_3F_5,2H_2O$ to those of $Sr(SnF_3)_2$ and PbF,SnF_3 shows the new compounds to contain the SnF_{3-} anion.[200]

One of the more significant papers of the year has compared the ^{119}Sn and ^{121}Sb spectra of a series of pairs of isoelectronic (although not isostructural) complexes.[201] These are K_2SnF_6 (SbF_6), SnF_4 (SbF_5), SnO_2 (Sb_2O_5), α-tin (InSb), β-tin (Sb in β-tin), SnO (Sb_2O_3), and SnF_2 (SbF_3). A plot of the chemical isomer shifts for tin against those for antimony is nearly linear and establishes that $\delta r/r$ for ^{121}Sb is opposite in sign to that for ^{119}Sn and nearly an order of magnitude larger (see Figure 6). The experimental data were combined with Hartree–Fock atomic SCF calculations to give some new values for nuclear constants: $\delta r/r(^{119}Sn)$ $(1{\cdot}2 \pm 0{\cdot}4) \times 10^{-4}$, $Q(\frac{3}{2}+{}^{119}Sn)$ $-0{\cdot}06 \pm 0{\cdot}02$ barn; $\delta r/r(^{121}Sb)$ $(-8{\cdot}5 \pm 0{\cdot}3) \times 10^{-4}$, $Q(\frac{5}{2}+{}^{121}Sb)$ $-0{\cdot}26$ barn. This is a further rebuttal of the postulate by Goldanskii *et al.* that the sign of $\delta r/r(^{119}Sn)$ is negative. However, the figures are probably only

[197] A. E. Jech and C. R. Abeledo, *J. Phys. and Chem. Solids*, 1967 **28**, 1371.
[198] J. J. Zuckerman, *J. Inorg. Nuclear Chem.*, 1967, **29**, 2191.
[199] C. G. Davies and J. D. Donaldson, *J. Chem. Soc.* (*A*), 1967, 1790.
[200] J. D. Donaldson and B. J. Senior, *J. Chem. Soc.* (*A*), 1967, 1821.
[201] S. L. Ruby, G. M. Kalvius, G. E. Beard, and R. E. Snyder, *Phys. Rev.*, 1967, **159**, 239.

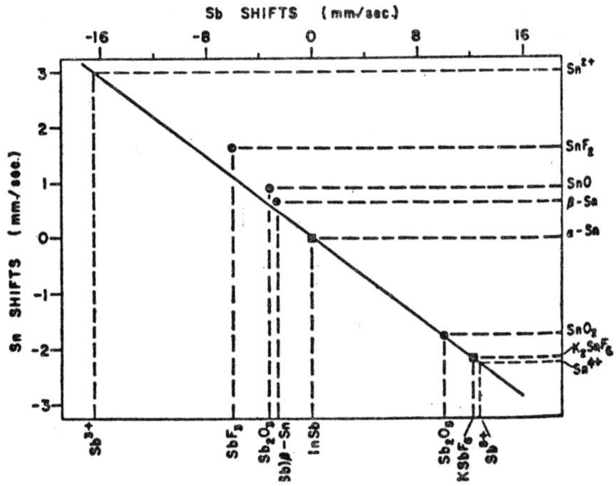

Figure 6 *Relation between the isomer shifts of similar compounds of tin and antimony. The points marked by squares are used to generate the straight line* (Reproduced by permission from *Phys. Rev.*, 1967, **159**, 239)

approximate as they still use only atomic wavefunctions and the theoretically unjustified tenet that α-tin has the configuration $5s5p^3$. Goldanskii *et al.* have published a new treatment of the chemical isomer shift in tin(IV) compounds which distinguishes between compounds which contain π bonding to the tin atom and those which do not.[202] $\delta r/r$ is assumed to be positive, and a decrease in shift in a π-bonding series of compounds is said to be a result of a decrease in 5*s*-orbital occupation, while in compounds without π bonds it indicates an increase in 5*s*-orbital occupation. However, the precise calculations given may be suspect because of the implicit assumption of α-tin having a 5*s*-orbital occupancy of unity.

Both the ^{119}Sn and ^{129}I resonances have been reported in SnI_4.[203] The 'effective temperatures' calculated from the Debye model are 166 and 85°K respectively for the two atoms, showing that the iodine atoms move more easily thermally. An enhancement of the intensities of the $\Delta m = 0$ transitions in the ^{129}I spectrum is explained on the basis of anisotropy of the recoil-free fraction. The Pasternak interpretation of ^{129}I chemical isomer shifts indicates a 'hole' concentration in SnI_4 of 0·87 electrons from the closed-shell iodide configuration.

Spectra of a wide range of six-co-ordinate tin(IV) complexes show no detectable quadrupole splitting even in those cases where the six directly bonded atoms are not identical, provided that all six have non-bonding p_π electrons.[204] When two of the attached groups have no such p_π electrons a sizeable splitting is seen. These observations were used to assign structural configurations to several compounds of uncertain

[202] V. I. Goldanskii, E. F. Makarov, and R. A. Stukan, *J. Chem. Phys.*, 1967, **47**, 4048.
[203] S. Bukshpan and R. H. Herber, *J. Chem. Phys.*, 1967, **46**, 3375.
[204] N. N. Greenwood and J. N. R. Ruddick, *J. Chem. Soc.* (*A*), 1967, 1679.

stereo-chemistry, *e.g.* $SnCl_4$(8-hydroxyquinoline), $SnCl_4$(8-hydroxyquinoline)$_2$, and $NO^+[SnCl_4NO_3]^-$.

A temperature-dependent asymmetry in the line intensities of the quadrupole-split spectra of Me_3SnCN and Me_3SnOH is interpreted in terms of the anisotropy of the mean-square displacement tensor in polymeric structures.[205] The temperature-dependence of the recoil-free fraction in a number of organo-tin compounds shows no systematic dependence on nearest-neighbour atom mass, ligand mass, molecular weight, co-ordination number of the tin, bulk chemical properties, chemical isomer shift, or quadrupole splitting.[206] Polymer formation may, however, be important.

o-Phenanthroline and α,α′-bipyridyl complexes of the type Bu_2SnX_2L (X = Cl, Br, or I) show large quadrupole splittings of the order of 4 mm./sec.[207] The dissimilarity to $SnCl_4$,bipy is consistent with the observations mentioned above.[204]

Several studies have given rather inconclusive results. These include an attempted correlation between the n.m.r. and Mössbauer data on the methyltin hydrides,[208] data for fifteen methyltin compounds which do not fit simple bonding models,[209] a series of five-co-ordinate complexes of R_3SnX with monoaza-aromatic bases,[210] and data for forty organo-tin compounds, most of which are without both quadrupole splittings and p_π non-bonding electrons.[211]

Both the ^{57}Fe and ^{119}Sn resonances have been observed and interpreted for the complex carbonyls $Me_4Sn_3Fe_4(CO)_{16}$ and $Me_4Sn_2Fe_2(CO)_8$.[212] The former shows evidence in the Mössbauer spectrum for the two different tin environments predicted by its stereochemistry.

The valence state and site symmetry of tin in the natural minerals arandisite, canfieldite, cassiterite, cylindrite, franckeite, herzenbergite, hulsite, nigerite, nordenskioldine, stannite, and teallite have been determined.[213] Similar studies have been made on ferrite spinels doped with tin,[214, 215] and on some quaternary oxides of the $BaSnO_3$ type.[216] The spectrum of

205 H. A. Stockler and H. Sano, *Phys. Letters*, 1967, **25**, *A*, 550.
206 H. A. Stockler, H. Sano, and R. H. Herber, *J. Chem. Phys.*, 1967, **47**, 1567.
207 M. A. Mullins and C. Curran, *Inorg. Chem.*, 1967, **6**, 2017.
208 L. May and J. J. Spijkerman, *J. Chem. Phys.*, 1967, **46**, 3272.
209 M. Cordey-Hayes, R. D. Peacock, and M. Vucelic, *J. Inorg. Nuclear Chem.*, 1967, **29**, 1177.
210 J. Nasielski, N. Sprecher, J. Devooght, and S. Lejeune, *J. Organometallic Chem.*, 1967, **8**, 97.
211 V. V. Khrapov, V. I. Goldanskii, A. K. Prokofev, and R. G. Kostyanovskii, *Zhur. obshchei Khim.*, 1967, **37**, 3.
212 M. T. Jones, *Inorg. Chem.*, 1967, **6**, 1249.
213 D. L. Smith and J. J. Zuckerman, *J. Inorg. Nuclear Chem.*, 1967, **29**, 1203.
214 G. N. Shlokov, P. L. Gruzin, and E. J. Kuprianova, *Doklady Akad. Nauk. S.S.S.R.*, 1967, **174**, 1144.
215 P. L. Grusin, G. N. Shlokov, and L. A. Alexeev, *Doklady Akad. Nauk S.S.S.R.*, 1967, **176**, 362.
216 M. V. Plotnikova, A. S. Viskov, K. P. Mitrofanov, V. S. Shpinel, and Yu. N. Venevtsev, *Izvest. Akad. Nauk S.S.S.R., Ser. fiz.*, 1967, **31**, 1112.

SnO_2 introduced into alkali–tin silicate and –borate glasses indicates six-co-ordination of the tin to oxygen. The disordered-lattice hypothesis for the structure of the glass is rejected as the evidence points to uniformity of the tin atoms in the glasses.[217] ^{119}Sn deposited on the surfaces of silica gels and zeolites is more strongly bound when the pore diameter is small.[218]

Several papers have appeared on investigations of alloy systems, chiefly using impurity doping methods; these are not discussed individually but are listed in the Bibliography at the end of the Chapter.

The pressure-dependence of the chemical isomer shift of ^{119}Sn in $SnMg_2$ is interesting in that non-linearity occurs as a result of a change from semi-conducting to metallic properties at about 50 kbar as a result of a decrease in the energy gap between the appropriate electron bands.[219] A decrease in the recoil-free fraction is found with decreasing particle size of metallic tin.[220] This is due to an increase in the fraction of atoms which are in 'surface states', where bonding is different from that in the bulk, rather than to a change in the properties of the bulk material within the particles themselves.

6 Rare Earths

Interest in rare-earth isotopes centres mainly on the determination of nuclear parameters and this section is therefore necessarily abbreviated to the more essential points of chemical interest and relevance.

Several groups have reported the 121·8 kev ($2^+ \rightarrow 0^+$) resonance in ^{152}Sm. The β-decay of ^{152}Eu in $CaF_2(Sm^{2+})$ and $Gd_2O_3(Sm^{3+})$, with absorbers of Sm_2O_3 and Sm^{2+} in CaF_2, show detectable chemical isomer shifts between the Sm^{2+} and Sm^{3+} electronic configurations.[221] The nuclear deformation parameter $\delta\beta/\beta$ was estimated to be $1{\cdot}0 \times 10^{-2}$. Virtually identical experiments gave a value of $1{\cdot}4 \times 10^{-2}$.[222] The magnetic moment of the 2^+ state has been determined to be 0·83 μ_N from the samarium–iron garnet magnetic spectrum.[223] The source of ^{152}Eu in Gd_2O_3 was shown to give a sharp single line even at 4·2°K, but the high energy of the γ-ray produces only a very low Mössbauer efficiency.

The chemical isomer shifts shown by two resonant isotopes of the same element in identical compounds should show a linear relationship. Earlier data for ^{151}Eu and ^{153}Eu contradicted this, but have now been shown to be in error. Comparisons of data using the 21·7 kev level of ^{151}Eu and the

[217] K. P. Mitrofanov and T. A. Sidorov, *Soviet Phys.—Solid State*, 1967, **9**, 693 (*Fiz. Tverd. Tala*, 1967, **9**, 890).
[218] I. P. Suzdalev, A. S. Plachinda, and E. F. Makarov, *Zhur. eksp. teor. Fiz.*, 1967, **53**, 1556.
[219] H. S. Moller and R. L. Mössbauer, *Phys. Letters*, 1967, **24**, *A*, 416.
[220] I. P. Suzdalev, M. Ya. Gen, V. I. Goldanskii, and E. F. Makarov, *Soviet Phys.—J.E.T.P.*, 1967, **24**, 79 (*Zhur. eksp. teor. Fiz.*, 1966, **51**, 118).
[221] P. Steiner, E. Gerdau, P. Kienle, and H. J. Korner, *Phys. Letters*, 1967, **24**, *B*, 515.
[222] D. Yeboah-Amankwah, L. Grodzins, and R. B. Frankel, *Phys. Rev. Letters*, 1967, **18**, 791.
[223] U. Atzmony, E. R. Bauminger, D. Froindlich, and S. Ofer, *Phys. Letters*, 1967, **26**, *B*, 81.

97 and 103 kev levels of ^{153}Eu for some intermetallic compounds,[224] and for the compounds EuSe, EuO, EuS, EuTe, EuB_6, $EuCO_3$, $EuCl_2$, and $EuSO_4$ [225] indicate a linear relationship, within experimental error. The influence of electronic structure on the chemical isomer shift of the ^{151}Eu transition has been considered.[226] The change from Eu^{3+} ($4f^6 5s^2 5p^6$) to Eu^{2+} ($4f^7 5s^2 5p^6$) produces a large negative shift because of the increased shielding of the 5*s* shell by the extra 4*f* electron. Variations of shifts within the Eu^{2+} compounds are due to partial occupancy of the 6*s* shell ($4f^7 5s^2 5p^6 6s^x$; $0<x<1$). The Eu^{3+} compounds are assumed to involve some measure of 4*f* covalancy with the ligands ($4f^{6+y} 5s^2 5p^6$; $0<y<1$). The increased shielding by the 4*f* electrons accounts for the observed negative chemical shifts with increasing covalency to the ligands. The unexpectedly high values found for S, Se, and Te ligands reflects the 4*f* π-bonding effects, which reduce the shielding and increase the chemical shifts. There is no evidence for π-bonding effects in the Eu^{2+} compounds, possibly because of the stabilisation effect of the half-filled 4*f* shell.

The europium–iron–gallium garnets ($Eu_3Ga_xFe_{5-x}O_{12}$) have been studied by ^{57}Fe and ^{151}Eu resonances at 4·2°K.[227] About 80% of the gallium in all samples occupies tetrahedral sites. About 88% of the exchange-field acting on an Eu^{3+} ion is produced by the two nearest iron neighbours in the tetrahedral sites although the Eu–O–Fe angle of 92° is often considered unfavourable for super-exchange interactions. The remaining 12% is produced by the four third-nearest neighbours in the octahedral sites. The Mössbauer spectrum of europium metal has shown that an observed anomaly in the specific heat at 16°K is not an intrinsic property of the metal but is due to an impurity, possibly EuH_2.[228]

Resonances in ^{155}Gd (86·5 kev, second excited-state), ^{156}Gd (89·0 kev), ^{158}Gd (79·5 kev), and the new isotope ^{160}Gd (75·3 kev) have all been reported after Coulomb excitation with a proton beam.[229] Neutron capture in ^{157}Gd populates the 79·5 kev level of ^{158}Gd and has been used to observe the electric-field gradients in GdF_3 and $GdCl_3$.[230]

The new 73·4 kev resonance in ^{164}Dy has been observed by a Coulomb-excitation method.[229] The 25·6 and 74·6 kev transitions in ^{161}Dy were used in a detailed study of the ferromagnetic, antiferromagnetic, and paramagnetic phases of dysprosium metal.[231] Population of the 81 kev (2+)

[224] U. Atzmony, E. R. Bauminger, I. Nowik, S. Ofer, and J. H. Wermick, *Phys. Rev.*, 1967, **156**, 262.
[225] E. Steichele, *Z. Physik.*, 1967, **201**, 331.
[226] F. A. Deeney, J. A. Delaney, and V. P. Ruddy, *Phys. Letters*, 1967, **25**, *A*, 370.
[227] I. Nowik and S. Ofer, *Phys. Rev.*, 1967, **153**, 409.
[228] O. V. Lounasmaa and G. M. Kalvius, *Phys. Letters*, 1967, **26**, *A*, 21.
[229] R. R. Stevens, jun., J. S. Eck, E. T. Ritter, Y. K. Lee, and J. C. Walker, *Phys. Rev.*, 1967, **158**, 1118.
[230] J. Fink, *Z. Physik*, 1967, **207**, 225.
[231] G. J. Bowden, D. St. P. Bunbury, and J. M. Williams, *Proc. Phys. Soc.*, 1967, **91**, 612.

state in ^{162}Dy by a ^{161}Dy(n, γ) ^{162}Dy reaction and of the 74·6 kev ^{161}Dy level by ^{160}Gd(n, γ) ^{161}Gd $\rightarrow$ ^{161}Tb $\rightarrow$ ^{161}Dy has been used to study the magnetic splitting of $DyCl_3,6H_2O$ and to derive excited-state nuclear parameters.[232]

One of the more chemically significant rare-earth studies has involved ^{161}Dy measurements at 4°K on $DyF_3,5H_2O$, $Dy(NO_3)_3,6H_2O$, $DyPO_4,5H_2O$, $Dy(MoO_4)_3$, $DyFeO_3$, $DyMn_2O_5$, DyAl garnet, Dy_2O_3, $DyCl_3,6H_2O$, and dysprosium oxalate, acetate, and ethyl sulphate.[233] The long electronic-relaxation times produce magnetic splittings in all cases. Calculations of theoretical relaxed spectra show that in all the compounds the Dy atom is in a position of axial or very close to axial symmetry. In all but the last three mentioned, the ground state was established to be a Kramer's doublet with J_z $\pm 15/2$. The results for the acetate contradict earlier e.s.r. data which implied a non-axial environment, and it is suggested that the majority of Dy^{3+} sites are non-resonant in the e.s.r. spectrum.

The new 91·5 kev ^{164}Er resonance following the β-decay of ^{164}Ho ($\tau_{\frac{1}{2}}$ 30 min.) has been reported in the form of a magnetically split spectrum in $ErCl_3,6H_2O$.[234] The ^{168}Er (79·8 kev) resonance was seen following Coulomb excitation,[229] as have magnetic effects on the Mössbauer spectra of ^{166}Er in metals.[235, 236] Non-equally spaced magnetic lines from a ($2^+ \rightarrow 0^+$) transition have been observed for the first time in a rare-earth salt in erbium ethyl sulphate, ^{166}Er $(EtSO_4)_3,9H_2O$, magnetically diluted with the yttrium salt.[237] The slowing down of the spin–spin relaxation causes a breakdown of the effective-field approximation. The possibility of a quadrupole effect is excluded.

The temperature dependence of the quadrupole splitting in ^{169}Tm metal has been used to derive crystal-field parameters appropriate to the Tm site-symmetry.[238, 239] Neutron diffraction has shown that Tm metal contains four inequivalent magnetisations with a magnetic structure repeating every seven lattice layers. However, the observed low-temperature Mössbauer spectrum is apparently inconsistent with this model. It has now been shown that the conflict can be resolved if the lattice and magnetic periodicities are no longer commensurate.[240] The magnetic periodicity is assumed to be $7+\varepsilon$ ($\varepsilon \ll 1$). The calculated spectrum is identical with that observed and is virtually independent of the magnitude of ε (see Figure 7). This type of behaviour could be induced by defects in the lattice.

[232] W. Henning, D. Heunemann, W. Weber, P. Kienle, and H. J. Korner, *Z. Physik*, 1967, **207**, 505.
[233] H. H. Wickman and I. Nowik, *J. Phys. and Chem. Solids*, 1967, **28**, 2099.
[234] E. Munck, D. Quitmann, and S. Hufner, *Phys. Letters*, 1967, **24**, *B*, 392.
[235] W. Wiedemann and W. Zinn, *Phys. Letters*, 1967, **24**, *A*, 506.
[236] R. A. Reese and R. G. Barnes, *Phys. Rev.*, 1967, **163**, 465.
[237] E. R. Seidel, G. Kaindl, M. J. Clauser, and R. L. Mössbauer, *Phys. Letters*, 1967, **25**, *A*, 328.
[238] D. L. Uhrich, D. J. Genin, and R. G. Barnes, *Phys. Letters*, 1967, **24**, *A*, 338.
[239] D. L. Uhrich and R. G. Barnes, *Phys. Rev.*, 1967, **164**, 428.
[240] R. L. Cohen, *Phys. Letters*, 1967, **24**, *A*, 674.

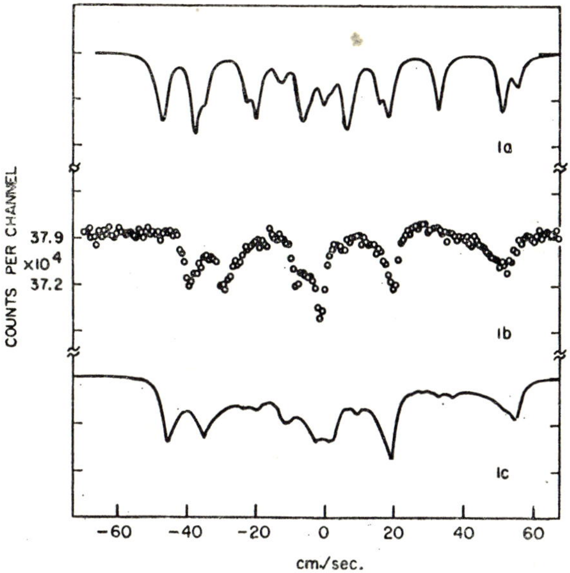

Figure 7 *Observed and calculated spectra of* Tm *metal at* 40°K; (1a) *calculated on seven-sub-lattice model;* (1b) *observed;* (1c) *calculated on assumption of incommensurate magnetic and crystal lattices*
(Reproduced by permission from *Phys. Letters*, 1967, **24**, *A*, 674)

The $0 \rightarrow 2$ resonances of ^{172}Yb (79 kev), ^{174}Yb (77 kev), and ^{176}Yb (82 kev) have been reported in Coulomb excitation experiments.[241] Relaxation effects in intermetallic compounds have been seen in ^{170}Yb (84·3 kev),[242] and in ytterbium–iron garnet and ytterbium–gallium garnet.[243] Replacement of Ga^{3+} by Fe^{3+} has a large damping effect on the spin-relaxation rates, and the similarity of the spectrum of ytterbium doped into yttrium–iron garnet shows the relative unimportance of rare-earth–rare-earth interactions. Chemical isomer shifts have been found in a small series of Yb^{2+} and Yb^{3+} compounds using the ^{170}Yb (84·3 kev) resonance at 4·2°K.[244] The compounds studied were $YbCl_2$, $YbSO_4$, $YbCl_3$, ytterbium–gallium garnet, $YbAl_2$, $YbSi_2$, and Yb metal.

The second excited-state resonance (75·9 kev) of ^{171}Yb has been studied using the 8·3 day electron-capture process in ^{171}Lu formed by the reaction $^{169}Tm(\alpha, 2n)^{171}Lu$. The $\frac{5}{2} \rightarrow \frac{1}{2}$ $E2$ decay gives a ten-line magnetic spectrum in $YbCl_3,6H_2O$ at 4·2°K because of long spin-relaxation times and the small electric-field gradient.[245]

[241] J. S. Eck, Y. K. Lee, J. C. Walker, and R. R. Stevens, jun., *Phys. Rev.*, 1967, **156**, 246.
[242] I. Nowik, S. Ofer, and J. H. Wernick, *Phys. Letters*, 1967, **24**, *A*, 89.
[243] S. Ofer and I. Nowik, *Phys. Letters*, 1967, **24**, *A*, 88.
[244] U. Atzmony, E. R. Bauminger, J. Hess, A. Mustachi, and S. Ofer, *Phys. Rev. Letters*, 1967, **18**, 1061.
[245] W. Henning, P. Kienle, and H. J. Korner, *Z. Physik.*, 1967, **199**, 207.

7 Other Elements

The new experimental technique of Coulomb-recoil-implantation has been applied to observe the ^{73}Ge (67 kev) resonance.[246] A thin layer of germanium on a chromium foil was used and the recoiling germanium atoms, excited by an oxygen-ion beam, pass into the foil and are arrested. A more intense unsplit emission line can be obtained in this way than by bombardment of an ordinary Ge target because the latter is reduced to an amorphous state during irradiation. Chemical isomer shifts of the order 1 mm./sec. were observed for Ge and GeO_2 absorbers, giving a crude estimate for $\delta r/r$ of $+1 \times 10^{-3}$. Krypton experiments using ^{83}Br decay in LiBr, NaBr, KBr, CsBr, NH_4Br, $KBrO_3$, and Kr-β-hydroquinone sources yielded only slightly broadened single unshifted lines.[247] There was no suggestion of the formation of a KrO_3 species, and the general lack of chemical isomer shifts shows the Kr outer-shell configuration to be unaffected by induced dipole interaction with its neighbours.

Further practical details of the ^{99}Ru (90 kev) resonance have appeared in the context of experiments to test for time-reversal invariance.[248] This is one of the few isotopes to show significant $E2$–$M1$ mixing in the Mössbauer resonance.

Some of the work on ^{121}Sb has already been mentioned in the section on ^{119}Sn (see page 359.) Similar studies have been published independently.[249] Quadrupole effects in ^{121}Sb are generally smaller than the natural linewidth and are unresolved. The magnetic hyperfine interactions recorded in the alloy MnSb are much more significant and the $\frac{7}{2}$ excited-state moment has been determined to be $+2{\cdot}35\ \mu_N$.[250] Magnetic fields have also been found at ^{121}Sb impurity atoms in iron and nickel.[251]

Quadrupole coupling data cannot be obtained for tellurium by n.q.r. methods, and ^{125}Te Mössbauer measurements have already been shown to be rather insensitive. A new method is proposed whereby ^{128}Te compounds are neutron-irradiated and the ^{129}I resonance of the decay product is observed.[252] If the β-decay does not alter the electric-field gradient significantly, the quadrupole coupling in tellurium can now be estimated more accurately. ^{129}I and ^{125}Te resonances were recorded in Te metal, TeO_2, and $Te(NO_3)_4$, and the ratio of the coupling constants was equal in the two compounds but not in the metal. The greater detail of the ^{129}I resonance suggests that TeO_2 has a chain structure like that of SeO_2 (see Figure 8) and that $Te(NO_3)_4$ is pyramidal with a three- or four-fold axis.

246 G. Czjzek, J. L. C. Ford, jun., J. C. Love, F. E. Obenshain, and H. H. F. Wegener, *Phys. Rev. Letters*, 1967, **18**, 529.
247 M. Pasternak and T. Sonnino, *Phys. Rev.*, 1967, **164**, 384.
248 O. C. Kistner, *Phys. Rev. Letters*, 1967, **19**, 872.
249 V. A. Bryukhanov, B. Z. Iofa, V. Kothhekar, S. I. Semenov, and V. S. Shpinel, *Zhur. eksp. teor. Fiz.*, 1967, **53**, 1588.
250 S. L. Ruby and G. M. Kalvius, *Phys. Rev.*, 1967, **155**, 353.
251 S. L. Ruby and C. E. Johnson, *Phys. Letters*, 1967, **26**, *A*, 60.
252 M. Pasternak and S. Bukshpan, *Phys. Rev.*, 1967, **163**, 297.

Anisotropy of the recoil-free fraction was observed in the metal and oxide, consistent with chain structures.

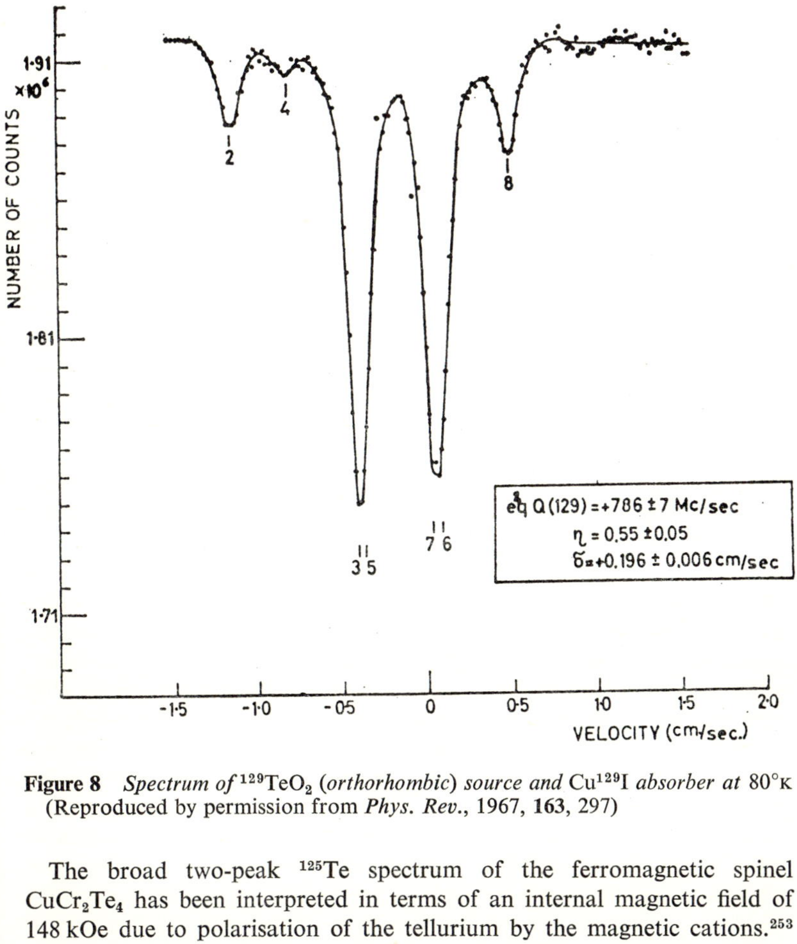

Figure 8 *Spectrum of* $^{129}TeO_2$ (*orthorhombic*) *source and* $Cu^{129}I$ *absorber at* 80°K (Reproduced by permission from *Phys. Rev.*, 1967, **163**, 297)

The broad two-peak ^{125}Te spectrum of the ferromagnetic spinel $CuCr_2Te_4$ has been interpreted in terms of an internal magnetic field of 148 kOe due to polarisation of the tellurium by the magnetic cations.[253] Chemical isomer shifts of a series of hexahalogeno-anions, TeX_6^{2-} (X = F, Cl, Br, or I), have been found to show a linear correlation with electronegativity of the ligand. The fluoride was not isolated and the data given refer to a frozen-solution spectrum. $\delta r/r$ is probably positive.[254, 255]

[253] J. F. Ullrich and D. H. Vincent, *Phys. Letters*, 1967, **25**, *A*, 731.

[254] V. A. Bryukhanov, B. Z. Iofa, A. A. Opalenko, and V. S. Shpinel, *Russian J. Inorg. Chem.*, 1967, 1044 (*Zhur. neorg. Khim.*, 1967, **12**, 1985).

[255] V. S. Shpinel, V. A. Bryukhanov, V. Kotkhekar, and B. Z. Iofa, *Zhur. eksp. teor. Fiz.*, 1967, **53**, 23.

The compounds FeTe, $FeTe_2$, ZnTe, CdTe, and HgTe have been studied.[256] Radiation-damage effects in $Pb^{125}Te$ and their change with time and with thermal annealing have been reported.[257]

The ^{129}I resonance in SnI_4 is referred to in the ^{119}Sn section on page 360.

A hyperfine field of 1040 kOe at xenon in iron metal is considered to arise from the well known 'transferred hyperfine structure' effect that leads to positive polarisation of the bound $5s^2$ shell.[258]

The hyperfine interactions in the resonance of ^{182}W ($0 \rightarrow 2$, 100 kev), ^{183}W ($\frac{1}{2} \rightarrow \frac{3}{2}$, 47 kev, and $\frac{1}{2} \rightarrow \frac{5}{2}$, 99 kev) and ^{186}W ($0 \rightarrow 2$, 122·5 kev) have been investigated in WS_2, WO_3, and tungsten alloys. Experimentation is difficult and the line splittings are only of the order of the linewidth.[259, 260]

^{189}Os has now been added to the list of Mössbauer isotopes. The 13·3 day ^{189}Ir isotope populates the third excited level of ^{189}Os at 69·6 kev, and a resonance was observed for an osmium metal absorber at 80°K.[261]

The same paper describes a new route to the hitherto unused 82·4 kev level of ^{191}Ir *via* the 3·0 day electron-capture decay of ^{191}Pt. A Ge(Li) detector system was used and the resonance was detected in an iridium metal absorber.[261] Both this and the already familiar 129 kev transition in ^{191}Ir, and 73 and 139 kev transitions of ^{193}Ir, have been studied in a small group of metals and compounds.[262] Some spectra showed resolved quadrupole effects. Plots of chemical isomer shifts for pairs of resonances in a given set of compounds are linear as predicted and permit relative values of the nuclear radii changes to be computed. The widest range of measurements, made with the 73 kev transition in ^{193}Ir, suggests that characteristic chemical isomer shift values will be given by different iridium oxidation states. The excited-state spin for the 73 kev ^{193}Ir transition has been confirmed to be $\frac{1}{2}$, and the $E2:M1$ mixing ratio is 0·37 [263] making an interesting comparison with ^{99}Ru. Well resolved chemical isomer shifts, internal magnetic hyperfine interactions, and quadrupole splittings are seen in a range of intermetallic compounds and salts. A parallel investigation of an iron–iridium ferromagnetic alloy gives $E2:M1$ as 0·31.[264] Details of independent measurements on the new ^{191}Ir resonance (82 kev) were also given in this latter paper. Hyperfine

[256] G. Albanese, C. Lamborizio, and I. Ortalli, *Nuovo Cimento*, 1967, **50**, *B*, 65.
[257] E. P. Stepanov and A. Yu. Aleksandrov, *J.E.T.P. Letters*, 1967, **5**, 83; *Z.E.T.F. Letters*, 1967, **5**, 101.
[258] D. A. Shirley, *Phys. Letters*, 1967, **25**, *A*, 129.
[259] D. Agresti, E. Kankeleit, and B. Persson, *Phys. Rev.*, 1967, **155**, 1342.
[260] B. Persson, H. Blumberg, and D. Agresti, *Phys. Letters*, 1967, **24**, *B*, 522.
[261] S. Jha, W. R. Owens, M. C. Gregory, and B. L. Robinson, *Phys. Letters*, 1967, **25**, *B*, 115.
[262] F. Wagner, J. Klockner, H. J. Korner, H. Schaller, and P. Kienle, *Phys. Letters*, 1967, **25**, *B*, 253.
[263] U. Atzmony, E. R. Bauminger, D. Lebenbaum, A. Mustachi, S. Ofer, and J. H. Wernick, *Phys. Rev.*, 1967, **163**, 314.
[264] F. Wagner, G. Kaindl, P. Kienle, and H. J. Korner, *Z. Physik*, 1967, **207**, 500.

magnetic fields and chemical isomer shifts in a range of rare-earth–iridium alloys have been studied using the ^{193}Ir 73 kev Mössbauer resonance.[265]

^{195}Pt resonances have been quoted in several compounds and alloys.[266, 267] Chemical isomer shifts in $PtCl_2$, PtO, $PtCl_4$, and PtO_2 were the same within experimental error, but the alloys show variation.[266] Hyperfine fields at the ^{195}Pt nucleus were of the order of 1200 kOe.

Au_4V is only the third alloy of non-magnetic metals which is known to be ferromagnetic. Below the Curie temperature the ^{197}Au spectrum shows magnetic interactions giving line broadening. The magnitude of the internal magnetic field indicates that the moment is localised on the vanadium and little if any is on the gold.[268]

The first report has been made of a Mössbauer resonance in ^{238}U following Coulomb excitation by 3 Mev α-particles.[269] The U_3O_8 absorber and uranium metal target were cooled to liquid helium or nitrogen temperature, and the 45 kev $2^+ \rightarrow 0^+$ transition was studied. This γ-transition has a broad linewidth and is highly converted, but a weak resonance absorption showing signs of quadrupole hyperfine structure was detected. The ^{237}Np resonance which is populated by α-decay from ^{241}Am has been known for some time. Preliminary data from other laboratories have now appeared of the ^{237}Np resonance in NpO_2 using sources of ^{241}Am in AmO_2 [271] and NpO_2.[270] Experiments with ^{233}U and ^{239}Pu were not successful.[271]

Acknowledgement is made to Mr. J. D. Cooper for translations of papers written in Russian.

8 Bibliography

The following is a list of those references on Mössbauer spectroscopy which have not been referred to in the main text.

G. Albanese, G. Fabri, C. Lamborizio, M. Musci, and I. Ortalli. Mössbauer effect in gallium arsenide, *Nuovo Cimento*, 1967, **50**, *B*, 149.

J. Baijal and U. Baijal. Mössbauer fractions and specific heat of tellurium, *J. Phys. Soc. Japan*, 1967, **22**, 1507.

L. H. Bennett and L. J. Swartzendruber. Ferromagnetic iron alloys lacking a hyperfine field at the iron site, *Phys. Letters*, 1967, **24**, *A*, 359.

V. G. Bhide and H. C. Bhasin. Mössbauer study of the $SrTiO_3$–^{57}Co system, *Phys. Rev.*, 1967, **159**, 586.

V. G. Bhide, H. C. Bhasin, and G. K. Shenoy. Observation of iron clusters in hydrogen-fired ^{57}Fe-doped $SrTiO_3$ by Mössbauer studies, *Phys. Letters*, 1967, **24**, *A*, 109.

265 A. Heuberger, F. Pobell, and P. Kienle, *Z. Physik*, 1967, **205**, 503.
266 D. Agresti, E. Kankeleit, and B. Persson, *Phys. Rev.*, 1967, **155**, 1339.
267 A. Buyrn, L. Grodzins, N. A. Blum, and J. Wulff, *Phys. Rev.*, 1967, **163**, 286.
268 B. D. Dunlap, J. B. Darby, jun., and C. W. Kimball, *Phys. Letters*, 1967, **25**, *A*, 431.
269 J. R. Oleson, Y. K. Lee, J. C. Walker, and J. W. Wiggins, *Phys. Letters*, 1967, **25**, *B*, 258.
270 V. A. Bryukhanov, V. V. Ovechkin, A. I. Peryshkin, E. I. Rzhekhina, and V. S. Shpinel, *Soviet Phys.—Solid State*, 1967, **9**, 1189.
271 G. N. Belozerskii, Yu. A. Nemilov, A. A. Chaikhorskii, and A. V. Shredchikov, *Soviet Phys.—Solid State*, 1967, **9**, 978.

V. G. Bhide and S. K. Date. Mössbauer study of intermetallic Fe_5Ge_3 system, *Solid State Comm.*, 1967, **5**, 435.

R. G. Borg, R. Booth, L. Himmel, D. N. Pipkorn, and C. E. Violet. Effective fields in dilute alloys of Mn in Cu and Au, *Phys. Letters*, 1967, **25**, *A*, 141.

F. Borsa, R. G. Barnes, and R. A. Reese. Nuclear magnetic resonance and Mössbauer effect study of ^{119}Sn in rare earth–tin intermetallic compounds, *Phys. Stat. Sol.*, 1967, **19**, 359.

H. Brafman, P. Gilad, P. Hillman, D. Lebenbaum, and Y. Wolfsohn. Use of the Mössbauer effect to measure small vibrations, *Nuclear Instr. Methods*, 1967, **53**, 13.

J. W. Burton and R. P. Godwin. Mössbauer effect in surface studies: ^{57}Fe on W and Ag, *Phys. Rev.*, 1967, **158**, 218.

D. C. Champeney. Theory of narrow-angle *X*-ray scattering and its measurement by the Mössbauer effect, *J. Appl. Phys.*, 1967, **18**, 549.

V. V. Chekin, L. E. Danilenko, and A. I. Kaplienko. Internal magnetic fields at ^{119}Sn nuclei in Heusler alloys, *Soviet Phys.—J.E.T.P.*, 1967, **24**, 472.

V. V. Chekin and V. G. Naumov. Isomer shifts on ^{119}Sn impurity nuclei in Pd–H alloys, *Soviet Phys.—J.E.T.P.*, 1967, **24**, 699.

J. Christiansen, P. Hindennach, U. Morfeld, E. Recknagel, D. Riegel and G. Weyer. Hyperfine structure investigations with the Mössbauer effect, following the reaction ^{56}Fe(*d,p*)^{57}Fe, *Nuclear Phys.*, 1967, **99**, *A*, 345.

A. Corciovei and E. Radescu. Calculation of the Debye–Waller factor for the Mössbauer effect in thin films, *Rev. Roumaine Phys.*, 1967, **12**, 177.

M. Cordey-Hayes and I. R. Harris. The magnetic susceptibilities and ^{119}Sn isomer shifts of some Pd–Sn alloys, *Phys. Letters*, 1967, **24**, *A*, 80.

L. Cser, J. Ostanevich, and L. Pal. Mössbauer effect in iron–aluminium alloys, *Phys. Stat. Sol.*, 1967, **20**, Part I, p. 581; Part II, p. 591.

I. I. Dekhtiar, B. G. Egiazarov, L. M. Isakov, V. S. Mikhalenkov, and V. P. Romansho. The influence of plastic deformation on the Mössbauer effect in Fe–Ni alloys of invar composition, *Doklady Akad. Nauk S.S.S.R.*, 1967, **175**, 556.

N. N. Delyagin. Interpretation of the isomeric shifts of the γ-radiation of impurity nuclei ^{119}Sn and ^{197}Au in metallic solid solutions, *Soviet Phys.—Solid State*, 1967, **8**, 2748 (*Fiz. Tverd. Tela*, 1966, **8**, 3426).

N. Y. Delyagin, H. El Sayes, and V. S. Shpinel. Magnetic hyperfine structure of the ^{155}Gd levels in metallic gadolinium and in the intermetallic compound $GdAl_2$, *Soviet Phys.—J.E.T.P.*, 1967, **24**, 64 (*Zhur. eksp. teor. Fiz.*, 1966, **51**, 95).

L. M. Epstein and A. Wachtel. Hyperfine structure in a paramagnetic ^{57}Fe Mössbauer absorber at room and high temperatures, *Appl. Phys. Letters*, 1967, **10**, 246.

R. B. Frankel, N. A. Blum, B. B. Schwartz, and D. J. Kim. Mössbauer evidence for a spin-compensated state in dilute Fe–Cu alloys, *Phys. Rev. Letters*, 1967, **18**, 1051.

G. Gorodetsky and S. Shtrikman. Studies of magnetic hardness in Vicalloy using the Mössbauer effect, *J. Appl. Phys.*, 1967, **38**, 3981.

S. M. Harris. Quantum-mechanical model of Mössbauer line narrowing, *Phys. Rev.*, 1967, **163**, 280.

A. Heilmann and W. Zinn. Method for determination of short-range order in Ni–Fe alloys by the Mössbauer effect, *Z. Metallk.*, 1967, **58**, 113.

A. A. Hirsch, Z. Eliezer, and D. Haimovici. Anhysteretic magnetization process and Mössbauer effect in iron powder, *Physica*, 1967, **35**, 29.

R. M. Housley and F. Hess. Analysis of Debye–Waller factor and Mössbauer thermal-shift measurements. Part II. Thermal-shift data on Fe, *Phys. Rev.*, 1967, **164**, 340.

D. G. Howard and J. G. Dash. Behaviour of Mössbauer *f* and thermal shift of ^{57}Fe in nickel near the Curie temperature, *J. Appl. Phys.*, 1967, **38**, 991.

G. P. Huffman and R. M. Fisher. Mössbauer studies of ordered and cold-worked Fe–Al alloys containing 30 to 50 at. % aluminium, *J. Appl. Phys.*, 1967, **38**, 735.

S. Hufner. Conduction-electron contributions to the hyperfine fields in metallic Eu and Gd, *Phys. Rev. Letters*, 1967, **19**, 1034.

R. Ingalls, H. G. Drickamer, and G. de Pasquali. Isomer shift of ^{57}Fe in transition metals under pressure, *Phys. Rev.*, 1967, **156**, 165.

Ya. A. Iosilevskii. The Mössbauer effect in two-component crystals, *Soviet Phys.—Solid State*, 1967, **8**, 2421 (*Fiz. Tverd. Tela*, 1966, **8**, 3032).

Ya. A. Iosilevskii. Mössbauer effect in complex multicomponent systems, *Phys. Stat. Sol.*, 1967, **24**, 749.

Y. Ishikawa and Y. Endoh. Antiferromagnetism of γ-Fe–Mn Alloys. Part II. Neutron diffraction and Mössbauer effect studies, *J. Phys. Soc. Japan*, 1967, **23**, 205.

A. P. Jain and T. E. Cranshaw. Anomalous temperature-dependence of the hyperfine fields at Sn in cobalt, *Phys. Letters*, 1967, **25**, *A*, 421.

A. P. Jain and T. E. Cranshaw. Mössbauer effect at ^{119}Sn in Au–Mn, Ag–Mn, Cu–Mn, and Au–Fe alloys, *Phys. Letters*, 1967, **25**, *A*, 425.

Yu. Kagan and A. M. Afanasev. Anomalous diffuse small-angle Mössbauer scattering in crystals, *J.E.T.P. Letters*, 1967, **5**, 40 (*Z.E.T.F. Letters*, 1967, **5**, 51).

R. Kamal, R. G. Mendiratta, S. B. Raju, and L. W. Tiwari. The recoil-less fraction and the thermal shift for ^{129}I (26·8 kev) transition in CsI, *Phys. Letters*, 1967, **25**, *A*, 503.

C. W. Kimball, J. K. Tison, and M. V. Nevitt. Hyperfine interactions at ^{57}Fe nuclei in intermetallic compounds of the Fe–Mn system with the β-manganese structure, *J. Appl. Phys.*, 1967, **38**, 1153.

S. L. Kordyuk, V. I. Lisichenko, O. L. Orlov, N. N. Polovina, and A. N. Smoilovskii. Mössbauer effect in a disperse system, *Soviet Phys.—J.E.T.P.*, 1967, **25**, 400 (*Zhur. eksp. teor. Fiz.*, 1967, **52**, 611).

L. Korecz and K. Burger. The isomeric shift in the Mössbauer spectra of iron complexes and the neutrality postulate of Pauling, *Magyar Kém. Folyóirat*, 1967, **73**, 68.

V. M. Krasnoperov, B. G. Lure, A. N. Murin, N. K. Cherezov, and I. A. Yutlandov. Using ^{83}Rb for Mössbauer effect investigations with ^{83}Kr, *At. Energ.*, 1967, **23**, 60.

M. A. Krivoglaz and S. P. Repetskii. Theory of diffusional broadening of the Mössbauer line in non-ideal crystals, *Soviet Phys.—Solid State*, 1967, **8**, 2325 (*Fiz. Tverd. Tela*, 1966, **8**, 2908).

R. N. Kuz'min, A. V. Kolpakov, and G. S. Zhdanov. Scattering of Mössbauer lines by crystals, *Soviet Phys.—Crystallography*, 1967, **11**, 457 (*Kristallografiya*, 1966, **11**, 511).

L. B. Kvashnina and M. A. Krivoglaz. Mössbauer spectra in crystals containing defects, *Fiz. Metall. i Metallov.*, 1967, **23**, 3.

P. Lutsch. Flow measurement in pipelines by means of the Mössbauer effect, *Wasser, Luft Bertr.*, 1967, **11**, 202.

E. F. Makarov, V. A. Povitskii, E. B. Granovskii, and A. A. Fridman. Study of Alnico 8 by Mössbauer spectroscopy, *Phys. Stat. Sol.*, 1967, **24**, 45.

H. Maletta, W. Heidrich, and R. L. Mössbauer. Dependence of concentration of the isomer shift of divalent ^{151}Eu in CaF_2, *Phys. Letters*, 1967, **25**, *A*, 295.

M. P. Maley, R. D. Taylor, and J. L. Thompson. Spin value and moment determination for the localised magnetic states of very dilute Fe impurities in Pt and Pd, *J. Appl. Phys.*, 1967, **38**, 1249.

H. L. Marcus, M. E. Fine, and L. H. Schwartz. Mössbauer-effect study of solid-solution and precipitated Fe-rich Fe–Mo alloys, *J. Appl. Phys.*, 1967, **38**, 4750.

H. L. Marcus and L. H. Schwartz. Mössbauer spectra of Fe–Mo alloys, *Phys. Rev.*, 1967, **162**, 259.

H. B. Mathur, A. P. B. Sinha, and C. M. Yagnik. Mössbauer spectra of cubic ferrites with spinal structure, *Indian J. Pure Appl. Phys.*, 1967, **5**, 155.

A. N. Murin, B. G. Lur'e, and P. Seregin. A study of the state of iron ions in silver halides using the Mössbauer effect, *Soviet Phys.—Solid State*, 1967, **9**, 1110 (*Fiz. Tverd. Tela*, 1967, **9**, 1424).

A. N. Murin, B. G. Lur'e, P. P. Seregin, and N. K. Cherezov. Study of the state of iron in AgCl single crystals using the Mössbauer effect, *Soviet Phys.—Solid State*, 1967, **8**, 2632 (*Fiz. Tverd. Tela*, 1966, **8**, 3291).

A. N. Nesmeyanov, A. M. Babeshkin, E. N. Efremov, and A. A. Bekker. Determination of recoil atoms of ^{119}Sn stabilising as metallic tin in tin oxide irradiated in an atomic reactor, *Vestnik Moskov. Univ.*, 1967, 107.

R. S. Preston. Temperature dependence of the isomer shift and the hyperfine field near the Curie point in iron, *Phys. Rev. Letters*, 1967, **19**, 75.

D. L. Raimondi and G. Jura. Effect of pressure on the Mössbauer effect of ^{57}Fe in nickel metal, *J. Appl. Phys.*, 1967, **38**, 2133.

S. Roth and E. M. Hörl. Decrease of Mössbauer recoil-free fraction in small tungsten particles, *Phys. Letters*, 1967, **25**, *A*, 299.

R. Segnan. Experimental investigation on exchange interaction in platinum–iron alloys, *Phys. Rev.*, 1967, **160**, 404.

G. A. Selyutin. On the shift and intensity of the Mössbauer line in metals, *Phys. Stat. Sol.*, 1967, **22**, K23.

J. S. Shier and R. D. Taylor. Temperature-dependent isomer shift and anharmonic binding of ^{119}Sn in Nb_3Sn, *Solid State Comm.*, 1967, **5**, 147.
E. Simanek, Nai Li Huang, and R. Orbach. 'Super' transferred hyperfine interactions for Fe^{3+} salts, *J. Appl. Phys.*, 1967, **38**, 1072.
O. I. Sumbaev. Small energy shifts of the X- and γ-ray lines and selection of wavelength standard within the ranges of X-ray and γ-radiations, *Izvest. Akad. Nauk S.S.S.R., Ser. fiz.*, 1967, **31**, 343.
W. L. Trousdale, G. Longworth, and T. A. Kitchens. The range of ferromagnetic exchange interactions, *J. Appl. Phys.*, 1967, **38**, 922.
C. C. Tsuei and E. E. Kankeleit. Mössbauer effect of ^{125}Te in simple cubic and amorphous tellurium-base alloys, *Phys. Rev.*, 1967, **162**, 312.
E. E. Vainshtein, P. M. Valov, G. P. Barsanov, and M. E. Yakovleva. Mössbauer spectra of iron in pyrites and marcasites of different origins, *Geokhimiya*, 1967, 876.
E. E. Vainshtein, P. M. Valov, F. A. Sidorenko, and T. S. Shubina. Mössbauer spectra of iron in the homogeneity region of ε-phase in the system iron-silicon, *Fiz., Metall. i Metallov.*, 1967, **23**, 367.
B. I. Verkin, V. V. Chekin, and A. P. Vinnikov. Effect of impurities on isomer shifts in metallic tin, *Soviet Physics—J.E.T.P.*, 1967, **24**, 16 (*Zhur. eksp. teor. Fiz.*, 1966, **51**, 25).
C. E. Violet and R. J. Borg. Magnetic ordering in dilute solid solutions of iron in gold. Part II. Electric hyperfine interactions, *Phys. Rev.*, 1967, **162**, 608.
G. K. Wertheim and J. H. Wernick. Mössbauer effect study of b.c.c. structure alloys, Fe–Al and Fe–Ti, *Acta Metallurgica*, 1967, **15**, 297.
I. R. Williams, G. V. H. Wilson, and B. Window. Hyperfine magnetic fields at cobalt and tin nuclei in dilute alloys, *Phys. Letters*, 1967, **25**, *A*, 144.
B. Window. Hyperfine fields at ^{119}Sn in dilute alloys, *Phys. Letters*, 1967, **24**, *A*, 659.
F. Wittman and L. Zitzlsperger. Mössbauer effect and colour measurements in natural minerals, *Optik*, 1967, **25**, 211.

Author Index